Encyclopedic Dictionary of Industrial Technology

Encyclopedic Dictionary of Industrial Technology

MATERIALS, PROCESSES AND EQUIPMENT

DAVID F. TVER
ROGER W. BOLZ

NEW YORK LONDON
CHAPMAN AND HALL

First published 1984
by Chapman and Hall
733 Third Avenue, New York, NY 10017

Published in Great Britain by
Chapman and Hall Ltd
11 New Fetter Lane, London EC4P 4EE

Printed in the United States of America

Library of Congress Cataloging in Publication Data

Tver, David F.
Encyclopedic dictionary of industrial technology.

Bibliography; p.
1. Manufacturing processes — Dictionaries.
2. Materials — Dictionaries. 3. Industrial equipment — Dictionaries. I. Bolz, Roger William. II. Title.
TS9.T84 1984 670'.3'21 84-1769
ISBN 0-412-00501-8

Preface

This volume has been prepared as a reference guide for all engineering, industrial and technical management personnel who are in any way involved in the manufacturing process, in product design, or in converting of raw materials to finished products. This *Encyclopedic Dictionary* covers a wide range of subjects from industrial materials, minerals, metals, plastics and synthetic fibers to machine tools, computers, lasers, robots and other production equipment as well as manufacturing processes.

Some of the materials reviewed are brass, steel, nickel, copper, bronze, cast iron, cements, clay, coal, coke, petroleum and petrochemicals, glass, limestone, rubber, paper, metal alloys, chemicals, synthetic fibers, textiles, plastics, resins, lubricants, and thermoplastics.

Various processes are reviewed such as metal casting, forming, machining, annealing, extrusion, heat treating, injection molding, papermaking and steel processing. In heat treating such areas as martempering, annealing, spheroidizing, tempering and austempering are included.

Different types of equipment related to the products are defined. In plastics such products are covered as nylons, polyesters, rayons, Teflon, Vinyon, Saran, acetates and acrylics.

Many of the manufacturing processes and equipment involved in the conversion of material to finished products are described along with products and their ultimate uses. Also, important associated manufacturing activities such as inspection, handling, and control are included to make the references as complete as is practicable.

This industrial materials, processes and equipment dictionary is an excellent hand guide in concise alphabetic form for easy reference but in sufficient depth to provide a clear understanding. In addition, there are tables showing the uses, strengths, characteristics and detailed composition of many materials.

It is believed that this book will be an invaluable quick reference source and should be a part of the library of every engineer, manufacturing supervisor, and management personnel involved in manufacturing as well as industrial processes.

The authors are indebted to and wish to acknowledge with special thanks to the American Society for Metals, The Materials Handling Institute, *Machine Design* magazine, and the companies referenced for their assistance in providing information, reference material, and illustrations for this work.

David F. Tver
San Diego, California

Roger W. Bolz, P.E.
Lewisville, North Carolina

abrasive belt grinding. An old processing method originally used in woodworking where power sanding is still widely used. Improvement in impervious plastic-bonded cloth belts makes precision abrasive belt grinding of surfaces practical using coolants. The so-called wet-belt process is a method of stock removal and surfacing performed on machines having a tensioned abrasive belt operating over precision pulleys at speeds between 2500 and 6000 sfpm (760 and 1830 smpm). Grinding takes place where the belt passes over a vertical support platen. The work is supported on a horizontal table or on a conveyor belt and contacts the abrasive belt with a motion against or normal to belt travel. Aluminum oxide or silicon carbide abrasive belts are used, generally with a water coolant, and finishes from around 40–50 μin. (0.001–0.0013 mm) to as fine as 3 μin. (8×10^{-5} mm) can be produced. Fixtures allow the grinding of cylindrical parts.

abrasive cut-off machine **(see cut off machines)**

abrasives. Except for diamonds, all abrasives used in large volume in industry are synthetic. Natural abrasives have some minor uses: garnet and flint are used for sandpaper; emery paper is used for polishing metals; rottenstone and pumice are used for wood finishing; and walnut shells are used for cleaning aircraft engine parts. Diamond, the only significant natural abrasive, is the hardest material known and is used for applications where the life of other abrasives is poor. Typical applications for diamond are in brick and concrete saws, wire drawing dies, drills for drilling hard rock, and cutters for machining hard metals. Next to diamond, the hardest abrasive in common use is silicon carbide, or "carborundum." In silicon carbide, sand, coke, and sawdust are mixed, and a high-temperature [4500°F (2480°C)] electric arc is passed through the mixture for a considerable period of time. Sawdust burns out to provide porosity for the escape of gases from the mass. At the conclusion of the process the center of the mixture is converted to silicon carbide, which is then crushed. Another abrasive is aluminum oxide, which is made from bauxite by fusion and, then, crushing. Aluminum oxide is not quite as hard as silicon carbide, but is tougher and more resistant to impact. It is preferable to silicon carbide for such applications as floor sanding machines. Aluminum oxide is the preferred material for grinding the harder metals, since it wears away faster than silicon carbide, thus exposing new edges for cutting. Silicon carbide is selected for cast iron and the softer types of nonferrous metals. In general, only these two abrasives are used in grinding wheels. The grit in the wheel is bonded either with a ceramic material such as sodium silicate or with an organic material such as a rubber or plastic cement.

abrasives, sand (see sand)

accuracy. (1) Quality, state, or degree of conformance to a recognized standard. (2) Difference between the actual position response and the target position desired or commanded of an automatic control system.

acetal (plastics). Acetal is an engineering-type thermoplastic and is referred to as a linear polymer of formaldehyde. Acetals compete with nylons and polycarbonates in many structural applications such as hardware components, gears, bearings, business machine assemblies, and housings. Like nylon and polycarbonate, stiffness and dimensional stability are improved by incorporating glass fibers and lubricity is enhanced by adding fluorocarbons. Acetals do not embrittle with long-term exposure at elevated temperature. Tensile strength is substantially the same after 1 yr in air at 240°F (116°C). Acetals cannot be dissolved by organic solvents, which precludes solvent bonding. They withstand staining by most common household foods and resist discoloration by industrial oils or greases. Moisture absorption is extremely low, leading to usage in shower heads, sprinklers, and pump assemblies. Acetals have limited resistance to strong acids and oxidizing chemicals, and they resist weak alkalies.

acetal copolymer (plastics). Acetal copolymer is an engineering thermoplastic that has a balance of mechanical properties and processing characteristics. The mechanical properties of the copolymer enable the resin to be used in applications in which design parameters are critical. In addition, it has wide latitude in processing. The copolymer is produced by polymerization of trioxane and ethylene oxide. The resulting product is an engineering resin with long-term retention of mechanical properties at high temperatures and in numerous chemical environments. The polymer is available in natural translucent white and in a wide range of standard colors. The copolymer is designated as an engineering resin because of its predictable design, processing, and end-use characteristics. It exhibits toughness, stiffness, and excellent impact resistance. These properties are maintained in hostile environments such as hot water, organic solvents, inorganic flat solutions, lubricants, and hot air. Acetal copolymer is one of the most creep-resistant crystalline thermoplastics. Other key properties are fatigue endurance and dimensional stability. In both unreinforced or glass-reinforced versions, acetal copolymer can be processed using standard types of injection-molding equipment.

acetal homopolymer (plastics). The chemical structure and high crystallinity result in an unusual combination of physical properties that bridges the gap between metals and plastics. These properties include high melt point, high strength and rigidity, excellent frictional properties, and resistance to fatigue. Acetal homopolymer products retain many of these desirable engineering properties over a wide range of useful service temperatures and humidities, as well as solvent exposures. It is an easily processed engineering thermoplastic especially designed for injection molding and extrusion. The high crystallinity of acetal homopolymer makes it possible to mold it at unusually short cycles. One of the strongest and stiffest thermoplastics, acetal homopolymer has a tensile strength at room temperature of 10,000 psi (6.9×10^4 kPa), and a flexural modulus of 410,000 psi (2.8×10^6 kPa). It has the highest fatigue limit of any of the commercially unfilled thermoplastics. At room temperature the fatigue endurance limit is 5000 psi (3.4×10^4 kPa) at 105°F (41°C), it is 3000 psi (2.1×10^4 kPa). Moisture has very little effect on mechanical properties. It has excellent creep resistance. The notched Izod impact strength is essentially constant from −40 to 212°F (−40 to 100°C). The material is excellent for snap-fit assemblies and integral springs.

acetylene (C_2H_2) (metallurgy). Acetylene is a colorless, combustible gas, with a characteristic odor. It forms highly explosive mixtures with air and oxygen, so that laws and regulations forbid the generation or use of acetylene fuel at pressures above 15 psi (103 kPa). Acetylene is produced by the chemical reaction between water and calcium carbide. When a piece of calcium carbide is dropped into water, bubbles will rise to the surface. With commercially pure oxygen and acetylene, the hottest known flame from gases can be produced, its estimated temperature being in the neighborhood of 6200°F (3425°C). Two volumes of acetylene together with five volumes of oxygen combine and react to produce four volumes of carbon dioxide and two volumes of water vapor. Therefore, for complete combustion of acetylene and oxygen, the ratio of oxygen to acetylene is $2\frac{1}{2}$ to 1. For oxyacetylene welding and heating, the most suitable mixture is generally obtained by using a 50–50 mixture of oxygen and acetylene through the torch. Such a mixture when burned at the tip of a properly designed torch, produces what is known as a neutral flame because its action is neutral in effect, being neither oxidizing nor carburizing in nature.

acid–Bessemer process. In steel making molten blast furnace iron (pig iron) of the correct composition is poured into a Bessemer converter, a pear-shaped container mounted on trunnions, and the converter is tilted to receive the molten metal. There is usually between 5 and 30 tons (4500 and 27,000 kg) of metal in the charge, depending on the custom and size of the plant. There is an acid refractory lining inside the converter so that the slag produced consists of more than 50% silica. After the converter is charged with molten pig iron, air is introduced from turboblowers, or blowing engines, from one of the two trunnions that support the vessel, into a blast box at the bottom of the converter. Air passes out of the box through many holes or tuyers that are $\frac{1}{2}$–$\frac{7}{8}$ in. (13–22 mm) in diameter into the interior of the converter; blast pressure of the air ranges from 20 to 35 psi (138 to 241 kPa). Automatic controls are used to ensure a uniform rate of air volume.

Acrilan fiber. Acrilan acrylic fiber is a copolymer of acrylonitrile, the raw materials of which are obtained from petroleum by the cracking process. Acetylene is formed and is reacted with hydrocyanic acid to give acrylonitrile, a liquid. The acrylonitrile is copolymerized with about 12% of other constituents, including the basic material, the resulting polymer being a white powder similar in appearance to talc. The polymer is dissolved, usually as a 20% solution in dimethyl acetamide, and the solution is spun into a bath made up of dimethyl acetamide and water that precipitates the fibers, which are stretched 350% and are permanently crimped. The fiber is sold with 72,000 filaments or more depending on the filament denier and is cut to staple of various lengths. The density of Acrilan is 0.04 lb/in.^{3} (1.17 g/cm^3) at 68°F (20°C). The dry tensile strength is 0.09 oz (2.5 g) per denier and extension break is 35%; corresponding figures for wet fiber are 0.07 oz (2.0 g) per denier and 44%. The fiber decomposes before it melts; under pressure it sticks to surfaces at about 473°F (245°C). Pure Acrilan is used mainly in sweaters, jersey knit outwear fabric, and blankets. In blends with cotton, it is used in work clothes, where its chemical resistance is useful. In blends with rayon, Acrilan is used to give good resistance to creasing and for permanent pleating. With wool, Acrilan gives dimensional stability. Acrilan is easy to wash and quick to dry.

acrylic plastics. The term acrylic is descriptive of a large class of resins, the most important of which is polymethyl methacrylate. At least 30 acrylate and methacrylate monomers of varying importance can be synthesized. The acrylics are used by such diverse industries as building, automotive, lighting, sign, appliance, and aircraft. These thermoplastic resins are marketed primarily as molding pellets or in such cast stock shapes as sheet, rod, tube, and block. Fine powders, films, and light-conducting filaments are also available. The major constituent of acrylic plastics is methacrylate. Unmodified acrylics are transparent and extremely stable against discoloration. They have almost unlimited color possibilities, exhibit superior dimensional stability, and offer desirable structural and thermal properties. They also demonstrate resistance to weather, breakage, and chemicals, and are light in weight. Physical blends of acrylic resins with other polymers are also in use. Colorless acrylic stock has a white light transmittance of 92%

and index of refraction of 1.49. Izod impact strength for cast acrylic is 0.4–0.5 ft-lb/in. of notch (0.22–0.27 J/cm); that of general-purpose molded acrylic range from 0.3 to 0.5 ft-lb/in. (0.16 to 0.27 J/cm). Tensile strengths for cast acrylics run from 800 to 11,000 psi (5500 to 7.6×10^4 kPa); for general-purpose molding material, from 7000 to 11,000 psi (4.8×10^4 to 7.6×10^4 kPa). Modulus of elasticity ranges from 350,000 to 470,000 psi (2.4×10^6 to 3.2×10^6 kPa). Unmodified acrylic compositions are not embrittled by temperatures of −40°F (−40°C). Impact resistance in most instances is a factor of thickness. Acrylics are not affected by alkalies, nonoxidizing acids, salt water and salt spray, photographic solutions, chemicals used in treating water, petroleum oils and grease, and household cleaning products; however, they are attacked by alcohols, strong solvents, and many aromatic hydrocarbons.

adaptive control. An adaptive-control system performs three basic functions: it measures on-line machine variables in a process in real time; it compares these measured outputs with established values; and it modifies machining activity by changing one or more variables to improve or optimize performance. The adaptive-control system performs these functions automatically.

additive assembly (see automatic assembly)

adhesives, acrylics. One of the chief advantages of modern acrylics is their tolerance of oily surfaces. The latest versions are said to be usable on many metals in their "as-received" condition. Acrylics get their tolerance through agents in the primer or adhesive that have the ability to penetrate or to react with the surface contaminants (oils) and make them harmless. Strength of the acrylics is comparable to that of some epoxies. Shear strengths of 6000 psi (4.2×10^4 kPa) have been reported. Although not formally called two-part adhesives, most first-generation acrylics are usually applied with a primer, so the effect is the same. Volume production requires automatic dispensing or coating equipment. Set time for acrylics can be 60 s and full cure can occur within 10 min. Although these adhesives cure at room temperature, cure time can be reduced by heating. Their gap-filling capability may range from 0.03 to 0.25 in. (0.76 to 6.35 mm). Some acrylics are sensitive to moisture. Operating temperatures may be as high as 350°F (177°C). The characteristic acrylic aroma can be objectional.

adhesives, anaerobics. This single-component adhesive cures in the presence of an active surface when oxygen is excluded. Anaerobics have become known chiefly for their use in threadlocking and machinery assembly. Typical applications are holding locking keys in place and securing bearings on shafts. They are also used to lock and seal pipe threads and, sometimes, to replace gaskets. The newer structural anaerobic adhesives are tough and impact resistant, and their shear strength is about 3000 psi (2.1×10^4 kPa). The operating temperature limit is in the 300–450°F (149–232°C) range. Because they are costly, anaerobics have been applied by the drop rather than the gallon and are usually limited to joining small parts. The use of heat or primers as surface-cure activators enables anaerobic adhesives to be used on nonmetallic parts with fast cures. Anaerobic threadlocking and sealing compounds can be controlled to develop required levels of removal strength. The most recent developments for anaerobic adhesives are combinations of urethanes and acrylics, which develop high-strength, flexible, and durable structural adhesives and gasketing compounds. Using surface activators, these single-component adhesives can cure in seconds and provide high tensile [4000 psi (2.8×10^4 kPa)], shear [3000 psi (2.1×10^4 kPa)], impact, and peel strengths on well-prepared metal surfaces. Since variations in surfaces can cause inconsistencies in strength and durability, parts should be carefully prepared for bonding. A unique new approach to anaerobic adhesive cure is the use of ultraviolet light for glass bonding or coating. The adhesive is sensitized only to light and will cure between transparent substrates in seconds when exposed to long-wavelength, ultraviolet, light.

Anaerobic structural adhesives also have limitations, including the fact that bond strengths may not be equal to those of other adhesives.

adhesives, cyanoacrylates. These adhesives are true single-component adhesives, with cure initiated by an alkaline surface. (Water vapor present on the surface of the part is often sufficient to initiate cure.) Because cyanoacrylates depend on joint spacing for cure, since they do not fill voids, they cure most quickly on smooth, close-fitting parts. The cyanoacrylate polymer forms strong, durable bonds to nearly every material in less than 10 s. As a result, such adhesives can be used for bonding plastics, metals, fabrics, paper, rubber, and, with some limitations, glass. The bonds provide good resistance to most environments. Limitations of cyanoacrylates are in both bond durability and application. Exposure to temperatures above 176°F (80°C), especially under conditions of high humidity, can cause bond failure. Cured bonds also offer limited resistance to impact and peel stresses. Finally, the hazard of the rapid cure, which can even bond skin, must be taken into consideration.

adhesives, epoxies. These are available as liquid pastes, films, tapes, microencapsulates, and powders. They can be cured either by mixing two or more components or by applying heat to a single-component adhesive. Cured epoxies adhere to many substrates; applications include structural bonding, potting, and coating. Through molecular modifications and additives, adhesives can be formulated for high tensile strength, good peel and impact strengths, heat resistance, or high durability. New formulations provide fast cures for two-component epoxies or low-temperature cures for single-component epoxies. Limitations still exist in application techniques for two-component adhesives, and heat-curing adhesives require refrigerated storage. Ultraviolet cures are available for some epoxy coatings, but these have limited use. Epoxy adhesives are generally used where large volumes of adhesive are required to develop strength and durability.

adhesives, polysulfone. Polysulfone is a tough, rigid, high-strength thermoplastic that maintains its properties over a temperature range from −150 to above 300°F (−101 to 149°C). It has found its widest commercial use in molding and extrusion applications for the electronic, electrical, appliance, automotive, aircraft, and general industrial markets. Even though polysulfone has been most extensively used in fabricated items, its ability to adhere to hot metals, its inherent high softening point, and its outstanding heat stability have aroused interest in the structural adhesive field. In addition to strong and ductile bonds, polysulfone adhesives combine the high strength, heat resistance,

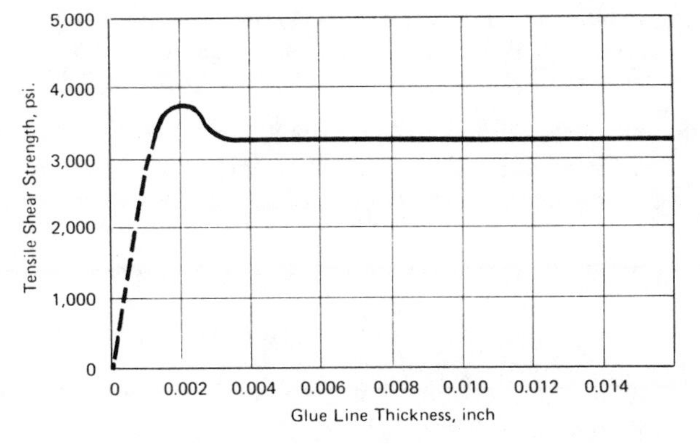

Courtesy Union Carbide Corp.

Effect of glueline thickness on joint strength.

and creep resistance of a thermosetting-type adhesive with the processing characteristics and toughness of a high-molecular-weight thermoplastic. Metal joints, when bonded, combine good shear strength with high peel strength, excellent creep resistance, and a broad use-temperature range. This combination of properties is outstanding for a structural adhesive, especially for one that is thermoplastic. Both films and solutions of polysulfone can be used to bond metal with equal success. Since polysulfone is thermoplastic, it is possible to use it also as a hot melt directly from an extruder. Extruded or molded parts can be adhered directly to metal by means of heat. By using polysulfone-clad metal, structural bonds can be obtained at room temperature by the use of solvent or at elevated temperatures by hot melt in less than 1 min. For aluminum, the cladding can be done simply by heating the metal to 700°F (371°C) and laminating polysulfone film to it by passing the composite through a nip roll. When carbon steel is used, it should first be primed with a diluted solution of the resin and baked for 10 min at 500°F (260°C). Following this, the film can then be ironed on at 500–600°F (260–316°C). Since the polysulfone is now strongly adhered to the metal substrate, it is no longer necessary to use heat to bond the clad metal pieces together. A 2–5% solution of polysulfone in methylene chloride can be used to achieve a strong bond at room temperature. The parts can be assembled most successfully by dipping them into the solution, air drying for 15 sec, and then assembling them in a jig and placing them under pressure of about 500 psi (3450 kPa) for 5 min.

adhesives, urethanes. These adhesives are available as one- or two-part systems and, like epoxies, form strong, durable bonds to many different substrates. Urethane adhesives are generally cured at room temperature as two-component systems. Elongation of urethane adhesives is excellent, making them suitable for bonding nonrigid materials and components that are subjected to severe shock. Two-part urethanes require dry storage and mixing. Environmental resistance is not as good as that of other types of adhesives, with moisture resistance being a particular limitation. The upper temperature limitation is approximately 212°F (100°C). They have excellent low-temperature qualities. Newly developed heat-cured one-component urethane adhesives feature a longer open time and heat resistance to above 392°F (200°C). These high performance, 100% solid adhesives require no mixing and demonstrate good gap-filling ability. They are suitable for most substrates that can withstand the cure temperature and require only a solvent wipe for surface preparation with many substrates.

admiralty metal (see brasses)

aging. Precipitation from solid solution resulting in a change in properties of an alloy, usually occurring slowly at room temperature (natural aging) and more rapidly at elevated temperatures (artificial aging).

age hardening. An aging process that results in increased strength and hardness.

age softening. The loss of strength and hardness at room temperature that takes place in certain alloys due to spontaneous reduction of residual stresses in the strain-hardened structure.

AI (see artificial intelligence)

alclad plate. Composite plate comprised of an aluminum-alloy core having on both surfaces (if on one side only, it is called alclad one-side plate) a metallurgically bonded aluminum or aluminum-alloy coating that is anodic to the core, thus electrolytically protecting the core against corrosion.

alcohol, ethyl (ethanol; CH_3CH_2OH). Sometimes called grain alcohol because starch from grain, when hydrolized to sugars and fermented by enzymes, produces ethyl alcohol and carbon dioxide. Starch from any source is a suitable starting material. Ethyl alcohol used in the laboratory for solvent purposes seldom is pure alcohol, but is usually a mixture of 95% alcohol and 5% water; 95% represents the maximum purity obtainable when alcohol is distilled because this is the constant-boiling-point composition. A constant-boiling-point mixture of liquids, called an azeotrope, cannot be separated by fractional distillation. In order to obtain absolute, or 100% ethyl alcohol, the water must be removed by methods other than fractionation, such as distilling a ternary mixture composed of alcohol, water, and benzene. These three liquids in a composition of 18.5%, 7.4% and 74.1%, respectively, also form an azeotrope with a constant-boiling-point temperature of 148.73°F (64.85°C). Therefore, if sufficient benzene is added to 95% alcohol and the mixture distilled, the water is removed in the distillate along with benzene and some alcohol, but pure alcohol is left in the still pot.

alcohols. Alcohols may be considered as hydroxyl-substituted hydrocarbons of the general formulas R—OH, and Ar—OH, respectively. The hydroxyl group (—OH) is the functional group that characterizes alcohols. Compounds that have hydroxyl groups joined to carbon atoms of alkyl groups are alcohols. The alcohols, like the alkyl halides, may be classified as primary, secondary, or tertiary according to the number of hydrocarbon groups attached to the carbon atom bearing the hydroxyl groups. However, the nomenclature of the alcohols is somewhat more extensive than that encountered in other families of substances. Common names usually are employed for the simpler members having one to four carbon atoms. Such names are formed simply by naming the alkyl group bonded to the hydroxyl function, followed by the word alcohol. Alcohols are named according to IUPAC rules by selecting and naming the longest carbon chain including the hydroxyl group. The terminal "e" of the parent hydrocarbon (alkane) is replaced by "ol". If more than one hydroxyl group appears in the chain, prefixes such as di, tri, etc., are used. Alkyl side chains and other groups are named and their position is indicated. The suffix "ol" is generic for compounds that contain hydroxyl groups. Although names such as cresol, glycerol, and cholesterol contain no clues to their structure, such names do indicate that each contains one or more hydroxyl groups.

aldehydes. Often referred to as carbonyl compounds. The carbonyl carbon of an aldehyde is always bonded to one hydrogen atom, the remaining bond being shared with an alkyl or an aryl group. Trivial names are commonly employed. The IUPAC system of nomenclature follows established rules. The longest carbon chain is named after the parent hydrocarbon with "al" added as a suffix to designate aldehyde. The carbonyl carbon atom of an aldehyde is always number one in the carbon chain and takes precedence over other functional groups that may be present. In many naturally occurring substances the "al" suffix is frequently employed in nonsystematic names to indicate the presence of aldehyde. The stem of the name frequently indicates the source of the substance. The aldehydes are practically insoluble in water. The lower-molecular-weight members of the aldehyde family have sharp, irritating odors, but the higher-molecular-weight members are fragrant.

alginate fibers. The Latin word for seaweed is alga, and it is from this that the word "alginate" is derived. One of the chief constituents of seaweed is alginic acid, and since this substance is a linear polymer, it is a valuable potential source of fibers. Alginic acid is a polyuronic acid, actually a polymer of *d*-mannuronic acid. Alginic acid is composed of very-long-chain molecules with reactive side chains (the carboxylic groups). Solutions of sodium alginate are readily soluble in water and, being very viscous, are highly suitable for spinning. In the manufacture of alginate fibers, seaweed, usually *Laminariae*, is collected, dried, and milled. The powdered seaweed is treated with a solution of sodium carbonate and caustic soda, which converts all the alginate in the seaweed to sodium alginate. The viscous, brownish solution of sodium alginate is suitably purified by sedimentation and is bleached and sterilized by the addition of sodium hypochlorite. The alginic acid is extracted from this solution by a series of chemical reactions and then it is purified and dried. In spinning, an 8–9% solution of sodium is made, and is sterilized by the addition of a bactericide. It is filtered and spun on a viscose spinning machine into a coagulating bath that contains normal calcium chloride solution, 0.02 *N* hydrochloric acid solution, and a small quantity of a cationic surface-active agent. As the sodium alginate issues from the jet, it is precipitated in filament form as calcium alginate. The filaments are drawn together, washed, lubricated, and dried and wound. Calcium alginate yarns have a dry strength comparable with that of viscose rayon, but their wet strength is low, their extensibility is sufficiently high to meet most textile requirements.

alkyds (plastics). Alkyd compounds are dry, granular, or nodular types formulated from polyester-type resins. These unsaturated resins are produced through the reaction of an organic alcohol with an organic acid. Selection of suitable polyfunctional alcohols and acids permits manipulation of a large number of repeating units. Formulating can provide resins that demonstrate a wide range of characteristics involving flexibility, heat resistance, chemical resistance, and electrical properties. Unsaturated polyester resins are dissolved in and reacted with unsaturated monomers such as styrene, diallyl phthalate, diacetone, acrylamide, methyl methacrylate, or vinyl toluene. A catalyst, usually of the peroxide type, is added to speed the reaction of cross-linking the resin and monomers, which results in a cured thermoset. To prepare the final compound, pigments, lubricants, and fillers are added to the resin.

allotropic metals. The atoms of some chemical elements, both metallic and nonmetallic, will crystallize on different space lattices under different conditions of temperature or pressure. Such materials are said to be allotropic. The principal allotropic metals are chromium, cobalt, iron, manganese, nickel, tin, and tungsten. Of these, the most important is iron. All of the major changes that take place in steel, and the other iron alloys, are related either directly or indirectly to the allotropic changes occurring in iron. Without these changes, iron and its alloys could not have their properties modified so drastically by heat treatment.

alloys. An alloy may be defined as a material composed of two or more elements, at least one of which is a metal, and possessing metallic properties. The addition of a second element to the first so as to form an alloy usually results in greatly changed properties. It is to obtain these changed properties that alloys are produced and used. Alloys may be formed by any one of three mechanisms. The first and probably the simplest type is where the two components are insoluble in each other in the solid state. In this case the base metal and the alloying element maintain their individual identities and properties. The lattice structures of both are unchanged. The alloying element and the base metal exist in the alloy as an intimate mixture. The second mechanism occurs when the two elements are soluble in each other in the solid state. They thus form a solid solution with the alloying element being dissolved in the base metal. The third mechanism of alloying is where the elements combine to form intermetallic compounds. In this case atoms of the alloying element replace atoms of the lattice of the base element in definite proportions and in definite relationships as to position. As a result of this type of mechanism the lattices of these intermetallic compounds have less symmetry and fewer planes of greater atomic density. These compounds are, therefore, more resistant to deformation and tend to be hard and less ductile, and of high strength. Normally, a useful alloy can only be produced if the elements concerned are soluble in each other in the molten state, that is, they form a single homogeneous solution in the crucible. Some molten metals do not dissolve in each other but instead form two separate layers, as do oil and water when an attempt is made to mix them. Thus, molten lead and molten zinc will not dissolve in each other completely [unless the temperature is in excess of 1468°F (798°C)]. Instead a layer of molten zinc (containing some dissolved lead) will float on top of a layer of molten lead (which will contain some dissolved zinc). Such a situation cannot be expected to give rise to the formation of a useful alloy. When cast, the lighter metal will tend to float to the surface before solidification is complete though there may be limited entanglement of the two component metals. Two such metals may be compounded successfully using the techniques of powder metallurgy. Generally, a prerequisite to the formation of a useful alloy is that all the components of the alloy shall be mingled intimately together in the liquid state. Certain materials with an excessive amount of nonmetals are not considered alloys but rather "cermets." The name is derived from the combination of ceramic materials and metals present.

alloy steels. These contain appreciable quantities of alloying elements in addition to carbon. They include (1) low-alloy, high-strength structural steels; (2) quenched and tempered low-carbon construction alloy steels; (3) AISI–SAE alloy steels; (4) alloy tool steels; (5) stainless steels; (6) heat-resisting steels; and (7) magnet steels.

alloy steels (AISI–SAE). Steels whose composition have been standardized by the American Iron and Steel Institute and the Society of Automotive Engineers. A numbering system using as many as five digits designates the composition of the alloy. The nominal carbon content is given in hundredths of a percent by the last two numbers when four digits are used and by the last three numbers when five digits are used. With the exception of the low-carbon, plain-carbon steels, the AISI–SAE steels are always used in the heat-treated condition. The AISI–SAE steels are used for applications such as carburized or through-hardened gears, steering mechanical parts, transmissions, shafting, and ordnance parts. The 52000 series are used mainly for ball and roller bearings.

alloy steels, austenitic. Steels that remain austenitic in structure (gamma iron) upon slow cooling from the temperature of solidification. These steels do not undergo any change in the condition of the iron and therefore exhibit no critical temperature upon cooling. These steels cannot be hardened by heat treatment, although they may be cold work hardened and annealed. If any precipitate occurs with these steels upon slow cooling from a high temperature, they may be reheated and quenched to redissolve the precipitate and keep it in solution. The austenitic steels exhibit great shock strength and low elastic strength, and are very ductile. They work harden very rapidly and develop great resistance to wear by abrasion. The chrome–nickel austenitic steels are very resistant to corrosion.

alloy steels, cementitic. Some alloying elements on being added to steel in increasing amounts fail to convert the steel into an austenitic type. A steel containing 18.0% of a special element and 0.6% of carbon, upon slow cooling from above its critical temperature, would have a structure of ferrite or martensite with numerous particles of cementite embedded in the ferritic or martensitic matrix. Such a steel has been referred to as a cementitic type of steel. The cementitic types of alloy steel are usually difficult to use, requiring special care in annealing to make it machinable. These steels are subject to such hardening heat treatments as to cause most of the cementite or carbides to be absorbed and retained in a martensitic structure. These steels are largely used in tools, particularly where hardness and resistance to wear are important.

alloy steels, cold work. Includes the water-, oil-, and air-hardening tool steels (W, O, A steels), all of which have a fairly high carbon content (0.60–2.25%) and varying degrees of hardenability as indicated by their names, and are controlled by the amount and kind of alloying elements they contain. The shock-resisting tool steels (S steels), considered as special cold-work tool steels, have a lower carbon content (0.50%) in order to improve their toughness. The high-carbon, high-chromium tool steels (D steels) have large amounts of chromium (12%) and other carbide formers. These alloy additions produce an air-hardening composition with excellent wear resistance, useful for blanking dies, thread-rolling dies, and brick molds. Since the strong carbide formers have such a great affinity for carbon, the carbon content is raised in these steels to ensure that there is enough uncombined carbon remaining in the austenite to yield a martensite matrix of adequate hardness upon hardening. The carbon tungsten tool steels (F steels) are similar to the oil- and water-hardening steels, but they have extra amounts of tungsten, resulting in improved wear resistance because of the tungsten carbide particles. The low-alloy special purpose tool steels (L steels) are similar to the W steels, but have higher amounts of strong carbide formers for improved wear resistance. The low-carbon mold steels (P steels) have the lowest carbon contents of all the tool steels and, after machining or pressing to shape, are carburized for improved wear resistance.

alloy steels, ferritic stainless. Contains chromium, no nickel, and tolerates only small amounts of austenite-stabilizing carbon. If the carbon content is increased, the chromium content must be increased in order to maintain balance and a ferritic structure. In this balanced condition these steels can be heated to the melting point without transforming to austenite. Thus, it is impossible to harden them by quenching and tempering.

alloy steels, high-speed. Contains either tungsten (T steels) or molybdenum usually with tungsten (M steels) as the principal carbide formers. Both T and M steels contain chromium and vanadium. The high carbon content is necessary to satisfy the carbide forming tendencies and to produce excellent wear resistance and hardness at red heat. At the same time the carbon content is not so high that toughness is lacking. Since molybdenum is a cheaper alloying element than tungsten and is about twice as effective as tungsten, the T steels have been almost entirely replaced by M steels. There is no significant difference between the performance of the two major classes of high-speed steels.

alloy steels, hot work. "H steels" contain fairly large amounts of strong carbide formers, and vanadium is found in these compositions. Their carbon content is below 0.65% carbon so that they exhibit moderately good toughness at high strength levels. They are used for forging dies, extrusion dies, and die-casting dies.

alloy steels, iron–chromium. Stainless steels containing from 14 to 18% chromium and a maximum of 0.12% carbon are magnetic and consist structurally of a solid solution of iron and chromium. They are a ferritic type of stainless steel used for general requirements where resistance to corrosion and heat is needed, but where service conditions are not too severe and slight discoloration of the surface during service can be tolerated. Steels of this type, which contain less than 0.10% carbon, with chromium from 14 to 20%, are not heat treatable except for an annealing treatment. They cannot be hardened by heat treatment because they do not transform into austenite when heated to an elevated temperature. These steels are susceptible to grain growth at elevated temperatures, and with a larger grain size they suffer a loss in toughness. The ferritic stainless steels may be cast and forged hot or cold, but they do not machine easily. They possess a high resistance to corrosion in an ordinary atmosphere providing they have been highly polished and are free from foreign particles. Their machinability can be improved by additions of molybdenum and sulfur, or phosphorus and selenium.

alloy steels, low-alloy, high-strength structural. Contains insufficient carbon and alloying elements to be hardened effectively by quenching to martensite. This is advantageous because it enables them to be welded without becoming brittle. At the same time, the alloying elements they contain alter the microstructure so that it resembles a higher-carbon steel cooled at moderately fast (air-blast quench) rates. In addition these steels contain slightly more phosphorus and silicon than do the carbon steels, thereby strengthening the ferrite network. These changes in microstructure raise the yield strength

Alloying Elements on Steel, Relative Effects

Element	*Influence on the properties of steel*	*Uses in steel*
Nickel	Stabilizes γ by raising A_4 and depressing A_3. It is the universal grain refiner in alloy steels (and many nonferrous alloys). Strengthens ferrite by solid solution. Unfortunately a powerful graphitizer	In amounts up to 5% as a grain refiner in case-hardening steels. Along with chromium and molybdenum in low-alloy constructional steels. In larger amounts in stainless and heat-resisting steels
Manganese	Like nickel it stabilizes γ but unlike nickel it forms stable carbides	Low-manganese steels are not widely used though in recent years it has been used to replace small amounts of more expensive alloying elements, e.g. nickel. The high-manganese (Hadfield) steel contains 12.5% Mn and is austenitic but hardens on abrasion
Chromium	Stabilizes α by raising A_3 and depressing A_4. Forms hard stable carbides. Strengthens ferrite by solid solution. In amounts above 13% imparts stainless properties. Unfortunately increases grain growth	In small amounts in constructional and tool steels. Also in ball bearings. In larger amounts in stainless and heat-resisting steels.
Molybdenum	Strong carbide-stabilizing influence. Raises the high-temperature creep strength of suitable alloys. Imparts some sluggishness to tempering influences	Reduces 'temper brittleness' in nickel-chromium steels. Increases red-hardness of tool steels. Now used to replace some tungsten in high-speed steels
Vanadium	Strong carbide-forming tendency. Stabilizes martensite and increases hardenability. Like nickel it restrains grain growth. Induces resistance to softening at high temperatures once the steel is hardened	Used in steels required to retain hardness at high temperatures, e.g. hot-forging dies, extrusion dies, die-casting. Also increasingly in high-speed steels
Tungsten	Has similar effects to chromium in stabilizing α. Also forms very hard carbides. Renders transformations very sluggish—hence, once hardened, a steel resists tempering influences	Used mainly in high-speed steels and other tool and die steels. particularly for use at high temperatures
Cobalt	Induces sluggishness to the transformation of martensite. Hence increases 'red hardness'	Super high-speed steels and 'maraging' steels. Permanent magnet steels and alloys
Silicon	A strong graphitizing influence — hence not used in high-carbon steels. Imparts casting fluidity. Improves oxidation resistance at high temperatures	Up to 0.3% in steels for sandcastings where it improves fluidity. In some heat-resisting steels (up to 1.0%)

Alloy Steels, Low-Alloy Constructional

Type of steel	Relevant spec'n: B.S. 970:	Composition (%)	Condition	Mechanical properties: Yield stress (N/mm^2)	Tensile stress (N/mm^2)	Elongation (%)	Izod (J)	Heat-treatment	Uses
Low manganese	150M28	0.28 C 1.50 Mn	Normalized	355	587	20	—	Oil-quench from 860°C (water-quench for sections over 38 mm diameter). Temper as required	Automobile axles, crankshafts, connecting rods, etc., where a relatively cheap steel is required
Nickel-manganese	503M40	0.40 C 0.90 Mn 1.00 Ni	Quenched and tempered at 600°C	494	695	25	91	Oil-quench from 850°C; temper between 550° and 660°C and cool in oil or air	Crankshafts, axles, connecting rods; other parts in the automobile industry and in general engineering
Manganese-molybdenum	608M38	0.38 C 1.50 Mn 0.50 Mo	28.5 mm bar, o.q. and tempered at 600°C	1000	1130	19	70	Oil-quench from 830°–850°C; temper between 550° and 650°C and cool in oil or air	A substitute for the more highly alloyed nickel-chrome-molybdenum steels
Nickel-chromium	653M31	0.31 C 0.60 Mn 3.00 Ni 1.00 Cr	28.5 mm bar, o.q. and tempered at 600°C	819	927	23	104	Oil-quench from 820°–840°C temper between 550° and 650°C. Cool in oil to avoid 'temper brittleness'	Highly stressed parts in automobile and general engineering, e.g. differential shafts, stub axles, connecting rods, high-tensile studs, pinion shafts
Nickel-chromium-molybdenum	817M40	0.40 C 0.55 Mn 1.50 Ni 1.20 Cr 0.30 Mo	O.q. and tempered at 200°C O.q. and tempered at 600°C	— 988	2010 1080	14 22	27 69	Oil-quenched from 830°–850°C; 'light temper' 180°–200°C; 'full temper' 550°–650°C—cool in oil or air	Differential shafts, crankshafts and other highly-stressed parts where fatigue and shock resistance are important. In the 'light tempered' condition it is suitable for automobile gears. Can be surface hardened by nitriding
	835M30	0.30 C 0.55 Mn 4.25 Ni 1.25 Cr 0.30 Mo	Air-hardened and tempered at 200°C	1470	1700	14	35	Air-harden from 820°–840°C; temper at 150°–200°C and cool in air	An air-hardening steel for aero-engine connecting rods, valve mechanisms, gears, differential shafts and other highly-stressed parts. Suitable for surface hardening by cyanide or carburising
Manganese-nickel-chromium-molybdenum	945M38	0.38 C 1.40 Mn 0.75 Ni 0.50 Cr 0.20 Mo	28.5 mm bar. o.q. from 850°C and tempered at 600°C	958	1040	21	85	Oil-quench from 830°–850°C; temper at 550°–660°C, and; cool in air	Automobile and general engineering components requiring a tensile strength of 700 to 1000 N/mm^2

Carbon Steel, Properties and Uses

	Relevant specifications	Typical compositions (%)	Heat-treatment	Typical mechanical properties. Y.P. (N/mm²)	T.S. (N/mm²)	Elong. (%)	Impact (J)	Hardness (Brinell)	Uses
Mild steel	B.S. 970:040A10	0.10 C 0.40 Mn	No heat treatment—except process annealing (9.5.1) to remove the effects of cold-work	—	300	28	—	—	Lightly stressed parts produced by cold-forming processes, e.g. deep-drawing and pressing
Structural steels	B.S. 15	0.20 C	No heat treatment	240	450	25	—	—	General structural steel
	B.S. 968	0.20 C 1.50 Mn	No heat treatment	350	525	20	—	—	High tensile structural steel for bridges and general building construction—fusion welding quality
Casting steel	B.S. 1504 161B	0.30 C	No heat treatment other than 'annealing' (9.5.4) to refine grain	265	500	18	20	150	Castings for a wide range of engineering purposes where medium strength and good machinability are required
Constructional steels	B.S. 970:080M40	0.40 C 0.80 Mn	Harden by quenching from 830°–860°C. Temper at a suitable temperature between 550° and 660°C	500	700	20	55	200	Axles, crankshafts, spindles, etc., under medium stress
	B.S. 970:070M55	0.55 C 0.70 Mn	Harden by quenching from 810°–840°C. Temper at a suitable temperature between 550° and 660°C	550	750	14	—	250	Gears, cylinders and machine-tool parts requiring resistance to wear
Tool steels		0.70 C 0.35 Mn	Heat slowly to 790°–810°C and quench in water or brine Temper at 150°–300°C	—	—	—	—	780	Hand chisels, cold sates, mason's tools, smith's tools, screwdriver blades, stamping dies, keys, cropping blades, miner's drills, paper knives
	B.S. 4659:BW1A	0.90 C 0.35 Mn	Heat slowly to 760°–780°C and quench in water or brine. Temper at 200°–300°C	—	—	—	—	800	Press tools; punches; dies; cold-heading, minting and embossing dies; shear blades; woodworking tools; lathe centers; draw plates
	B.S. 4659:BW1B	1.00 C 0.35 Mn	Heat slowly to 770°–790°C and quench in water or brine. Temper at 150°–350°C	—	—	—	—	800	Taps; screwing dies; twist drills; reamers; counter sinks; blanking tools; embossing, engraving, minting, drawing, needle and paper dies; shear blades; knives; press tools; center punches; woodworking cutters; straight edges; gouges; pneumatic chisels; wedges
	B.S. 4659:BW1C	1.20 C 0.35 Mn	Heat slowly to 760°–780°C and quench in water or brine. Temper at 180°–350°C	—	—	—	—	800	Engraving tools, tiles; surgical instruments; taps; screwing tools

Stainless and Heat-Resisting Steels

Type of steel	Relevant specification	Composition (%)	Condition	Typical mechanical properties: Yield stress (N/mm²)	Tensile strength (N/mm²)	Elongation (%)	Hardness (Brinell)	Heat-treatment	Uses
Stainless iron	B.S. 970:403S17	0.04 C 0.45 Mn 14.00 Cr	Soft	340	510	31	—	Non-hardenable except by cold work	Wide range of domestic articles — forks, spoons. Can be spun, drawn, and pressed
Cutlery steel	B.S. 970:420S45	0.30 C 0.50 Mn 13.00 Cr	Cutlery temper Spring temper	— —	1670 1470	— —	534 450	Water- or oil-quench (or air-cool) from 950–1000°C. Temper: (for cutlery) — at 150–180°C; (for springs) — at 400°–450°C	Cutlery and sharp-edged tools requiring corrosion resistance. Circlips, etc. Approximately pearlitic in structure when normalized.
18/8 stainless	B.S. 970:302S25	0.05 C 0.80 Mn 8.50 Ni 18.00 Cr	Softened Cold-rolled	278 803	618 896	50 30	170 —	Non-hardening except by cold-work. (Cool quickly from 1050°C to retain carbon in solid solution)	Particularly suitable for domestic and decorative purposes. An austenitic steel
18/8 stainless (weld decay proofed)	B.S. 970:321S20	0.05 C 0.80 Mn 8.50 Ni 18.00 Cr 1.60 Ti	Softened Cold-rolled	278 402	649 803	45 30	180 225	Non-hardening except by cold-work. (Cool quickly from 1050°C to retain carbon in solid solution)	A weld-decay proofed steel (fabrication by welding can be safely employed). Used extensively in nitric acid plant and similar chemical processes

Type of steel	Relevant specification	Composition (%)	Condition	Testing temperature (°C)	0.1% P.S. (N/mm²)	Tensile strength (N/mm²)	Elongation (%)	Maximum working temperature (°C)	Uses
Heat-resisting steel	AISI Series 311	0.15 C 20.00 Cr. 25.00 Ni	Forged or rolled	—	—	—	—	820	Conveyor chairs and skids, heat-treatment boxes, recuperator valves, valves and other furnace parts
Heat-resisting steel	AISI Series 302B	0.10 C 1.50 Si 1.00 Mn 19.00 Cr 11.00 Ni	Air-cooled from 1100°C	20 400 600 800 900	347 248 217 139 85	698 527 465 248 155	55 40 37 45 50	1000 (air) 950 (flue gases)	A fairly cheap grade of heat-resisting steel with a good combination of properties

Type of steel	Relevant Specification	Composition (%)	Heat-treatment	Uses
'60'-carbon-chromium	B.S. 970:526M60	0.60 C 0.65 Mn 0.65 Cr	Oil-quench from 800°–850°C. Temper: (i) for cold-working tools at 200°–300°C (ii) for hot-working tools at 400°–600°C	Blacksmith's and boilermaker's chisels and other tools. Mason's and miner's tools. Vice jaws. Hot-stamping and forging dies
1% carbon-chromium	B.S. 970:534A99	1.00 C 0.45 Mn 1.40 Cr	Oil-quench from 810°C; temper at 150°C	Ball- and roll-bearings. Instrument pivots. Cams. Small rolls
High carbon-high chromium	B.S. 4659:BD3	2.10 C 0.30 Mn 12.50 Cr	Heat slowly to 750°–800°C and then raise to 960–990°C. Oil-quench (small sections can be air cooled). Temper at 150°–400°C for 30–60 minutes	Blanking punches, dies and shear blades for hard, thin materials. Dies for moulding abrasive powders, e.g. ceramics. Master gauges. Thread-rolling dies
1% vanadium	—	1.00 C 0.25 Mn 0.20 V	Water-quench from 850°C; temper as required	Cold-drawing dies; etc.
4% vanadium	—	1.40 C 0.40 Mn 0.40 Cr 0.40 Mo 3.60 V	Water-quench from 770°C; temper at 150°–300°C	Cold-heading dies; etc.
Hot-working die steel	B.S. 4659:BH12	0.35 C 1.00 Si 5.00 Cr 1.50 Mo 0.40 V 1.35 W	Pre-heat to 800°C, soak and then heat quickly to 1020°C and air cool. Temper at 540°–620°C for 1½ hours	Extrusion dies, mandrels and noses for aluminum and copper alloys. Hot-forming, piercing, gripping and heading tools. Brass-forging and hot-pressing dies
High-speed steels 18% Tungsten	B.S. 4659:BT1	0.75 C 4.25 Cr 18.00 W 1.20 V	Quench in oil or air blast from 1290°–1310°C. Double temper at 565°C for 1 hour	Lathe, planar and shaping tools; millers and gear cutters. Reamers; broaches; taps; dies; drills and hacksaw blades. Bandsaws. Roller-bearings at high temperatures (gas turbines)
12% Cobalt	B.S. 4659:BT6	0.80 C 4.75 Cr 22.00 W 1.50 V 0.50 Mo 12.00 Co	Quench in oil or air blast from 1300°–1320°C. Double temper at 565°C for 1 hour	Lathe, planar and shaping tools, milling cutters, twist drills, etc. for exceptionally hard materials. Has maximum red-hardness and toughness. Suitable for severest machining duties, e.g. manganese steels and high-tensile steels, close-grained cast irons
Molybdenum '562'	B.S. 4659:BM2	0.83 C 4.25 Cr 6.50 W 1.90 V 5.00 Mo	Quench in oil or air blast from 1250°C. Double temper at 565°C for 1 hour	Roughly equivalent to the standard 18–4–1 tungsten high-speed steel but tougher. Drills, reamers, taps, milling cutters, punches, threading, dies, cold-forging dies
9% Mo–8% Co	B.S. 4659:BM42	1.00 C 3.75 Cr 1.65 W 1.10 V 9.50 Mo 8.25 Co	Quench in oil or air blast from 1180°–1210°C. Triple temper at 530°C for 1 hour	Similar uses to the 12% Co–22% W high-speed steel

about 10–15% above that of carbon structural steel. Lighter secretions, of lower cost, can therefore carry a given load. Since rust and corrosion become increasingly important as section sizes become smaller, copper and phosphorus are used to provide improved corrosion resistance. These steels are used for railroad cars and transportation equipment. They are also useful for bridge structural members and for columns in steel-frame buildings.

alloy steels, maraging. A special type of high-nickel steel. These compositions develop martensite upon cooling from the austenitizing temperature, but the martensite formed in these steels is, unlike the martensite of AISI alloy steels, ductile and tough. The ductility and toughness of this martensite results from its low-carbon content, which is below 0.03% carbon. In the martensitic condition these steels can be cold worked and can be hardened by precipitation at temperatures below the austenitizing temperature, e.g., 900°F (480°C). Hardening is believed to result from precipitation of compounds such as Ni_3Mo and Ni_3Ti. The hardened maraging steels have yield strengths up to 300,000 psi (2.1×10^6 kPa) and Charpy V-notch impact strengths well over 10 ft-lb (13.6 J). The impact strength for maraging steels with 200,000 psi (1.4×10^6 kPa) yield strength is in the 50–60 ft-lb (67.8–81.3 J) range. These steels are useful in the manufacture of large structures having critical strength requirements such as space vehicle cases, hydrofoil struts, and extrusion press rams.

alloy steels, martensitic. Contains balanced amounts of chromium (ferrite stabilizer) and carbon and nickel (austenite stabilizers), so that on heating the steel becomes austenitic, but upon cooling tends to revert to ferrite. These compositions can be heated to the austenitic range of temperatures and will transform to martensite upon cooling at suitable rates. The carbon content is sufficient to produce a martensitic hardness that is adequate for cutlery and surgical instruments. The properties of martensitic alloy steel are not unlike those of fully hardened and martensitic carbon steel, and, therefore, these steels are used for tools and dies that are subjected to heating. These steels are usually hardened by air cooling from above their critical temperature and annealed by unusually slow cooling in a furnace.

alloy steels, pearlitic. The pearlitic type of alloy steel contains a relatively small percentage of pearlite, although with a low percentage of carbon the amount of this element may be as high as 6%. The structure and characteristics of pearlitic alloy steels are similar to carbon steels, and the microscopic analysis may reveal all pearlite, or a mixture of pearlite and free cementite, all depending on the amount of carbon and pearlite. Generally, this element lowers the eutectoid ratio of carbon to iron, and it therefore requires less than 0.85% carbon in a special steel to produce 100% pearlite or eutectoid steel. The lower-carbon-alloy steels of this group usually are used for structural purposes; they also may be hardened by case-hardening methods and used for tools. The medium-to-high carbon steels of this group are subjected to heat treatments and are used in highly stressed parts of structures and machines or for many applications in tools.

alloy steels, precipitation-hardenable stainless. Have austenitic, martensitic, or semiaustenitic and ferrite stabilizers, principally chromium and nickel. Lowering the chromium/nickel ratio tends to stabilize the austenitic condition; raising it promotes transformation to martensite. Hardening is accomplished by precipitation of titanium or copper from martensitic types by precipitation of aluminum from semiaustenitic types and by precipitation of carbide from austenitic types. Precipitation may result from simple heating and aging cycles, perhaps following a subzero treatment or cold-working-equipment treatment to transform an austenitic structure to a martensitic one. The precipitation-hardenable steels were developed for applications such as aircraft structural members where the size and shape of the structure prevented hardening by cold work or by conventional quenching and tempering.

alloy steels, quenched and tempered low-carbon construction. Known as low-carbon martensites. They are similar to low-alloy, high-strength structural steels, except that their alloy content permits quenching to bainite or martensite. The low-carbon martensite of these steels retains toughness to −50°F (−46°C), and these compositions produce welded joints fully as strong as the unwelded base metal. The low-carbon martensites tend to have slightly higher alloy contents than the low-alloy, high-strength structural steels and in addition may contain alloying elements such as boron, vanadium, and molybdenum, all of which contribute to hardenability, and vanadium, molybdenum, and titanium, which form persistent carbides that resist softening upon tempering. These steels are used in the form of plate for the construction of welded pressure vessels and as structural members for large steel structures, mining equipment, and earth movers.

alloy steels, silicon (see silicon steels)

alloy steels, stainless. Rely primarily upon the presence of chromium for the achievement of their stainless qualities. In general, the higher the chromium content, the more corrosion resistant is the steel. There are three common classes of stainless steels: (1) austenitic, (2) ferritic, and (3) martensitic. The names of these classes reflect the microstructure of which the steel is normally composed. The alloying elements in steel can be classed as austenite stabilizers and ferrite stabilizers. The austenite stabilizers of importance are carbon, nickel, nitrogen, and manganese. These elements enhance the retention of austenite as the steel is cooled. When 12% or more manganese is present, or when 20% or more nickel is present, it is impossible to cool steel slowly enough to allow austenite to transform to ferrite. The ferrite stabilizers of importance are chromium and the strong carbide formers. The ferrite stabilizers tend to prevent transformation of steel to austenite upon heating. Whether a steel is austenitic, ferritic, or martensitic depends on the balance between the amounts of austenite and ferrite stabilizers present, and the heating–cooling cycle to which steel has been subjected.

alloy steels, tool. Plain carbon steels if used for cutting tools lack certain characteristics necessary for high-speed production, such as red hardness, abrasive resistance, and hot strength. Carbon is a most important alloy and determines many of the mechanical properties that may be made use of in steel. As the carbon content increases, the hardness, strength, wear resistance, and red hardness increase, and the toughness and impact strength decrease. In general, their high-temperature strength and hardness are poor when compared with alloy tool steels. They deform during hardening so that cracking and distortion take place. The low-carbon tool steels (up to 0.35%) are used for structural steels, bars, sheet steels, etc. Medium-carbon steels (up to 0.50%) are used for forgings and high-strength-steel castings. High-carbon steels are used for forgings, high-strength wire, high-strength tools, and cutting tools. The AISI-SAE designations list 13 different classifications of which the major groups are:

Designation	Type
W2-W7	Water-hardening, cold-work tool steels
S1-S5	Shock-resisting tool steels
O1-O7	Oil-hardening, cold-work tool steels
A2-A7	Air-hardening, cold-work tool steels
D1-D7	High-carbon, high-chromium, cold-work tool steels
H11-H43	Hot-work tool steels
T1-T15	Tungsten high-speed tool steels
M1-M36	Molybdenum high-speed tool steels
L1-L7	Low-alloy, special-purpose tool steels
F1-F3	Carbon-tungsten tool steels
P1-P20 PPT	Low-carbon mold steels

allyl plastics. The allyl family of esters has as its basis the allyl radical, usually formed from reaction with allyl alcohol. Allyl esters based on mono- and diabasic acids are available as low-viscosity monomers and thermoplastic prepolymers. Allyl monomers and prepolymers find wide utility as a cross-linking agent for unsaturated polyester resins and in the preparation of reinforced thermoset molding compounds. Allyl resins can be processed by all modern thermoset techniques. The resins are excellent for their maintenance of electrical properties under conditions of high temperature and high humidity. Reinforced molding compounds are characterized by their combination of dimensional stability, chemical resistance, good mechanical strength, and exceptional heat resistance. The purity of the resin minimizes ionic contamination and corrosion of metallic inserts.

Alnico magnet alloy (see magnet steels)

alternating current (ac) motors. The basic principles of magnetism and electromagnetic induction are the same for ac and dc motors, but the application of the principle is different because of the rapid reversals and changes of magnitude found in alternating current. There are three principal types of ac motors. These are, the universal motor, the induction motor, and the synchronous motor. A universal motor is identical with a dc motor and may be operated on either alternating or direct current. The induction motor has a wide variety of applications because of its operating characteristics. It does not require special starting devices or excitation from an auxiliary source and will handle a wide range of loads. It is adaptable to almost all loads when an exact and constant speed is not required. Synchronous motors rotate at a speed that is synchronized with the applied alternating current. These motors have some features in common with induction motors and a construction similar to that of alternators.

alumina cements. Made by fusing limestone and bauxite in an electric furnace or blast furnace and grinding the clinker so produced. They have the property of high strength relatively soon after setting. They eventually attain a strength about twice that of Portland cement, but are not as rapid in setting. Alumina cements are hydraulic and are highly resistant to attack by seawater as well as to high temperature. They set with a high evolution of heat. They are used in mortars for fire brick and for construction in locations to be subjected to high temperatures.

aluminum. Aluminum and aluminum alloys are characterized by lightness, high strength–weight ratio, good corrosion resistance, high thermal and electrical conductivities, nontoxicity, and ease of fabrication. They are used in both the wrought and case condition in many applications, such as in aircraft construction, in chemical and process industries, in building industries, and in domestic uses. The strength of pure aluminum is increased by cold work and alloying, but this also reduces its corrosion resistance. One of the main limitations of aluminum is its low melting point [1220°F (660°C)] and its inability to be used at temperatures above 572°F (300°C) (because of creep) in spite of its superior oxidation resistance. However, aluminum, having a face-centered-cubic lattice, maintains its ductility at a very low temperature. High-purity aluminum (99.9% Al) is very soft and ductile, but it is considerably strengthened by the addition of even slight amounts of alloying elements such as silicon, manganese, iron, and copper. Commercially pure aluminum and some of its solid-solution alloys can be hardened only by cold work, but many age-hardened alloys have been developed that contain alloying elements such as copper, zinc, or magnesium. The most widely used is an aluminum–copper alloy known as Duralumim, containing from 2.5 to 5.5% copper and small amounts of magnesium and manganese. Aluminum alloys containing magnesium attain their optimum properties by natural aging at room temperatures, whereas those without magnesium require artificial aging at elevated temperatures.

aluminum, Alclad alloys. Aluminum alloy sheets, plates, tubes, and wire may be clad with high-purity aluminum or aluminum alloys to improve corrosion resistance. In this case, the alloys are identified by preceding the designation of the core alloy with the designation Alclad, for example, Alclad 2024-0 sheets. Sheet and plate may be clad on one or both sides. The strength of the cladding material is usually less than that of the core alloy. Consequently, the strength of the Alclad metal is slightly less than that of unclad materials, in proportion to the amount of cladding applied.

aluminum alloys. Aluminum can be hardened by solid-solution hardening, by cold working, and by precipitation hardening. Among the elements added to aluminum for precipitation hardening are copper, manganese, nickel, and silicon. However, silicon is used mainly to improve castability. Zinc is used as a solid-solution hardener, and magnesium is added to improve corrosion resistance. Room-temperature mechanical properties of aluminum alloys are, in general, inferior to those of steel, almost equal to those of copper alloys, and superior to those of magnesium alloys. Dividing the strength values by the specific gravity yields a number known as the specific strength. Comparison of specific strengths shows that most, but not all, aluminum alloys are superior to most steel compositions. Specific strengths have an important bearing on payloads and dead weight. When structural rigidity is required, the additional bulk of aluminum alloys is advantageous. Aluminum alloys, in general, are considered easily machined, although in some cases good surface finish is difficult to obtain. Joining of aluminum alloys is accomplished successfully by a variety of welding and brazing methods. Aluminum alloys are readily mechanically worked cold or at elevated temperatures, since they are relatively soft and have good ductibility.

aluminum, anodizing. In the anodizing method, a light or heavy oxide coating is produced on aluminum by making the aluminum part of the anode in an acid electrolytic bath. The anodic coating is porous and must be sealed by immersion in boiling water or linseed oil. Such coatings are applied to aluminum windows and doors, aluminum trim on buses, and aluminum scales and other parts of drafting machines. The thickness of the oxide film is controlled basically by the anodizing time. There is a practical maximum thickness, because while the film grows from the inside out, as a conversion coating must, the electrolyte dissolves the film already formed. As a result of this chemical dissolution, the rate of film growth steadily decreases, until, finally, the rate of film formation and dissolution become equal, and no further increase in thickness can occur. For most applications a film of 0.5–2 mils (0.013–0.05 mm) is developed. The electrolyte for anodizing may be sulfuric acid, chromic acid, oxalic acid, or boric acid. The latter two acids are used only for special effects. Oxalic acid produces a yellowish tint in the oxidized surface. Boric acid is largely restricted to the production of thin and impermeable dielectrical condenser foil. The sulfuric acid process is the usual one employed, in an acid concentration of 15–24%. "Hard anodizing" or "hard coating" is the production of an anodized coat, specifically for wear resistance. The hard coat is the same oxide produced by any other oxidizing processes and has the same hardness, but the coating is thicker and less porous.

aluminum bronze (see bronzes)

aluminum casting alloy series. Formulated for sand casting, permanent mold casting, and die casting. These alloys are designated by a three-digit number, such as 108, 359, etc. There is little system in these numbers, and the problem is complicated further by the fact that casting alloys have more than one numbering system. The most used system is probably that of the Aluminum Association. In this system, silicon casting alloys have numbers up to 99; silicon–copper alloys, 100 to 199; magnesium–casting alloys, 200 to 299; and silicon–manganese, 300 to 399. Copper alloys are preferred for their hardness when machinability is a factor; silicon alloys are useful when fluidity to follow machinability is a factor, and when fluidity to follow intricate shapes is desired. The high-magnesium alloys require care in melting to prevent excessive oxidation loss of the magnesium content.

aluminum chloride. A white solid when pure. On heating it sublimes and, in the presence of moisture, anhydrous aluminum chloride partly decomposes with the evolution of hydrogen chloride. It is used in the petroleum industries and various phases of organic chemistry. Aluminum chloride is a catalyst in the alkylation of paraffins and aromatic hydrocarbons by olefins and also in the formation of complex ketones, aldehydes, and carboxylic acid derivatives. Anhydrous aluminum chloride is manufactured primarily by the reaction of chlorine vapor on molten aluminum.

aluminum, electrical and thermal properties. The electrical conductivity of aluminum is second to copper among common conductor materials. The electrical resistivity of aluminum and its alloys ranges from about 17 ohms per cir mil ft for high-purity grades to 20–38 ohms per cir mil ft for other alloys. The thermal conductivity in Btu/hr/sq ft/°F/ft ranges from a high of about 135 for high-purity aluminum grades to a low of about 67 for several 5000 and 7000 series alloys. Aluminum has excellent corrosion resistance in many environments, its good performance being due to the formation of a protective, tightly adherent, invisible coating dioxide film that develops in air, oxygen, or oxidizing media. This film is generally stable in solutions having a pH between 4.5 and 8.5; however, because the film is soluble in certain strong alkaline or acid solutions, aluminum can be severely attacked in these media. Only those high-strength aluminum alloys that precipitate separate phases in the microstructure are susceptible to stress corrosion.

aluminum oxide (see abrasives)

aluminum production. Bauxite ore, from which aluminum is obtained, contains as impurities oxides of iron, silicon, and titanium. It is inexpensive, and the largest portion of the cost is for transportation of the ore and for electricity used in the production process. Because of the high melting temperatures of alumina, 3720°F (2050°C), it cannot be reduced by the usual furnace techniques used for copper and iron. Since aluminum is separated by a reducing process, the impurities present in the ore would also be reduced along with the aluminum unless they were first separated. This separation is usually accomplished by the Bayer process wherein the bauxite is digested by a caustic soda solution under pressure. The alumina is then dissolved out as a solution of sodium aluminate. Aluminum hydrate is separated from the solution of sodium aluminate by precipitation and converted to the oxide, Al_2O_3, by calcination. The electrolysis is carried out in cells made of steel shells lined with carbon, which acts as the cathode. The cells are filled with molten cryolite into which is dissolved about 16% alumina. Carbon anodes dip into the electrolyte and introduce the current. The aluminum is separated from the electrolyte and is deposited on the bottom since it is heavier than the molten cryolite. It is removed periodically as it collects. Powdered alumina is added to the bath to replace the aluminum removed. Aluminum has a specific gravity of 2.7 as compared with 7.85 for steel.

alums and aluminum sulfate. An alum is a double sulfate of aluminum or chromium and a monovalent metal (or a radical, such as ammonium). Alums are used in water treatment and sometimes in dyeing. They have been replaced to a large extent in these applications by aluminum sulfate, which has a greater alumina equivalent per weight. Pharmaceutically, aluminum sulfate is employed in dilute solutions as a mild astringent and antiseptic for the skin. An important application of it is in clarifying water and in the sizing of paper. It reacts with sodium resinate to give the insoluble resinate. A small amount is consumed by the dye industry as a mordant. Soda alum, aluminum sulfate, is used in some baking powders.

Aluminum-Base Alloys

Relevant specifications	Composition (%) (balance Al)	Condition	Typical mechanical properties: 0.1% P.S. (N/mm^2)	Tensile strength (N/mm^2)	Elongation (%)	Characteristics and uses
Wrought alloys — not-heat-treated						
B.S. 1470 7:N3	1.2 Mn	Soft Hard	45 170	110 200	34 4	Metal boxes, milk bottle caps, food containers. cooking utensils, roofing sheets, panelling of transport vehicles
B.S. 1470 7:N4	2.5 Mg	Soft 3/4 hard	75 215	185 265	24 4	Marine superstructures, life boats, panelling for marine atmospheres, chemical plant, panelling for land-transport vehicles
B.S. 1470 7:N6	5.0 Mg	Soft 1/4 hard	125 215	265 295	18 8	Ship-building and applications requiring high strength and corrosion resistance
Cast alloys — not-heat-treated						
B.S. 1490:LM4	5.0 Si 3.0 Cu	Sand cast Chill cast	70 80	150 170	2 3	Sand castings; gravity and pressure die-castings. General purpose alloys where mechanical properties are of secondary importance.
B.S. 1490:LM6	11.5 Si	Sand cast Pressure die cast	55 85	170 215	7 4	Sand castings; gravity and pressure die-castings. Excellent foundry properties. One of the most widely useful aluminum alloys ('modified'). Radiators, sumps, gear boxes and large castings.
Wrought alloys — heat-treated						
B.S. 1470 7:HT14	4.0 Cu 0.8 Mg 0.5 Si 0.7 Mn	Solution treated at 480°C; quenched and aged at room temperature for 4 days	280	400	10	General purposes — stressed parts in aircraft and other structures. The original 'duralumin'
B.S. 1470 7:HT30	1.0 Mg 1.0 Si 0.7 Mn	Solution treated at 510°C; quenched and precipitation hardened at 175°C for 10 hours	150	250	20	Structural members for road, rail and sea transport vehicles; architectural work; ladders and scaffold tubes. High electrical conductivity — hence used in overhead lines
DTD 5074	1.6 Cu 2.5 Mg 6.2 Zn 0.3 Ti	Solution treated at 465°C; quenched and precipitation hardened at 120°C for 24 hours	590	650	11	Highly stressed aircraft parts such as booms. Other military equipment requiring a high strength mass ratio. The strongest aluminum alloy produced commercially
Cast alloys — heat-treated						
B.S. 1490:LM8	4.5 Si 0.5 Mg 0.5 Mn 0.15 Ti	Solution treated at 540°C for 8 hours; quenched in oil, precipitation hardened at 165°C for 10 hours	—	280	2	Good casting properties and corrosion resistance. Mechanical properties can be varied by heat-treatment
B.S. 1490:LM14	4.0 Cu 0.3 Si 1.5 Mg 2.0 Ni 0.2 Ti	Solution treated at 510°C; precipitation hardened in boiling water for 2 hours or aged at room temperature for 5 days	215	280	—	Pistons and cylinder heads for liquid and air-cooled engines. General purposes. The original 'Y alloy'

amino resins (plastics). The amino resins are condensation-type thermosetting polymers produced by the reaction of formaldehyde with amino compounds, containing the $-NH_2$ groups. The most important members of this class of resins are urea formaldehyde and melamine formaldehyde. Other amino resins of only minor importance now include those based on benzoguanamine, ethylene urea, and aniline. Amino resins have found applications in the fields of industrial and decorative laminating, adhesives, protective coatings, textile treatment, paper manufacture, and molding compounds. Their clarity permits products to be fabricated in virtually any color. Finished products having an amino resin surface exhibit excellent resistance to moisture, greases, oils, and solvents; are tasteless, and odorless; are self-extinguishing; offer excellent electrical properties; and resist scratching and marring.

amorphous metals. Also called glassy metals, these are noncrystalline in structure, unlike ordinary metals. Amorphous metals are produced by squirting a stream of molten metal at a rapidly spinning wheel. The metal solidifies instantly into thin ribbons, the metal cooling so rapidly that the metal atoms remain disordered like those of glass, hence the term "glassy metals." This gives the metal a variety of valuable properties such as greater strength, increased corrosion resistance, electrochemical stability, and improved magnetic characteristics.

angle press. A hydraulic molding press equipped with horizontal and vertical rams, and specially designed for the production of complex moldings containing deep undercuts.

annealing (metals). A heat-treating process to achieve one or more of the following results: (1) the removal of stress; (2) the softening of the steel; (3) the refinement of the grain structure; (4) the production of a specific type of microstructure; (5) the alteration of certain physical, mechanical, and electrical or magnetic properties; and (6) the removal of gases. All of these effects produce stability or equilibrium in steel. The most common type of annealing is called full annealing, which consists of heating to the range of temperatures indicated as general annealing, followed by a soaking period during which the temperature is maintained constant until the object is heated through. A slow cooling period then follows, either in the furnace or in some medium that promotes slower cooling than does air. This produces a coarse but stable structure in the pearlite. The steel is soft, ductile, easily machinable, and low in strength. Sheet and wire are usually subjected to a process anneal. The temperature does not exceed the lower critical temperature and is then slowly reduced. The purpose is to relieve the stress in the grain structure that has been deformed by cold working. Other annealing processes are known as patenting, bright, solution, and close annealing.

annealing, cycle (metals). Done by cooling austenitized steel at a rate to reach a desired zone on the S-curve. The metal is held at the chosen temperature until transformation is complete. Then the work may be cooled in any way practicable, by quenching or in air, because no more transformation occurs. The main advantage is a short cycle time, 4–8 h as compared to 5–30 h for conventional annealing. The end structure may be pearlite or spheroidite (or a mixture of both) depending on selections of temperature and time. Spheroidizing is the name given to the process when the carbon is collected into coarse round carbide particles, especially in high-carbon steels. This is a desirable structure to machine because the hard particles in a soft ferrite matrix are readily pushed aside by a cutting tool.

annealing, full (metals). Consists of heating an iron base alloy to 100°F (38°C) above the critical temperature, holding it there for uniform heating and cooling at a controlled slow rate to room temperature. The work may be held in a heavily insulated furnace with heat cut off or it may be buried in an insulating material such as ash or asbestos.

annealing, normalizing. An annealing process in which the object is heated to a temperature of 100–150°F (38–66°C) above the upper critical and then cooled in still air. Unlike full annealing, there is no extended soaking period, and the cooling is more rapid. Its purpose is to produce a more-uniform structure and to reduce hardness. In low-carbon steels, the "banding" that results from full annealing is reduced by normalizing. Banding is a marked segregation of carbide and ferrite in pearlite and is not desirable for machining. In high-carbon steels, normalizing tends to reduce the network of excess cementite around the pearlite grains and to prevent the formation of large carbide plates within the grains. This improves machinability. Steel is stronger, harder, and less ductile when normalized than when fully annealed.

annealing, process or commercial. Consists of heating iron base alloys to a temperature a little below the critical [commonly 1020–1200°F (550–650°C)] and cooling for desired results. Some softening occurs, but the main benefit is stress relief. The operation is sometimes called stress relieving. The advantage of the process is that warpage of thin sections and surface corrosion and scaling are slight because the temperature can be kept low. The process has wide application in preparing steel sheets and wire for drawing or redrawing, for stress relief of weldments and castings, and to remove the embrittling effects of heavy machining and flame cutting. Large work may be heated locally by a torch, smaller pieces in a furnace.

annealing, spheroidization. In any heat treatment that results in the formation of cementite in spherical form. This type of structure imparts greater ductility, machinability, and breaking strength than are present in pearlitic structure of the same hardness. Spheroidizing is an annealing process in which the material is heated to a temperature at or below the lower critical, held for relatively long periods of time, cooled slowly to about 1000°F (538°C), and then cooled more rapidly.

anodizing. A process for the development of protective and decorative films by anodic oxidation processes. **(See also aluminum, anodizing.)**

arc furnaces. The basic feature of a direct-arc electric furnace is that the top may be lifted or swung off so that the charge may be introduced. Heating is provided by lowering the electrodes and striking an arc between the electrodes and the metal charge. The current path is from one electrode across the arc to the metal, through the metal, and back across the arc between the metal and another electrode. Once the furnace has been started, the current flow is controlled automatically by apparatus that controls the arc length by raising or lowering the electrodes as required. The direct-arc furnace may be used to melt metal or to heat and hold molten metal at a desired temperature. The indirect-arc type of electric furnace is useful for melting small quantities of metals. In this type the arc is maintained between two horizontal electrodes with the metal being heated by radiation. The barrel-shaped furnace may be rocked about its longitudinal axis in order to stir the metal. It is rolled over farther for pouring.

arc welding. In arc welding the heat source is an electric arc maintained between the work and an electrode, or between two electrodes. This results in a temperature in excess of 7000°F (3870°C), which is more concentrated than in the case of gas-flame welding. For many years nearly all arc welding was done with direct current. In recent years there has been a great increase in the use of alternating current. When direct current is used, if the work is made positive (the anode of the circuit) and the electrode is made negative, straight polarity is said to be used. When the work is negative and the electrode is positive, the polarity is reversed. When bare electrodes are used, greater heat is liberated at the anode. Certain shielded electrodes, however, change the heat conditions and are used with reverse polarity. **(See also Welding and particular type.)**

artificial aging (see aging)

artificial intelligence (AI) (electronics). A computer program that draws conclusions from a set of rules that make up its data or knowledge base. It allows a computer to operate on its own, for example, to learn, adapt, or correct and improve operations. AI is being applied in oil exploration, electronic system testing, etc.

argon. A monatomic inert gas, which does not dissolve in liquid metals. These characteristics make it an especially useful gas. Its boiling point at atmospheric pressure is −302.4°F (−185.8°C), and its specific gravity compared to air is 1.0–1.38. Argon serves a wide range of uses in incandescent and fluorescent lamps, Geiger–Müller counter tubes, and inert-atmosphere heating furnaces. Its largest use is in the welding and thermal cutting of metals that must be protected from oxidation, such as zirconium, aluminum, and titanium.

argon oxygen decarburization. Process intended to function as an auxiliary to electric arc furnace melting and is most applicable to low-carbon stainless steels. Argon is an inert gas derived as a by-product when oxygen is extracted from the air. It is used in conjunction with oxygen in a furnace quite similar to the basic oxygen furnace. Molten metal from the electric arc furnace is poured in the argon-oxygen-decarburization furnace; it is tilted back upright and blown with the mixture of gases the required length of time. The process results in accelerated decarburization and refining, saving considerable time in the electric arc furnace. It is most beneficial for steels containing large amounts of chromium and allows less expensive chromium additions.

asbestos. About 95% of the world's asbestos production consists of the mineral chrysotile, a fibrous form of serpentine, a hydrous magnesium silicate. It is found in veinlets, mostly less than 1 in. (2.54 cm) in width, in dark-green massive serpentine. Chrysotile occurs as silky fibers, which are called cross fiber when they extend from wall to wall of the veinlets, and slip fiber when they form strands more or less parallel to slickensided vein walls. Chrysotile in the mass is green or yellowish green, and has a pearly luster; but when disaggregated into its individual fibers makes a white fluffy mass like cotton. Electron micrographs shows that each chrysotile fiber is a hollow cylindrical tube. Asbestos fibers are noninflammable, and heat resistant up to about 500°F (260°C). Spinning fiber includes the longer grades of chrysotile and crocidolite and is made into asbestos, thread, yarn, tape, and cloth. A few of the products in which these are used are brake linings, clutch facings, fireproof theater curtains and scenery, gaskets, safety clothing, blankets and draperies, chemical filters, and heat-resistant conveyor belts. Yarn is twisted and braided to form various types of packing. Nonspinning fiber consists of the shorter grades of chrysotile and crocidolite and all grades of amosite and anthophyllite. Most of it is used in compressed, molded, or cast products in which asbestos fibers make up felted mass in a binder. Roofing shingles, millboard, and corrugated paneling are made from asbestos and Portland cement. Other uses include asbestos-cement pipe, molded brake linings and clutch facings, and paper for a variety of uses.

AS/RS (see automated storage/retrieval system)

assembly robot. Designed, programmed, or dedicated to putting together parts into subassemblies or complete products. **(See also robot.)**

Aston process (metals). In the first step of this process, pig iron is tapped into ladles where it is desulfurized. It is then purified to a high state in a Bessemer converter. Simultaneously, the second step for preparation of the slag is carried out in an open-hearth furnace, by melting together oxide and siliceous materials. In the third step the refined iron is poured into a ladle containing the molten slag. The iron solidifies rapidly, with the liberation of dissolved gases. The gas liberation is sufficient to shatter the solidified metal into small fragments. These fragments cohere as they settle to the bottom of the ladle and are impregnated with liquid slag. The surplus slag is poured off and the metal mass is squeezed in the same manner as the puddler's balls were handled. From 6000 to 8000 lb (2725 to 3630 kg) sponge balls are obtained by the Aston process at the rate of one every 5 min.

asynchronous computer. A computer in which each event of the performance of each operation starts as a result of a signal generated by the completion of the previous event or operation, or on the availability of the parts of the computer required by the next event or operation.

ATE (see automatic test equipment)

austempering. A process in which a piece of metal is quenched in a molten salt bath having a temperature in the lower bainite region. Unlike martempering, in austempering the piece is maintained at this temperature until transformation to bainite is complete. It may then be either quenched or slowly cooled to room temperature. This may be followed by tempering to reduce hardness. Austempered steel has about the same strength and hardness as the same steel quenched to form martensite and tempered in a conventional manner, but it has greater ductility. Because of certain hardenability requirements in the steel, the austempering of plain carbon steel is largely confined to thin-section materials, such as wire, sheet, and strip. This restriction does not apply to alloy steels and cast irons. **(See also steel, austempering.)**

austenite (metals). One of the important high-temperature phases in steel is austenite, a homogeneous phase consisting of a solid solution of carbon in gamma form of iron.

Austenite is formed when steel is heated to the relatively high temperature of more than 1450°F (788°C), with the limiting temperatures for the formation of austenite varying with composition.

austenitic stainless steels (metals). Essentially chromium–nickel–iron alloys, generally of a composition varying from 16 to 25% chromium, and 6 to 22% nickel, with a maximum of up to 0.25% carbon. Other alloying elements such as molybdenum, vanadium, and titanium are also added to develop or intensify specific properties. The most widely used grade of austenitic stainless steel is the type "18-8" containing from 17 to 19% chromium, 8 to 10% nickel, and up to 0.20% carbon. When 2–4% molybdenum is added to the basic composition, types 136 and 317 are obtained, which show an improved corrosion resistance against pitting and increased high-temperature strength.

automated storage/retrieval system (AS/RS). A combination of equipment and controls that handles, stores, and retrieves materials with precision, accuracy, and speed under a defined degree of automation. Systems vary from relatively simple, manually controlled order-picking machines operating in small storage structures to giant, computer-controlled storage/retrieval systems totally integrated into the manufacturing and distribution process. **(See figure on p. 26.)**

automatic assembly. Often referred to as automated assembly or mechanized assembly, automatic assembly is basically the assembly of products on a machine. Two types of assemblies are handled on these machines: (1) an additive assembly is one in which a series of discrete parts are added to one another in a specific sequence, i.e., a ball point pen; (2) a multiple-insertion assembly is one in which a series of discrete parts are assembled in different locations on a common base. In the first, any failure to insert a component stops the operation for corrective action. In the second, any failure to insert a component has no direct affect on the subsequent inserting operations, i.e., a circuit board.

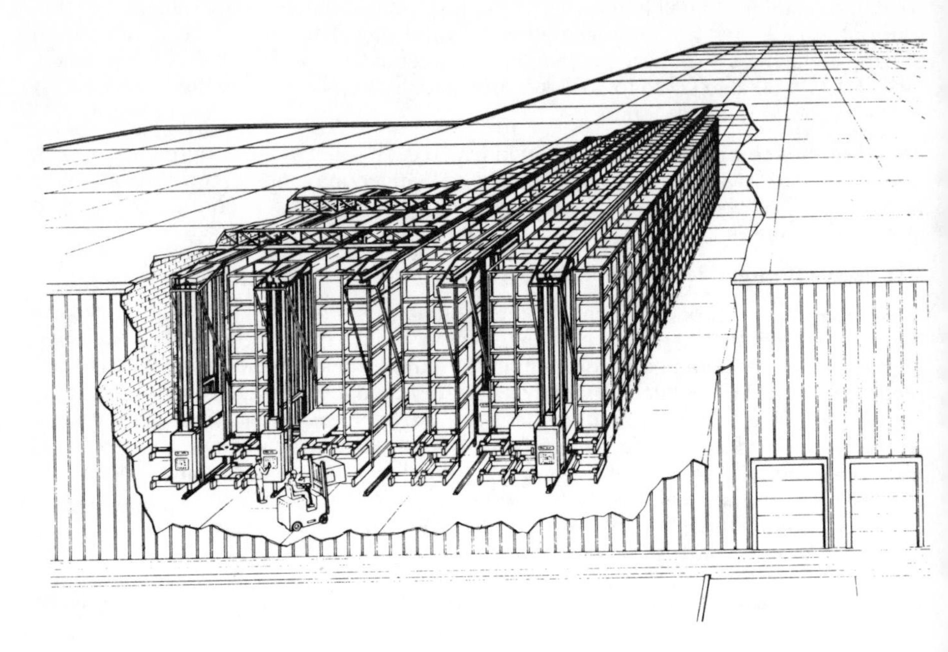

AS/RS in freestanding building.

automatic bar and chucking machines. Originally designed for the rapid, automatic production of screws and similar threaded parts, the automatic screw machine has long since exceeded the confines of this narrow field and today plays a vital role in the mass output of precision parts in endless variety. Capable of producing turned and formed parts of both simple and intricate design, the basic machine utilizes a variety of tool slides, automatically sequenced, each of which performs a portion of the necessary machining on a rotating bar and cuts off finished pieces in rapid succession as the bar is continually fed out at each cutoff. By no means limited to the production of threaded parts, the wide range of possibilities afforded by this method of production both enhances its usefulness and makes imperative its consideration as a means of manufacturing machine parts in quantities. Generally, automatic screw machines fall into several categories: single-spindle automatics, multiple-spindle automatics, and automatic chucking machines. Single-spindle machines are usually designed to produce parts in rapid succession from a length of bar stock fed through the machine spindle, whereas the multiple-spindle machine is available both as a bar machine and as a chucker. On the chuckers automatically operating chucks grip cast, forged, or other single parts and carry them through the cycle, ejecting the parts on completion. A wide variety of models are available in each basic type of machine as well as special adaptations.

automatic crane. A crane that when activated operates through a preset cycle or cycles. It can be powered by hardwire remote control or radio control; it can also function automatically. Remote or radio controls are used when the operator may not, or cannot, get near the load, or when required to operate several cranes from a remote point. Automatic operation reduces or completely eliminates the need for operators.

automatic test equipment (ATE) (electronics). Term used for quality and acceptance testing in the manufacture of electronic chips, circuit boards, and systems.

automation. The word ''automation'' is a contraction of the words automatic–operation. It implies the process of doing things automatically. It is not synonymous with any other word. It does not merely mean mass production; mass production is volume manufacture of interchangeable products. In the 1840s, Robbins, Kendall, and Lawrence of Windsor, Vermont, mass produced rifles on a truly interchangeable basis completely by hand methods. Automation is based on but goes a long step beyond mere mechanization. Mechanization simply means doing things with or by machines, not *necessarily* automatically. True automation implies continuous or cyclic arrangement for manufacturing, processing, or performing services as automatically as is *economically practical* or necessary. Actually, it is possible to automate a single operation, a sequence of operations, a whole department, or a plant. Automation can be segregated into several types. It is possible to create an automatic batch system or an automatic continuous system. One chemical process, for instance, may be more economically carried out in batches while another is most economical when produced continuously. Process characteristics as well as production economics dictate the best method. Thus, job-lot-type operations may require batching arrangement with amenity to continuous change while mass production operations may be more economically carried out on a continuous basis. The total system consists of three major subsystems that may be inextricably interwoven into the final whole. The three subsystems are: processing or making, handling the work from one step to another, and controlling the actions as required to keep it continuous and to maintain quality of output. **(See also robot.)**

B

bagasse. A refuse of cane from which the sugar juice has been extracted. It has a fibrous structure like that of wood, and also a similar analysis. The most variable item is moisture, which normally ranges from 40 to 60%. The relatively high ash in bagasse is due to silt picked up in the harvesting of the cane.

bag molding. A method of applying pressure durng bonding or molding, in which a flexible cover, usually in connection with a rigid die or mold, exerts pressure on the material being molded, through the application of air pressure or drawing of a vacuum.

balancing (see dynamic balancing)

ball broaching (machining). A method of securing bushings in gears or other components without the need for keys, pins, or splines. A series of axial grooves, separated by ridges, is formed in the bore of the workpiece by cold plastic deformation of the metal. A tool, having a row of free rotating balls around its periphery, is pressed through the part. When the bushing is pressed into the broached bore, the ridges displace the softer material of the bushing into the grooves, thus securing the assembly. The balls can be made of high-carbon chromium steel or carbide, depending on the hardness of the component.

ball mills (see coal)

band saw (see cut-off machines)

bar. A solid product that is long in relation to cross section, which is square or rectangular (excluding plate and flattened wire) with sharp or rounded corners or edges, or is a regular hexagon or octagon, and in which at least one perpendicular distance between parallel faces is $^3/_8$ in. (9.5 mm), or greater.

bar code. An identification code formed by an array of rectangular bars and spaces which are arranged in a predetermined pattern following unambiguous rules in a specific way to represent elements of data that are referred to as characters. The code, used to identify products, boxes, etc., in production and to control their movement throughout manufacturing, is designed to be read optically.

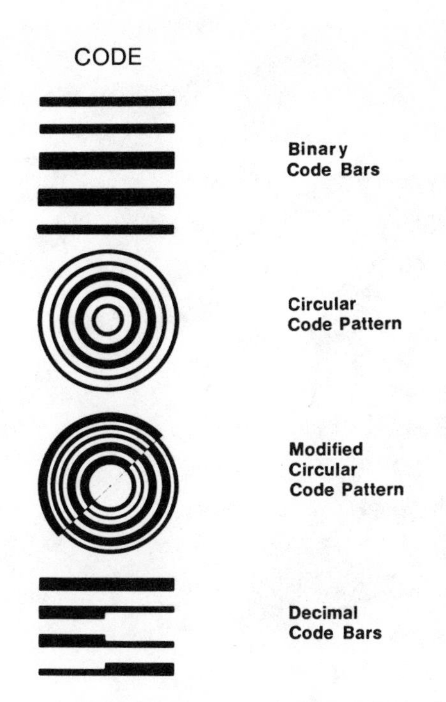

Examples of bar codes.

bar code reader. A device used to identify and read bar code symbols in manufacturing process. **(See also scanner, moving beam bar code reader, and figure on p. 30.)**

bar, cold-finished. Bar brought to final dimensions by cold working to obtain improved surface finish and dimensional tolerances.

bar, cold-finished extruded. Cold-finished bar produced from extruded bar.

Bar, cold-finished rolled. Cold-finished bar produced from rolled bar.

Barex (nitrile barrier resin). Barex 210 resin (A Sohio Petroleum Company trademark) is a lightweight, transparent, impact-resistant rigid copolymer of high acrylonitrile content. It combines the advantages of plastics with gas, aroma, and flavor barrier characteristics approaching those of metal and glass. The resin is generally approved for the packaging of all foods and beverages by FDA regulations. The containers can be disposed of in normal waste channels without adversely affecting incinerator operation or causing any change in effluent composition. Barex 210 resin has a good melt processing stability and impact resistance and has a rather broad rubbery range, which adapts it to blow molding and thermoforming. It has been processed on a wide variety of commercial equipment with cycle time comparable to rigid PVC, ABS, or, in some cases, impact polystyrene. The resin has a relatively high melt viscosity and adequately powered properly designed equipment is required.

bar, extruded. Bar brought to final dimensions by hot extruding.

barite (minerals). Barite is natural barium sulfate, with a specific gravity of 4.3–4.6, which is exceptionally high for a nonmetallic mineral; it is commonly white to light gray, but may be bluish, brown, or nearly black. It is generally opaque and has a pearly

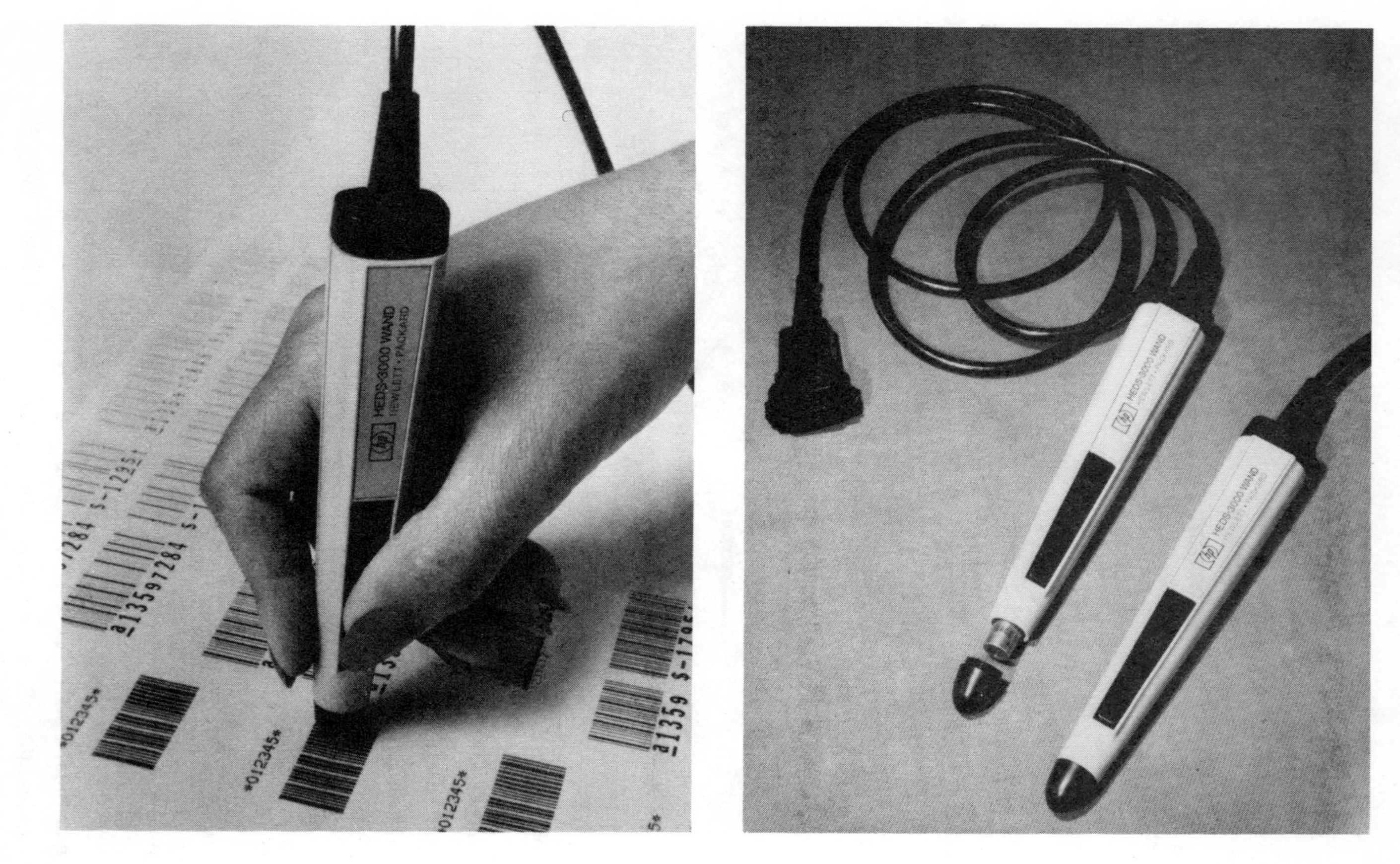

Digital bar code wand.

to vitreous luster. Deep drilling for oil and gas by the rotary method, in regions where very high gas pressures are likely to be encountered, requires a heavy circulating fluid to control the reservoir pressure and prevent blowouts. Pulverized barite, added to normal clay–water mixture, serves this purpose because it is heavy, inert, clean, and relatively inexpensive. Barite is also a source of barium chemicals. Precipitated barium sulfate is a paint extender and a filler in paper, rubber, and linoleum. Barium chloride is used in leather and cloth manufacture; the carbonate is a component of ceramic glazes and enamels; the oxide is used in glassmaking and in electric furnace metallurgy; the hydroxide is utilized to recover sugar from molasses; and the nitrate is an ingredient of signal flares and detonators. In ground form, barite is used chiefly in rubber and as an extender in paint.

bar, rolled. Bar brought to final dimensions by hot rolling.

barium salts. The most common naturally occurring barium compounds are the mineral carbonates, or witherite. The application of barium compounds are varied. Barium carbonate is employed as a neutralizing agent for sulfuric acid and, because both barium carbonate and barium sulfate are insoluble, no contaminating barium ions are introduced. Application is found in the synthetic dyestuffs industry, the glass industry, and bricks and clay. Witherite is used chiefly to prepare other compounds. Barium sulfate is a white pigment. It is used as a filler for paper, rubber, linoleum, and oilcloth. Because of its opacity to x-rays, barium sulfate, in a purified form, is important in contour photographs of the digestive tract. Barium sulfide and zinc sulfate solutions are mixed to give a precipitate of barium sulfate and zinc sulfide, which is given a heat treatment to yield the cheap but good pigment lithopone. Barium chlorate and nitrate are used in pyrotechnics to impart a green flame.

basic-oxygen process, steel making. Molten pig iron and steel scrap are charged into a large, cylindrical, open-mouth furnace, which has a basic lining. The furnace is returned to the upright position and burned lime and flux are poured in through a chute. A water-cooled oxygen lance is then lowered into the mouth of the furnace to a position several feet above the surface of the charge. High-purity oxygen is then blown, under considerable pressure, onto the surface of the bath. The impinging oxygen blows the slag aside and reacts violently with the exposed iron to form iron oxide. Part of the oxide reacts with the flux to form a basic slag, while the remainder is mixed with the bath through turbulence, and oxidizes impurities. The oxygen treatment requires about 27 min, and a complete cycle requires about 50 min. The resulting product is virtually the same as open-hearth steel. An oxygen-process furnace having a capacity of 100 tons (91 tonnes) is about 26 ft (8 m) high and 18 ft (5.5 m) in diameter. The major disadvantage is that the charge must include a considerable quantity of molten pig, thus limiting the amount of scrap that can be used.

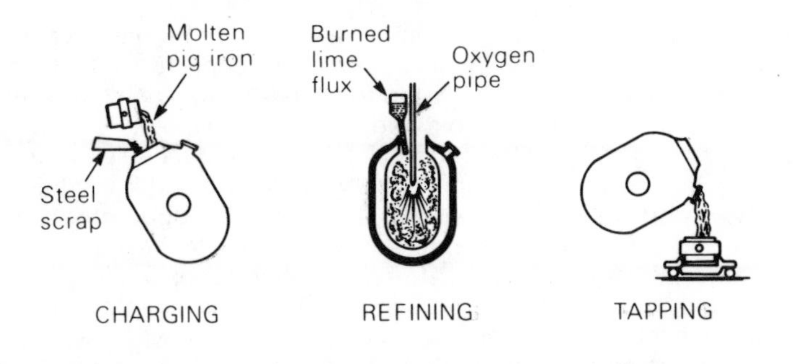

Schematic diagram of the basic-oxygen steel-making process.

bath furnaces, metal heat treatment. Bath furnaces may be gas or oil fired or electrically heated. Gas- and oil-fired salt bath furnaces have low first and operating costs and are versatile. They can be restarted easily, and pots can be interchanged in one furnace to use a variety of salts. The bath may be heated externally or by immersed radiant tubes. Temperature control is not uniform in externally heated furnaces as in others. There are several kinds of electrically heated salt bath furnaces. All are surrounded by insulated casing. Heating may be done by resistance elements around the pot in

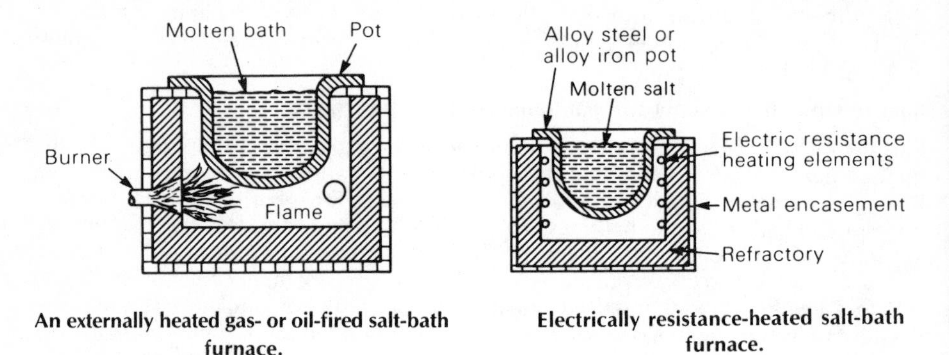

An externally heated gas- or oil-fired salt-bath furnace.

Electrically resistance-heated salt-bath furnace.

the externally heated type. Well-insulated heating elements are put directly into the bath in the immersion-heating-element type of furnace. Temperatures are usually limited to 1100°F (593°C) for satisfactory resistor life. Higher temperature can be held by passing electricity through the bath between electrodes. The immersed-electrode salt bath furnace has electrodes immersed in a metal pot. The submerged-electrode furnace has water-cooled electrodes extending through the sides into a ceramic brick pot. The molten salt penetrates the refractory material until it reaches a zone cooled enough to freeze it, thereby sealing the pot. Electrode furnaces use alternating current transformed to low voltage (5–15 V) because direct current decomposes the liquid salt. Temperatures are easy to control within 5°F (2.8°C).

beam power tube (electronics). A beam power tube is a special kind of tetrode with the performance characteristics of a pentode. It is highly efficient as a power amplifier. The electrodes are arranged so that secondary emission from the plate is suppressed by the negative space charge without the use of a suppressor grid. A beam-conforming electrode, connected internally to the cathode, concentrates the transient electrons in the vicinity of the plate.

beryllium bronze (see bronzes)

beryl minerals. Beryl is beryllium aluminosilicate. It commonly occurs in simple, prismatic hexagonal crystals, which range from microscopic to 1 ft (0.3 m) or more in diameter and several feet in length. Crystals frequently show a marked tapering toward one or both ends. The mineral is also found in anhedral masses. Beryl occurs in commercial amounts only in pegmatites, and is valuable chiefly as the sole commercial source of beryllium metal. More than 90% of the beryllium consumed is used as a hardening agent in alloys, mostly those of copper. It is used in small proportions up to 4%. Addition of 2.5% beryllium to copper produces an alloy that, after heat treatment, is harder than pure copper and several times as strong. Beryllium–copper alloys have high electrical and thermal conductivity, are nonsparking and nonmagnetic, and resist fatigue, corrosion, and wear. They are made in many special forms, which are used at points of abnormal wear in engines and motors, as current-carrying springs in instruments, gages, and switches, and in other specialized applications. The alloys are also used in the manufacture of nonsparking tools. Beryllium is also alloyed with zinc, nickel, and other metals. Beryllium oxide is a high-temperature refractory.

Bessemer process. Steels made in a Bessemer furnace are designated as either acid or basic steel, depending on the nature of the refractory linings of the furnace. Silica is an acid lining, while dolomite and magnesite are basic linings. The nature of the linings controls the slag since a basic slag could rapidly dissolve an acid lining and an acid slag would have the same effect on a basic lining. With a basic lining, a large percentage of the phosphorus and some of the sulfur can be removed, but the greater amount of iron oxide left in the steel renders basic steel inferior to acid steel. In making Bessemer steel, molten iron direct from the blast furnace is poured into the converter or vessel. In the bottom of the vessel are a number of holes through which air is blown. The air first oxidizes the silicon and manganese, which, together with some iron oxide, rise to the top and form a slag. The carbon then begins to burn, and the blowing is continued until all but about 0.05% of the carbon has been eliminated. The progress of the blow can be determined from the flame coming from the vessel. The oxidation of the impurities has raised the temperature of the metal to the point where it can be cast conveniently. When the blow is completed, the amount of carbon necessary to bring the carbon content to the specified percentage, together with manganese to counteract the influence of sulfur and silicon to degassify, are added to the molten metal. The finished steel is then poured into a ladle by tipping the vessel, and from the ladle it is poured into molds for subsequent rolling or forging.

bilateral manipulator. A master–slave manipulator with symmetric force reflection where both master and slave arms have sensors and actuators such that in any degree of freedom a positional error between the master and slave results in equal and opposing forces applied to the master and slave arms. The term, two-armed manipulator, can refer to two arms performing a task in cooperative movements, or can refer to two arms in the sense of a master–slave manipulator.

billet. A hot-worked semifinished product suitable for subsequent working by such methods as rolling, forging, or extruding.

binnable materials. Those small items stored either on shelves or directly in containers.

bituminous cements. Bituminous materials used for cementing are either asphalts or tars. Asphalts may be classified as either native or pyrogenous. Native asphalts are found in surface deposits and result from the weathering of petroleum. They occur in pure or fairly pure states or are associated with substantial amounts (up to about 40%) of such mineral materials as sand, clay, shale, sandstone, and limestone. Pyrogenous asphalts are produced from petroleum refining; actually they are the residues from the fractional distillation of petroleum. They usually have a smooth, glossy surface as compared to the duller appearance of the native asphalts. Both range from dark brown to black in color. Gilsonite is a hard, brittle, practically pure native asphalt that is found in rock crevices or veins from which it is removed by mineral mining methods. It is used to improve the properties of other asphalts that are low in bitumen. Asphalt cement made from gilsonite is tough and rubbery. Grahamite, which is found in various localities, is similar to gilsonite, although it varies considerable in purity. Pyrogenous asphalts are also called petroleum asphalts; they are the residues from the fractional distillation of petroleum. They are almost entirely soluble in carbon disulfide. Some crude oils contain relatively large amounts of asphaltic materials, whereas others contain very little. Oils of a paraffinic base may contain from trace amounts to appreciable quantities. The mixed-base crudes are higher in asphalts. Those of a naphthenic base are divisible into high-gasoline–low-asphalt and low-gasoline–high-asphalt classes. Selected naphthenic crudes are used to obtain the petroleum for roadways. Tars may be defined as "black to dark-brown bituminous condensates which yield substantial quantities of pitch when partially evaporated or fractionally distilled, and which are produced by destructive distillation of organic material such as coal, oil, lignite, peat, and wood." Tars used for cementing purpose originate as coal tar or water-gas tar. Coal tar is a by-product of the manufacture of coke from bituminous coal. Water-gas tar is a by-product of the manufacture of water gas from coal or coke. Tars and asphalts are quite similar in general appearance, but tars have a characteristic odor that distinguishes them from the asphaltic materials. Tars have lower liquefaction temperatures than do asphalts and are more sensitive to small temperature changes. They have better penetrating quality or ability than asphalts but do not make as good a paving material. A combination of tar undercoat and asphalt surface is often used for roads.

black annealing. Box annealing or pot annealing ferrous alloy sheet, strip, wire. **(See also box annealing.)**

black oxide. A black finish on a metal produced by immersing it in hot oxidizing salts or salt solutions.

blank. A piece of metal cut or formed to regular or irregular shape for subsequent processing such as by forming, bending, or drawing.

blanking. The cutting of flat sheet stock to shape by striking it sharply with a punch while it is supported on a mating die. Punch presses are used. Also called die cutting. **(See also die cutting.)**

blanking, fine. A German development for producing extremely accurate finished parts in a single stroke, fine blanking is gaining in use in the United States largely because of cost savings in production. Costs can be reduced as much as 50% on some designs. The process uses tools with virtually no clearance between mating parts of the punch and die [actually held to 0.0003 in. (0.008 mm) or less]. Although similar to ordinary blanking, the key to success is in an impingement ring in the die that presses into the perimeter immediately around the outline of the part to be blanked. Thus, the punch extrudes the blank through the material under controlled pressure creating a fine finish and straight edges.

blanking, steel-rule die. Often referred to as "skeleton" dies, steel-rule dies are those widely used in blanking paper, rubber sheet, plastic foam, and fiber forms. These dies are built-up by bending or fitting together, usually by screwing to a wood base form, steel-rulelike cutting blades as required by the blank to be cut. Steel-rule dies are extremely low in cost and easy to obtain. A wide variety of materials can be blanked, including thin-gage soft or half-hard sheet metals. These dies can be used in power presses fastened to the upper platen or merely set in place over the material for blanking as in a checker- or dinker-type cutting press or between rollers.

blast cleaning. This process uses a pneumatic blast gun and abrasives to clean or impart a matte finish to metals, plastics, etc. Blast cleaning machines are in standard, special-purpose, and fully automated types. Media used include: glass beads for decorative, blending, light deburring, peening, general cleaning, texturing; aluminum oxide for fast cutting, matte finish, descaling, and cleaning; garnet for noncritical cleaning, cutting, and texturing; crushed glass for fast cutting at low cost with short life; steel shot for general-purpose rough cleaning in foundry operations and peening; steel grip for rough cleaning with coarse textures, foundry welding applications, and some texturing; cut plastics for deflashing thermosetting plastics, cleaning, and light deburring; crushed nut shells for deflashing plastics, cleaning, and very light deburring, and for use in fragile parts.

blast furnace. A blast furnace is a large, round steel structure of varying diameter, from 90 to 110 ft (27.4 to 33.5 m) tall. There are four major parts to a blast furnace:

the top, the stack, the bosh, and the hearth. The hearth is a storage region for the molten metal and molten slag. Above the hearth are the combustion zone and the fusion zone. These two zones taken together comprise a region called the bosh. The bosh is the region in which melting takes place. The temperature is about 3000°F (1650°C) that as a result of combustion, aided by a continuous hot-air blast that furnishes the oxygen necessary to support combustion. On top of the bosh zone is the stack. This region is made up of the heat absorption zone and a reduction zone. The temperatures range from about 2200°F (1200°C) at the bottom of the heat-absorption zone to 400°F (205°C) at the top of the reduction zone. The temperatures are continuous, and there are no distinct dividing lines between the various zones. At the top of the blast furnace rests the "top." To ensure an even distribution of the charge, the top houses two inverted cone-shaped bells. The gaseous products of combustion are carried off through four ducts spaced around the top of the furnace. The charge—successive layers of iron ore, scrap, coke, and limestone—is dumped into the top of the furnace. As the coke burns, aided by the air forced into the furnace, the ore melts and collects in the hearth. As the melting process proceeds, the entire mass settles and thus makes room for the addition of charge at the top. While the melting is going on, the limestone forms a slag with the impurities. The iron ore that is most frequently used is in the form of hematite. The ore is commonly about 50% iron and 50% earthy matter, known as gangue. This is composed of silica, alumina, calcium oxide, magnesium oxide, water, phosphorus, and often some sulfur. These elements in the iron ore affect the end product. The average temperature of molten iron in the hearth of a blast furnace producing basic iron is approximately 2700°F (1480°C), and the average temperature of the molten slag is slightly higher than that.

blast furnace gas. Results from various reactions occurring throughout the blast furnace. This gas contains a relatively high percentage of carbon monoxide, along with carbon dioxide, nitrogen, and water vapor. Its heating value varies between 90 and 110 Btu/ft^3 (3170 and 3870 kJ/m^3) and is dependent principally on the quality of coke used, the speed of combustion, and the character of the ore treated. The combustible in blast furnace gas consists of about 30% carbon monoxide, 2–3% hydrogen, and a trace of methane. The inert gas consists of about 9.0% carbon dioxide and 59% nitrogen; when washed, it carries a considerable amount of water in suspension as fine droplets as well as water vapor. An average figure of the amount of gas produced in a blast furnace is 150,000 ft^3/ton of iron (4950 m^3/tonne), or about 160,000 ft^3/ton of coke (5280 m^3/tonne) charged. Approximately 10% is lost through leakage, and 30% is required to preheat the blast. As it leaves the "top" this gas is hot and contains considerable dust having a high iron oxide content. Much of the dust is removed and then sintered for return to the blast furnace as part of the ore charge.

blast furnace products. Iron ore, coke, limestone, and air are the essential products going into a blast furnace. The products that come out are: (1) Pig iron—the main purpose of the blast furnace is to produce pig iron of suitable quality and grade for the purpose intended. (2) Slag—skimmed off into cars or into granulating troughs for subsequent use. (3) Gas—which has the following approximate composition: CO, 22–27%; CO_2, 10–17%; N, 60–62%. The calorific power of the gas coming from the furnace is about 0.1 Btu/ft^3 (900 cal/m^3). (4) Flue dust—in the form of very fine particles of charge blown out of the blast furnace by the gas current, together with small quantities of volatile material. The coke fines make the flue dust rather high in carbon.

bleaching (papermaking). The term bleaching in the pulp and paper industry is defined as the destruction or solubilizing of coloring matter in the fibers and the subsequent washing out of these coloring substances in order to obtain white fibers. While pure cellulose is white, the digestion or cooking of wooden chips in pulp manufacturing never completely removes the impurities or encrusting substances around the fibers of liberated cellulose. The result is that these chemical pulps are not white, but rather brownish in color. Bleaching may be thought of as a further step in the purification of fibers, this step being a continuation of the purification of the fibers begins in the cooking of pulps. The purpose is to liberate the whiteness that pure cellulose has without chemically harming or degrading the cellulose. There is no one bleaching procedure used universally in the industry. Owing to the differences in chemical nature and the extent of the color-producing substances associated with the fibers, different chemical treatments are used. Different species of trees bleach differently. The chemical used and the manner in which the wood was cooked determine to a great extent what bleaching procedure will be used. There are several fundamental reactions involved in bleaching. The first chemical reaction is one resulting from the treatment of pulp with chemicals so as to render water soluble the impurities in the fiber. These chemically altered impurities now being water soluble can be washed out from the pulp. The second reaction results from the treatment of the pulp with a chemical such as alkali to dissolve those water-insoluble reaction products from the first reaction. These first two reactions represent steps taken to solubilize and remove as much as possible those substances contributing to the dark color before actual bleaches are used. The third fundamental reaction is one of actual bleaching whereby strong oxidants such as bleaching powders are used to either destroy or render colorless the remaining substances causing the dark color.

bleaching, hypochlorite (papermaking). Calcium hypochlorite is used in hypochlorite bleaching. Bleaching is done in one stage (single-stage bleaching) by adding the bleach liquor to the pulp fibers prior to papermaking under proper conditions of temperature, pH, etc., and allowing it to react until the fibers bleach to the desired brightness. Washing usually follows. Hypochlorite bleaches mainly by oxidizing the impurities to a colorless form and also making them water soluble.

bleaching, multistage (papermaking). Bleaching in more than one step in which bleaching of sulfate pulp is accomplished by using from five to ten stages. A five-step bleaching of sulfate pulp might be done as follows: Pulp is first treated with chlorine as a purifying step. In this chlorination stage, the chlorine combines with or alters the lignin and other colorizing substances so that they may become soluble in water or alkali. The chlorinated pulp is now washed with water. The second stage is one of alkali extraction whereby the chlorinated pulp is treated with an alkaline solution so as to dissolve the reaction products left from the first stage. Washing with water follows again. In the third stage the pulp is treated with hypochlorite to decolorize by bleaching action those colored residues remaining after the first two stages. After washing with water, stage four follows, which consists of another alkali extraction and washing. The fifth stage consists of bleaching again with hypochlorite followed by washing with water. These five steps constitute a five-stage bleaching and are the very minimum in producing bleached sulfate. Often there may be several additional stages to the five described.

bloom. A semifinished hot rolled product, rectangular or square in cross section, produced on a blooming mill.

blow molding. A method of fabrication in which a parison (hollow tube) is forced into the shape of the mold cavity by internal air pressure.

blow molding, extrusion (plastics). A process that makes use of many types of equipment to produce containers ranging in size from fractional ounces to 55 gal (208.5 liters) and greater, as well as industrial and consumer parts. Any thermoplastic can be blow molded with varying degrees of success. There are three steps basic to all blow molding: (1) formation of a hollow tube of molten resin called a parison, (2) positioning of the

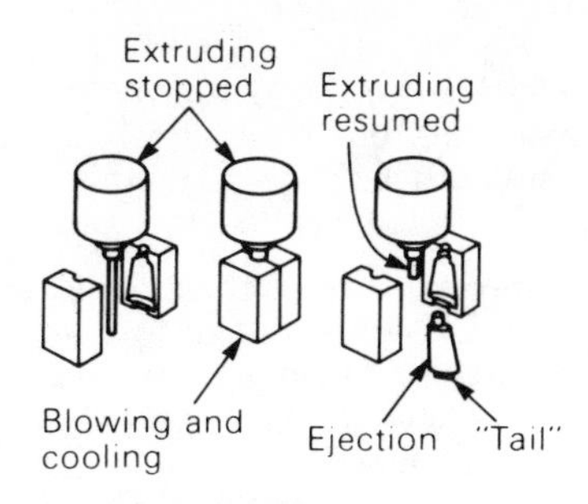

Intermittent extrusion blow molding

parison between mold halves, and (3) blowing the parison so that it takes the shape of the mold. There are four basic means by which a parison can be formed. (1) Continuous extrusion. In this process the extruder screw revolves continuously at a uniform rate to force the melt through an annular die and form the hollow tube or parison. (2) Intermediate extrusion. The extruder for this process is equipped with a reciprocating screw. Melt, instead of being discharged continuously through a nozzle to the die, is trapped in front of the screw, which is forced backward by the pressure of the accumulating melt. When the required quantity of melt has been built up, the path to the die is opened and the screw moves forward, forcing the melt through the die to form the desired length of parison. (3) Neck injection tube extrusion. A process that combines intermittent extrusion with the injection molding process. Melt is first forced into the injection mold where the neck of the container is completely formed. (4) Injection blow molding. A parison with neck finish completely formed is produced in an injection mold, then transferred to a blow mold where the process is completed.

blow molding, injection (plastics). The process of injecting molten plastic into a cavity where the neck of a container is formed and material is placed around a hollow core pin. The core pin is then transported to the blow station where the neck section remains the same while the hot plastic surrounding the core pin is blown into the final shape of the container by introducing air through the hollow core pin. Several types of injection blow molding machines are used.

blow molding, preform (plastics). A cold parison blow molding system, with the blow molding operation separated into two basic processes: (1) The extrusion of a perfect parison, requiring a high degree of skill in plastic processing; (2) the blow molding of a bottle from a perfect parison, merely a mechanical function.

bluing. Subjecting the scale-free surface of a ferrous alloy to the action of air, steam, or other agents at a suitable temperature, thus forming a thin blue film of oxide and improving the appearance and resistance to corrosion. Note: This term is ordinarily applied to sheet, strip, or finished parts. It is used also to denote the heating of springs after fabrication to improve their properties.

board binding (papermaking). Board commonly found in the stiff binding of books. It is made on a wet machine, of mixed papers and moderately sized. It ranges in thickness from 0.020 to 0.200 in. (0.51 to 5.1 mm). Important properties are smoothness, high density, and strength. Shoe counter board is often made with the same facilities as binder board. Raw materials are chosen for maximum strength, and long fibered stocks such as cordage, kraft, and sulfite are commonly employed. Hard sizing is required, as the counter is subject to moist conditions.

board, Bristol (papermaking). Term covers a class that includes file folders, calling cards, business record cards, and tabulating cards. These are generally high-grade products having a minimum thickness of 0.006 in. (0.152 mm). They may be made on either Foudrinier or cylinder machines, but in either case are of the same stock throughout or nearly so. They are stiff and have very smooth surfaces. One grade, known as "Wedding Bristol," is made by pasting together two sheets of Bristol board.

board, container (papermaking). Most container board is Foudrinier board of solid unbleached kraft. Much care is exercised to get maximum strength and durability. Kraft pulp provides strong fibers, and refining of the pulp is designed to give strength. The board is sized with rosin for water resistance and usually has a surface application of wax size and starch to make it abrasion resistant. If the containers are to be used for overseas military shipments, the outer layer of the box is treated with wet-strength resin; the board is also calendered to provide a good printing surface.

board, corrugating (papermaking). Used as the cushioning layer in corrugated containers. It is traditionally 0.009 in. (0.23 mm) thick and is sometimes referred to as "nine-point." It was formerly made of straw pulp and later chestnut semichemical pulp. In addition to these two sources, kraft pulp is much used today. It must have resilience and high resistance to crushing. It is normally not sized so that it may readily be softened with steam for corrugating and adhere readily to the backing board when adhesive is applied to the crests of the flutes.

board, folding boxboard (papermaking). This grade includes boards used for making boxes, which are shipped folded flat to the point of use and are then opened and assembled to receive merchandise. The quality may vary widely, but it is in general cylinder board which carries a liner containing some virgin wood pulp together with high-grade waste papers. The filler and back may contain mixed paper or news. Thickness varies from 0.013 to 0.052 in. (0.33 to 1.32 mm) and weighs from 60 to 180 lb/1000 ft^2 (293 to 878 kg/1000 m^2). It must have a fair degree of strength, be capable of bending without cracking, and possess a good surface for printing. Some are clay coated for the last-named purpose.

board, food container (papermaking). A rather arbitrary class embracing products that vary widely. They have in common the fact that they are used to package food and include among other uses mild bottle board, butter cartons, ice cream board, and cottage cheese containers. In most cases the board is heavily waxed before use. The base stock is generally bleached chemical wood pulp, but groundwood has also been used. It may be made on either a cylinder or Foudrinier machine. The board is well sized with rosin, sometimes in conjunction with wax. Even though the board is waxed, sizing of the base stock as a second line of defense is considered necessary. Good appearance and freedom from odor are important characteristics of these boards. Falling into this class, but of different construction, is can stock. This board is used to make containers of which the body consists of a cylinder made of spirally wound strips of board. It is commonly made on a cylinder. It is very hard sized, whether or not it is intended to be waxed. The board must be stiff and yet flexible enough so that the bottom disk can be rolled in.

board, setup boxboard (papermaking). This grade is used in boxes that are set up at the time of fabrication. The board is made on a cylinder machine but may be solid (having the same stock in all plies). It must be strong to resist rough handling, but bending quality is less necessary than in folding boxboard. It is made in thicknesses of 0.016–0.065 in. (0.41–1.65 mm).

borates (minerals). Elemental boron is unknown in the native state because of its strong affinity for oxygen, but exists in a large number of combinations. The minerals of commercial importance are a small group of borates, which contain boric oxide, sodium or calcium oxide, or both, and water of crystallization; borax; kernite; colemanite; and ulexite. Borax commonly occurs as glassy, clear to translucent monoclinic crystals and crystalline aggregates. In dry air it turns white by dehydration to tin-calconite, a dull white powdery mineral. Hardness is 2.0–2.5; the specific gravity is 1.72. It is easily fusible and highly soluble in hot water. A little borax added to a batch from which ordinary glass is made improves melting conditions, and imparts clarity and brilliance to the product. Twelve to fourteen percent of borax in a glass mix produces heat-resistant borosilicate glass, such as Pyrex. Porcelain enamels for bathtubs, stoves, refrigerators, and metal signs may contain as much as 20% borax. Borax is readily soluble in water, producing a solution with mild antiseptic and detergent properties. Not only are borax solutions used as disinfectants and preservatives, but are applied to prevent growth of mildew in starches, mold on citrus fruit, and fungi in lumber. Borax is used in washing and cleansing directly or in detergents. Casein, used in the manufacture of coated paper, playing cards, and plywood, is dissolved in a borax solution during processing. Borax is utilized as a flux in brazing, welding, and soldering, and as a deoxidizer in metallurgy. Borax and other boron compounds are among the most versatile intermediate chemicals in the manufacturing industries.

boring (machining). A machining process in which internal diameters are generated in true relation to the centerline of the spindle by means of single-point cutting tools, and is most commonly used for enlarging or finishing holes or other circular contours. Although most boring operations are done on simple, straight through holes, the process is also applied to a variety of other configurations. Tooling can be designed for the boring of blind holes, holes with "bottle" configurations, and circular contoured cavities, and bores with numerous steps, undercuts, and counterbores. The process is not limited by length-to-diameter ratio of holes; with the workpiece properly supported, holes having diameters that exceed length (or vice versa) by a factor of 50 or more have been successfuly bored. Boring is sometimes used after drilling to provide drilled holes with greater dimensional accuracy or improved finish. It is more widely used, however, for finishing holes too large to be produced economically by drilling, such as large cored holes in castings.

boring machines. Available boring equipment may be divided into two distinct categories, the horizontal and the vertical. With horizontal boring machines, the work is held stationary on a table having in and out as well as back and forth movements. A vertically adjustable head carries the boring spindle, which can be fed into the work. Large, intricate, or bulky parts are normally machined on the large universal horizontal machines. These are usually of the planer type, table type with standard or revolving tables, floor type, or multiple-head crossrail type machines. Many adaptations of these models and new numerically controlled machining centers are also available, extending the scope and capacity range of hole boring. For convenience, where length or height is less than the diameter, the usual practice is to machine in a vertical boring mill. Vertical boring mills, however, revolve the work on a circular table while the tools, much as in a vertical turret lathe, are fed into and away from the work. Clamped to the rotary table of these mills, parts such as plates, turbine bucket wheel forgings, and flywheels, can be turned and faced as well as bored.

box annealing. Annealing a metal or alloy in a sealed container under conditions that minimize oxidation. In box annealing a ferrous alloy, the charge is usually heated slowly to a temperature below the transformation range, but sometimes above or within it, and is then cooled slowly; this process is also called close annealing or pot annealing. **(See also annealing and black annealing.)**

brasses. Brasses contain from 5 to 45% zinc with or without smaller additions of one or more of other elements. Zinc forms with copper a series of solid solutions known as alpha brasses containing varying amounts of zinc, from 5 to 37%. The alpha brasses have the same face-centered-cubic crystal structure as copper, which accounts for their ductility and the ease with which they may be cold worked. The alpha brasses can also be hotworked, but the presence of impurities such as lead and bismuth causes hot shortness. When the zinc content exceeds 37.5% a new beta phase appears that is hard and brittle. Alloys of composition from 37.5 to 44% zinc are composed of two phases, alpha and beta, whereas those with a zinc content of from 44 to 46% consist of a beta phase. The single-phase alloys are not used in engineering applications because of their brittleness, but the two-phase alloys have good hot-working properties over a wide range of temperatures and are extensively used as hot-working materials.

brasses, admiralty metal. A copper–zinc alloy containing approximately 1.0% tin and 0.03% arsenic. Its resistance to corrosion is superior to that of ordinary brasses that run free from tin. This composition is used for tubing condensers, preheaters, evaporators, and heat exchangers in contact with freshwater, salt water, oil, steam, and other liquids at temperatures below 500°F (260°C).

brasses, alpha–beta (Muntz metal). When the content of zinc in brass is increased from 45 to 50% zinc, the alloy is called Muntz metal. This alloy contains both the alpha and beta constituents in its structure. Muntz metal may be hot worked even when it contains a high percentage of lead. Such brass is useful for screws and machine parts, where ease of working and, particularly, ease of machining are more important than strength. Some of the applications for Muntz metal include sheet form for sheathing, condenser heads, perforated metal, condenser tubes, valve stems, and brazing rods.

brasses, red. There are four types of wrought brasses in this group: gilding metal, 95–5; commercial bronze, 90–10; rich low brass, 85–15; and low brass, 80–20. (The second number gives the percentage of zinc.) These brasses are workable both hot and cold. The lower the zinc content, the greater the plasticity and workability. These brasses are superior to the yellow brasses or high brasses for corrosion resistance and show practically no bad effects from dezincification or season crackling. The applications for red brass include valves, fittings, rivets, radiator cores, detonator fuse caps, primer caps, plumbing pipe, bellows and flexible hose, stamped hardware, and screen cloth. These alloys may be shaped by stamping, drawing, forging, spinning, etc. They have good casting and machining characteristics and are weldable.

brasses, yellow or cartridge. The yellow alpha brass is the most ductile of all the brasses, and its ductility allows the use of this alloy for jobs requiring the most severe cold-forging operations such as deep drawing, stamping, and spinning. Basically, the structure of annealed yellow or cartridge brass is copper containing from 27 to 35% zinc. This alloy is used for the manufacture of sheet metal, rods, wire, tubes, cartridge cases, and many other industrial shapes. A rolled and annealed brass has a tensile strength of about 28,000 psi (1.93×10^5 kPa). By cold rolling, however, its tensile strength may be increased to 100,000 psi (6.9×10^5 kPa). Brass is obtainable in varying degrees of hardness by cold rolling usually designated quarter-hard, half-hard, and hard. **(See also ball broaching.)**

brazing. A group of welding processes that join solid materials together by heating them to a suitable temperature and by using a filler metal having a liquidus above 840°F (450°C) and below the solidus of the base materials. The filler metal is distributed between the closely fitted surfaces of the joint by capillary attraction. **(See also welding.)**

brazing sheet. Sheet of a brazing alloy, or sheet clad with a brazing alloy on one or both sides.

Composition, Properties, and Uses of Some Brasses

Name	Composition (%) Cu	Zn	Other	Color	Temper	Tensile strength psi	Elongation in 2'' %	Typical uses
Gilding metal	95	5		Copper	Light anneal	35,000	42	Drawing, spinning, forming jewelry, forgings
					Soft anneal	34,000	44	
					Hard	55,000	4	
					Spring	65,000	3	
Commercial bronze	90	10		Bronze	Light anneal	41,000	42	Stamped hardware, forgings, screws, rivets, screen cloth, jewelry, bullet jackets
					Soft anneal	38,000	45	
					Hard	64,000	4	
					Spring	73,000	3	
Red brass	85	15		Red	Light anneal	45,000	48	Hardware, radiator cores, plumbing pipe, condenser tubes, flexible hose
					Soft anneal	40,000	43	
					Hard	71,000	5	
					Spring	83,000	3	
Rich low brass	83	17		Red-gold	Soft	42,000	43	Corrosion-resistance applications; pipe, tags, dials, wire, jewelry
					Hard	75,000	4	
Low brass	80	20		Red-gold	Light anneal	47,000	47	Drawing, forming, corrosion-resistance applications; jewelry (for gold plating), bellows, flexible hose
					Soft anneal	43,000	55	
					Hard	75,000	7	
					Spring	90,000		
Brazing brass	75	25		Typical brass color	Light anneal	52,000	50	Drawing, spinning, and for fabrication by brazing; springs
					Soft anneal	45,000	62	
					Hard	76,000	7	
					Spring	95,000		
Cartridge or spinning brass	70	30		Typical brass color	Light anneal	53,000	54	Cartridge cases and spun or drawn articles; eyelets, pins, rivets, radiator tubes
					Soft anneal	46,000	64	
					Hard	76,000	7	
					Spring	97,000	3	
High brass	66	34		Typical brass color	Light anneal	53,000	54	Stamping, blanking, drawing, spinning, springs, screws, rivets, grillwork, chains
					Soft anneal	46,000	64	
					Hard	76,000	7	
					Rivet	60,000	30	
					Spring	93,000		
Extruded rivet metal	63	37		Typical brass color	Rivet	60,000	30	Rivets, screws
Muntz metal	60	40		Reddish	Hot rolled	54,000	45	Architectural work, welding rod, condenser tubes, valve stems, ship sheathing
					Cold rolled	80,000	5	
Architectural bronze	56	41.25	2.75 Pb	Bronze	Plain	60,000	25	Extruded shapes, forgings, interior ornamental bronze; free cutting
					Extruded			
Silicon brass	82	17	1 Si	Brass	Soft	55,000	60	Refrigerator evaporators, fire extinguisher shells; high strength, weldable
					Hard	90,000	8	
Aluminum brass	76	22	2 Al	Brass	Soft	52,000	65	Condenser tubes, corrosion- and erosion-resistant
					Hard	85,000	10	
Admiralty metal	71	28	1 Sn	Brass	Soft	52,000	65	Condenser tubes; especially resistant to corrosion by sea water
					Hard	85,000	10	
Naval brass	60	39.25	0.75 Sn	Brass	Soft	54,000	45	Tube heads, marine shafting, bolts, nuts, forged parts; resists sea water
					Hard	65,000	15	
Manganese bronze	59	39	0.75 Sn 1.25 Fe 0.05 Mn	Brass	Soft	60,000	35	Welding rod, perforated coal screens, extruded parts; wear- and corrosion-resistant
					Hard	75,000	5	

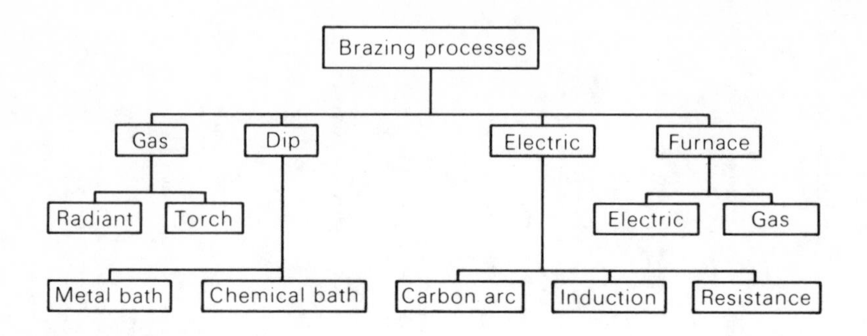

General breakdown of basic brazing processes and their various offshoots.

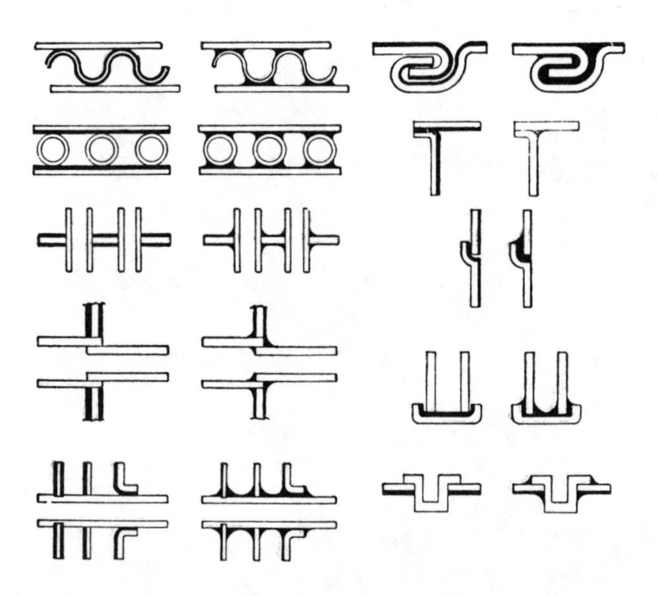

Typical joints with braze-clad sheet, before and after brazing, are shown at the right.

brazing wire. Wire for use as filler metal in joining by brazing.

brick and terra cotta. In the manufacture of brick and terra cotta, the raw materials (consisting of clays, shale, and other materials) are first washed and ground. The mixture is then placed in a pug mill, which consists of a horizontal cylindrical container with a shaft to which a group of blades are attached. The rotation of the shaft makes it possible for the blades to produce a uniform plastic mass of the clay. When the pug mill has served its purpose, the clay is transferred to molds if special shapes are desired. In the stiff mud process used for brick, hollow tile building blocks, and similar shapes, the plastic clay is extruded through dies. The brick is then dried and is subsequently fired in kilns to produce partial vitrification. The rate of cooling used influences the properties. Too rapid cooling will produce warping and brittleness. Paving brick of maximum quality is obtained by a cooling or annealing period of about a week. Shapes such as sewer pipe and paving brick require a hard impervious surface; a coating is added that vitrifies the surface. This is accomplished by including with the clay materials such as magnesia, lime, and ferric oxide that will fuse with the clay during the burning process.

brick, building. Standard building brick is 2¹/₄ × 3³/₄ × 8 in. (5.7 × 9.5 × 20.3 cm) and glazed brick is ¹/₈ in. (0.3 cm) wider. The compressive strength varies from 3000 to 9000 psi (2.1×10^4 to 6.2×10^4 kPa) and the modulus of elasticity from 1.4 million to 4 million psi (9.7 million to 27.6 million kPa). Building brick is graded as red, salmon, or arch brick. Red brick is well burned, high-grade construction material. Salmon brick is soft and incompletely burned; it is used where strength is not important, as in masonry fill. Arch brick comes from the high part of the kiln, is warped and deformed from overburning, and is usually used where it cannot be seen in the finished structure.

Brinell hardness test (see hardness test)

broaching (machining). A machining process in which a cutting tool that has multiple transverse cutting edges is pushed or pulled through a hole or over a surface to remove metal by axial cutting. Because broaches are multitoothed cutting tools produced to close tolerances, they are expensive; consequently, the process usually is employed only for high production. Because several edges of a broach are cutting at once, forces are much greater than in other machining methods, and broaching is rated as the most severe of all machining operations. Usually a broach is a tapered bar into which teeth have been cut so as to produce a desired contour in a workpiece by a single pass of the tool.

bronzes. The phosphor bronzes are actually tin bronzes containing 1–11% tin. Phosphorus is used in small amounts as a deoxidizer. These bronzes are used in sheet and bar form for their high tensile strength, toughness, formability, fatigue strength, and corrosion resistance. Tensile strength may be as high as 100,000 psi (6.9×10^5 kPa). Phosphor bronzes are frequently specified for instrument parts such as diaphragms, bellows, springs, and electrical contacts. Cast bronze plumbing fittings are a common article in domestic and commercial plumbing lines strung in copper pipe. A commonly used alloy for casting these fittings is a tin bronze, analyzing 81% copper, 3% tin, 7% lead, 9% zinc. The tin and zinc provide castability while the lead gives machinability. Silicon bronzes are more familiarly known by their trade names, such as Everdur and Herculoy. These also are high-strength alloys, with strengths as high as 110,000 psi (7.6×10^5 kPa). The cupronickels contain up to 30% nickel. Aluminum bronzes contain 4–11% aluminum. The alloys containing over 8% aluminum are heat treatable to give microstructures similar to steel. Slow cooling gives a lamellar soft structure, like pearlite; fast cooling produces a martensitic structure. The aluminum bronzes are used for a wide range of articles, including electrical hardware, marine hardware, propellers, tubing, and pumps. Tensile strength may exceed 100,000 psi (6.9×10^5 kPa) after heat treatment, but ductility is then minimal. Beryllium copper is another heat-treatable bronze, giving the highest strength of all the copper alloys, together with excellent fatigue resistance, formability, corrosion and wear resistance, and formability. Zirconium copper and chromium copper are heat treatable and used chiefly for instrument work and electrical hardware such as resistance welding tips and switch parts. **(See table p. 45.)**

Refractory Brick, Characteristic Properties

Type	Class	Characteristics
High-Alumina	99% alumina 90% alumina 80% alumina 70% alumina 60% alumina 50% alumina	High refractoriness, which increases with increasing alumina content. High mechanical strength at high temperatures. Excellent-to-fair resistance to spalling. Greater resistance than fire clay brick to corrosion by most basic slags and fluxes.
Fire clay	Super-duty	Stability of volume and high mechanical strength at high temperatures. Excellent resistance to thermal spalling. Fair resistance to highly acid slags; lower resistance to basic slags.
Fire Clay	High duty Intermediate duty Low duty	The chemical and physical properties of fire clays vary between wide limits; hence, fire clay brick with widely varying combinations of properties are available. This fact accounts for their suitability for service under widely different operating conditions. Most fire clay brick have relatively good spalling resistance and thermal insulation value. Fair resistance to acid slags and fluxes, lower resistance to basic slags and fluxes.
Silica	Super-duty Conventional	High refractoriness and resistance to abrasion. High mechanical strength at high temperatures. Thermal conductivity at high temperatures about 25% greater than that of fire clay brick. High resistance to corrosion by acid slags. Fair resistance to attack by oxides of lime, magnesia, and iron. Readily attacked by basic slags and fluorine. Not subject to thermal spalling at temperatures above 1200°F (649°C); poor resistance to thermal spalling at low temperatures
Magnesite	Fired 85–90% magnesia More than 90% magnesia Chemically bonded. metal encased	Extremely high refractoriness and high thermal conductivity. Great resistance to corrosion by basic slags; poor resistance to slags containing high percentages of silica. The chemically bonded and metal encased brick have marked resistance to spalling.
Chrome	Fired	High resistance to corrosion by basic and moderately acid slags and fluxes. In general, basic slags do not adhere to chrome brick. Under certain unusual conditions, iron oxide is absorbed and causes a damaging expansion. Thermal conductivity lower than that of magnesite brick but higher than that of fire clay brick.
Magnesite-chrome	Chemically bonded Chemically bonded, metal encased Fired	Mechanical strength and stability of volume at high temperatures. Excellent resistance to spalling. High resistance to corrosion by basic slags.
Chrome-magnesite	Chemically bonded Fired	Mechanical strength and stability of volume at high temperatures. Excellent-to-good resistance to spalling. High resistance to corrosion by basic slags.
Fosterite	Fired	High refractoriness. Excellent strength at high temperatures. Marked resistance to corrosion by alkali compounds. Fair resistance to most basic slags; attacked by acid slags.

Bronze Alloys

	Cu[a]	Sn	Si	Zn	Ni + Co	Other
Phosphor bronze	90+	1.25–10				
Silicon bronze	98		1.53–3			
Architectural bronze	57	3		40		
Cupronickel	70				30	
Aluminum bronze	90+					4–11 Al
Nickel silver	65			17	18	
Beryllium bronze	97.9				0.2	1.9 Be
Chromium copper	99.1					0.9 Cr
Zirconium copper	99.85					0.15 Zr

[a]Values in percent.

bronzes, aluminum. These are classified into two main groups, single-phase and duplex-phase alloys. The alloys are ductile and corrosive resistant. Since a 5% aluminum alloy has a color similar to that of 18 carat gold, it is used largely for the manufacture of imitation jewellery. The 10% aluminum alloy can be heat treated in a manner similar to carbon steel. Slow cooling will give a structure of primary (analogous to ferrite) in a eutecotoid of a + ²/₂ (analogous to pearlite), and quenching produces a hard martensitic structure (β'). Subsequent tempering modifies this structure with results parallel to those obtained in carbon steel. Difficulties in casting these alloys arise due to the rapid oxidation of aluminum at the casting temperature, and this has limited the popularity of aluminum bronzes.

bronzes, beryllium. Contains up to 2% beryllium and can be strengthened by precipitation treatment to give tensile strengths in the region of 2×10^5 psi (1400 N/mm^2). It is used as a tool material in the gas industries and explosive factories where sparks, generated from steel tools, might have disastrous results.

bronzes, leaded. Lead does not alloy with copper but may be mixed with copper, by agitation or mechanical mixing, while the copper is in the molten state, and under suitable conditions this may be satisfactorily cast in a mold, resulting in the lead being well distributed throughout the casting in small particles. Lead may be added to both bronze and brass for the purpose of increasing the machinability and acting as a self-lubricant in parts that are subjected to sliding wear, such as a bearing. The lead particles, with their soft, greasy nature, reduce the frictional properties of the alloy. Lead is a source of weakness and is usually kept below 2% although some bearing bronzes may contain as high as 50% lead.

bronzes, phosphor. From 1.5 to 10% phosphorus is added to bronze for deoxidation purposes during melting and casting operations. The phosphorus also increases the fluidity of the molten metal, thereby increasing the ease of casting into fine castings and aiding in the production of sounder castings. With the higher phosphorus content, a hard compound, Cu_3P, is formed, which combines with the Cu_3Sn present in these alloys, increasing the hardness and resistance to wear. The alpha constituent appears as dendritic crystals in a matrix consisting of a mixture of alpha phase and the hard, brittle compound Cu_3Sn. These alloys are used largely for gears and bearings.

bronzes, tin. Fall into two groups, the single-phase a-alloys are tough, ductile, strong, and corrosion resistant. They are more heavily cored in the cast condition than are the brasses. Diffusion of tin in copper is slow, so that cast copper–tin alloys with as little as 6% tin may contain some of the brittle intermetallic compounds S ($Cu_{31}Sn_8$). In industrial practice cooling is never slow enough to permit transformation to E

(Cu_3Sn) to occur. The a bronzes are cold rolled to sheet and drawn to wire and other sections. Bronzes containing in excess of 10% tin may show considerable amounts of S due to excessive coring of the a. They are too brittle, therefore, to permit cold working and are used for the manufacture of corrosion-resistant castings and bearings in which the presence of particles of the hard S phase, cushioned in a tough matrix of a, provides a hard-wearing low-friction surface. Tin bronzes are generally deoxidized with small amounts of phosphorus left in the alloy. The addition of phosphorus in excess of this amount and up to 1.0% is made when increased strength, corrosion resistance, and casting fluidity are necessary. These alloys are known as phosphor bronzes.

brushing, power. Rotary brushes are used for conditioning surfaces to achieve specific degrees of smoothness or roughness; to remove burrs and lightly attached inclusions of particles, scale, and oxides; to finish edges by rounding or blending; to remove surface film or encrustations; and, generally, to affect the surface itself or cause the removal of materials from or along a surface. Brushing tools do not find efficient use to remove metal or other materials from a surface to achieve a dimensional size or tolerance, although recent advancements with abrasive filament brushes do remove metal uniformly to a small degree and within specified tolerances. Production yield of a brushing operation is influenced by the design of the brushing machine used to operate the brush. Completely automatic equipment designed to feed and move parts through one or more brushing stations is capable of finishing at the rate of 1000–2000 parts per hour. On the other hand, a simple work fixture, such as a handheld tool for holding the part in front of a rotating brush, would result in a much lower production yield per hour.

brush plating. Generally used for touching up defective electroplating, for rebuilding worn parts, and for localized plating where conventional processes are unsuitable. Materials generally can be deposited faster and thicker, and they adhere better where brush plated. But brush plating is a handheld, one-at-a-time process. The process, however, has been automated where both anode contact area and solution flow are suitably controlled.

buffing. A mechanical finishing operation in which fine abrasives are applied to a metal surface by rotating fabric wheels for the purpose of developing a lustrous finish.

burning. (1) Permanently damaging a metal or alloy by heating to cause either incipient melting or intergranular oxidation. **(See also overheating.)** (2) In grinding, getting the work hot enough to cause discoloration or to change the microstructure by tempering or hardening.

burnishing. Steel burnishing is essentially a method of improving the surface finish or sizing a part by pressure application of a roll or blunt tool to the surface. Modifications of the burnishing process include roller burnishing, bearingizing, ballizing, ball broaching, and roll planishing. Internal and external cylindrical surfaces (such as the bores of internal combustion engine cylinders, the fillets of shafts subjected to torsion and bending, and the bearing surfaces of axles for railway rolling stock) have been finished by cold rolling by pressing a suitably shaped roller against the rotating part. The cold-working action may serve to increase the fatigue resistance of highly stressed parts, or the hardness and wear resistance of bearing surfaces. Such burnishing is usually done by the pressure from hardened and polished single rollers or attachments having rollers at the front and rear to balance the pressure. Metal is not removed in roller burnishing, rather the metal in the surface crests is displaced into the microscopic valleys to produce a high-quality finish.

burr. A thin ridge of roughness left by a cutting operation such as slitting, trimming, shearing, blanking, or sawing.

burred edge. A thin turned-down edge of sheet or foil resulting from shearing.

butyl rubber. The copolymer of isobutylene with about 2% isoprene is known as butyl rubber. The isoprene imparts sufficient unsaturation to the molecule to permit curing or vulcanization. Polymerization is carried out in solution at extremely low temperatures, followed by precipitation of the rubber by the addition of water, separating the rubber as crumb, which can be filtered and dried with conventional equipment. Butyl rubber is uniquely impermeable to air and has found its major use in inner tubes and as liners of tubeless tires. Butyl rubber is also inert to oxidation and is useful for weatherstripping and other similar exposure conditions.

C

CAD (see computer-aided design or computer-aided drafting)

CAD/CAM (see computer-aided design/computer-aided manufacturing)

CAD/D (see computer-aided drafting and design)

cadmium. Used for the plating of steel hardware and fasteners and as an alloy in bearing metals. The uses of tin are similar to those of cadmium. Since cadmium compounds are toxic, this metal is not permitted in food-processing equipment. Because of its high neutron-capture cross section, cadmium is used in the control of nuclear reactions.

CAE (see computer-aided engineering)

calendering. Process used to produce thermoplastic sheet, particularly of PVC and polythene, or rubber. The calendering machine consists of two or more heated rolls into which the plastic material is fed as a heated dough. The formed sheet is cooled on a chill roll. Calendering machines are also used to coat paper, fabric, and foil with a film of plastic material.

CAM (see computer-aided manufacturing)

cantilever truck. A self-loading counterbalanced or noncounterbalanced industrial truck, equipped with cantilever load-engaging means such as forks.

capacitors (electronics). A capacitor is formed when two conductors are brought within close proximity to each other without touching. In such an instance, air would be the medium of separation between two such conductors and the air thus acts as an insulator. Other types of insulation can also be employed, such as mica, paper, ceramic, and plastic. The insulation between the two conductors that form a capacitor is called the dielectric. Capacitors take many forms, some utilizing metal plates for the conducting surfaces, while others use metal foil. A capacitor effect is also formed between two wires that are close together or between a length of wire and another metal surface, such as the circuit chassis. Such capacity effects are often undesirable. The capacitors that are desired are usually of the physical type specifically for application in electronic circuits.

carbides (refractories). Carbides are characterized by very high melting points, but lack oxidation resistance at high temperatures. The most important refractory materials are carbides of silicon and boron and interstitial carbides of the transition elements such as zirconium and titanium. Of all the refractory carbides silicon carbide (SiC) refractories have been used longest. They are in several grades depending on the amount and type of binding used. Four major types of bonds have been used, resulting in four types of products. The oldest is the clay-bonded material produced by adding refractory clay as a bonding agent. This considerably lowers the refractories of the product since the bond softens at 2190–2730°F (1200–1500°C). The other types of bond used are silicon nitride, and silicon. Silicon-nitride-bonded materials give a product of high strength and superior thermal shock resistance and make it possible for closer dimensional tolerances to be maintained. Silicon-bonded silicon carbide materials find limited use because the melting point of silicon is only 2600°F (1427°C). Other highly refractory compounds are borides, nitrides, silicides, and sulfides. Boride groups containing 30 binary borides with melting points greater than 3615°F (1990°C) can support substantial loads under oxidizing conditions over a temperature range of 2910–3990°F (1600–2200°C).

carbide, sintered (see sintered carbide)

carbon. A naturally abundant nonmetallic element in some inorganic and all organic materials. It is found in amorphous (coal), graphitic, and diamond forms. It is capable of bonding itself chemically to form a large number of long-chain molecules. It sublimes above 6332°F (3500°C). Coal is its most common form. In industry coal is used to manufacture carbon coal electrodes for electric arc furnace use. Coke for heating fuel and other metallurgical uses is derived from coal by destructive distillation of the volatiles.

carbon black. A very finely divided, essentially nonporous type of carbonaceous material, which is produced in a precisely controlled pyrolytic petrochemical process. The best rubber-grade blacks are spherical carbon particles fused together in grapelike clusters. Each particle is composed of several thousand microcrystallite bundles, which are stacked together in a random order. Each bundle consists of three to five micrographic planes or platelets, elliptically shaped and with axis lengths of about 0.02–0.04 in. (0.5–1 mm), which are stacked in a not quite parallel manner. About 50–80% of the carbon atoms in carbon are thought to be present in the microcrystallite bundles, while the remainder is disorganized carbon. Most of carbon black consumption is in rubber products especially tires, shoe heels, and mechanical goods, where it is used to improve the wearing qualities by imparting toughness. Carbon black is significantly responsible for the adoption of synthetic rubber [especially styrene–butadiene (SBR) copolymer] for automobile tires during peacetime. High-abrasion furnace blacks, when combined with "cold-polymerized" SBR, produce tire treads that consistently give better wear than controls compounded with natural rubber and channel blacks. The rubber compound in tires contains about 35% carbon black. Most printing inks are also based on carbon black. The pigments in black plastics, paints, lacquers, and enamels are usually carbon black. Color-grade blacks used for inks and paints have been replaced by furnace blacks, particularly chemically and air-oxidized types.

carbon dioxide. Carbon dioxide gas, and its solid phase known as "dry ice," is one of the most versatile chemicals in industry. It is not recovered from air, but is usually produced by burning carbon compounds. Carbon dioxide gas is 53% heavier than air. At atmospheric pressure (760 torr) the phase change from dry ice to CO_2 gas occurs at −110°F (−78.5°C).

carbon fibers. Carbon fibers are polycrystalline and consist of a large number of small crystallites. These are about 4×10^{-7} in. (10^{-8} m) thick and 10^{-6} in. (2.5×10^{-8} m) in diameter and consist of parallel planes of carbon atoms grouped together. Examination by the electron microscope shows that crystallites are grouped together in units known as fibrils, which are subunits of the fiber. Carbon fibers are produced by the pyrolysis of polyacrylonitrile filaments. The polymer material is heated at a high temperature in an inert atmosphere so that decomposition takes place and polymer chains are, so to speak, stripped of all other atoms and groups of atoms leaving a skeleton of carbon atoms in the structure of graphite. The fibers are kept in tension during pyrolysis in order to prevent curling and loss of desirable properties. The final treatment temperature, which may be in excess of 3630°F (2000°C), to some extent governs the mechanical properties of the resultant carbon fiber. Two varieties of fiber are obtained by different heat treatments, a high-strength fiber and a high-modulus fiber. Carbon-fiber reinforcement of plastics has been used in the manufacture of such diverse products as fan blades of aero-engine compressors, pressure bottles, racking kayaks, molded gear wheels in carbon-fiber reinforced nylon, and racing car bodies.

carbonitriding. A case-hardening process in which a suitable ferrous material is heated above the lower transformation temperature in a gaseous atmosphere of such composition as to cause simultaneous absorption of carbon and nitrogen by the surface and, by diffusion, create a concentration gradient. The process is completed by cooling at a rate that produces the desired properties in the workpiece.

carbon steels. Steels in which carbon is the alloying element that essentially controls the properties of the alloys and in which the amount of manganese cannot exceed 1.65% and the copper and silicon contents must each be less than 0.60%. The carbon steels can be subdivided into those containing between 0.08 and 0.34% carbon, those containing between 0.35 and 0.50% carbon, and those containing more than 0.50% carbon. These are known, respectively, as low-carbon, medium-carbon, and high-carbon steels. Low-carbon steel is relatively soft and ductile and cannot be hardened appreciably by heat treatment. It is used for tin plate, automobile body sheet, fencing wire, light and heavy structural members (auto frames, I beams, etc.), and hot and cold-finished bars used for machined parts. Cold finishing improves surface finish, mechanical properties, and machinability of these compositions. Medium-carbon steel is used for high-strength-steel castings and for forgings, such as railroad axles, crankshafts, gears, turbine bucket wheels, and steering arms. Medium-carbon steel can be hardened by heat treatment, but it cannot be through-hardened in sections whose thickness is greater than 0.5 in. (12.7 mm). High-carbon steel is used for forgings such as railroad wheels and for hot-rolled products such as railroad rails and concrete reinforcing bars. High-strength wire products such as piano wire and suspension bridge cable are made from high-carbon steel. High-carbon steel tools are among the most useful general-purpose tools for applications such as blanking dies, sledges, chisels, and razors.

carburizing, gas. Accomplished by subjecting the steel to the action of a carbonaceous gas. The carbon in the gas is dissolved in the austenite or combines with iron to form ferric carbide. The gases used are natural gas, methane, ethane, propane, and butane. Carburizing is done in an oven at about 1700°F (925°C). Gas carburizing produces a case depth equal to that of pack carburizing but in about one-half the time.

carburizing, liquid. Makes use of salts that are solid at ordinary temperatures but liquid at carburizing temperatures. The salts contain the cyanogen radical and are usually sodium or potassium cyanide. The process is also known as cyaniding. The heated piece to be hardened is immersed in a bath of the molten salt at temperatures of 1400–1600°F (760–870°C) for 15 min to 1 h, the average time being about 20 min. The cyanides yield carbon monoxide and nitrogen, both of which react with iron. The case thus formed is relatively thin, but the time for carburizing is greatly reduced over that for pack carburizing.

carburizing, pack. A method for carburizing in which the parts to be heated are packed with a carburizing material into steel containers and heated to a temperature of 1650–1750°F (900–955°C). They are held at this temperature for from 6 to 20 h, the average being from 8 to 10 h. During this period carbon from the carburizing material combines with oxygen to form carbon monoxide, which reacts with the iron of the surface by giving up carbon that dissolves in the austenite. This carbon then moves inward by a process of diffusion. The case depth is a function of the concentration of carbon at the surface, temperature, time, and steel composition. The surface carbon concentration varies with the type of carburizing material and the temperature. The depth of the case ranges from less than 1/16 in. (1.6 mm) to about 1/8 in. (3.2 mm). Heat treatment following carburizing varies with the piece being manufactured.

cartridge brass (see brasses)

case hardening (metals). The process of obtaining a hard surface but retaining most of the properties of the original metal in the core. Hardness in steel can be increased by raising the carbon content. The process of increasing the carbon content of the surface while maintaining the original carbon content in the interior is called case carburizing. The surface layer in which the carbon content has been appreciably raised is called the case and the extent of carbon penetration is known as the case depth.

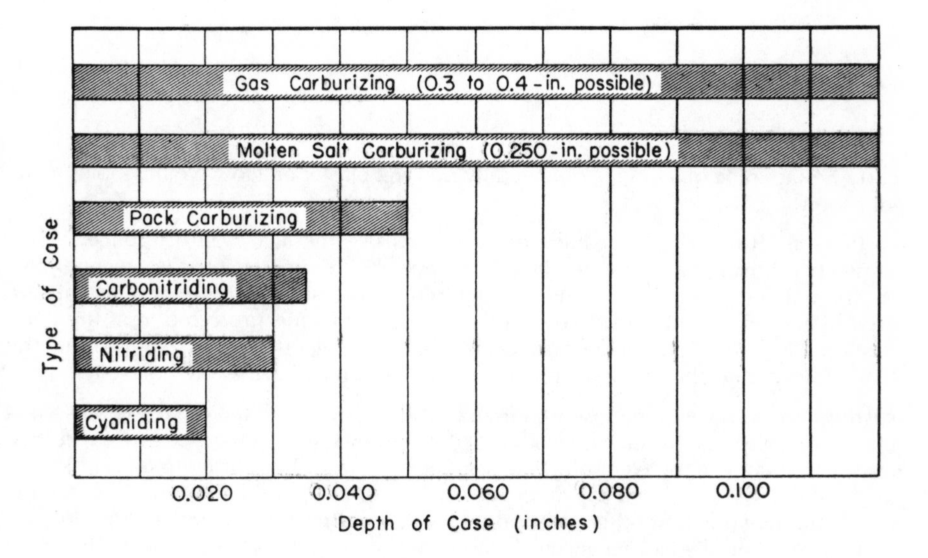

General ranges in case depth possible by different methods of case hardening.

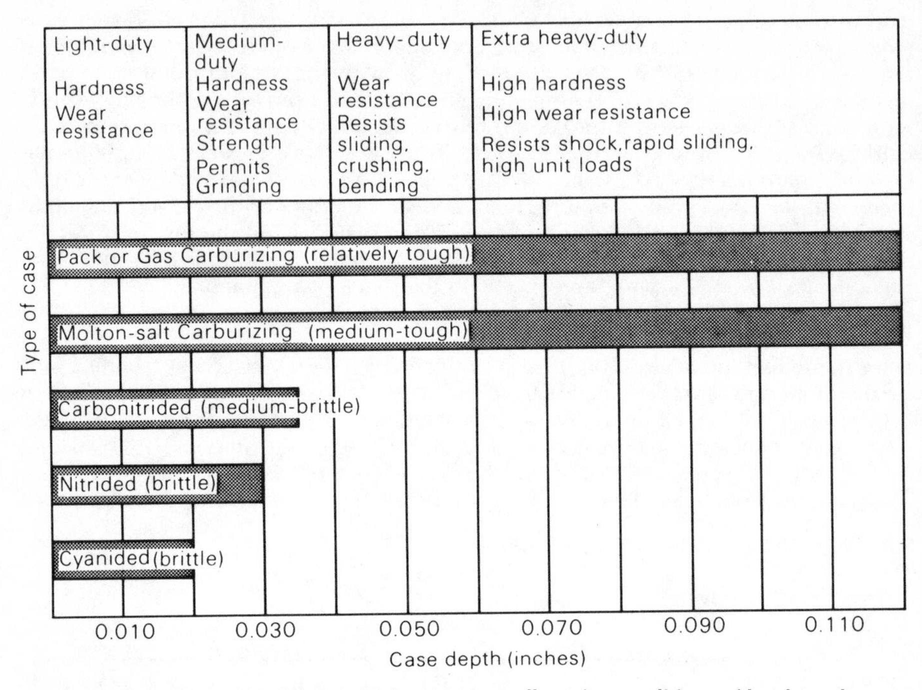

Chart showing relationship of case thickness to overall service conditions of hard-tough parts.

cast. A facsimile of a forging obtained by pouring plaster or a low-melting-point metal into the die cavity.

casting, air furnace. Malleable-iron foundries use the acid-hearth furnace. While somewhat similar to a small open-hearth furnace, it differs in that it has not regenerative equipment and receives its charge through removable bungs. Air furnaces usually are fired by fuel oil or pulverized coal, and the molten metal is protected from the combustion flames by slag. The air furnace has the advantage that the rate of heating, the temperature, and the chemical composition of the metal are easily controlled.

casting, arc furnace. In most arc furnaces the top may be lifted or swung off so that the charge may be introduced. Heating is provided by lowering the electrodes and striking an arc between the electrodes and the metal charge. The current path is from one electrode across the arc to the metal, through the metal, and back across the arc between the metal and another electrode. Once the furnace has been started, the current flow is controlled automatically by apparatus that controls the arc length by raising or lowering the electrodes as required. The direct-arc furnace may be used to melt metal or to heat and hold molten metal at a desired temperature. Fluxing materials are added usually through a side door to provide a protective cover over the molten metal. Since the metal is thus protected, the temperature may be controlled as desired, and the metal maintained at a given temperature as long as desired, high-quality metal of any desired composition can be obtained. The very high temperatures obtainable make high melting rates possible. Furnaces of this type in capacities up to 100 tons (91 tonnes) are in use, but the more common sizes are below 25 tons (23 tonnes). Linings may be either acid or basic, the former being more frequently used.

casting, centrifugal. In centrifugal casting centrifugal force is utilized to force the molten metal to conform to the shape of a mold. The centrifugal force is obtained by rotating the mold about an axial centerline at speeds of from 300 to 3000 rpm while the molten metal is being introduced. In true centrifugal casting, the mold rotates about either a horizontal or vertical axis. In this process either a dry sand or metal mold is used and it determines the outer surface of the casting. The outer shape of such castings usually is round, but hexagonal or other symmetrical but slightly out-of-round shapes may be used. In this type of centrifugal casting no mold or core is needed to form the inner surface of the casting. When a horizontal axis is used, the inner surface is always cylindrical. If a vertical axis is used, the inner surface is a section of a parabola, the exact shape being a function of the speed of rotation. Since in true centrifugal casting the metal is forced against the walls of the mold with considerable pressure, and solidifies first at the outer surface, a dense structure is obtained and all impurities tend to collect at the inner surface because they are lighter than the metal. They can easily be removed by a light machining cut if required. This process is used for mass production of such item as pipe, large gun barrels and liners, cylinder liners, brake drums, and aircraft engine cylinders. Semicentrifugal casting utilizes centrifugal force to cause the molten metal to flow from a central feeding reservoir into one or more sand molds that are rotated about their central axis. It is used primarily for such items as wheels.

casting, continuous. Accomplished by pouring molten metal into the mold and by keeping the mold filled at all times. The metal at the lower end of the mold is cooled so that it solidifies. Rollers grip the solid metal and pull it out from the end of the mold at a controlled speed. As the metal is withdrawn, a cutting torch or saw is used to cut the continuously moving metal to length. Materials such as brass, zinc, copper and its alloys, aluminum and its alloys, and carbon and alloy steels may be cast by one of several continuous-casting processes. When steel is cast, a brass mold or thin-steel-wall mold is used. The outside of these molds must be cooled. Shapes such as circles, tubes, squares, and other geometric shapes may be continuously cast. Ingots, slabs, and sheet may also be continuously cast.

casting, die. A metal mold casting process in which molten metal is forced under considerable pressure into a steel mold or die in the die-casting process. The molten metal is shot through a runner and gate to fill the die. Vents and overflow wells are provided for escape of air. The metal is pressed into all the crevices of the die, and the pressure is held while the metal freezes to ensure density. Many dies are water cooled to hasten freezing. After the metal has solidified the die is pulled open, and the part is ejected by pins actuated by a special mechanism. The production is called a die casting. Metals and their alloys that are die cast are zinc, aluminum, magnesium, copper, lead, and tin. Die casting is mainly found in the high-production industries. Two basic types of machines are used, the gooseneck and cold chamber. Gooseneck-type machines are used for the low-melting-point materials. They are fast in operation, but because the gooseneck is in contact with molten metal continuously they cannot be used for high-melting-point materials such as brass or bronze, and are not suitable for use with aluminum. Thus their use is confined to zinc- and tin-base alloys. In the cold-chamber type of die casting machine the metal for each casting is fed into the cold chamber. The plunger then forces the metal out of the chamber and into the die. The advantage of this type of machine is that there is very little pick-up of iron because the chamber remains cool, and high pressures are exerted on the metal during solidification and early stages of cooling so as to produce denser structures. Cores can be used successfully in die castings, but they increase the cost of the dies. The dimensional accuracy which can be obtained in die castings is one of their outstanding characteristics. Because of the excellent dimensional accuracy and the smooth surfaces which can be obtained, most die castings require no machining except the removal of small amounts of fin, or flash, around the edge and possibly drilling or tapping of holes.

casting, horizontal molds. For metals having low melting points such as zinc, lead, and copper, horizontal-tray-type molds may be used. These produce slabs suitable for rolling operations. This type of ingot tends to have a fairly heavy oxide coating on its surface, along with a concentration of other impurities. This requires removal of the top portion of the ingot. Flat, traylike molds produce ingots in which there is an absence of pipe formation and very little gas porosity, because freezing is progressive from the bottom of the mold. As the columnar crystals grow upward, gas is evolved at the liquid–solid interface and rises to the surface of the molten metal where it escapes into the air; thus blow-holes and gas porosity are almost eliminated. Shrinkage, instead of being concentrated in a single pipe, is distributed across most of the surface area.

casting, induction furnace. There are two types of induction furnaces. The high-frequency type consists of a crucible surrounded by a coil of copper tubing. When a high-frequency electrical current is passed through the copper coil, a rapidly alternating magnetic field is established, which induces secondary currents in the metal contained in the crucible. These secondary currents heat the metal very rapidly. Cooling water is passed through the copper tubing. Low-frequency induction furnaces are used particularly in connection with die-casting. This type consists of a primary coil with a loop of molten metal acting as the secondary coil. Because the secondary has only one turn, a low-voltage, high-amperage current is induced in it. This high-amperage current provides the desired heating. Furnaces of this type require molten metal in order to start them operating, but the temperature can be controlled readily, so that they are useful as holding furnaces where it is desired to maintain molten metal at a constant temperature for an extended period of time, as in die-casting machines.

casting, investment. In investment casting wax patterns are produced in precision metal molds and then fixed to a tree. The wax is then "invested" with the mixture of finely powdered silimanite and ethyl silicate. The action of water and heat on the latter cause it to form a strong silica bond between the silimanite particles. The heating process also melts out the wax pattern leaving a mold cavity which will receive the charge of metal. To improve the accuracy of the mold impression, metal is cast into the mold under pneumatic pressure or by centrifuging. One advantage of the process is that extremely complex shapes can be cast since the pattern is not withdrawn.

casting, low-pressure. Molten metal is forced by gas pressure upward through a stalk to fill a mold. The metal cools inwardly in the mold to the stalk and freezes when pressure is held. Then the pressure is released and the still molten metal in the stalk returns to the pot. The process is used mostly to cast aluminum in plaster, cast-iron, and steel molds. Low-pressure casting is at a stage between hydrostatic casting and high-pressure die casting. It gives moderately thin sections and intermediate accuracy and surface finish, density, and details.

casting, permanent-mold. Refers to the process of die casting in which the molten metal is fed into the cavity by gravity. The head pressure of the molten metal forces the metal into the mold. The molds are coated with a refractory material and lampblack. The dies are closed and the metal is poured into the dies and allowed to solidify. The dies are opened, and the casting is ejected. The dies are then cleaned and the process is repeated. Cores may be positioned in the dies before they are closed and removed as soon as solidification starts, or sand cores may be used and knocked out of the casting after solidification has taken place. This process is used for making steel and cast-iron castings. It is also used to make alloy castings having a copper, aluminum, or magnesium base. Typical products are refrigerator compressor cylinder blocks, heads, and connecting rods; flat iron sole plates; washing machine gear blanks of cast iron; automotive pistons and cylinder heads; kitchenware; and typewriter parts of aluminum. Castings range in weight from a few ounces to 300 lb (136 kg) and even more, with most under 50 lb (22.7 kg).

casting, plaster-mold. Disposable or semipermanent plaster molds or cores for metal molds are made from plaster of Paris (gypsum) with added talc, silica flour, asbestos fiber, and other substances to control setting time and expansion. A slurry is poured over a pattern and allowed to harden. The mold is stirred to remove water and prevent the formation of steam upon exposure to hot metal. Metals commonly cast in plaster molds are yellow brass, manganese and aluminum bronzes, aluminum, magnesium, and their alloys. Typical products are aircraft parts, plumbing fixture fittings, aluminum pistons, locks, propellers, ornaments, and tire and plastic molds. Plaster molds have low heat conductivity and provide slow and uniform cooling. Castings can be made in intricate shapes with varying sections and particularly with extremely thin walls.

Typical Properties of Various Cast Metals

	Zn-12		Zn-27		Zn-8		Bronze (85-5-5-5)	Aluminum (356-T6)	Cast-Iron (Class 30)
	Sand	Perm Mold	Sand	Sand Ht[a]	Sand	Perm Mold	Sand	Sand	Sand
Tensile Strength, psi × 10^3	40–45	45–50	58–64	45–47	36–40	32–37	37	33	30–34
Yield Strength, psi × 10^3, .2% offset	30	31	53	37	28	30	17	24	—
Elongation, % in 2 in.	1–3	1–3	3–6	8–11	1–2	1–2	30	3.5	—
Hardness, BHN	105–120	105–125	110–120	90–100	90–100	85–90	50–65	70	179+
Young's Modulus, psi × 10^6	12	12	10.9	10.9	12.4	12.4	13.5	10.5	13
Density, lb cu in.	0.218	0.218	0.181	0.181	0.23	0.23	0.318	0.097	0.26
Melting Range, F	710-780	710-780	708-903	708-903	707-759	707-759	1570-1849	1076-1130	2100+
Electrical Conductivity, % IACS	28.3	28.3	29.7	29.7	27.7	27.7	15	40	—
Normal Composition	11 Al, 0.75 Cu, 0.02 Mg, bal Zn		27 Al, 2.2 Cu, 0.015 Mg, bal Zn		8.4 Al, 1 Cu, 0.02 Mg, bal Zn		—	—	—

[a]Heat Treatment—3 hr at 610°F and slow furnace cool.

casting, plaster-mold (Shaw process). Uses plaster compounds for molds. In this process a slurrylike mixture of a refractory aggregate, hydrolyzed ethyl silicate, and a jelling agent is poured over a pattern. This mix forms a jell over the patterns. The jell mold has adequate strength and flexibility so that it can be stripped from the pattern and then returned to the exact shape it had while fitted to the pattern. The mold is then ignited to burn off the volatile elements in the mix. It next is brought to a red heat in a furnace. This firing makes the mold rigid and hard, but at the same time causes a network of microscopic cracks to form. This microcrazing produces fissures that provide excellent permeability and a good degree of collapsibility to accommodate the shrinkage of the solidifying metal. The Shaw process can be used for castings of all sizes and results in excellent surface finish, detail, and dimensional accuracy.

casting, sand. In casting, the metal is brought to a fluid state by heating, and is then poured or forced into a mold. After cooling to a solid condition, it is removed from the mold and, when necessary, machined to final dimensions, or used in the as-cast condition. Casting is used to produce many small as well as fairly large pieces when shaping by machining would be too expensive. The product may be formed in the shape and size close to that ultimately required, making it unnecessary to remove large quantities of stock by machining. Casting is also used for making large objects such as frames or bases of machine, although here welding offers serious competition. To produce a mold for sand casting, sand is packed around a pattern that has the shape of the desired casting. Except in the full-mold processes, after the sand has been packed firmly, the pattern is removed, leaving a cavity of the desired shape. An opening, called a sprue hole, is provided from the outside of the mold through the sand and connected to the cavity by a system of channels, called runners. The molten metal is poured into the sprue hole and enters the mold cavity through the runners and an opening, called a gate which controls the rate of flow.

casting, shell mold. In ordinary sand casting the molding sand contains sufficient natural clay to act as a binding material for the silica patterns. In shell molding processes this natural clay bond is replaced by a synthetic bonding material of the thermosetting resin (phenol formaldehyde) type. The molding mixture consists of clean, clay-free silica and sand mixed with 5% of the plastic bonding agent. Each half of the shell mold is made on a metal pattern plate, which is heated to about 480°F (250°C) before being placed on top of a "dump box." The box is then inverted so that the plate is covered with the sand/resin mixture. The resin melts and in about 30 s the hot pattern plate becomes coated with a shell of resin-bonded sand. This shell hardens as the resin sets and the box is turned to its original position so that surplus sand/resin mix falls back into the box. The shell is then stripped from the pattern plate with the help of ejector pins. The two halves of the mold thus produced are then clamped or cemented together to receive their molten charge. The main advantage of this process lies in the high degree of dimensional accuracy obtained. Surface finish is also superior to that associated with ordinary sand castings.

casting, slush. Molten metal is poured into a metal mold. After the skin has frozen, the mold is turned upside down or slung to remove the still liquid metal. The thin shell that is left is called a slush casting. Toys and ornaments are made in this way from zinc, lead, or tin alloys. A slush casting is usually bright and well suited for plating, but is weaker and takes longer to make than a die casting.

casting, steel (bottom-pressure method). The principle consists of a ladle filled with molten steel placed in a pressure vessel, the latter being covered with a lid in which has been inserted a pouring tube that dips down into the molten steel almost to the bottom of the ladle. A gooseneck connects the pouring tube to the mold in the casting position. When air pressure is applied to the pressure vessel, the pressure of the air on the molten steel causes it to rise in the pouring tube and gooseneck and to enter the mold. The height at which the metal must be raised is that of the highest point to be filled with metal. The highest point depends on the length of the mold and riser, the height above the level of the metal in the ladle, and the angle above the horizontal at which the mold is set. Rate of casting is controlled by regulating the pressurized equipment to increase air pressure in the desired increments.

casting, vertical molds. Cross sections of vertical molds show wide variations. Vertical molds have square, rectangular, or round cross sections. The square and rectangular molds usually have rounded corners. The sides may be smooth or vertically corrugated. Corrugations in the mold produce more rapid cooling by increasing the ratio of the surface to mass, thus reducing the tendency to form large columnary crystals. Rounding the corners of the molds reduces the tendency toward formation of planes or weakness where the ends of columnar crystals meet. Square vertical molds are designed as big-end-up or big-end-down molds.

cast iron. An alloy of iron, carbon, and silicon. There are the following types of cast iron: pig iron, gray cast iron, white cast iron, malleable cast iron, and ductile cast iron. The generic term cast iron is applied to all these materials because they are always utilized by melting and then casting them into their final form. With the exception of ductile iron, they do not possess the necessary plasticity, even at elevated temperatures, to be formed by hammering, rolling, or forging. Most cast irons containing from 2.5 to 3.7% carbon, and silicon is always present in amounts ranging from 0.5 to 3%.

cast iron, ductile. In ductile iron the graphite is formed in spherical or nodular shapes instead of the flat flakes formed in gray iron. By compacting the graphite in this form the notch effect of the flat flake is eliminated. The resulting material combines many of the desirable properties of cast iron and steel. Ductile iron is not a single material but a group of alloy irons containing magnesium, nickel, and other elements. It is the presence of the magnesium that causes the graphite to form in nodules.

cast iron, gray. Receives its name from the color of a freshly made fracture. The gray color is caused by free carbon in the form of graphite flakes. Carbon is soluble in iron (austenite) up to 2%. The eutetic, ledeburite, is a mixture of saturated austenite and cementite that transforms at the eutectoid temperature into cementite and pearlite. Fractures occur mostly through the flakes of graphite, leaving part on each face. These can be brushed off; a silvery white surface remains. The large number of graphite flakes leave a reduced area of iron, and since graphite has no tensile strength, the tensile strength of cast iron is low. The strength of gray cast iron can be improved by controlling the composition and the rate of cooling.

General Engineering Grades of Gray Iron—Mechanical and Physical Properties[a]

Class	30	35	40	45	50	60	70	80
Tensile strength, psi (min)	30,000	35,000	40,000	45,000	50,000	60,000	70,000	80,000
Compressive strength, psi	105,000	115,000	125,000	135,000	150,000	175,000	200,000	225,000
Torsional strength, psi	40,000	45,000	54,000	60,000	67,000	76,000	85,000	90,000
Modulus of elasticity,[b] k psi × 10^{-6}	14	15	16	17	18	19	20	21
Torsional modulus, psi × 10^{-6}	—	—	5.5	6.6	7.0	8.0	8.1	—
Impact strength, Izod AB (1.2-in. diam. unnotched), ft-lb	23	25	31	36	65	75	120+	120+
Brinell hardness	180	200	220	240	240	260	280	300
Endurance limit, psi								
Smooth	15,000	17,500	19,500	21,500	25,500	27,500	29,500	31,500
Notched	(13,500)[c]	(15,500)	17,500	(19,500)	21,500	23,500	25,500	27,500
Damping capacity	Excellent	Excellent	Excellent	Good	Good	Good	Fair	Fair
Machinability	Excellent	Excellent	Excellent	Good	Good	Fair	Fair	Fair
Wear resistance	Fair	Fair	Good	Good	Excellent	Excellent	Excellent	Excellent
Pressure tightness	Fair	Fair	Good	Good	Excellent	Excellent	Excellent	Excellent
Specific gravity:								
g per cc	7.02	7.13	7.25	7.37	7.43	7.48	7.51	7.54
lb per cu in.	0.254	0.258	0.262	0.267	0.269	0.270	0.272	0.273
Thermal coefficient of expansion: (in. per in.) × 10^{-6} (50–200°F)	6.5–6.7	—	6.6–6.8	6.4–6.4	—	—	—	—
(50–500°F)	6.9–7.2	—	7.1–7.3	6.8–7.0	—	—	—	—
(50–800°F)	7.4–7.6	—	7.4–7.6	7.0–7.2	—	—	—	—
Magnetic properties	Mag.	Mag.	Mag.	Mag.	Mag.	Mag.	Mag.	Mag.
Pattern shrinkage, in. per ft	1/10–1/8	1/8	1/8	1/8	1/8	1/8–3/16	1/8–3/16	1/8–3/16
Coefficient of friction (against steel)	—	—	—	(0.19)	(0.195)	(0.20)	—	—

[a]Adapted from material of the ASTM by Babcock & Wilcox Company.
[b]At 25% of tensile strength.
[c]Values in parentheses are estimated.

cast iron, malleable. Malleable iron casting consists almost entirely of ferrite and temper carbon. Temper carbon is graphite in a nodular form rather than in the flaky form found in gray cast iron. The carbon appears in the nodular form rather than flaky form because the graphite was separated from the cementite, Fe_3C, while the alloy was in the solid state, thereby preventing the growth of flakes. While not strictly malleable, castings are soft and can be bent without breaking. They are of two kinds, known as white heart (European) and black heart (American), the terms referring to the differences in the process and the products and countries of origin. In either type, the molten metal must have a composition that when cast at the desired temperature into castings of the required size, they will still be free from graphite carbon and flakes will be absent from the melt. Malleable cast iron is usually manufactured in the reverberatory or air furnace, as it is sometimes called. The electric furnace is also used and in either type of furnace the metal is charged either with solid metal or, in some cases, in the liquid state from the cupola, where the duplexing process is utilized. Duplexed malleable iron is melted in the cupola and refined in electric arc or air furnaces.

cast iron, nodular. Nodular cast iron, also referred to as ductile iron, spherulitic iron, and spherulitic graphite iron, is essentially cast iron in which graphite is present in the form of tiny balls of spherulities, rather than as flakes as in gray iron or compacted aggregates (nodules) as in malleable iron. The chemical composition of nodular irons is much like that of gray irons, somewhat as follows: total carbon, 3.20%; silicon, 1.80–2.80%; manganese, up to 0.80%; sulfur, 0.03% maximum; phosphorus, 0.10% maximum. In order to produce the spheroidal graphite structure, one or more alloying elements are added to the molten metal, usually magnesium, cerium, lithium, sodium, barium, or other elements. Although some nodular iron castings are used in the as-cast condition, in most cases they are given a heat treatment in the nature of a ferritizing anneal, or a normalizing treatment with quenching and tempering being used in some instances. Nodular or ductile iron castings have a high shock resistance, stiffness, and modulus of elasticity with the properties of large generators made from this material as follows: tensile strength, 65,000–85,000 psi ($4.5–5.9 \times 10^5$ kPa); elongation 3.5–4.5% in 2 in. (5.1 cm); hardness, 163–228 Brinell; modulus of elasticity, 21.5–23.5 million psi ($1.5–1.6 \times 10^8$ kPa).

cast iron, white. Iron in which most of the carbon is in combined form as cementite, Fe_3C, and little if any is present as graphite. The constitution is cementite and pearlite, and the lack of free graphite gives it the silver–white color. Cementite is hard and brittle and white iron is the same. White cast iron is produced by slowly cooling a cast iron that is low in silicon. It is also produced by rapidly cooling a cast iron that would normally be gray if cooled slowly. In this case the iron is called chilled cast iron.

CCD camera (computers). A solid-state camera that uses CCD (charge-coupled device; also called a bucket-brigade device) to transform a light image into a digitized image. A CCD camera is similar to a CID camera, except that its method of operation forces readout of pixel brightnesses in a regular line-by-line scan pattern. There is only one readout station and charges are shifted along until they reach it.

cellophane. Made from a viscose solution. The spinning solution is extruded through a narrow slot (instead of through a number of small orifices) as a film into a coagulating bath and is then purified, softened, and dried. In practice, the alkaline viscose solution is pumped into a rectangular casting head (the extrusion of the film is called "casting" instead of "spinning") which has a narrow slit running along the bottom; the slit is positioned just beneath the surface of the acid-coagulating bath, so that as the viscose is pumped through the slit it congeals to a film. The coagulated film is then drawn from the bath by a roller, which delivers the film to the processing baths. In these, the film is carried by rollers through a number of treating baths in which it is successively regenerated, washed, dyed if desired, and finally (now transparent) softened with a little glycerine. The film is dried by passing over and under heated rollers and is finally wound on to a metal core and is then called the "mill roll."

cellulose. A large, complex molecule ($C_6H_{10}O_5$) that forms the major part of living plants, grass, and trees. Commercially, purified cellulose is obtained from cotton-gin operations or is extracted from wood pulp. Long cotton fibers called staple, ranging from less than one to several inches long, are first removed from cotton seeds to textile yarns and cordage. Shorter lengths called linters, which are too short for spinning, are used for plastics. Removed from the seed, they are cleaned, filtered, and chemically purified to form the basis for cellulose plastics. Purified cellulose called alpha-cellulose is furnished to manufacturers of phenolic, urea, and melamine molding compounds, as a filler and for production of translucent panels. Cellulose is a highly crystalline polymer having strong intermolecular bonds. The presence of hydroxyl groups permits direct substitution of ester or ether groups to form a number of cellulose plastics.

cellulose acetate. Commonly made by mixing cellulose with acetic anhydride, acetic acid, and sulfuric acid in a stainless-steel Werner–Pfleiderer type of mixer for about 3 h. If the triacetate is desired, it is precipitated with dilute acetic acid. If secondary acetate is desired, the triacetate is partially hydrolyzed by added water before precipitating. The acetate is centrifuged and dried. In another method the reaction between the cellulose and the acetic anhydride is carried out in boiling methylene chloride, using perchloric acid instead of the sulfuric acid as a catalyst.

cellulose nitrate. Because of its flammable nature cellulose nitrate is made in small batches, 50 lb (22.7 kg) or less. High alpha-cellulose is stirred slowly with a mixture of approximately 55% sulfuric acid, 25% nitric acid, and 20% water for about 30 min. A much acid as possible is removed by rapid centrifuging. It is then washed thoroughly with water and finally boiled in water for several hours to degrade to a more stable and more soluble product of lower nitrate content. The boiling, formerly a batch process, is generally carried out continuously in a long tube. After draining, the remaining water is displaced by ethyl alcohol, and a plasticizer, a stabilizer such as urea, and colorants are added in a dough-type mixture. Camphor, the original plasticizer, is excellent. Its keto groups form strong bonds with the nitrate groups while its bulk molecule spaces the cellulose chains far enough apart to ensure flexibility. Sometimes part of the camphor is replaced by orthonitrobiphenyl.

cellulose triacetate. A product of complete acetylation of cellulose with all three hydroxyl groups in each glucose residue acetylated. In manufacturing, the raw material is either purified cotton linters or specially pure grades of wood pulp, in either case a pure form of cellulose. The reactivity of the cellulose is enhanced by pretreatment with acetic acid. Two methods of acetylation can be used. (1) Nonsolvent process in which the activated cellulose is esterified with acetic anhydride in the presence of a nonsolvent such as benzene, which has a slight swelling action on the esterified cellulose. An acid catalyst such as sulfuric acid, toluene sulfonic acid, or perchloric acid is used; this acid catalyst must subsequently be removed from the fiber by heating in a nonsolvent medium with acetic acid, thus purifying the solid cellulose triacetate, which is then dried. (2) Solvent process, in which the activated cellulose is esterified with acetic anhydride and acetic acid, using sulfuric acid as a catalyst. Alternatively, methylene chloride can be used instead of acetic acid. In wet spinning, cellulose triacetate is dissolved in glacial acetic acid, and this is extruded into either water or dilute acetic acid.

cellulosic plastics. Produced by chemical modification of cellulose, a natural polymer. Cellulose itself is not a thermoplastic since it will not melt. Nevertheless, by the viscose process it is made into film (cellophane) and into fiber (rayon). Organic cellulose ester plastics are triacetate, acetate, butyrate, and propionate. Cellulose acetate esters are prepared by the reaction of chemical cellulose with acetic acid and acetic anhydride, with sulfuric acid as a catalyst in most processes. In the standard synthesis it is necessary to allow the reaction to proceed to the triester, which may be recovered if this is the desired product. It has an acetyl content in excess of 45%. For the production of plastic-grade cellulose acetate, some acetyl groups are removed by hydrolysis. The plastic-grade cellulose acetate contains 38–40% acetyl. The synthesis of cellulose acetate butyrate and cellulose propionate resembles that of cellulose acetate except that butyric or propionic anhydride is substituted for part of the acetic anhydride. In the case of butyrate and propionate, the triesters are of no commercial utility. Plastic grades of cellulose acetate butyrate analyze about 26–30% butyrl and 20–25% acetyl. Cellulose propionate for plastic production contains about 39–47% propionyl and 2–9% acetyl.

cementing materials. Cementing materials of engineering interest are of three types: compounds of calcium, bituminous materials, and glues. Calcium cements are of two classes. One is gypsum, which is obtained from calcium sulfate ($CaSO_4$), while the other is lime, which is obtained from calcium carbonate ($CaCO_3$). Gypsum cements include plaster of Paris, wall plaster, and hard-finish plaster. Calcium carbonate cements include lime, natural cement, Portland cement, and high-early-strength cement. Calcium cements have the property of hardening and developing adhesive properties when combined with water. They are used in layers to bind together brick or stone in the forming of masonry. With rock and similar materials, they are used to make concrete, and in thin layers they are used as a decorative coating.

cementite (metals). An intermetallic compound, iron carbide, having the formula Fe_3C. It is very hard and brittle and its tensile strength is not known, although it is believed to be low. Cementite appears white and in eutectoid steel is found mixed with dark-appearing ferrite to form pearlite. In hypereutectoid steels, it may also be present as boundaries around the grains or as needles within the colonies of pearlite. It appears in hypoeutectoid steels only as part of pearlites.

cement, Portland. Materials used in the manufacture of Portland cement are combinations of clay-bearing (argilaceous) and lime-bearing (calcareous) materials. In the manufacture of Portland cement the raw materials are first crushed and finely ground. Next they are mixed in their correct proportions and ground again. They are then fed into a rotating kiln where burning takes place. After removal from the kiln the cement is finely ground and then usually placed in sacks or bags containing 94 lb (43 kg) or about 1 ft^3 (0.03 m^3) of cement. When water is added to Portland cement, it does not set within 1 h, but fully sets within 10 h. The strength of Portland cement mortars increases with time after mixing. The rate of hardening or the rate of the chemical action that occurs depends on the size and chemical composition of the cement constituents, the temperature, and the amount of available moisture for combination with cement.

cements. Cements are composed primarily of oxides of silicon, calcium, and aluminum, namely, silica (SiO_2), lime (CaO), and alumina (Al_2O_3). There are also small amounts of FeO_3 and MgO present in cements. The properties of cement will depend on the relative proportions of silica, lime, and alumina present. Portland cements, for example, have about 60% lime, 20% silica, and 20% alumina plus impurities. The cements include hydraulic lime, Puzzolan cement, slag cement, and natural cement. Hydraulic lime is produced by heating limestone that contains sufficient silica to yield a cement that will harden under water. Puzzolan cement is made from volcanic ash, sand, and slaked lime mixed together and ground. Slag cement is finely ground blast-furnace slag and hydrated lime. Slag cement is used in building construction where strength is not too important. Natural cements are made heating limestone that contains alumina and silica. **(See also bituminous cements.)**

centerless turning machines. In principle not greatly unlike the Swiss automatic, the Taber centerless turning machine performs turning operations without the use of centers. Bar stock is continuously fed through the drive spindle and is supported by a carriage-mounted rigid collect assembly. The machine turns shafts with contours by means of a template and direct-tracer stylus or numerical control in one pass. At any desired point the carriage can be stopped to permit facing, grooving, or cutoff from a six- or eight-station tool turret. The centerless turning machine thus extends the capabilities of the Swiss method to larger diameter shafts—up to 2.25 in. (5.7 cm) in diameter and up to 50 in. (127 cm) in length.

ceramic fiber insulation. As a substitute for asbestos, high-temperature insulation is made by vacuum forming high-alumina ceramic fiber and inorganic bonds. The material is lightweight and resilient, with minimum shrinkage. It has good dielectric strength, excellent insulating qualities, and good resistance to thermal shock. It can be produced as moist moldable felt, as flat and curved boards, and in vacuum-formed shapes.

ceramic magnets. A class of magnet materials has been developed, which now shows considerably higher usage than the alnicos. These materials have become popular owing to the lower cost of their basic materials and the ease with which these magnets can be fabricated into stator housings, permitting lower manufacturing costs. This class of permanent-magnet material is identified as ceramics. The general ingredients of this kind of magnet material are iron oxide and barium or strontium elements, which are relatively inexpensive and abundant in supply. A number of rare earths, when used with cobalt, have powerful magnetic properties, including samarium, lanthanum, cerium, praseodymium, and the natural mixture known as Misch metal. Samarium–cobalt magnets have been produced with a maximum energy product of 17 million gauss-oersteds as compared to a maximum energy product of about 5–10 million for conventional magnets. A praseodymium–samarium–cobalt alloy has produced a magnet with an energy product of 22.6 million gauss-oersteds.

ceramics. The term "ceramics" is applied to a range of inorganic materials of widely varying uses. Generally, these materials are nonmetallic and in most cases have been treated at a high temperature at some stage during manufacture. The word "ceramic" is derived from the Greek, *keramos* or "potter's clay," although the group of materials now so described includes glass products; cements and plasters; some abrasives and cutting tool materials; building materials such as bricks, tiles, and drain pipes; various electrical insulation materials; refractory linings for furnaces; porcelain and other refractory coatings for metals; as well as the more traditional uses in pottery, crockery, and sanitary ware. From the point of view of composition and structure, ceramics can be classified conveniently into four main groups: (1) Amorphous ceramics — these are substances referred to generally as glasses. They include those such as obsidian that occur naturally and manmade glasses used for the manufacture of bottles, windows,

Typical Properties of Representative Engineering Ceramics

Property	Alumina (dense, sintered)	Boron carbide (hot pressed)	Boron nitride (hot pressed)	Silicon carbide/ silicon nitride	Silicon nitride (hot pressed)	Silicon nitride (reaction bonded)	Silicon carbide (hot pressed)	Silicon carbide (re-crystallized)	Glass-bonded mica
Flexural strength (10^3 psi)									
At 70°F	46–50	44	—	115	130	30–35	110	16–65	10–13
At 2250°F	21–44	—	—	45	45	40–45	80	20–40	—
Compressive strength (10^3 psi),	>340	420	40	400	500	100	500	100	27–29
Hardness, Knoop	2,500	2,800	—	2,200	2,200	—	2,500	2,400	90(Rc)
Modulus of elasticity, tension (10^6 psi)	45–55	65	7	52	45	25	62	35	6
Maximum use temperature (°F)	3,092	1,100	3,000	—	3,000	2,600	3,360	3,200	750–1,300
Coefficient of thermal expansion 70–2000°F (10^{-6}in./in.-°F)	4.2–4.3	3.2	4.3	2.4	1.7	1.7	2.7	2.7	6.0–6.5
Thermal conductivity 70°F (Btu-in./hr-ft^2-°F)	120–225	104–197	100–200	—	101	101	290	290	—
Density (lb/in.^{3})	0.13–0.14	0.087	0.076	—	—	—	—	—	0.13
Electrical resistivity (ohm-cm)									
At 70°F	$>10^{14}$	0.5–1.0	$>10^{13}$	10^6	10^{11}	10^{15}	10	10^2	10^{14}
At 1500°F	10^8–10^{10}	—	$>10^4$	10^6	10^4	10^7	10^{-1}	10^{-1}	—

Note: These property data are for general reference only, not for design use. Variations in material structure, composition, orientation, and other characteristics, as well as in testing methods used, can produce wide differences in reported values.

and lenses. (2) Crystalline ceramics—these may be single-phase materials like magnesium oxide or aluminum oxide, or various mixtures of materials such as these. In addition some carbides and nitrides belong to this group. (3) Bonded ceramics—materials in which individual crystals are bonded together by a glassy matrix, as in a large number of products derived from clay. (4) Cements—a number of these are crystalline but some may contain both crystalline amorphous phases. Ceramics are far less ductile than metals and tend to fracture immediately if any attempt is made to deform them by mechanical work. They are often of complex chemical composition and their structure may be relatively complex. These structures fall into two main groups: (1) Simple crystal structures—these include structures in which the bonding is either ionic or covalent. Silicon carbide is also a ceramic material of simple composition and structure. The bonding is covalent and the structure is tetrahedral such that each silicon atom is surrounded by four carbon atoms and, conversely, each carbon atom by four silicon atoms. As a result the structure is very like that of diamond. (2) Complex silicate structures—the majority of ceramic materials, and particularly those derived from clay, sand, or cement, contain the element silicon in the form of silicates.

Melting Points of Ceramic Materials, °F (°C)

Hafnium carbide	7520 (4160)	Silicon carbide	4890 (2699)
Tantalum carbide	7025 (3885)	Zirconia	4850 (2677)
Carbon	6600 (3649)	Calcium oxide	4660 (2571)
Zirconium carbide	6370 (3521)	Beryllia	4570 (2521)
Tantalum nitride	6050 (3343)	Titania	3870 (2132)
Titanium carbide	5660 (3127)	Alumina	3722 (2050)
Thoria	5630 (3110)	Nickel oxide	3560 (1960)
Zirconium boride	5530 (3054)	Silica	3110 (1710)
Magnesia	5070 (2799)		

ceramic lubricants. Ceramic materials are used as lubricants under conditions of either extreme high pressure or high temperature. Graphite is probably best known of these lubricants. Lime is used as a lubricant in wire drawing and molten glass in the extrusion of steel shapes. Mica, soapstone, and talc are occasionally used as lubricants. Molybdenum disulfide, MoS_2, is a ceramic lubricant. This material is gray in color, with a high specific gravity of 4.9 and a melting point of 2200°F (1204°C). Its coefficient of friction is very low, but it is capable of sustaining high bearing pressures.

cermets. Combinations of metals and ceramics, bonded together in the same manner in which powder-metallurgy parts are produced. They combine some of the high refractories of ceramics and the toughness and thermal-shock resistance of metal. Except for a few varieties they are still in the experimental stage. The oxide-based cermets are primarily chromium–alumina based or chromium–molybdenum–alumina based. Carbide-based cermets are based on tungsten or titanium and tantalum carbides. To produce parts from cermet materials, the powders are pressed in molds at pressures ranging from 10,000 to 40,000 psi (6.9×10^5 to 2.8×10^6 kPa). They are then sintered in controlled-atmosphere furnaces at about 3000°F (1650°C). Cermets are used principally as nozzles for jet and other high-temperature devices, crucibles, and cutting tools.

Charpy test. A pendulum-type single-blow impact test in which the specimen, usually notched, is supported at both ends as a simple beam and is broken by a falling pendulum. The energy absorbed, as determined by the subsequent rise of the pendulum, is a measure of impact strength or notch toughness. (Contrast with **Izod test.**)

chatter. In machining or grinding, (1) a vibration of the tool, wheel, or workpiece producing a wavy surface on the work and (2) the finish produced by such vibration.

chatter marks. Intermittent transverse marks due to vibration during machining, rolling, extrusion, or drawing, resulting in nonuniform surface reflection and brightness.

chlorinated polyether (plastics). Pentacrylthritol is chlorinated and the resulting 3.3-bis(chloromethyl) oxetane is polymerized to produce a linear, crystalline polyether containing about 45% chlorine. The average molecular weight is about 300,000. Chlorinated polyether is thermoplastic with a melting point of 358°F (181°C) and a low melt viscosity, allowing it to be readily formed by injection molding or extrusion into strain-free, close tolerance, dimensionally stable final products. It has low creep characteristics. It can be used up to 275°F (135°C) in contact with many chemicals. Certain fillers such as graphite, glass, asbestos, silica, and polymolybdenum disulfide are used or are under study to improve certain properties such as mold shrinkage and thermal expansion. Thermal conductivity is low and electrical properties are good. It is used largely where chemical resistance is important. These uses include injection-molded valves and pipe fittings; extruded pipe and tubing; and extruded sheets for lining tanks and other chemical equipment. Chlorinated polyether may also be applied as a coating to metal surfaces by the fluidized-bed method, by flame spraying, or by spraying as a solution in an organic solvent or a suspension in water. Equipment such as pumps and meters may be made either of metal coated with the resin or solid molded resin parts.

chromium steels. To steel, chromium imparts hardness, wear resistance, and useful magnetic properties, permits a deeper penetration of hardness, and increases the resistance to corrosion. Chromium content in steels varies from a small percentage to approximately 35%. Chromium is the principal alloying element in many of the alloy or special steels. The low-carbon or pearlitic-chromium steels are the most important steels, containing chromium from 12% to 35%. When chromium is used in amounts up to 2% in medium- and high-carbon steels, it is usually for the hardening and toughening effects and for increased wear resistance and increased resistance to fatigue failure. Chromium, when added to steel, raises the critical temperature and slows up the rate of transformation from austenite to pearlite during the hardening process. With additions of less than 0.50% chromium to the steel, there is a tendency to bring about a refinement of grain and to impart slightly greater hardenability and toughness. Chromium is added to the low-carbon steels to effect greater toughness and impact resistance at subzero temperatures, rather than to increase hardenability. Chromium is used quite frequently in tool steels to obtain extreme hardness. Applications of such include files, drills, chisels, roll thread dies, and steels commonly used for ball bearings, which contain approximately 1% carbon and 1.5% chromium.

chucking lug. A lug or boss added to a forging so that "on center" machining and forming may be performed with one setup or chucking. This lug is then machined away.

CID camera (computers). A solid-state camera that uses a charge-injection imaging device (CID) to transform a light image into a digitized image. The light image focused on the CID generates minority carriers in a silicon wafer, which are then trapped in potential wells under metallic electrodes held at an elevated voltage. Each electrode corresponds to one pixel of the image. To register the brightness of one pixel of the image, the voltage on the electrode that corresponds to that pixel is changed to inject the charge stored under that electrode into the substrate. This produces a current flow in the substrate that is proportional to the brightness of the image at that pixel location, and is therefore capable of producing a gray-scale image. In a CID camera, pixels of the image can be read out in an arbitrary sequence. This is not possible with a CCD camera. In come CID cameras, the same image can be read out hundreds or thousands of times (nondestructive readout capability).

CIM. Acronym for computer-integrated manufacturing. **(See also computer-aided manufacturing.)**

clad metals (see laminated metals)

CLA process. An acronym for the inventors Chandley-Lamb and Air. This investment-casting process overcomes many of the disadvantages of the historical method of pouring metal into investment molds. The open end of a mold is placed through the bottom of a partable vacuum chamber, and the chamber is closed by the machine which also lowers the chamber until the open end of the mold is submerged in molten metal and applies a vacuum in the chamber. This causes the metal to fill the mold, one row at a time, in a controlled nonsplashing fashion. The vacuum level and the time to reach that level can be preset to achieve optimum results. The vacuum is held until the castings solidify and on its release, most of the gating, which is still liquid, returns to the melt. The cast mold is removed and the ceramic is blasted from the castings, which emerge only with short gate stubs ready for finishing. **(See also investment casting.)**

clays. Clays and soils have their origin in the mechanical and chemical disintegration of rock. Materials most commonly found in natural clays are the hydrated silicates of alumina, commonly in the form $Al_2O_3{\cdot}2SiO_2{\cdot}2H_2O$, or kaolite. Impurities such as lime, potash, magnesia, and soda are usually present in clay. Clay is found in the form of residual, glacial, sedimentary, or loess clay. They are also obtained from shales or slate. Residual clays occur in deposits at the site of rock from which they were formed. Glacial clays are those that have been transported and deposited in banks by the action of glaciers. Clays deposited by sedimentation and transported by waters are sedimentary clays, while loess clays are those deposited by winds. If only alumina, silica, and water are present in clay, the clay is white after firing. Most clays contain iron and manganese oxides, which give color to the clay and reduce the refractoriness of pyrometric cone equivalent of the clay. A wide range of products is produced from clays, including pottery, chemical stoneware, electrical porcelain, brick, tile, drilling mud for oil wells, filters, Portland cement, and catalyst for the "cracking" of petroleum. There are basically three broad groups of soils from an engineering viewpoint: (1) granular or cohesionless soils, including sands and gravels; (2) fine-grained soils, including silts and clays; and (3) organic soils, including peat, muskeg, and muck. Clays are plastic when blended with suitable water content, while silts cannot be thus plasticized. Silt particles are in the range of 1.97×10^{-4}–1.97×10^{-3} in. (0.005–0.05 mm), while clay particles are even smaller.

clays, adsorbent. Adsorbent clays have the highly useful property of being able to remove coloring matter from oils, fats, and waxes. Decolorizing (bleaching) is a selective adsorption, such as occurs in dyeing fabrics. The clays that have adsorbent properties in the crude state are called naturally active clay or fuller's earth (the latter term refers to the practice of fulling, or cleansing, wool by removing grease from the fibers). There are, in addition, a number of clays that are not active in the natural state but can be made highly adsorptive by treatment with sulfuric acid. These are termed activable clays in the crude form and activated clays after treatment. Adsorbent clay finds its chief use in the petroleum industry. It is used to decolorize and remove gums from gasoline and acid sludges from lubricating oils, and to clarify paraffin and other waxes.

clays, ball. A sedimentary clay that has good plasticity, high refractoriness, and good bonding ability. It burns white or creamy white and is used chiefly for making white earthenware and porcelain and as a binder in chinaware.

clays, brick. The cheapest type of clay from which satisfactory bricks can be made. The presence of liberal amounts of iron causes these clays to burn red. A good brick clay produces bricks uniform in color, density, hardness, strength, and regularity of form. Silica is a critical ingredient, since increasing amounts reduce shrinkage in burning but excessive amounts destroy cohesiveness and cause crumbling. Calcium and magnesium carbonates act as fluxes, reducing the fusing temperature and influencing the color. Iron oxide tends to increase strength and exerts a strong influence on color.

clays, ceramic. About 75% of all clay produced is used in manufacturing formed and fired ceramic products. These products may be divided into two major groups. The first include pottery, chinaware, chemical stoneware, sanitary ware, high-grade tile and porcelain, and certain forms for use in firing these products in the kiln. The second includes what are commonly termed the structural clay products, that is, common brick, paving brick, drain tile, construction tile, sewer pipe, etc. Ceramic ware other than structural clay is made from clays high in kaolinite. One variety is the white clay called kaolin. A second variety is ball clay, a light-colored to white kaolinitic clay. Whiteware clay, China clay, and pottery clay are informal terms for clays with the indicated uses. Structural clay products are made from a variety of common clays and shales, which are designated in production tables as miscellaneous clays, and also from refractory fire clay. The clay minerals are kaolinite and illite, and there is generally a considerable fraction of nonclay minerals. Ferric oxide is desired because it acts as a flux, allowing a low temperature of vitrification. Consequently, most brick and tile clays have a red or yellow burned color.

clays, China. A residual or sedimentary form, pure white and relatively uncontaminated by iron compounds. It is also called kaolin, and is used for high-grade chinaware, for porcelain, and for floor, wall, and electrical tile. Because of its low plasticity, kaolin often requires mixing with other clays such as ball clay.

clays, fire. A residual or sedimentary form, the marine or lacustrine forms being more common. These clays are used in making refractory materials or as refractory cements. Their temperature of fusion ranges from 2700°F to 3200°F (1480°C to 1760°C). Fire clays are called plastic if they can be replasticized, that is, if they can be made plastic by grinding and mixing with water after being burned hard. Clays that cannot be replasticized are called "flint" clays.

clays, illite group. A general term for micalike clay minerals. In electron micrographs, illite is seen to occur as small, poorly defined flakes, sometimes grouped in irregular aggregates. Its basic unit is a three-sheet packet like that of montmorillonite, but with some substitution of aluminum for silicon in the tetrahedral layers and with considerable iron and magnesium in the intervening layer. The units are firmly bonded by potassium atoms and the lattice is nonexpanding. When illite loses most of its potassium from between the crystal packets, water may enter and the lattice may acquire the capacity to expand.

clays, kaolinite group. Of the several minerals in this group, kaolinite itself is of chief economic importance. Its basic structure is a two-layer lattice, consisting of a gibbsite sheet and a silica tetrahedral sheet, repeated indefinitely parallel to the plane of the *a* and *b* axis. Kaolinite does not expand with increasing water content, and no appreciable isomorphous replacement of alumina by other metals occurs. The formula is $(OH)_8Al_4Si_4O_{10}$. Electron micrographs show that kaoline occurs as flaky particles, or as hexagonal outline. Exceptional concentrations of pure white kaoline are formed by hydrothermal, weathering, and sedimentary processes. Such deposits are sources or high grade clay (kaolin) for china, paper and refractories, sanitary ware rubber, firebrick.

clays, montmorillonite group. Minerals of the montmorillonite group tend to occur as irregularly shaped or lath-shaped particles. They differ from kaolinite in that (1) the basic structure does not consist of two layers but of three, a gibbsheet being enclosed sandwichlike, between two sheets of tetrahedral layers; (2) magnesium may replace much of the aluminum in the gibbsite layer; (3) each three-layer unit is loosely bound to its neighbors in the *c* direction by water, the amount of which may vary considerably; (4) between the units, in addition to water, there may also be ions of calcium, sodium, or potassium; montmorillonite shows the most pronounced "ion-exchange'' capacity of any of the clay minerals. The general formula of montomorilonite is $(OH)_4Al_4Si_8O_{20}{\cdot}nH_2O$.

clays, refractory. Clays suitable for the manufacture of fire brick, clay crucibles, and pots for molten glass, and for binder in molding sands are termed fire clays. Their properties are similar to those of the ceramic clays, except that they withstand higher temperatures before melting. The refractoriness of a clay is measured by its pyrometric cone equivalent (PCE). Clays with a PCE below 19 [2760°F (1515°C) at temperature rise of 36°F (20°C) per hour] are not considered refractory. There are not exact industry-wide standards, but, in general, low-duty fire clays have a PCE ranging from cone 19 to cone 28 [2760–2940°F (1515–1615°C)]; intermediate duty, up to cone 30 [3000°F (1650°C)]; high duty, up to cone 32 [3090°F (1700°C)] and superduty, up to cone 37 [3230°F (1775°C)]. Many fire clays have high plasticity, others practically none. Nonplastic clay is known as flint clay. Flint clay must be mixed with plastic clay to become usable. Clays with intermediate degrees of plasticity are termed semiflint and semiplastic. The chief clay minerals are kaolinite and illite. Some kaolin is also used in refractories and large tonnages of betonite (clay mineral, montmorillonite) are utilized as bonding agents in synthetic foundary sands.

cleaning, alkaline. A process that utilizes a combination of alkalies and surface-active agents to remove excess oil, lift solids from the surface of metal, and suspend soil particles. Alkaline cleaners often contain sodium hydroxide, carbonates, silicates, and phosphates. Sodium hydroxide provides required alkalinity. Sodium carbonate softens water, adds electrolytic salts that assist surface-active agents, and buffers the alkalinity of the hydroxide. Silicates disperse and suspend solids. Phosphates—added as orthophosphates, pyrophosphates, or polyphosphates—have detergent properties similar to, but not as good as, silicates. Pyrophosphates prevent formation of insoluble calcium and magnesium soaps. Polyphosphates are effective water softeners; but, the unstable polyphosphates revert to orthophosphates, limiting the life of the cleaner. Stable chelating agents, such as ethylene–diaminetetraacetic acid, are used to extend water-softening properties.

cleaning, blast (see blast cleaning)

cleaning, emulsion. A process for soil removal by solvent action. A protective film is left on metal, by emulsion cleaning—a less-expensive process than solvent cleaning. Soaps and surface-active agents combined with a petroleum solvent produce a concentrate that, with water, may form a stable emulsion, an unstable emulsion maintained by agitation, a diphase emulsion of oil floating on water, or an emulsifiable concentrate that, as it emulsifies on the workpiece, is sprayed away with water. Stable emulsions are used to remove light soils and provide several weeks of rust protection. Unstable emulsions are used to remove heavily deposited oil-based shop soils. Diphase systems are used to remove organic films and caked buffing compounds. Emulsifiable concentrates are used when a heavily soiled part requires prolonged soaking.

cleaning, fluidized bed. The principle of fluidized-bed cleaning involves forcing a gas (usually air) upward through a heated container filled with finely divided, inert, aluminum oxide particles, setting the mass of particles into turbulent motion in a condition in which the particles are microscopically separated from each other by the gas and the whole mass behaves much like a liquid. This turbulent motion and rapid circulation of the particles provide an extremely high rate of heat transfer from the particles to the object being cleaned, thus removing organic materials by a combination of pyrolysis and partial oxidation reactions. A uniform temperature gradient is maintained throughout the chamber so that metal parts do not distort. The aluminum oxide particles are nontoxic and nonreactive and will not abrade parts being cleaned. In addition to cleaning dies, the system may be used for cleaning screen packs, screws, and injection molding equipment and to strip paints and coatings from metal parts. For highly flammable materials such as cellulosics, nitrogen may be used as the fluidizing gas.

cleaning, solvent. Aliphatic hydrocarbons, chlorinated hydrocarbons, alcohols, and ketones are commonly used for industrial cleaning. A few basic types are perchloroethylene (nonflammable); kerosene [flash point of 145°F (63°C)]; Stoddard solvent [flash point of 105°F (41°C)]; and methyl ethyl ketone, MEK [flash point of 35°F (2°C)].

cleaning, ultrasonic. A method for cleaning small- to medium-sized parts without surface or dimensional change. Ultrasonic-frequency sound waves are used in a suitable solvent bath to remove molecular-size particles to provide uncontaminated surfaces.

clean room, inspection. An area that incorporates high standards of control of humidity, temperature, and all forms of particulate matter and contamination.

closed loop. In computer programming, a loop that has no exit and whose execution can be interrupted only by intervention from outside the computer program in which the loop is included.

closed-loop control. Control achieved by a closed feedback loop, i.e., by measuring the degree to which actual system response conforms to desired system response, and utilizing the difference to drive the system into conformance.

cloud and pour point (lubricants). The cloud point is the temperature at which the water, wax, and some other substances in an oil begin to crystallize and separate out from an oil. A low cloud point is important when the oil is wick fed and there is danger of exposure to low temperature. The pour point is the temperature at which an oil just barely flows. It is indicative of the suitability of an oil to be used in a gravity-fed lubricating system and of the channeling potentialities of the oil.

CLV process. Many alloys containing reactive elements are melted and investment cast at very low pressures, less than 10 microns of mercury [2×10^{-4} psi (1.33 Pa)], to reduce contamination. In order to achieve these advantages for vacuum cast alloys this complex process was developed. In the CLV machine the metal is melted under high vacuum in a large lower chamber. When the metal is at the proper temperature, a hot mold is loaded in the top chamber, which is closed and evacuated. During loading, a valve above the melt keeps the vacuum over the melt. Argon is flooded into the system, the valve is opened, and the melt is raised until the open end of the mold is submerged. A higher vacuum is then applied to the top chamber, which fills the mold. After the castings and the gate stubs are solidified, the upper chamber is flooded with argon and most of the gating returns to the melt. The melt is then lowered, the valve closed, the mold removed, and the unit is ready for the next cycle.

CNC (see computer numerical control)

coal, anthracite. Pennsylvania anthracite is dense, shiny black in color, and homogeneous in structure, with no marks or layers. It is hard, and can be handled with very little breakage. It burns with a short, clear, bluish flame. It is principally used for heating homes and in gas producers. The so-called western, and particularly the Arkansas, anthracites are really semianthracites. They are dense, but softer than the Pennsylvania anthracite, shiny dark gray in color, and somewhat granular in structure. The grains have a tendency to break off in handling lumps and produce a coarse sandlike slack. They burn with a short, clear, bluish flame. The granular structure has been produced by small vertical cracks in horizontal layers of comparatively pure coal separated by very thin partings. The cracks are the result of heavy downward pressure and probably shrinkage of the pure coal due to a drop in temperature.

coal, bituminous. Low-volatile bituminous coals, also called semibituminous, are of grayish black color and distinctly granular in structure. The grain breaks off very easily, and handling reduces the coal to slack. Any lumps that remain are held together by thin partings. Because the grains consist of comparatively pure coal, the slack is usually lower in ash than the lumps. The coal cakes in fire, burns with a short, clear flame, and is usually regarded as smokeless.

coal, bituminous, high-volatile A. Bituminous coals that come entirely from the Appalachian region. They are mostly homogeneous in structure, with no indication of grains, but some do show distinct layers. They are hard, and stand handling with little breakage. The moisture, ash, and sulfur content is low, and the heating value is high. They include the best steaming and coking coals. They cake in fire, and smoke when improperly burned.

coal, bituminous, high-volatile B. Bituminous coals of distinct laminar structure, with thin layers of black, shiny coal alternating with dull, charcoallike layers. They are hard and stand handling well. Breakage occurs generally at right angles and parallel to the layers, so that the lumps generally have a cubical shape. They make good steam coal, and some of them are good coking coals. They cake in fire and smoke when improperly burned.

coal, bituminous, high-volatile C. Bituminous coals that occur in the Illinois coal fields. They are of distinct laminar structure, are hard and stand handling well. They generally have high moisture, ash, and sulfur content. They are fair steaming coals, some of them make good coke, although under certain conditions the coals will cake in fire. They are considered as free-burning coals.

coal, bituminous, medium-volatile. Medium-volatile bituminous coals come principally from the Appalachian region, and include the best steam coals. They are the transition from high-volatile to low-volatile coal and, as such, have the characteristics of both. Many of them have the granular structure, are soft, and crumble easily. Some of them have homogeneous structure, with very faint indication of grains or layers. Others are of more distinct laminar structure, are hard, and stand handling well. They cake in fire and smoke when improperly burned.

coals, caking. Caking coal, when heated in a furnace, passes through a plastic state during which the individual pieces fuse together into large masses of semicoke, which is impervious to flow of air. The air can pass only through cracks formed between the masses of semicoke. The very active combustion in these cracks widens them into craterlike fissures, which allows passage of a large volume of air, much of which is not used in combustion, while, in the centers of the masses of coke, the fuel remains inactive, owing to the lack of air. The masses of coke must be broken with fire tools or by the action of the stoker, which has moving rams or other means of breaking the masses of semicoke. Caking coals cannot be burned successfully on traveling grate stokers, because these stokers have no means for breaking the masses of semicoke.

coal crushers. The type most commonly used for smaller capacities is the swing hammer type of crusher. This consists of a casing enclosing a rotor to which are attached pivoted hammers or rings. Coal is fed through a suitable opening in the top of the casing and crushing is effected by the impact of the revolving hammers or rings directly on or by throwing the coal against the liners or spaced grate bars in the bottom of the casing. The degree of size reduction depends on the hammer type, speed, wear, and bar spacing. These crushers produce a uniform coal sizing and break up pieces of wood and foreign material with the exception of metallic items. They are provided with pockets to catch foreign material that is too hard to crush.

coal, lignite. Lignites are of brown color and of a laminar structure in which the remnants of woody fibers may be quite apparent. In some of the lignite seams of North Dakota, preserved stumps of trees may be found which can be whittled into curly shavings with a pocket knife. Freshly mined lignite is tough, although not hard, and requires a heavy blow with a hammer to break the large lumps. However, on exposure to air it loses moisture rapidly and disintegrates. Even when it appears quite dry, the moisture content may be as high as 30%. Owing to the high moisture and low heat value, it is not economical to transport long distances. It can be burned quite efficiently on traveling-grates and spreader stokers, and in pulverized form. Because of the tendency of the lignite to disintegrate, the fuel must not be agitated, since agitation speeds up the disintegration.

coal pulverizers (attrition mills). In an attrition mill the grinding elements consist of pegs and lugs mounted on a disk rotating in a chamber, the periphery of which is lined with wear-resistant plates and the walls of which contain fixed rows of lugs within which the rotating lugs mesh. The fan rotor is mounted on the pulverizer shaft. Instead of using an external classifier, a simple shaft-mounted rejector type is used. This mill type exhibits all the characteristics of the impact mills. No true attrition mill is used for coal pulverizing because of the high degree of wear on parts that would result. A high-speed mill that utilizes a considerable amount of attrition grinding along with impact grinding is, however, used for direct firing of pulverized coal.

coal pulverizers (ball mills). Basically a hollow horizontal cylinder, rotated on its axis, whose length is slightly less to somewhat greater than its diameter. The inside of the cylindrical shell is fitted with heavy cast liners and is filled to a little less than half with forged steel or cast alloy balls varying from 1 to 2 in. (2.54 to 5.08 cm) in diameter. Rotating slowly, 18–35 rpm, or about 20 rep for an 8-ft (2.4-m) diameter mill, the balls are carried about two-thirds of the way up the periphery and then they continually cascade toward the center of the mill. Coal is fed into the cylinder through hollow trunnions, and intermingles with the ball charges. Pulverization, which is accomplished through continual cascading of the mixtures, results from (1) impact of the falling balls on the coal; (2) attrition as particles slide over each other as well as over the liners; and (3) crushing as balls roll over each·other and over the liners with coal particles between them. Larger pieces are broken by impact, and the fine grinding is done by attrition and crushing as the balls roll and slide within the charge.

coal pulverizers (impact mills). Consists primarily of a series of hinged or fixed hammers or lugs revolving in an enclosed chamber lined with cast wear-resistant plates. Grinding results from a combination of impact of the hammers on the larger particles and attrition of the smaller particles on each other and across the grinding surfaces. An air system with the fan mounted either internally or externally on the main shaft, induces a flow through the mill. Either an internal or external type of classifier may be used. The ability to handle high inlet temperatures plus the return of dried classified rejects to the incoming raw feed make this mill type an excellent dryer.

coal pulverizers (ring roll and ball race mills). Normally of medium speed and utilizes primarily crushing and attrition of particles plus a very small amount of impact to obtain size reduction of the coal. The grinding action takes place between two surfaces, one rolling over the other. The rolling element may be either a ball or a roll, while the member over which it rolls may be either a rack or a ring. If a ball, the diameter is usually from 9 to 24 in. (23 to 61 cm) for the larger pulverizers and somewhat less for the smaller ones. When the rolling elements are balls, they are confined between races. In the majority of designs the lower race is the driven rotary member, while the upper race is stationary. Some designs also utilize a rotating upper race. The required grinding pressure is obtained by forcing the races together either with heavy springs or with pneumatic or hydraulic cylinders. Some additional grinding pressure is obtained from centrifugal forces of the rotating balls.

coal stokers, multiple-retort. An extension of the single-retort stoker. It is nothing more than a series of single-retort stokers built into the same machine with appropriate mechanisms provided to operate the various components in unison. The fuel bed is of two distinct types. In the underfeed section, there are parallel rows of hills and valleys, extending from front wall to the discharge end of the retorts. The hills occur over the relatively inactive retort areas, because they have no provisions for air admission. The coal is supplied through a reciprocating feed, and this produces a certain amount of segregation in the fuel bed. The coarse coal finds its way to the tuyeres near the front. The fines travel the length of the retort.

coal stokers, single-retort. Fuel is introduced in the retort, and the incoming coal progressively forces fuel out of the retort and onto the side grates. This feeding action from retort outward places the entire fuel bed under compression and automatically closes any holes that may tend to form in the fuel bed, thus overcoming one of the commonest obstacles to efficient firing.

coal stokers, spreaders. Uses the combined principles of pulverized coal and stoker firing in that fines are burned in suspension and the larger particles are burned on the grate. Feeding and distributing mechanisms continually project coal into the furnace above an ignited fuel bed. With this method of firing, the chemical characteristics of the coal have little effect on the fuel bed. Strongly caking coals show little tendency to mat and burn with the same ease as the free-burning variety. Flash drying of the incoming fuel, rapid release of volatile, and suspension burning of the fuel are factors that make this method of firing so widely applicable. Practically all types of coal have been burned successfully on spread stokers, as have cellulose fuels, including bagasse, wood chips, bark, hogged wood, sawdust, shavings, coffee grounds, and furfural residue.

coal stokers, traveling-grate. Every type of fuel that is mined, with the exception of caking bituminous coals, can be burned successfully on various grate stoker designs. Traveling-grate cokers may also be employed as a part of chemical process operations to produce coke and carbon dioxide. The two types commonly used are known as chain grate and bar grate, both performing the same function of carrying the fuel fed by gravity from a hopper; supporting it throughout the combustion period, and depositing the refuse in the ashpit. A chain-grate surface consists of a series of links strung on rods in a staggered arrangement and moved by sprockets or drums. Bar-grate surface consists of rows of keys strung on bars which are in turn carried by chains driven by sprockets. The chain grates tend less to retain siftings and clinker accretions; bar grates are less affected by wear and easier to repair.

coal stokers, underfeed. Single or multiple retort consists essentially of a trough or troughs into which coal is pushed by rams or screws. Part of the combustion air is introduced into the fuel bed through tuyeres or grate bars. Movement of the fuel discourages the formation of large coke masses. Volatile matter is distilled off the coal in these troughs or retorts and burns above the incandescent fuel bed. The partly coked and somewhat caked coal then falls into the air-admitting tuyeres or grate bars where the fixed carbon is burned out. Progressively the fuel is pushed sideways or forward until the refuse is discharged to the ashpit.

coal, subbituminous. Occurs mainly in the Rocky Mountains, the Great Plains, and Pacific Coast Provinces of Canada. These coals are brownish black or black in color. Most of these coals are of a homogeneous structure with smooth surface and with no indication of layers. They have high-moisture content, although appearing dry. When exposed to air, they lose part of the moisture and crack with an audible noise. On long exposure to air, they disintegrate. In the fire they have no caking property and crumble into small pieces. They are free-burning coals.

coal tar. A by-product in the carbonization of coal. The tar compounds are extremely complex and number in the hundreds. The solid material, which is insoluble in benzene, is contained as colloidal and coarse dispersed particles and is known as "free carbon." The composition of the tar is dependent on the temperature of carbonization and, to a lesser extent, on the nature of the cooking coal. The following is an example of coal tar analysis and physical properties:

carbon	89.9%
hydrogen	6.0%
sulfur	1.2%
oxygen	1.8%
nitrogen	0.4%
moisture	0.7%
gravity	1.18 deg. Baumé
viscosity	900 SSF
flash point	156°F (69°C)
heating value	16,750 Btu/lb (40,585 kJ/kg)

coal tar/distillate. A mixture of many chemical compounds, mostly aromatic, which vary widely in composition. It is a coproduct of the destructive distillation or pyrolysis of coal. Most of the tar in the United States is produced by steel companies as a coproduct from blast-furnace coke production. The end product of the distillation of coal tar is pitch, usually more than 60% of the crude tar. The objects of the distillation are to produce a salable end product with a separation of the valuable products into useful cuts. The general methods of distillation of coal tar are (1) the 3000–10,000 gal batch still, (2) the continuous still, with a single distillation column, using side streams, (3) the continuous unit, using multiple columns with reboilers. The fractions obtained in an ordinary continuous distillation, which will vary with the coal and with conditions are (1) light oils, with cuts up to 390°F (199°C); (2) middle oils, or cresote oils, generally the fractions from 390 to 480 or 520°F (199 to 249 or 271°C), which contains naphthalene, phenol, and cresols; (3) heavy oils, representing the fraction from 480 to 570°F (249 to 299°C); (4) anthracene oil is usually the fraction from 520 to 570°F (271 to 299°C) up to 660 or 750°F (349 or 399°C). It is washed with various solvents to remove phenanthrene and carbazole; the remaining solid is anthracene. Coal tar is used as a fuel, roads, and roofs.

cobalt. Weakly magnetic with about the same mechanical properties as iron. With respect to corrosion resistance, the behavior of cobalt is more nearly like that of nickel and superior to iron. Cobalt is used in high-speed steels where it contributes to the hardness by solid-solution hardening of ferrite. It is used in magnet steels and ferrous alloys such as the Alnicos, some of which are alloys containing aluminum, nickel, cobalt, and iron. The Stellites are cobalt-base metals, containing smaller amounts of chromium and tungsten. They are particularly well known for good high-temperature properties such as creep strength and oxidation resistance. Because of the difficulty of forging and machining, precision casting of parts is the common practice.

coextrusion. For use in plastic bottle manufacture, six layers are extruded simultaneously. Ethylene vinyl alcohol (EVOH) is sandwiched between two layers of adhesive and two layers of polypropylene (PP), with an additional regrind layer. American Can produces such bottles on proprietary rotary blow molders using a coextrusion system consisting of five extruders mounted piggyback on a common base.

cogging. The first operation on a tool-steel ingot is to convert it into a billet by a process known as cogging. This may be done by hammering, rolling, or pressing. In cogging, the ingot is carefully heated to a predetermined temperature and the cross section is reduced with a corresponding amount of elongation. Thus a 9-in. (22.7-cm) square ingot would yield a 4-in. (10.2-cm) square billet, which would be about six times as long as the original ingot. As the hot billet comes from the mill, hammer, or press, the top is "cropped" to remove the pipe. This crop goes back to the melting department as scrap for future heats.

coining (see cold forging)

coke. The residue remaining following the destruction heating or carbonization of certain bituminous coals in the absence of air. During the carbonization process, the water and most of the volatile matter in the coal is driven off, leaving a solid, porous residue consisting essentially of carbon and ash. Coke is produced through two processes: high-temperature carbonization, employing temperatures of 1800–2200°F (982–1200°C), and low-temperature carbonization which is carried out at temperatures of 1000–1400°F (538–760°C). The resultant cokes are quite different in both composition and physical characteristics. High-temperature coke is hard, finely cellular, carbonaceous mass of high strength and containing 3% volatile matter. Low-temperature coke is a noncoherent, granular structure with a volatile matter content of 7–20%. The properties of the low-temperature coke render it unsuitable for metallurgical purposes; however, owing to its higher reactivity, it makes an excellent smokeless fuel. Nearly all the coke produced in the United States is carbonized at high temperatures in byproduct ovens of the slot type. The volatile products of this process, consisting essentially of gas, ammonia, tar, and light oil, are recovered and separated into their various fractions.

Coke, Analysis of Typical U.S. Coke as Fired

	Proximate analysis percent				Ultimate analysis percent							Heating value Btu per lb			
	Moisture	Vol. matter	Fixed carbon	Ash	Moisture H_2O	Carbon C	Hydrogen H_2	Sulfur S	Oxygen O_2	Nitrogen N_2	Ash	Higher	Lower	Atmos. air of zero excess air, lb/10^6 Btu	CO_2 at zero excess air, percent
High temperature coke	5.0	1.3	83.7	10.0	5.0	82.0	0.5	0.8	0.7	1.0	10.0	12200	12095	796	20.7
Low temperature coke	2.8	15.1	72.1	10.0	2.8	74.5	3.2	1.8	6.1	1.6	10.0	12600	12258	763	19.3
Beehive coke	0.5	1.8	86.0	11.7	0.5	84.4	0.7	1.0	0.5	1.2	11.7	12527	12453	805	20.5
Byproduct coke	0.8	1.4	87.1	10.7	0.8	85.0	0.7	1.0	0.5	1.3	10.7	12690	12613	801	20.5
High temperature coke breeze	12.0	4.2	65.8	18.0	12.0	66.8	1.2	0.6	0.5	0.9	18.0	10200	9950	805	20.1
Gas works coke:															
Horiz. retorts	0.8	1.4	88.0	9.8	0.8	86.8	0.6	0.7	0.2	1.1	9.8	12820	12753	807	20.6
Vertical retorts	1.3	2.5	86.3	9.9	1.3	85.4	1.0	0.7	0.3	1.4	9.9	12770	12659	810	20.4
Narrow coke ovens	0.7	2.0	85.3	12.0	0.7	84.6	0.5	0.7	0.3	1.2	12.0	12550	12493	802	20.6
Petroleum coke	1.1	7.0	90.7	1.2	1.1	90.8	3.2	0.8	2.1	0.8	1.2	15060	14737	773	19.5
Pitch coke	0.3	1.1	97.6	1.0	0.3	96.6	0.6	0.5	0.3	0.7	1.0	14097	14036	813	20.7

coke breeze. In the manufacture of coke, considerable quantities of fine results. These fines are separated from the larger and more valuable coke by screening. Coke breeze is readily burnable. High capacities with very good efficiency can easily be maintained with properly designed furnaces and forced-draft traveling-grate stokers. Generally all coke passing through a 3/4 in. (19 cm) round-hole screen is called breeze. The value of one breeze as compared to another will depend on the percentage of undersize contained. The undersize content, where possible, should be held above 20% and below 30% for best performance.

coke oven gas. Obtained during the high-temperature carbonization of bituminous coal to make coke. Substantially all volatile matter in the raw coal is distilled off, and the vapors thus obtained contain fixed gases, liquid, and solids. Various methods of cooling, separation, and extraction are used to recover the liquids and solids. Some of the fixed gases are then used to heat the coke ovens, while the remainder is available for boiler, process, or domestic fuel. The quality of coke oven gas depends on the character of the coal processed, the duration of the coking operation, the maximum temperature reached, and, to some extent, the conditions of operation. At very high carbonizing temperatures the volatile matter of the coal is thermally cracked to a greater extent than at normal temperatures [1800–2100°F (982–1149°C)], thereby lowering the heating value of the gas discharged, but simultaneously increasing the thermal yield of gas.

Coke Oven Gas

Constituents	Percent by volume						
CO_2	1.8	2.3	1.0	3.13	0.75	1.4	2.6
CO	6.3	9.4	4.8	11.93	6.0	5.1	6.1
O_2	0.2	0.6	—	—	—	0.5	0.6
H_2	53.0	49.3	53.5	42.16	53.0	57.4	47.9
N_2	3.4	6.5	3.7	—	12.1	4.2	3.7
CH_4	31.6	28.4	34.0	37.14	28.15	28.5	33.9
C_2H_4	3.7	3.5	3.0	5.64	—	2.9	5.2
Higher heating value Btu per cu ft	580	550	557	645	466	526	588

coking, beehive. The beehive oven consists of a beehive-shaped brick chamber provided with a charging hole at the top of the dome and a discharging hole in the circumference of the lower part of the wall. The coal is introduced through the hole in the dome and spread over the floor. The heat retained in the oven is sufficient to start the distillation. The gases given off from the coal mix with the air entering at the top of the discharge door and burn; the heat of combustion is sufficient for pyrolysis and distillation.

coking, coproduct. The coproduct coke oven is a narrow chamber, usually about 38–40 ft (11.6–12.2 m) long, 13 ft (4 m) high, and tapering in width from 17–18 in. (43.2–45.7 cm) at one end and to 15 or 16 in. (38.1 or 41 cm) at the other. The ovens hold from 16 to 24 tons (14.5 to 21.8 tonnes) of coal. These ovens are used for carbonizing coal only in large amounts and are built in batteries of 10–100 ovens. The general arrangements for the operation of a coproduct coke oven with its various accessories, is elaborate and costly of masonry structure and is erected with the closest attention to engineering details, so that it can withstand the severe strains incurred in its use and remain gastight, even after the great expansion during heating up. The oven block is built of refractor brick, with heating flues between the coking ovens.

coking, delayed (petroleum coke). In this process the reduced crude oil is heated rapidly and flows to isolated coking drums where it is coked by its own contained heat. The process requires several drums to permit removal of the coke in one drum while the others remain on stream. The residual product that solidifies in these drums is termed "delayed coke." When first removed from the drum, it has the appearance of run-of-mine coal, except that coke is dull black. The analysis of the coke varies with the crude from which it is made. The components range as follows: moisture, 3–12%; volatile matter, 10–20%; fixed carbon, 71–88%; ash, 0.2–3.0%; Btu/lb (dry), 14,100–15,600 (34,160–37,800 kJ/kg).

coking, fluid (petroleum coke). Two large vessels are used in fluid coking. One is known as a reactor vessel, and the other as a burner vessel. In this process fluid coke is both the catalyst and the secondary product. The used coke is first heated in the burner vessel, either by adding air and burning a portion of the coke or by burning an extraneous fuel such as oil. The heated seed coke then flows into the reactor vessel where it comes in contact with the preheated residual oil, and the lighter fractions of the oil are flashed off. The coke that is produced both deposits in uniform layers on the seed coke and forms new seed coke. Thus, there is a constantly accumulating coke reservoir, which is tapped off and available as a boiler feed. The coke thus formed is a hard, dry, spherical solid resembling black sand. It is composed of over 90% carbon with varying percentages of sulfur and ash, depending on the source of the crude oil. Typical analysis are: fixed carbon, 90–95%; volatile matter, 3–6%; ash, 0.2–0.5%; sulfur, 4.0–7.5%; Btu/lb, (higher) 14,100–14,600 (34,160–35,375 kJ/kg).

cold drawing (high-energy forming). This process uses a high explosive, such as dynamite or gases. Very-high-energy shock waves result, fan out in all directions, and force the metal into the preformed die cavity. Because of the high energies employed, very-high-strength materials may be formed by this method. High-strength materials, when deformed, will spring back excessively. However, as a result of the high energies employed, very little springback takes place. In addition to the powder explosives and the high-velocity expanding gases employed, shock waves generated from a capacitor discharges across a gap have also been employed. The shock waves fan out through a nonconducting liquid in all directions and impact the workpiece causing it to deform. The magnetic fields generated as the high voltage is discharged induce a current in the workpiece. The forces generated in the workpieces as the current builds rapidly cause the metal to deform. This deformation is controlled to give the desired shape to the workpiece.

cold drawing (metalworking process). Round, rectangular, square, hexagonal, and other shapes of bars up to about 4 in. (10.2 cm) across or in diameter, wire of all sizes, and tubes are commonly finished by cold drawing. Wire cannot be hot rolled economically smaller than about 0.2 in. (5.1 mm) in diameter and is reduced to smaller sizes by cold drawing. Steel, aluminum, and copper and its alloys are cold drawn in large quantities. In cold drawing the leading end of a piece is tapered for insertion through the die. A piece is pulled through a hole of smaller size and emerges correspondingly reduced in size. Drawing pressure against a die must exceed the yield strength of the work material and commonly is as much as 100,000–300,000 psi (6.9×10^5–2.1×10^6 kPa) for steel.

cold drawing (stretch forming). A process of cold working that uses the principle of stretching and wrapping a metal around a form. A form mounted on a ram is forced up, as the grippers move out and down. The metal is stretched above its elastic limit and wrapped at the same time. Multiple curvatures may be achieved with this method since there is practically no springback.

cold drawing (wire drawing). Accomplished by pulling a wire through a hardened die, usually carbide. Small-diameter wires are drawn through a diamond die. The wire is cleaned by pickling and then washed. A pointed or reduced cylindrical diameter at the end of the wire is pushed through the die, gripped with a pair of tongs, and pulled through until it can be attached to a power driven reel. The reel continues to pull the wire through the die as it is being rolled onto the reel. The material is lubricated to aid in the drawing operation and to impart a good surface finish to the wire. The dies themselves are usually water cooled because of the heat generated by the severe cold working. Annealing in a controlled-atmosphere furnace is required to restore the ductility lost during the cold working. Reductions in each pass through a die range from about 10% for steel to 40% for more-ductile materials. The cold drawing of tubes, seamless or welded, is accomplished by drawing the tube over a mandrel and through a reduce die opening.

cold extrusion (steel). The forcing of unheated metal to flow through a shape-forming die. It is a method of shaping metal by plastically deforming it under compression at room temperature while the metal is within a die cavity formed by the tools. The metal issues from the die in at least one direction with the desired cross-sectional contour, as permitted by the orifice created by the tools. Cold extrusion is always performed at a low enough temperature that the metal never reaches the recrystallization temperature [about 1100–1300°F (593–704°C) for steel], and work hardening always occurs. Extrusion differs from other processes, such as drawing, in that the metal is always being pushed in compression and never pulled in tension. The most important single advantage of cold extrusion is the substantial savings that can be realized, the amount saved depending on the part design, its previous method of manufacture, and the production volume requirements. Another advantage is that cheaper, less-critical raw materials can sometimes be used. Cold extrusion is generally limited to the production of cylindrical parts having solid or hollow cross sections, or minor variations of such shapes, such as hexagonal or square.

cold forging, coining. Requires that the work be confined while the impression is being forced into the surface or surfaces. The thickness of the metal is being changed as the process progresses. The coining operation requires high pressures. Since the length of the stroke is fixed, the volume of the metal must be accurately calculated. There is no way in which the metal can escape. Small amounts of metal flow are involved. The materials must have good ductile qualities. This process may be used to force impressions into metals for coins, medals, jewelry, etc.

cold forging, heading. A method of gathering metal in certain sections along lengths of wire or rod stock by causing the metal to plastically flow between dies without preheating the material. The process employs the same principles as hot upset forging, the main differences being that cold material is used and forming is done cold. With the blanks held in stationary dies on horizontal presses, usually called headers, the end protruding from the face of the dies is struck axially by upsetting tools. In this way the work length is shortened and increased in diameter over a portion of its length. Cold heading can also be used to bend or flatten the material on selected portions of the blank. However, this is usually done on only one end of the blank unless double-end headers are employed.

cold forging, hubbing. Used to form impressions or patterns in annealed steel. The desired pattern is machined into the punch, which is then hardened. The punch is next forced into the annealed steel. The annealed steel block is reinforced with a steel retainer ring while the hubbing is being done. Any excess metal from the plastic flow of the metal is machined away. Since the pressures are high (60 to 80 tons/in.^{2}) and the working severe, several operations may be required with annealing operations in between.

Complicated forms may be successfully hubbed with high finish. Duplicate cavities are easily made by this process. The cavities because of the finish required, may be used as molds for plastic molding or die casting. This process is often improperly called hobbing.

cold forging, riveting. Riveting and staking are operations for fastening two parts together. In riveting the upsetting or cold heading is applied to both ends of a precalculated length of material. The slug of metal is placed in aligned holes in the parts to be fastened. The forms are machined into the punch and die. Both ends of the slug are struck to form the heads of the rivet. Punching the holes in the pieces to be fastened, inserting the rivet, and cold heading may all be accomplished in one operation in specially designed machines. Staking may be used to fasten two pieces by forming the metal against the sides of punched or drilled holes into grooves in the upper pieces. The pressures required are small in comparison with the pressures required in riveting.

cold forging, swaging. A process of reduction, closely related to, but a specialized precision application of, mechanical forging, for shaping round bars, tubes, or blanks by means of numerous, uniformily spaced, short hammer blows applied in rapid succession by rotating dies. Rotating backers contact rolls to bring the dies together. The dies reciprocate thousands of times per minute as the spindle in which they are mounted rotates, thus displacing metal and making the blank conform to the shape of the dies used and to a predetermined size. The blows overlap each other to produce a smooth surface on the workpiece. Blanks for rotary swaging can be practically any symmetrical cross-sectional shape—round, square, oval, hexagonal, etc. Basic uses of swaging include pointing or tapering, forming (internal or external), straight reducing, and assembling. Straight reductions can be made down to diameters as small as 0.005 in. (0.127 mm), and on parts of unlimited length. The amount of reduction per pass depends on the material being swaged, and the maximum diameter is limited only by the size or capacity of the machine. **(See also rotary swaging.)**

cold rolling (metals). Sheets, strips, bars, and wires may be rolled cold. The material is cleaned by pickling in a dilute sulfuric acid solution and washed with a hot lime–water solution. The material is then rolled in the same kind of rolling mill as the one used in hot rolling. The rolling process gives directional properties to the material and imparts a surface skin which has been work hardened. Excessive working of the material causes embrittlement, and may require annealing. After annealing, the material is sized with additional passes through a finishing mill.

cold rolling, Roto-Flo process. A cold-rolling process for forming serrations and splines is performed on Roto-Flo hydraulically operated machines made by Michigan Tool Company. Workpieces are usually held between centers and rolled to full depth in a single pass of a pair of traversing, rack-type forming tools. Separating forces in excess of 100,000 lbf (445 kN) have been encountered in rolling hard materials, and as little as 3000 lbf (13,350 N) on softer materials. Metal on the periphery of the parts is displaced by plastic deformation to form the required shape. Positive control of all factors involved in accurate displacement of the metal ensures forming the correct number of teeth properly spaced around the periphery. The teeth can be formed at any desired position lengthwise on the part.

cold saw (see cut-off machines)

cold working (embossing). Uses a matching punch and die with the impression machined into both surfaces. However, this process differs from coining process in that the material thickness remains constant. It is actually a shallow drawing process rather than a squeezing process. The metal flow in the coining process is lateral, whereas the metal flow in embossing is in the direction of applied force.

cold working (metals). Cold-worked metal is formed to shape by the application of pressure at temperatures below the critical temperature and for the most part nominally at room temperature. It is preceded by hot working, removal of scale, and cleaning of the surface, usually by pickling. Cold working is done mostly to hold close tolerances and produce good surface finishes and to enhance the physical properties of the metal.

cold working (shot peening). A process of cold working in which large quantities of steel shot are fired at the surface of a workpiece. The fibers in the surface of the workpiece stretch and place the subsurface fibers in tension. These subsurface tensile forces place the outer fibers in compression. The compression forces counteract failure, which might occur owing to tensile forces applied to the workpiece, such as fatigue failure. The surface of the workpiece is work hardened by the process.

collator (computer). Can be used to perform the following types of operations: (1) Sequence checking. After the sorter has been used to place a file of cards in a desired sequence, the file can be checked on the collator to determine if the sequence is correct. The collator does this by comparing each card with the one ahead of it. (2) Merging. In merging, two files of cards already in sequence can be combined into one file. (3) Matching. In matching, instead of merging the two files, cards in either file that do not match the other can be separated and cards that do match remain in the two original groups. (4) Card Selection. The collator also has the ability to select certain types of cards from a file without disturbing the sequence of the others. The selecting task of the collator is similar to that of the sorter, except that the collator can select on more than one card column.

colloids (crystals). Colloids cover nearly all the materials of ordinary life, such as fine suspension of inorganic compounds and metal soaps, proteins, and organic high polymers. Any rigid classification of colloidal systems is extremely difficult and remains incomplete because a considerable overlapping of properties occurs among various colloidal substances. Colloidal systems that use liquid as an external phase are considered under two main groups—lyophobic ("liquid-hating") and lyophilic ("liquid-loving") colloids. When water is a dispersing medium, the terms hydrophobic and hydrophilic are used, respectively. Lyophobic or hydrophobic colloids are dispersions of the particles composed of many atoms or molecules that are of sufficient size [0.39–39 μin. (0.01–1 μm)] and refractivity to be visible in the ordinary ultramicroscope. Because of this size they are considered as two-phase systems. Lyophobic colloids are mostly inorganic compounds such as metal particles and certain hydroxides. Lyophilic colloids are usually macromolecular dispersions of organic high polymers, proteins, and soaps. The size of dispersed particles (macromolecules) is much smaller than that of lyophobic colloids, ranging from 0.039 to 3.9 μin. (0.001 to 0.1 μm). Such dispersions are not visible in ultramicroscopes and are usually regarded as one-phase systems.

coloring. (1) The overall application on foil of colored lacquer or ink for decorative purposes. (2) A type of buffing.

compression molding (plastics). Normally, compression molding of thermoplastics is not economically competitive with the more rapid injection-molding process, because a mold requires heating to soften the resin, followed by cooling to solidify the molded shape. One molding cycle could take as much as 30 min, but the injection cycle takes only seconds. Large, thick slabs are compression molded from specified thermoplastics, which are then machined into shapes not otherwise obtainable with molds. Typical applications for compression-molded parts include appliance bases, knobs, ashtrays, missile components, battery cases, pot handles, bottle caps, rocket nozzles, dinner ware, wall switch plates, gears, cams, pulleys, washing machine agitators, and electrical and electronic components. The basic process consists of placing a predetermined weight of molding compound into a heated mold cavity, closing the mold, and exerting pressure by means of a compression press. The press and mold remain closed until the compound becomes rigid enough to maintain its molded shape. Hardening (cure) is caused by a chemical change induced by the heat of the mold. When the cure, which is first determined by trial, is completed, press and mold are opened, and the part is removed while hot. Part removal is manual or automatic, depending on mold design.

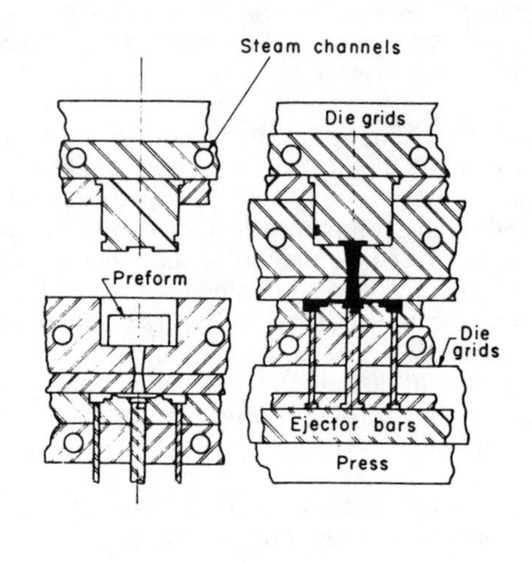

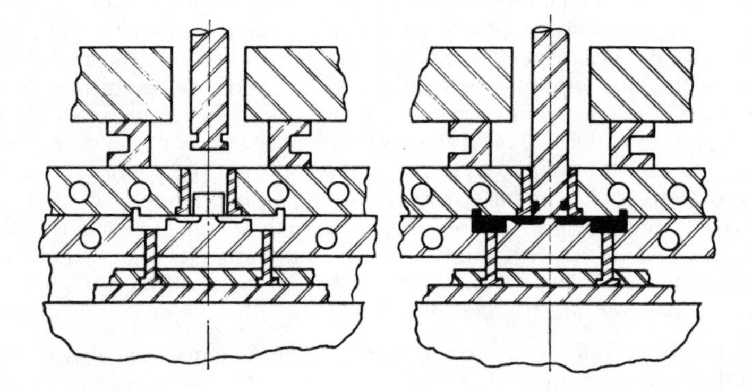

Die setup for transfer molding (top); duplex or plunger arrangement for transfer molding (bottom).

computer-aided design (CAD). Some computer systems can be used for design while others are limited to the drawing and editing of lines. [**See also computer-aided drafting and design (CAD/D).**]

computer-aided design/computer-aided manufacturing (CAD/CAM). The true meaning of CAD/CAM is a system that actually integrates the design and manufacture of something, a feat very few CAD/D systems are capable of. CAD and CAM are largely separate phenomena that are just beginning to be linked.

computer-aided drafting and design (CAD/D). A descriptive term applied to systems that assist not only in the production of drawings, but also in the development of the intelligence of the drawing. Most CAD/D systems have six major components (four hardware and two software): (1) Central Processing Unit (CPU). This is the brain part of the computer. (2) Storage. Typically fixed, Winchester, floppy disks, or sometimes tape; the storage portion of the CAD/D system is where drawings are kept electronically. (3) Workstation. This is the console where the CAD/D operator communicates with the computer, and where the computer responds or communicates with the user. (4) Plotter Station. The plotter station is where images stored in the computer memory are printed on drafting media. (5) Operating System (OS). The OS is the master control program that coordinates the activities of all four of the hardware components of the system. (6) Applications Program. The applications program or user software is the most important component of a CAD/D system. User software creates the working environment for making drawings or creating designs.

computer-aided engineering (CAE). This new term pertains to computer systems that assist in the engineering or design cycle, but have little, if any, drafting capability. CAE systems are becoming common personal workstations to CAD/D systems, allowing design and engineering calculations to be passed to a CAD/D system for drafting.

computer-aided manufacturing (CAM). The utilization of computer technology in the management, control, and operations of the manufacturing facility through either direct or indirect computer interface with the physical and human resources of the company or organization. Developments in CAM are in four main areas: numerical control, robotics, process planning, and factory management. A key concept of CAM is that these individual functions are not only computerized but are tied together through a shared database in computer memory.

computer graphics. Just as engineering graphics is a tool for traditional design, computer graphics is a tool for computer-aided design. [**See also computer-aided design (CAD) and computer-aided drafting and design (CAD/D).**]

computer-integrated manufacturing (CIM.) The logical organization of individual engineering, production, and marketing/support functions into a computer-integrated system. Functional areas—such as design, inventory control, physical distribution, cost accounting, planning, purchasing, etc.—are integrated with direct materials management and shop-floor-data acquisition and control. Thus the loop is closed between the shop floor and its controlling activities. Shop-floor machines serve as data-acquisition devices for the control system and often its direct command. Strategic plans smoothly give way to tactical operations, at known cost.

computer numerical control (CNC). A numerical control system wherein a dedicated, stored-program computer is used to perform some, or all, of the basic numerical control functions in accordance with a control program stored in the computer. Considered to be the fourth generation of numerical control. **(See also numerical control.)**

computer process control. Manufacturing process controlled by digital computer. This involves a multistage development of the manufacturing process as a whole. The computer controls the entire manufacturing process, accepting input measurements from instruments and controlling accordingly the operations through the drive controls, drive motors, power transmission devices, valves, etc., to keep the process at preset norms. The computer is the key element in the three major subsystems involved: (1) processing or making steps or stages; (2) handling operations for the materials being processed; and (3) the control system of which the computer is central.

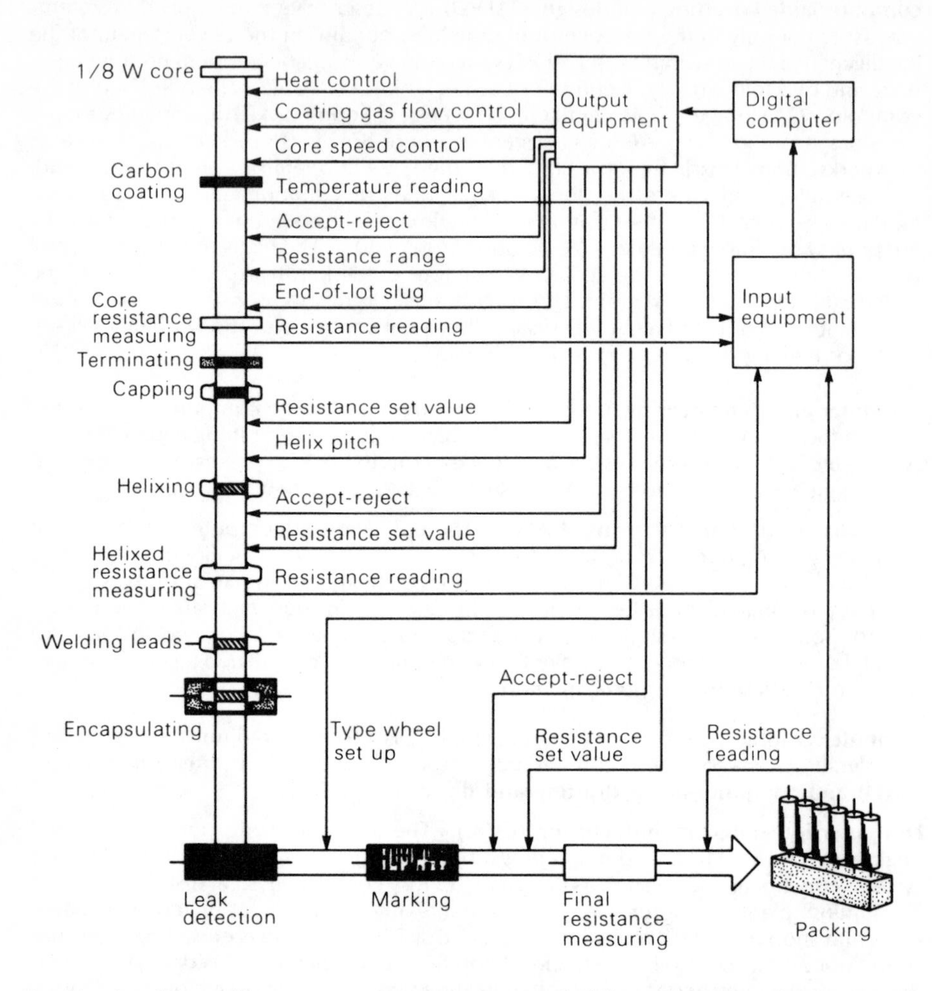

Combination diagram illustrates computer process control of flow and routing of major control signals for computer-controlled deposited-carbon resister production system.

computer unit (computer). The part of a computer system that effects the retrieval of instructions in proper sequence, the interpretation of each instruction, and the application of the proper signals to the arithmetic unit and other parts of the system in accordance with the interpretation. The performance of these operations requires a vast number of "paths" over which data and instructions may be sent. Routing data over the proper paths in the circuitry, opening and closing the right "gates" at the right time, and establishing timing sequences are major functions of the control unit. All of these operations are under the control of a stored program.

contact scanning. In ultrasonic inspection, a planned systematic movement of the beam relative to the object being inspected, the search unit being in contact with and coupled to this object by a thin film of coupling material.

control punch (computer). The most common method used to differentiate master cards from detail cards when the decks are run through punched card equipment is by means of a control punch. The majority of the machines are designed so that with proper board wiring they can recognize an 11-punch in a particular card column of a numeric field. The 11-punch, more commonly known as an x-punch, is usually used to signal the machine that a master card is passing through as opposed to a detail card (a NO X card). Board wiring then causes the particular machine to initiate one set of operations for the master card and another set for the detail card.

controller (robot). An information processing device whose inputs are both the desired and the measured position, velocity, or other pertinent variables in a process and whose outputs are drive signals to a controlling motor or actuator. The controller has a threefold function: first to initiate and terminate motions of the manipulator in a desired sequence and at desired points; second to store position and sequence data in memory; and third, to interface with the "outside world." Robot controllers run from simple step sequences through pneumatic logic systems, diode matrix boards, electronic sequencers, and microprocessors to minicomputers. The controller may be either an integral part of the manipulator or housed in a separate cabinet. The complexity of the controller both determines and is determined by the capabilities of the robot. Simple nonservo devices usually employ some form of step sequence. Servo-controlled robots use a combination of sequencer and data storage (memory). This may be as simple as an electronic counter, patchboard, or diode matrix and series of potentiometers or as sophisticated as a minicomputer with core memory. Other memory devices employed include magnetic tape, magnetic disk, plated wire, and semiconductor (solid-state RAM). Processor or computer-based controller operating systems may be handwired, stored in core memory, or programmed in ROM (read-only memory). The controller initiates and terminates the motion of the manipulator through interfaces with the manipulator's control valves and feedback devices and may also perform complex arithmetic functions to control path, speed, and position. Another interface with the outside world provides two-way communications between the controller and ancillary devices. The interface allows the manipulator to interact with whatever equipment is associated with the robot's task.

conveyor, accumulation. Any conveyor designed to permit accumulation of packages, objects, or carriers. May be roller, live roller, roller slat, belt, vibrating, power-and-free, or tow conveyors. **(See also conveyor, minimum pressure accumulation.)**

conveyor, belt. An endless fabric, rubber, plastic, leather, or metal belt operating over suitable drive, tail end, and bend terminals and over belt idlers or slider bed for handling bulk materials, packages, or objects placed directly on the belt. **(See figure p. 82.)**

conveyor, bucket. Any type of conveyor in which the material is carried in a series of buckets. **(See also elevator, bucket.)**

conveyor, controlled velocity roller. A roller conveyor having means to control the velocity of objects being conveyed. **(See also conveyor, roller.)**

conveyor, en masse. A conveyor composed of a series of skeleton or solid flights on an endless chain or other linkage which operates in horizontal, inclined, or vertical paths within a closely fitted casing for the carrying run. Bulk material is conveyed and elevated in a substantially continuous stream with a full cross section of the casing. **(See figure p. 82.)**

conveyor, flight. A type of conveyor comprised of one or more endless propelling media, such as chain, to which flights are attached, and a trough through which metal is pushed by the flights.

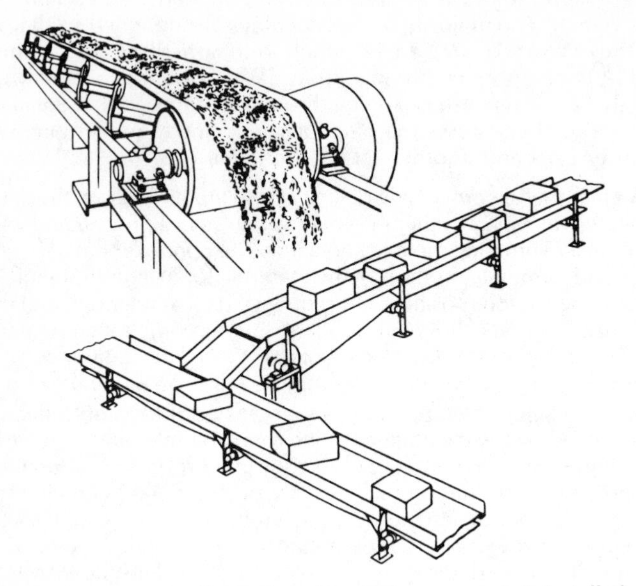

Illustration of typical conveyor belt installations.

Illustration of an en masse conveyor.

conveyor, live roller. A series of rollers over which objects are moved by the application of power to all or some of the rollers. The power-transmitting medium is usually belting or chain.

conveyor, minimum pressure accumulation. A type of conveyor designed to minimize build-up of pressure between adjacent packages or cartons. **(See also conveyor, accumulation.)**

conveyor, monorail. This is an overhead method of conveying in a manufacturing plant that employs a single overhead rail on which roller carriers with suitable hooks travel. A monorail system is primarily designed to operate overhead and not interfere with ground-level activities. Carriers for monorail conveyors are either hand propelled or power driven. When power driven, they may be controlled by a pendant or cab or from a remote station. In some cases, loads are automatically moved for a predetermined distance, then moved under manual control at load or unload areas. Loads can be moved nonstop, or they may be programmed to stop at intermediate stations along the route. Automation in a monorail system can be as basic as simply picking up and delivering loads without operator assistance. An automatic monorail also can be programmed to raise, lower, and rotate the load; automatically store finished or semifinished goods; or route coded products to appropriate points on the assembly line. Automatic monorail systems have been built with multiple routes and multiple elevations. All types of hoisting mechanisms can be used with monorails. A wide variety of below-the-hook lifting devices designed to handle many different sizes and types of loads can be suspended from the hoist. **(See also conveyor, trolley.)**

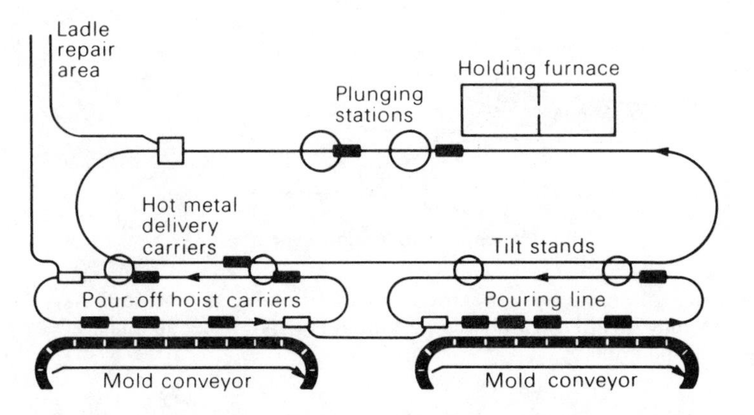

A total of three monorail loops are used in this system. Three 5-ton (4.54-tonne) monorail carriers distribute hot metal on the large loop. The metal is poured into "pour-off" ladles, which are suspended from carriers operating on the two smaller monorail loops.

conveyor, pneumatic. An arrangement of tubes or ducts through which bulk material or objects are conveyed in a pressure and/or vacuum system.

conveyor, power-and-free. A conveying system wherein the load is carried on a trolley or trolleys that are conveyor propelled through part of the system and may be gravity or manually propelled through another part. This arrangement provides a means of switching the free trolleys into and out of adjacent lines. The spur or subsidiary lines may or may not be powered. **(See also conveyor, monorail and figure p. 84.)**

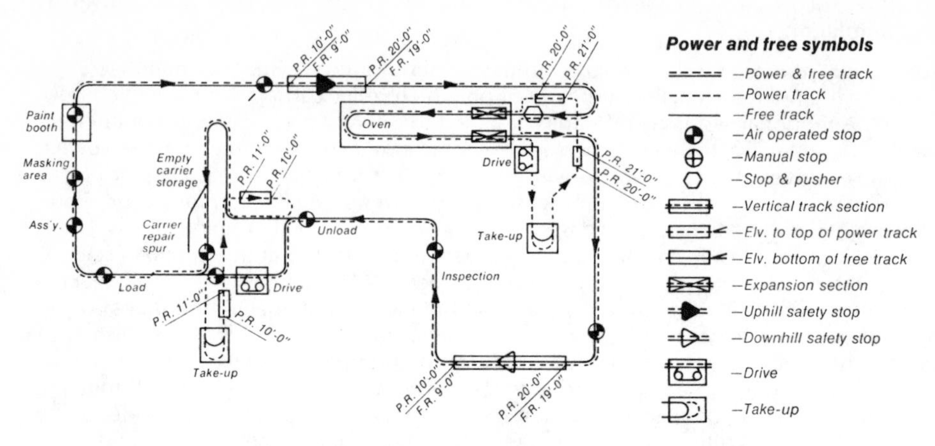

Typical schematic for power-and-free conveyor layout.

conveyor, roller. A series of rollers supported in a frame over which objects are advanced manually, by gravity, or by power.

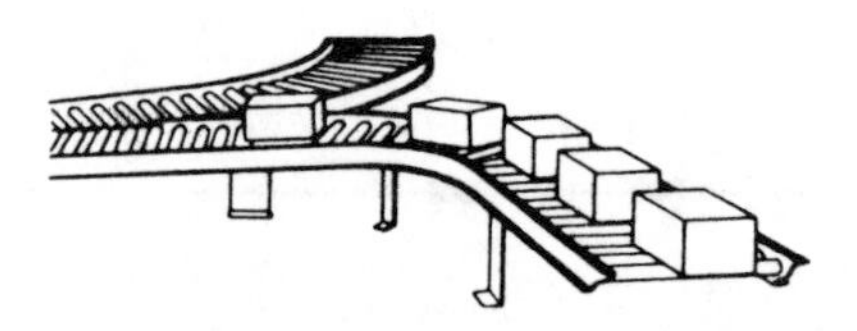

Illustration of a roller conveyor.

conveyor, screw. A conveyor screw revolving in a suitably shaped stationary trough or casing which may be fitted with hangers, trough ends, or other auxiliary accessories for moving bulk materials.

Illustration of a screw conveyor.

conveyor, skate-wheel. A type of wheel conveyor making use of series of skate-wheels mounted on common shafts or axles, or mounted on parallel space bars on individual axles.

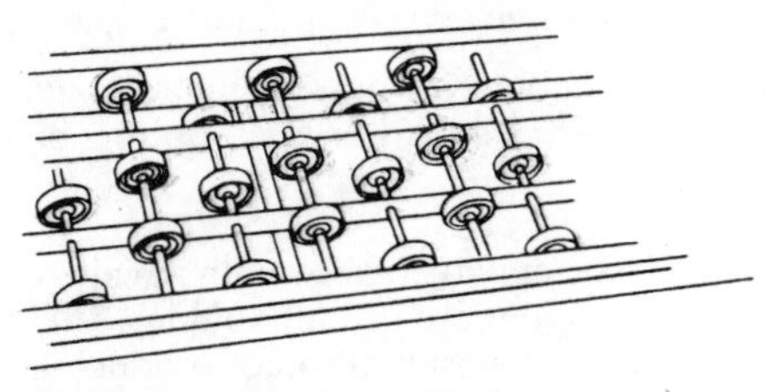

Illustration of a skate-wheel conveyor.

conveyor, steel belt. A belt conveyor using a steel band belt as the conveying medium.

conveyor, tow. An endless chain supported by trolleys from an overhead track or running in a track at the floor with means for towing floor-supported trucks, dollies, or carts.

conveyor, trolley. A series of trolleys supported from or within an overhead track and connected by an endless propelling means such as chain, cable, or other linkage with loads usually suspended from the trolleys.

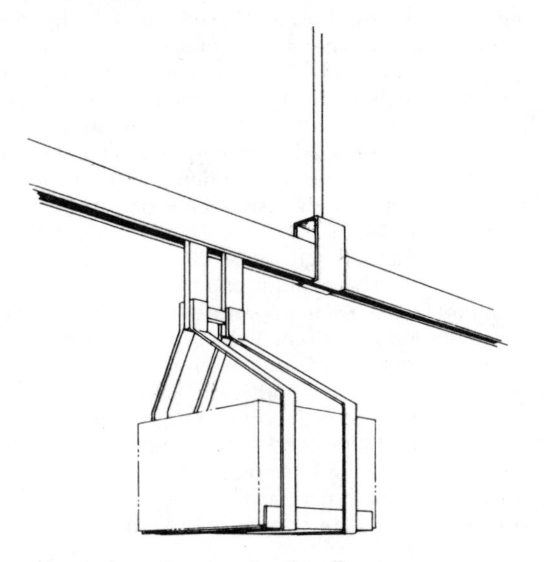

Illustration of an overhead trolley conveyor.

conveyor, troughed belt. A belt conveyor with the conveyor belt edges elevated on the carrying run to form a trough by conforming to the shape of the troughed carrying idlers or other supporting surface.

conveyor, troughed roller. A roller conveyor having two rows of rollers set at an angle to form a trough over which objects are conveyed.

conveyor, tubular drag (see conveyor, en masse)

conveyor, vertical articulated platform. A type of vertical conveyor in which sections of articulated slat conveyor aprons form rigid platforms for vertical movement in continuous flow. The platforms are flexible in only one direction, and they assume a vertical position on the noncarrying run to minimize space requirements.

conveyor, walking-beam. A conveyor employing a horizontal beam or beams with a reciprocating motion that moves a load on a supporting bed, rails, or rollers a given distance with each stroke.

copper. Similar to nickel in its ability to alloy with a wide range of metals and is face-centered cubic. The ores of copper are usually sulfides. The nature of copper ore and its copper content determines the best method of producing metallic copper from the ore. This is done for the most part by three different methods: (1) Ores in which copper is present as sulfides are usually concentrated, then smelted from the molten slag and matte. (2) The second method of obtaining copper from its ores is used when copper is present in the ores in the form of oxides, or when copper is present as native or nearly pure copper. In either case, these may be smelted in the blast furnace or open-hearth-type smelter, then fire refined, in some cases, electrorefined. (3) The third method applies to the low grades of oxidized ores, and are usually leached with sulfuric acid solutions and then concentrated and classified, and the concentrates of approximately 90% copper are ready for the smelter. Lower grade of oxidized ores of copper are usually leached with sulfuric acid solutions and then electrodeposited in an electrolyte onto a pure copper plate. Several types of copper are produced, depending on the source from which they were obtained and on the method used. These include cathode copper, electrolytic tough pitch (ETP) copper, and oxygen-free high conductivity (OFHC) copper. Copper is noted for its high conductivity for both heat and electric current, with a large amount used in busbars and other types of electrical conductors. The terminal conductivity of copper, about 2700 Btu/°F/in. (655 kJ/°C/cm), is almost 10 times that of steel. Copper is used for chills, continuous molds, and other equipment that must remove heat. Pure copper is almost unweldable because of the difficulty of retaining sufficient welding heat in the area of the weld. It is rarely resistance welded. Pure copper has a modulus of elasticity value of 16×10^6 psi (1.1×10^8 kPa), a yield strength of 9000 psi (6.2×10^4 kPa), and ultimate strength of 32,000 psi (2.2×10^5 kPa). Elongation is about 50% in 2 in. (5.1 cm). The melting point of pure copper is 1981°F (1083°C). However, copper cannot be used at temperatures very much above room temperature owing to loss of strength and oxidation. The current-carrying grades of copper must be relatively pure, since alloying additions, even in minute amounts, greatly increase the resistivity. Electrolytic tough pitch copper (ETP grade) containing 0.04% oxygen is used for conductors. Oxygen-free high-conductivity copper (OFHC grade) is the best available. Copper is not used in the food-processing industries because many copper compounds are toxic. Copper alloys are numerous; zinc, tin, aluminum, and silicon all serve as solid-solution hardeners in these alloys. In general, brasses are alloys of copper and zinc, and bronzes are alloys of copper and elements other than zinc.

copper alloys. Copper can be hardened and strengthened by cold working or by solid-solution alloying with zinc, tin, aluminum, silicon, manganese, and nickel. It forms a precipitation-hardenable alloy when small amounts of beryllium are present. The ductility of many of the solid-solution alloys is greater than that of copper. The room-temperature mechanical properties of copper alloys are approximately intermediate between those of aluminum alloys and steel. Copper alloys are known for attractive appearance and corrosion resistance, accounting for large-scale use in hardware, particularly marine hardware. Copper–zinc alloys are called brasses. Alpha-phase brasses with up to 36% zinc include ductile, easily worked compositions such as cartridge brass for deep drawing. The beta phase appears with a large proportion of zinc. It is more brittle and less easily cold worked but more corrosion resistant and machinable. Most bronzes are alloys of copper with up to 12% tin. Typical applications are for bearings and worm gears.

copper–nickel alloys. Copper–nickel alloys, called cupronickels, are copper-base alloys containing nickel from 5 to 30%. Because of the mutual solubility of copper and nickel in the solid state, cupronickels form a series of solid-solution alloys that can be easily cold worked. Their mechanical properties as well as their corrosion and heat resistance increase progressively with increasing nickel content.

Copper-Zinc Alloys; Composition, Properties And Uses

Common name	Composition—Percent						Condition	Tensile strength psi	Elongation in 2" %	Typical uses
	Cu	Zn	Sn	Pb	Ni	Mn				
Muntz metal	59	41					Hot rolled	54,000	45	Architecutural work, condenser tubes, valve stems.
							Cold rolled	80,000	5	
High brass	66	34					Annealed sheet	46,000	64	Cold forming, radiator cores, springs, screws.
							Hard	76,000	7	
Cartridge brass	70	30					Annealed sheet	53,000	54	Cartridges, musical instruments, tubing.
							Hard	76,000	7	
							Spring	92,000	3	
Low brass	80	20					Annealed sheet	47,000	47	Drawing, architectural work, jewelry.
							Hard	75,000	7	
							Spring	91,000	3	
Commercial bronze	90	10					Soft sheet	38,000	45	Screen wire, hardware, forgings, screws, jewelry.
							Hard sheet	64,000	4	
							Spring	73,000	3	
Manganese bronze	59	39	0.7			0.05	Bars, half hard	72,000	20	High tensile strength propellers, valve stems, worm wheels, brazing rods, castings.
							Bars, hard	80,000	10	
Admiralty metal	70	29	1				Soft	45,000	60	Corrosion-resistant to sea water, condenser tubes, heat exchangers.
							Hard	95,000	5	
Screw stock	61.5	35.5		3			Soft	47,000	60	Screw-machine products.
							Hard	62,000	20	
Nickel silver	65	17			18		Soft sheet	50,000	40	Ornamental metal work, base for plated ware, marine hardware.
							Hard sheet	93,000	3	

copper production. Since much of the copper ore used has such a low percentage of copper, the first step in production is a concentration process normally yielding a product that is about 50% copper. The concentrate is next roasted to reduce the sulfur and arsenic. It is then smelted with coke and flux in a reverberatory furnace to produce matte, which is a mixture of copper and iron sulfides, being about 40% copper. The smelting also removes part of the iron usually present in the ore. The molten matter is then put into a converter, which is similar to those used in making Bessemer steel. Air blown through the molten matte oxidizes and blows out the sulfur. The iron is oxidized and goes into the slag. This processing converts the matte into blister copper, which contains 97–98% copper plus copper oxide. Blister copper is put into a reverberatory furnace for further refining and then cast into anode bars. These bars are made the anodes in an electrolytic refining process. During the electrolysis spongy copper forms on the cathodes. This copper from the cathodes is melted and treated to reduce the oxides prior to casting into molds. A common method of reducing the oxides is by stirring the molten copper with green logs. The resulting product contains only about 0.07% oxygen and is called electrolytic tough pitch copper. If the molten copper is not poled, it is used for copper-base alloys where very low oxygen content is not necessary.

Copper Base Alloys

Relevant specifications	Composition (%) (Balance Cu)	Condition	Typical mechanical properties: 0.1% P.S. N/mm²	Tensile strength N/mm²	Elongation (%)	Hardness (VPN)	Characteristics and uses
Brasses							
B.S. 2870/5:CZ106	30 Zn	Annealed Hard	77 510	325 695	70 5	65 185	Cartridge brass: deep-drawing brass, having maximum ductility of the copper-zinc alloys
B.S. 2870 5:CZ106	37 Zn	Annealed Hard	95 540	340 725	55 4	65 185	'Common brass': a general purpose alloy suitable for limited forming operations by cold-work
B.S. 2870 5:CZ109	40 Zn	Hot-rolled	110	370	40	75	Hot-rolled plate used for tube plates of condensers. Also as extruded rods and tubes. Limited capacity for cold-work
B.S. 2870/5:CZ114	Mn, Al, Fe, Sn } up to 7% total 37 Zn	Grade A Grade B	230 280	465 540	20 15	— —	High-tensile brass: wrought sections for pump rods, etc. Cast alloys; marine propellors, water turbine runners, rudders. Locomotive axle boxes
Tin bronzes							
B.S. 2870:PB101	3.75 Sn 0.10 P	Annealed Hard	110 620	340 740	65 5	60 210	Low-tin-bonze: good elastic properties combined with corrosion resistance. Springs and instrument parts
B.S. 1400:PB1 c	10 Sn 0.5 P	Sand-cast	125	280	15	90	Cast phosphor bronze: mainly bearings—cast as sticks for machining of small bearing bushes
B.S. 1400:G1 c	10 Sn 2 Zn	Sand-cast	125	295	16	85	'Admiralty gunmetal': pumps, valves and miscellaneous castings, particularly for marine purposes because of good corrosion resistance. Also statuary because of good casting properties
Aluminum bronzes							
B.S. 2870/5:CA101	5 Al Ni up to Mn 4.0% total	Annealed Hard	125 590	385 775	70 4	80 220	Imitation jewellery, etc. Excellent resistance to corrosion and to oxidation on heating, hence used in engineering particularly in tube form
B.S. 1400:AB1/c	9.5 Al 2.5 Fe Ni } up to Mn } 1.0% each	Cast	185	525	30	115	The best-known aluminum bronze for both sand- and die-casting. Corrosion-resistant castings
Cupro-nickels							
B.S. 2870/5:CN105	25 Ni 0.25 Mn	Annealed Hard	— —	355 600	45 5	80 170	Mainly for coinage, e.g. the current British 'silver' coinage
B.S. 3073/6:NA13	68 Ni 1.25 Fe 1.25 Mn	Annealed Hard	215 570	540 725	45 20	120 220	Monel Metal: Combines good mechanical properties with excellent corrosion resistance. Mainly in chemical engineering plant

coreless electric induction furnace. Operates on the principle of an electric transformer. The primary is a high-frequency alternating current, and this is passed through a water-cooled copper coil that surrounds the melting crucible. The crucible is filled with carefully selected steel scrap, which acts as the secondary of the transformer. When the high-frequency current passes through the copper coil, currents are generated in the pieces of scrap, causing them to become very hot so that in relatively few minutes the entire charge is melted. One advantage of this furnace is that after the charge is entirely molten, the induced electric currents cause a stirring action in the bath, producing uniformity in both temperature and composition. **(See also induction furnace.)**

corrosion. Corrosion of common metals is an electrochemical phenomenon associated with the flow of electric currents over finite distances. It is the gradual chemical (direct) or electrochemical attack on metal by its surroundings, in such a manner as to convert the metal to oxide, salt, or some other compound, thereby losing strength, ductility and other desirable properties. Corrosion occurs by two fairly well recognized mechanisms: (1) Direct chemical attack. Includes all types of corrosion in which there is little or no percipitable flow of current through the metal for measurable distances. (2) Electrochemical attack. Applies to all those cases where there are definite anodic and cathodic areas that are separated by finite distances resulting in the flow of current between anode and cathode. Corrosive attacks manifest themselves in many forms, frequently occurring in combination with other mechanisms of failure such as corrosion fatigue, erosion–corrosion, and stress corrosion. It may be brought about by an almost unlimited variety of corrosive media such as air, industrial atmospheres, soil, acids, bases, and salt solutions.

corrosion, acidity. As acidity increases, the number of hydrogen ions in a solution increases. This permits a greater amount of metal to enter solution as a replacement for the hydrogen, in other words, to combine with the negative ions or cations. Thus, acidity acts to promote corrosion; and although the concentrations may not be great enough to produce a continuous reaction, the direction of the force is such as to assist any other factors that increase corrosion rates.

corrosion, dezincification. In some metals and alloys and under certain environmental conditions, localized areas may be attacked in such a manner that one or more constituents are dissolved, leaving that particular area devoid of that certain element. This is known as dezincification, a term originally applied to the removal of zinc from brasses by means of existing corrosive environment. The term now applies to any condition of corrosion in which any specific element is removed from an alloy. The dezincification of brasses results in a transformation from its usual yellow color to that of copper, indicating a predominance of copper and a lack of zinc, accompanied by a severe loss of strength because of this loss of zinc from the alloy.

corrosion, electrolytic action. Electrolytic action starts with the dissolving of a metal in water or other dissolving medium. When dissolved, the metal is ionized and thereby possesses a positive electrical charge. It then either gives up its charge and is deposited in some other place or combines with another ion to form a new compound. The place from which the metal was dissolved is called the anode, and the place at which it is deposited is the cathode; these two form the electrodes of an electrical circuit. The water or solution is the electrolyte, and the driving force is the solution pressure of the metal. Solution pressure may be roughly defined as the electromotive force that exists between a metal and its ions in a solution. Solution pressure differs in all metals.

corrosion, fatigue. A type of stress corrosion. A safe endurance limit may be nonexistent if repeated stressing occurs under even mildly corrosive conditions such as exposure to ordinary tap water. In the case of fatigue failure, the strained crystals become anodic and therefore are susceptible to corrosive attack. Once a crack starts it may act as an artificial pit, and its anodic corrosion will be accelerated because of the relative area effect. The type corrosion is the combined action of corrosion and repeated stresses and is much more serious than the sum of these factors (stress and corrosion), acting individually.

corrosion, intergranular attack. Occurs when the grain boundaries of metal are selectively corroded. Probably caused by galvanic action resulting from the composition difference existing between the grain boundary and the grain center of the metal being attacked. It has been observed in some kinds of chromium–nickel stainless steels, particularly following improper welding or heat-treating practices, and in some copper-base alloys such as the brasses. The growth of intergranular cracks is favored by establishment of oxygen concentration cells and the relatively small anode area.

corrosion, oxidizing agents. An uncombined metal has zero valence; oxidizing agents tend to induce a positive valence in it. With a positive valence the metal tends to form a compound by combining with a negative ion. The act of forming this compound is a form of corrosion, since the change is from an uncombined state to a combined state, with the properties of the compound being generally undesirable for engineering purposes. The greatest oxidizing agent is oxygen, which may be present in the air or dissolved in a liquid. A further effect of oxygen is its combination with evolved hydrogen to form water, thereby removing the coating of hydrogen that would inhibit further corrosion. Oxidation also occurs without the presence of oxygen. Such chemical corrosions as the formation of sulfides by sulfur or sulfur compounds are examples of oxidation reactions. Dry ammonia and some organic compounds produce similar types of corrosion.

corrosion protection, anodizing. A method of coating aluminum and magnesium and their alloys, usually by the use of an electrolyte of 3% solution of chromic acid at a temperature of about 100°F (38°C). In the process the voltage is raised from 0 to 40 V at the rate of about 8 V/min and is maintained at full voltage for 30–60 min; this produces a current density of about 1–3 A/ft^2 (10.8–32.4 A/m^2) of surface coated. The resulting coating is very thin, and its principle use has been as a base for paint on aluminum and magnesium parts.

corrosion protection, chromizing. A cementation process much like sheradizing in parts to be coated are packed in a container with a mixture of 55 parts of chromium or powdered ferrochromium and 45 parts of alumina by weight. This mixture is then heated either under vacuum or in a protective atmosphere (hydrogen preferred) at 2370–2560°F (1299–1404°C) for 3 or 4 h. Coatings by this process are generally about 0.004 in. (0.1 mm) thick and contain about 40% chromium at the surface of the part being coated.

corrosion protection, metal cementation. A method of metal protection against corrosion in which the metals zinc, chromium, aluminum, and silicon are successfully applied by actually alloying them into the surface to be protected. This includes several procedures, such as sheradizing, chromizing, anodizing, metal cladding, aluminum cladding, and vacuum-metallized coating, depending on the metals used.

corrosion protection, metal spraying. A process in which the coating metal is usually drawn into wire and fed through a specially constructed spray gun that is operated with compressed air and a fuel gas. The gases at the nozzle are ignited, the wire is melted as it is fed into a nozzle and projected against the surface to be coated at a speed of over 500 ft/s (152.4 m/s). Even though the molten metal particles are cooled instantly to a temperature of about 80°F (27°C), the impact causes them to adhere firmly to the steel surface, if it is thoroughly cleaned. This process of metal spraying is used effectively for building up surfaces such as large drive shafts that have become worn, and also for the application of thin coatings as a protection against serious corrosion.

corrosion protection, sheradizing. Consists of thoroughly cleaning parts to be coated by pickling or sandblasting, then packing them into metal drums with fine zinc dust, usually containing 5–8% zinc oxide. The contents of the drum are heated from 650 to 750°F (343 to 399°C) for several hours during which time the drums are rotated in the furnace. The resulting coating is thin and consists of intermetallic compounds of iron and zinc, ranging from an iron-rich alloy adjacent to the steel base to almost pure zinc at the surface.

corrosion, stress. Likely to occur when static surface tensile stresses act in combination with a corrosive medium. The stresses can be residual stresses resulting from previous steps in the fabrication or use of the metal. The stresses can also be the result of applied loads, but they must in both cases be at the surface and tensile in nature. Failure is believed to start by corrosion at anodic areas in the surface of the metal, and surface tensile stresses are intensified at the bottom of minute corrosion notches. These stresses cause continual breakdown of any passive film that might tend to form, and so corrosion proceeds at the root of the fissure. As the crack grows deeper, stresses from applied loads are intensified at accelerating speeds and corrosion proceeds faster. When applied loads are present, final fracture may occur with sudden violence. Examples of stress corrosion failures are the season cracking of brass and the caustic embrittlement of boilers.

corrosion test, metal. The ability of metals to resist atmospheric corrosion, or corrosion by liquids or gases, is of primary importance. Common types of corrosion are the pitting or localized type, the direct chemical or solution type, and electrolytic or galvanic corrosion. Accelerated corrosion tests have been devised whereby the behavior of the materials in actual service may be deduced from conditions applied for weeks, instead of years as in actual use. Corrosion may be measured by determining the loss of tensile strength of specimens, by loss of weight in materials that dissolve in the corroding medium, or by gain in weight when a heavy coating of rust is formed. There are no standard tests, but it is of importance to see that corrosion test conditions nearly duplicate the conditions of service.

corrosion test, metal (intergranular corrosion). Attacking a metal with the so-called Strauss solution and microscopic examination for precipitation of carbides are two of the tests employed in looking for intergranular corrosion. In the former test, the steel is boiled in the Strauss solution for about 24 h. The solution usually consists of 47 ml of concentrated sulfuric acid and 0.5 oz (13 g) of copper sulfate crystals per liter. If intergranular corrosion is present, the attacking reagent will cause the metal to lose its metallic resonance, cause a considerable decrease in the electrical conductivity, or cause the metal to develop cracks following a bend test. In the test employing microscopic examination for precipitation of carbides, one side of the specimen is polished and then electrolytically etched in an oxalic acid or sodium cyanide solution. Such an etch causes precipitated carbides to appear as black particles. If carbides are found, the specimen is given an embrittlement test.

corrosion test, metal (mercurous nitrate). Brass and bronze objects are tested by dipping in a 40% solution of nitric acid and 1% mercurous nitrate. After a few minutes the specimens are removed, cleaned, and dried. Upon low-power microscopic examination, if no cracks are visible, the specimens will probably not crack because of internal stresses.

corrosion test, metal (salt spray). In this test, specimens that have been thoroughly cleaned are placed in a specially prepared box or chamber and exposed to a fog or spray of salt water for a designated number of hours. The concentration of the spray and the exposure period varying depending on the severity of the service required. Such a test is suitable for testing the corrosion resistance of all coatings used on steel. Coated steel will withstand the attack for varying periods depending on the protection the coating affords.

corundum (see emery)

coupon. A piece of metal from which a test specimen may be prepared—usually produced as an integral extra piece as on a casting or forging or as a separately cast or forged piece.

CPU (central processing unit) (computer). The center of the computer through which the information flows. It is the place where the program is realized; it is the place where the information is obtained through programming instructions.

cranes. A machine for lifting, lowering, and/or moving a load with the hoisting mechanism an integral part of the machine. Cranes may be traveling, portable, or fixed. Traveling cranes can be of various types: underhung crane with a movable bridge running on the lower flanges of an overhead fixed runway structure and carrying a movable or fixed hoisting mechanism; overhead crane with a movable bridge running on the top surface of rails of an overhead fixed runway structure and carrying a movable or fixed hoisting mechanism; gantry crane is similar to an overhead crane, except that the bridge for carrying the hoisting mechanism is rigidly supported on two or more legs running on fixed rails or other runway; semigantry crane with one end on the bridge supported on one or more legs running on fixed rails or other runway and the other end of the bridge supported by a truck running on an elevated rail or runway; wall crane having a jib with a movable or fixed hoisting mechanism and operating on a runway attached to the side walls or columns of a building. Portable cranes have a revolving superstructure with power plant, operating machinery, and boom mounted on a fully mobile carriage not confined to a fixed path. Crawler, locomotive, and truck cranes are classified as portable cranes. Fixed cranes are nonmobile. Derricks and jib cranes are classified as fixed cranes.

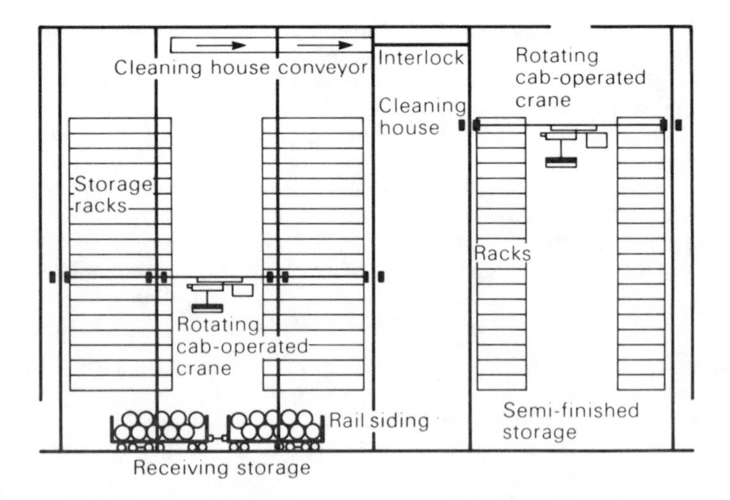

In this use, a rotating cab-operated underhung crane unloads railcars and places wire bundles in vertical racks until they are scheduled for cleaning.

cranes, top riding. Designed for loads greater than those for underhung cranes, these are the heavy-duty rectilinear traveling cranes that carry the main beam over the rails. In addition to the lifting motion, two horizontal movements at right angles to each other allow a load to be deposited at any point within the rectangle covered by the movement. **(See also cranes.)**

creep, metal. The deformation of materials under the combined effects of temperature, time, and constant tensile stress. The long-term effect of temperature is manifest as creep. If a tensile specimen is subjected to a fixed load at an elevated temperature, it will continue to elongate continuously until rupture finally occurs. While the rate of elongation is small, it is sufficient to be of great importance in mechanisms that operate at high temperatures for long periods of time. Such mechanisms as steam or gas turbines and high-temperature pressure vessels are cases where creep is very important. In general, the higher the temperature, the higher the rate of creep. Likewise the rate of creep depends on the applied stress. Creep takes place in three stages. In general, the alloying elements—nickel, manganese, molybdenum, tungsten, vanadium, and chromium—are helpful in lowering the rate of creep of steel. At high temperatures, coarse-grained steels seem to be more creep resistant than fine-grained steels, while below the lowest temperature of recrystallization the reverse is true. Steels having about 0.4% carbon are more creep resistant than those having higher or lower percentages of carbon.

critical path technique. This is a project management system involving engineering, construction, and maintenance of manufacturing facilities and installation. The critical path technique integrates all of the factors or building blocks of a project—manpower, money, time, materials, equipment, etc.—to allow management to develop a balanced, optimum, time-cost schedule ensuring timeliness and a minimum use of resources, and the means for management by exception. The critical path technique differs from the traditional methods in two fundamental ways: It separates planning from scheduling—planning consists in determining what tasks must occur in a project and their order of performance. Scheduling is the act of translating the plan to a timetable, and it relates time and costs directly. This shows that the minimum costs and the related time of an activity in a project can be shortened at some sacrifice in cost. The critical path technique starts with arrow diagramming, which incorporates all elements of a project. Operations, methods, and resources (time, money, manpower, equipment, and material) plus imposed conditions (design, delivery, approval, budget, completion date, decisions, etc.) are molded into a coordinated plan and model. Each activity, task or operation, is represented as an arrow. Each arrow indicates the existence of a task. The length or direction of the arrow has no meaning. Time flows from the tail to the head of each arrow. The arrows will then interconnect to show the sequences in which the tasks will be performed. The result is the arrow diagram. Three basic questions are asked and answered about each arrow: (1) What must be done before one can start this activity? (2) What can be done concurrently? (3) What must immediately follow this activity? With a thorough knowledge of the job, only these three questions need be answered to develop a complete network, which is a plan for a project based on logic. **(See also PERT and figure p. 94.)**

crucible process (steel making). High-grade tool steels and some alloy steels are still made by the crucible process, although the electric furnace is capable of making steel equal in quality to crucible steel. In the crucible process, wrought iron or good scrap, together with a small amount of high-purity pig iron, ferromanganese, the necessary alloying metals, and slagging materials are placed in a clay–graphite crucible covered with an old crucible bottom and melted in a gas or coke-fired furnace. After the charge is entirely molten, with sufficient time allowed for the gases and impurities to rise to the surface, the crucible is withdrawn, the slag is removed with a cold iron bar, and

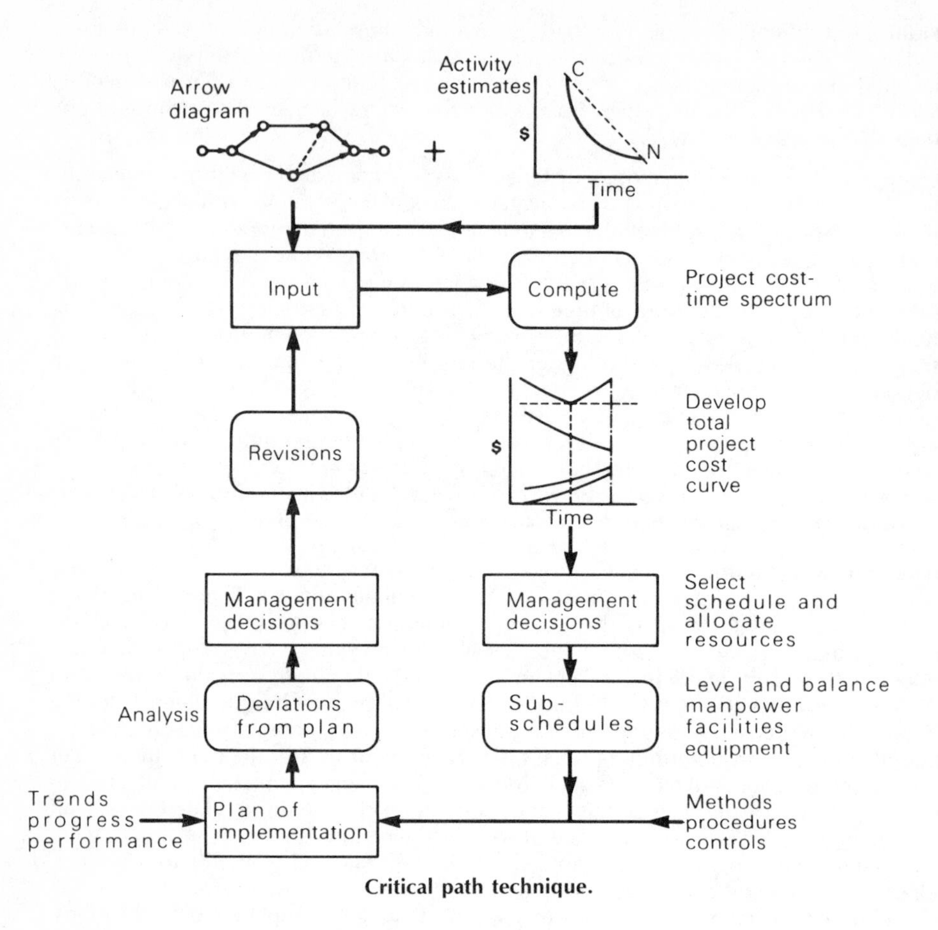

Critical path technique.

the resulting 50–100 lb (23–46 kg) of steel is poured into a small ingot, which is subsequently forged to the desired shape. The crucible process differs from other steel making in that little or no refining is included; the purity of the metal is dependent almost entirely on the purity of the material charged.

crusher. A device to break up large pieces of friable materials into smaller pieces. **(See also coal, crushers.)**

crystal diode. Consists of a tiny wafer of silicon or germanium and a platinum catwhisker, housed in a sealed capsule. The crystal acts as the cathode of the diode and the catwhisker acts as the anode. The diode characteristic depends on the unique property of a semiconductor that permits a relatively large flow of electrons with a small applied voltage in the forward direction, and only a very small flow of electrons in the reverse direction, even at much large applied voltage.

crystal formation, *n*-type. In order to make the pure intrinsic crystal a slightly better conductor, certain impurities are added to the crystal during the crystallization. Both germanium and silicon have four valence electrons and in their crystalline-lattice structure; these electrons are held in covalent bond. Impurities are selected which have five valence electrons. These include phosphorus (P), arsenic (As), bismuth (Bi), and antimony (Sb). These are classified as pentavalents. The process of adding these impurities is called "doping" and the impurities are "dopants." Only a small amount of dopant is used, on the order of one atom of impurity to 10 million atoms of germanium. The crystal now becomes an extrinsic crystal and the conduction characteristics of the crystal will change.

crystal formation, *p*-type. Beginning with a pure crystal of germanium or silicon, a minute quantity of a trivalent impurity is added during crystallization. Common trivalents that have three valence electrons are indium (In), gallium (Ga), and boron (B). The three valence electrons of the indium join in covalent bond with the germanium atoms. But the germanium atom is not satisfied; another electron is needed to complete the covalent bonding structure. There is, therefore, a positive site or hole that has a strong attraction for an electron, if one happened to be in the vicinity. The trivalent dopant creates many of these positive holes and conduction through this crystal is by holes. The crystal becomes a *p*-type crystal and its majority carriers are holes. A few electrons freed by electron-hole pairs will permit a very small conduction and the minority carriers in a *p* crystal are electrons. Since this dopant creates holes which will accept electrons, it is called an acceptor impurity.

cubic boron, nitride (CBN). Offering extreme edge durability, chemical and thermal stability on ferrous materials, and a low coefficient of friction, CBN is a near-ideal precision abrasive. Manmade, CBN is twice as hard as conventional abrasives. It is most suitable for applications where standard abrasives do not perform satisfactorily. Effective use of CBN wheels requires rigid machines, in good condition. Minimum vibration at any speed is essential to efficient CBN performance.

cupolas, metal casting. Essentially a vertical cylindrical steel shell with a refractory lining. It usually is supported on heavy steel or iron legs so that its lower end is 3-5 ft (0.9-1.5 m) above the floor. The lower end is closed by hinged doors, which may be opened to permit cleaning the inside. A sand bottom, 4-6 in. (10.2-15.2 cm) deep and sloping downward to the pouring spout, is prepared on top of the closed bottom doors. The pouring spout has a taphole 1-2 in. (2.54-5.08 cm) in diameter. After the sand bottom has been prepared, kindling wood is placed on the bottom and a bed of coke on top of this. The wood is ignited and the coke bed is increased until it is 18-48 in. (45.7-122 cm) in depth, depending on the size of the cupola. Natural draft is supplied through the tuyeres, which are openings around the periphery of the cupola about 15-20 in. (38.1-50.8 cm) above the top of the sand bottom. When the coke bed has been built up to the desired depth and is thoroughly ignited, alternate charges of coke and iron are added through the charging door located about half the distance up the height of the cupola. The metal charge may be either pig iron or scrap. About one part of coke, by weight, is used to eight or ten parts of iron. In addition, small amounts of limestone, or other materials, may be added to act as a flux and increase the fluidity of the metal. After the iron and coke charges have been built up to the level of the charging door, the charge is heated for 30 min or longer under natural draft before the forced draft is started. The natural draft openings are then closed and the blower is started. Under forced draft the temperature in the cupola rises rapidly, and the iron charge starts to melt and run down through the coke. Actual melting takes place in the zone just above the tuyeres. As the metal melts, it collects in the lower portion of cupola between the sand bottom and the slag spout, located slightly below the tuyeres. When sufficient amount of molten metal has collected, the air blast is turned off temporarily, the clay plug is punched out, and the metal is permitted to flow out into a pouring or holding ladle. When sufficient metal has been removed, the taphole is again plugged and the air blast is turned on.

curing ultraviolet. To cure decorative coatings on such materials as plastic tubes, paperboard for cartons, plastic containers for motor oil, beverage cans, metal lid sheet stock, and flexible packaging, ultraviolet energy (wavelength range 10–380 nm) is used. An ink or coating formulated for ultraviolet curing consists entirely of pigments and a resin monomer or prepolymer, most commonly acrylic. It does not need solvent. When the wet ink passes under ultraviolet lamps, the radiated energy polymerizes the acrylic resin almost instantaneously, forming a durable solid film. The method thus eliminates both the solvent and the gas problem with conventional inks and coatings. Ultraviolet curing also reduces total energy consumption. The amount of electrical energy required to run an ultraviolet oven is only 5–25% of the Btu equivalent in natural gas needed to run a baking oven.

cut-off machines. A variety of machines are used for cutting off bar stock and piece parts in manufacturing operations. Cold saw machines utilize a revolving metal saw blade mounted on an arbor with advance and retract movement, either hand operated or powered. A duplex cold saw cuts off both ends of shafts, axles, etc., to a given length. A multiple cold saw is used to cut a number of short lengths simultaneously. Automatic machines of this type are usually special. A vertical cold saw has the saw blade horizontal and arbor vertical. With special tables these machines can remove gates and risers from castings or do long, straight cuts on steel plates. An abrasive cut-off machine uses an abrasive wheel where a cold saw is impractical. High-speed machines, low-speed machines, wet and submerged machines, and portable machines are used in this classification. Organic bond wheels are employed and metal bar stock and tubing, plastics, glass, porcelain, etc., are readily cut off. Hardened tool steels, Stellite, etc., can be cut. Resinoid wheels are used for dry cutting on large sections where burn and burring are unacceptable. Carbon, plastics, and tile are usually cut wet at up to 16,000 sfpm (4877 smpm). Rubber-bonded wheels are used for cutting off glass tube, porcelain, light wall metal tube, and plastics, either wet or submerged, at speeds as low as 4000 sfpm (1219 smpm). The original cut-off machine was merely a special-purpose lathe that rotated the bar stock and severed it to desired length by means of a cut-off or parting tool. These are still made and used regularly where the economics dictate. Also, other machines used for cutting off are the friction saw, which cuts metals by virtue of creating friction between a large saw blade or disk to meet the material so the teeth or notches removes it; the band saw, which uses a continuous band blade operating over a pair of large wheels with hand or power feed through the blade and horizontal self-feeding or vertical styles; and the common hacksaw, which uses straight heavy blades with reciprocating motion and power feed.

cyaniding metal heat treatment. Imparts a file-hard and wear-resistant case to steel by immersing it in a molten cyanide salt bath for a time and then quenching it. A cyanide case is seldom over 0.010 in. (0.254 mm) thick, and carburized cases are usually thicker. However, cyaniding requires lower temperatures [usually below 1600°F (870°C)] and less time (30–60 min) than carburizing.

D

data bank (computer). A centralized computer storage facility containing extensive and detailed data on individuals, groups, corporations, etc. A collection of libraries of data. Specifically, one line of an invoice may form a record; a complete set of such records may form a file; the collection of inventory control files may form a library; and the libraries used by a business organization are known as its data bank.

DDAS (see digital acquisition system)

DDC (see direct digital control)

decarburization (see argon–oxygen decarburization)

deep drawing, metals. A forming process used to shape metals into bowls, cups, panels, and similar shapes. In deep drawing the metal is deformed by flowing radially inward and then turning the corner to become the wall of the cup. The amount of drawing in a single operation should be such that the ratio of the height to diameter of the article formed does not exceed 1. If this ratio exceeds 1, the article must be made by redrawing in one or more steps. By reverse drawing, or turning a cup inside out on the redraw operation, extraordinarily large reductions can be made. High-pressure and very-high-speed processes of formation have been introduced. Under very high pressure the ductility of metal considerably increases, permitting the combination of otherwise separate operations. High-velocity-impact forging machines are used to form thin-wall parts in one blow from a single slug of metal. Such machines can exert up to roughly 2×10^6 ft-lb (2.7×10^6 J) of energy with every blow with cycle times from 5 to 15 s. They are also used for producing close-tolerance no-draft forgings in alloy steels, stainless steels, maraging steels, titanium alloys, and some superalloys. Explosive forming involves the application of very high pressures ranging from 100,000 to over 1,000,000 psi (6.9×10^5 to 6.9×10^6 kPa), over a very short period of time (microseconds). In most industrial processes this is accomplished by producing an intense pressure pulse in a liquid medium caused by detonation of an explosive or explosion of a wire by a sudden passage of very high currents. The shock wave generated in the liquid forces a metal blank against a shaping die, causing the required deformation. **(See also explosive forming.)**

deep lane storage. Storage depth that is greater than two loads deep on one or both sides of the aisle.

degreasing, spray. If insoluble matter cannot be removed by immersion in boiling solvent, spray degreasing may be necessary. In this procedure the work is first lowered into the vapor to remove oil and grease, then sprayed with warm solvent from a lance, and finally given a vapor rinse to remove all traces of oil and grease. To minimize solvent loss, it is essential that spraying be done below the vapor level of the degreaser in a manner that does not disturb the vapor level, and baffles or screens must be placed to prevent the rebound or ricochet of droplets of solvent into the area above the vapor level.

degreasing, vapor-phase. A vapor-phase degreaser is a tank containing a quantity of solvent that is heated to its boiling point by either electricity or steam. The solvent vapor rises and fills the tank to an elevation determined by the location of a cold-water condenser. The vapor condenses at this point and returns to the liquid sump. The tank extends above the condenser to minimize air currents inside the tank. The work is lowered into the vapor. The solvent condensing on the cool metal surface dissolves and washes away oil and grease film. This action is continued until the metal reaches the temperature of the solvent vapor, whereupon it becomes dry. The length of time required depends on the shape, size, surface area, temperature, and specific heat of the parts to be cleaned. If the material is not allowed to remain long enough to reach the temperature of the vapor, considerable solvent will be dragged out with the dripping parts. Variations of the vapor-phase degreaser are available for special purposes. In one type both a liquid immersion tank and a vapor phase are available. This equipment is useful when articles to be degreased have intricate shapes and are heavily contaminated with dirt and grease. Immersion of the part in boiling solvent prolongs solvent action and affords mechanical scrubbing by the boiling liquid. A number of solvents are used with vapor-phase degreasers including trichloroethylene, perchloroethylene, methyl chloroform, ethylene dichloride, and certain Freons. The degreaser must be designed and applied in relation to the specific solvent in use.

denier. A unit of yarn number. The number of unit weights of 0.05 g per 450 m length. Denier is equal numerically to the number of grams per 9000 m. It is not a measure of size. It is a measure of density.

detector through-beam. A method of object detection in which a source is mounted on one side and a detector on the other. The object is detected when it "breaks" the beam. **(See also photoelectric control.)**

detergents, soaps and syndets. Cleansing agents include both soaps and syndets (synthetic detergents). The molecular structure of synthetic detergents and soaps are similar, being characterized by the presence of a long nonpolar, hydrophobic ("water-hating") hydrocarbon tail and a polar, hydrophilic ("water-loving") head. Soaps when used in hard water precipitate as insoluble calcium, magnesium, and iron salts of the long-chain fatty acids, but synthetic detergents are not precipitated in the presence of these ions. Solubility in hard water is one of the principal advantages that a synthetic detergent has over a soap. On the other hand, the large-scale use of syndets within the past decade has presented a serious disposal problem. Many of these detergents have an aromatic ring in the hydrocarbon segment of the molecule. This type is generally classified as alkyl benzene sulfonate (ABS) detergents. The benzene ring is very resistant to oxidation.

dew point. In many industrial processes the dew point is a more significant measurement than relative humidity. The dew point is the temperature at which a given sample of moist air is fully saturated and begins to deposit dew. A typical measuring instrument used is a gold-plated mirror surface, which is bonded to a copper thermistor holder. This assembly is chilled by a Peltier effect thermoelectric cooler. The air or other gas being measured for dew point is passed by the mirror. A neon lamp is beamed on the mirror, which reflects the beam toward a photoelectric resistor. As dew forms on the mirror and clouds it, there is a change in the amount of light reflected. This change is detected by an optical-sensing bridge. The result is a change in the electrical resistance, which becomes the input for the amplifier. This effects a proportional change in the power supplied to the cooler. The thermistor in the cooling assembly measures the dew-point temperature. The heat sink serves as a thermal capacitance.

diamagnetic materials (electronics). Term used for materials that become only slightly magnetized, even though under the influence of a strong magnetic field. When a diamagnetic material is magnetized, it is in a direction opposite to that of the external magnetizing force. Diamagnetic materials have a permeability of less than 1 and include such metals as copper, silver, gold, and mercury.

diamond (see abrasives)

diamond pyramid hardness test (see Vickers hardness test)

diamonds, industrial. Diamond is used extensively for grinding tungsten carbide tooling, glass, and nonferrous metals, and for sawing and drilling concrete and stone. Tungsten-filament wire for electric lights and millions of miles of other kinds of wire are drawn through dies. Three natural types are employed: (1) crystalline and cleavable diamonds, more or less off-grade and off-color; (2) bort, translucent to opaque, gray or dark brown, with a radiated or confused crystalline structure, and (3) carbonado, frequently known as black diamond or carbon, which occurs in an opaque, tough, crystalline aggregate without cleavage. An extensive use of industrial diamonds is in the grinding of shaped carbide or ceramic tips for turning and boring tools. Bonded-diamond wheels for grinding hard, abrasive materials such as cemented carbide tools, ceramics, and glass are made from synthesized diamond powder or crushed bort, held by a matrix of resin, vitreous material, or sintered metal. Tools set with industrial diamond are a practical necessity in truing and dressing worn abrasive wheels in all industries where precision grinding is performed. Diamond lathe tools, because of the hardness and high heat conductivity of diamond, allow greatly increased machining speeds in turning nonferrous metal parts to close tolerances and fine finishes. Since the continuous drawing of wire through reduction dies causes much wear, dies for accurate wire drawing are made of diamond. Diamond bits for drilling through hard rock formations for oil, gas, or water, for boring blast-charge holes in mines and quarries, and for drilling holes in and taking cores from concrete structures are set with small whole diamonds in a sintered metal bond.

diatomite. A light-colored, lightweight, friable sedimentary rock composed of siliceous shells of microscopic aquatic plants called diatoms. Diatomite may range from yellowish tan to brown but is usually light in color. The material is soft, finely granular, and more or less chalky. The pore space is partly or wholly filled with mechanically held water. Commercial grades of diatomite have the following range of chemical composition: SiO_2, 85–92%; Al_2O_3, 4–10%; Fe_2O_3, 0.8–2.0%; CaO, 0.1–2.0%; MgO, 0.1–2.0%; alkalies, 0.2–1.5%. The alumina comes from a small amount of clay. A large percentage of powdered diatomite's use is as a filter aid. Pulverized diatomite is extensively used as a filler in such products as paper, molded plastics, and synthetic rubber. It also finds wide application in paint. The low thermal conductivity of diatomite makes it useful as a heat insulator. Bricks may be cut directly from the quarry face, and used after drying and sizing. Usually the material is crushed, blended, pressed into bricks, and calcined; or diatomite powder may be combined with inorganic binders and asbestos fibers to produce blocks and pipe insulation. Diatomite insulation is applied to ovens, furnaces, kilns, and similar equipment.

die casting (see casting)

die cutting. Blanking. Also, cutting shapes from sheet stock by striking it sharply with a shaped knife edge known as a "steel-rule die." Clicking and dinking are other names for die cutting of this kind. **(See also blanking.)**

dielectric constant. The ability of a material to store electrostatic energy relative to the storage ability of a vacuum. A capacitor is an electrical device for the temporary storage of electrons, and is essentially two parallel conducting plates with a dielectric between them. The electrons collect on either plate under the pressure of a voltage. The storage capacity of a capacitor is termed the capacitance. One of the variables determining capacitance is the dielectric constant of the dielectric material of the capacitor, the capacitance varying directly with the dielectric constant. A vacuum is assigned a dielectric constant of 1.0, just as the magnetic permeability of a vacuum is 1.0. Like most physical constants, the dielectric constant is a somewhat variable constant, generally decreasing at very high frequencies. The dielectric constant for most plastics such as polyethylene is in the range of 2–3.0. For most ceramic materials, values lie between 4 and 10.

dielectric heating. The heating of a nominally insulating material in an alternating electric field due to its internal losses.

dielectric strength. Since a dielectric is an insulator, dielectric strength means the insulating capacity of a material against high voltages. A sufficiently high-voltage field across an insulator will cause it to break down and conduct. Dielectric strength is usually expressed as volts per mil thickness. Dielectric strength of ceramic insulators, such as the alumina of a spark plug body or high-voltage porcelain, ranges from 50 to 400 V/mil thickness. Refrigerants are required to have dielectric strengths of over 25 V/in.; for most of the Freons it is over 400 V/mil. The highest values of dielectric strength are given by such plastics as polyethylene and Teflon, with the value in the range of 1000–5000 V/mil.

die threading. A machining process for cutting external threads in cylindrical or tapered surfaces by the use of solid or self-opening dies. Die threading is a slower method of producing external threads than thread rolling, but is faster than single-point threading in a lathe. The hardness of the work metal limits the application of die threading. Attempting to die-thread metals harder than Rockwell C 36 usually causes excessive wear or breakage of tools. Single-point threading or thread grinding is recommended for metals harder than Rockwell C 36. The size of workpiece seldom limits application of the process. Die threading has been used to cut 280 threads per inch in rods 0.017 in. (0.432 mm) in diameter. At the other extreme, 3 threads per inch have been die cut into 24 in. (61 cm) pipe for the petroleum industry.

diffusion (crystallization). The occurrence of many mass-transport phenomena within a solid and between solid–solid, solid–liquid, and solid–gas are controlled by diffusion. Diffusion is a movement of atoms, ions, or molecules as the result of thermal agitation. Although thermal agitation may cause a random movement of atoms, there is a certain drift in the direction of their concentration gradient. Diffusion is very rapid in gases and liquids, but it is a slow process in solids, where the movement of atoms is greatly restricted. In spite of this, diffusion is responsible for most solid–solid state reactions. Diffusion in solids occurs by the relative displacement of atoms or molecules within the solid. For any atom or molecule to move from one position to another within the crystal lattice, it must overcome the potential-energy barrier and must have a site into which it can go. This requires a certain activation energy that originates in the thermal vibration of atoms or molecules, and the presence of vacancies or other defects in the crystal. In the case of interstitial atoms, they jump from one interstitial site to the next by squeezing between the atoms of the crystal lattice. Thus the number of diffusing atoms will correspond to the number of activated atoms passing over the energy barrier in a unit time. Since diffusion in solids occurs by means of such defects as vacancies and interstitials, the diffusion mechanisms are classified as vacancy mechanism, interstitial mechanism, and interstitialcy mechanism. Vacancy mechanism involves the diffusion atom moving by jumping into a neighboring vacant lattice site. Interstitial mechanism occurs when an atom passes from one interstitial site to one of its nearest-neighbor interstitial sites without permanently displacing any of the matrix atoms.

digital computer. A computer in which discrete representation of data is mainly used. (1) A digital computer uses 1 and 0 as symbols and operates on the fact of the direct relationship between these numbers and the on and off condition at a specific place in the computer circuitry. An ''analog'' computer operates on a wide range of conditions, one of which is the amount of voltage. (2) The digital computers operate on representation of real numbers or other characters coded numerically. The digital computer has a memory and solves problems by counting precisely, adding, subtracting, multiplying, dividing, and comparing. The ability of digital computers to handle alphabetical and numerical data with precision and speed makes them best suited for business applications.

digital data acquisition system (DDAS). Used in plant-floor automation to collect digitized data, occasionally from limit switches and scanners, but most often from information inserted in workstations by means of punched cards and readable tags, as well as information inserted by the operator, using dials or a keyboard. The collected information is used on dynamic inventory control, for work-flow monitoring and control, for work measurement, and for pay determination. The DDAS is different from the data acquisition system (DAS) in that the input–output system is not noise limited but is bandwidth limited. A single installation may be spread over many buildings including thousands of input stations and many miles of cabling. Data-handling requirements of 100K bit/s to several megabits/s are not unusual. Another difference is that the DAS application is based on a set of semistandardized operations; for example, signal scaling and correction limit comparison, logging, and recording. In the DDAS applications the type of data collected and the use to which they are put are less well defined, thus demanding complete freedom in programming. As a result while the local workstations are very simple special-purpose digital devices, a general-purpose central computer is clearly needed.

diode. A diode is a two element unilateral conductor. A diode has two connections, anode and cathode, with current flowing in only one direction. It has a high resistance to a current in its reverse direction. There are several methods of forming a diode and they are used individually or sometimes in combination. The desired result is to produce a surface which separates the *n*-type and *p*-type crystals. Diodes are made by alloying an *n*-type germanium crystal with a metal base such as aluminum or indium. In the rate growth method, a crystal is drawn from a molten semiconductor material containing *n*- and *p*-type impurities. The concentration of impurities in the drawn crystal depends on the rate at which the crystal is formed, so that *pn* junctions can be made by changing the growth rate periodically. Another method is by diffusion in which a solid crystal is heated in an atmosphere containing the desired impurity. The gas diffuses gradually into the base crystal to form a layer of the opposite type. This method is used in the manufacturing of integrated circuits. Commonly encountered diodes include vacuum tubes, gas-filled rectifiers, such as the mercury arc rectifier, x-ray tubes, photocells, and welding arcs.

diodes, solid-state. In solid-state *pn* diodes current flows in one direction with forward bias but very little flows with reverse bias. The difference in resistance between the two conditions is sufficiently great (particularly for silicon diodes) to permit the solid-state diode to perform virtually all the functions of the vacuum tube diode in electronic circuitry.

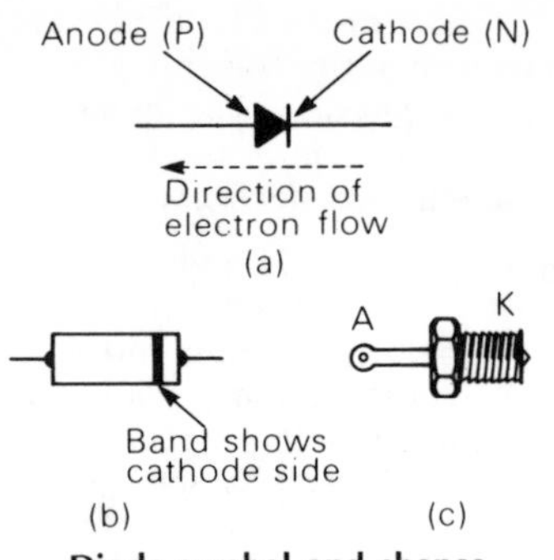

Diode symbol and shapes.

diode, tunnel. The tunnel diode has a region of negative resistance in its current–voltage characteristics. The tunnel diode differs from the regular *pn* junction diode in that the semiconductor used is much more heavily "doped" and has therefore a relatively large number of charge carriers. For small forward biases, the narrowness of the depletion layer allows charges to "tunnel" across the junction even though they do not have sufficient energy to overcome the counter emf of the junction. As the forward bias increases, the tunnel-conduction mechanism becomes less effective until eventually the potential at which the normal conduction process occurs is reached. When the diode is forward biased about 150 mV, the negative-resistance characteristic can be utilized. The characteristics of various tunnel diodes will differ significantly only in the magnitude of the peak current.

dip plating (see immersion plating)

dip soldering, soldered connections. The process whereby assemblies are brought into contact with the surface of molten solder for the purpose of making soldered connections.

direct digital control (DDC). Use of a digital computer to provide the computations for the control functions of one or multiple control loops used in process control operations.

direct digital controller. A special-purpose machine that replaces an analog set-point controller, such as a flow-rate or temperature controller. Such controllers compare a measured value to a set value and compute a correction signal. The desirability of a digital implementation is indicated by the need for accuracy, for a means to change set points dynamically, and for the ability to modify algorithms easily. Since the cost of even a special-purpose digital machine would be greater than that of the analog device, all designs for DDCs have been based on the capability of one processor to handle at least 16 and as many as a thousand control loops.

direct labor materials handling ratio (DLMH). This is a measure of the proportion of direct labor time spent in performing materials handling. The DLMH ratio = (materials handling time spent by direct labor) ÷ (total direct labor time).

direct numerical control (DNC). In this case the program is stored in the computer memory rather than on tape. The tape reader is eliminated along with the significant cost of operating and maintaining a tape system. The large cabinets of electronic gear used for processing the program no longer occupy valuable manufacturing space. These items, which were previously required for each machine tool, are replaced by a single computer that controls a group of a dozen or more machine tools directly via cable connections. DNC not only offers higher operational reliability, but it also opens the door to many new and promising manufacturing innovations.

disk memories (computer). Millions of bits can be stored on horizontally stacked disks that look like records in a juke box. Each disk contains a number of tracks in each of which thousands of bits can be recorded. A comb of vertical arms moves in and out of every disk simultaneously, stopping at the desired track, and reading or writing information from whatever disk is specified.

display console. In computer graphics, a console consisting of at least one display device (CRT) and usually one or more input devices such as an alphanumeric keyboard, functions key, a tablet, a joy stick, a control ball, or a light pen.

distillate, heavy. Furnishes lubricating oils, heavy oils for various purposes, and waxes. Heavy distillates are also hydrocracked to lighter distillate fuels and to gasoline.

distillate, intermediate. Includes gas oil, heavy furnace oil, cracking stock, diesel fuel oil, absorber oil, and distillates cracked and reformed to produce gasoline of adequate quality. Often distillates are blended with heavy tar to reduce the viscosity of asphalt so that it can be readily applied as road oil; such material is known as cutback asphalt.

distillates, light. These embrace naphthas and refined oils, aviation gasoline, motor gasoline, petroleum solvents, and jet-fuel kerosene. Gasoline is the most important product. Aviation gasoline with an octane number of 100 or higher is composed of about one-third alkylate derived from isobutane and gaseous olefins, blended with catalytically cracked gasoline and suitable crude distillate to which several cubic centimeters of TEL per gallon is added to reduce the knocking tendency.

dolomite. A sedimentary rock containing more than 50% of the minerals calcite and dolomite, in which the mineral dolomite is more abundant than calcite. There are all gradations between limestone and dolomite. Dolomite rock with 10% or more of calcite is calcitic dolomite. A commercially valuable stone, high-purity dolomite contains more than 97% total carbonates, all but 3 or 4% being mineral dolomite. Dolomite has many uses including (1) as crushed stone, for concrete aggregate, road material, railroad ballst, and sewage filter beds; and in finer form, for poultry grit, stucco, manufactured sand, coal-mine dust, and whiting; (2) as a fluxing agent in the smelting and refining of iron and other metals; (3) as a soil conditioner, to correct soil acidity and promote plant growth; (4) as a source of lime, a basic chemical of prime importance with a long list of industrial applications; (5) as a chemical raw material in glassmaking, acid neutralization, and other processes; (6) as dimension stone.

down-ender. A device used to rotate an object from a position on its end to a position on its side. **(See also up-ender.)**

double deep storage. Loads that are stored two deep on each side of the aisle.

double-shot molding. A means of turning out two-color parts in thermoplastic materials by successive molding operations.

drawing. (1) In forging, an operation of working metal between flat dies to reduce the cross section and increase length. (2) In stamping the process of pulling material through a die to reduce the size and change the cross section or shape.

drawn product. A product formed by pulling material through a die.

drilling. Where holes 1-in. (2.54-cm) or under are needed, production with spiral or straight fluted drills is common unless the method by which the part is processed originally permits production of the hole along with other features, as with die casting, molding, etc. With drilled holes the major consideration is hole depth. Single-operation drilling with ordinary drills allows the production of hole depths up to about five times the drill diameter. Where a series of drilling spindles can be used successively, holes can often be drilled to depths of eight to ten times diameter. Such holes, however, often may not be sufficiently straight to be satisfactory. By using chip breaker attachments or internal oil hole drills with black oxide treatment, holes to as much as 20–25 diameters in depth have been produced. Conventionally, however, holes over $^1/_8$-in. (3.2 mm) diameter to 50 diameters or more in depth are produced on deep hole drillers with deep-hole, crankshaft, or carbide gun drills. Such drills run up to 2 in. (5.08 cm) in diameter and can produce holes 60–70 diameters in depth using high-pressure coolant flushing. Maximum depth for the larger sizes has run to 12 ft (3.7 m). Common spiral fluted drills are available in five series: By decimal size in flat drills (from 0.002 to 0.080 in. in half-thousandth steps); by numbers from 80 to 1 (0.0135 to 0.228 in. diameters); by letters from A to Z (0.234 to 0.413 in. diameters); by fractions from $^1/_{64}$ to $3^1/_2$ in. (in steps of $^1/_{64}$ in. from $^1/_8$ to $1^3/_4$ in. diameters, steps of $^1/_{32}$ in. from $1^{25}/_{32}$ to $2^1/_4$ in. diameters, and steps of $^1/_{16}$ in. from $2^5/_{16}$ to $3^1/_2$ in. diameters); by millimeters (available in steps of 1 mm from 1 to 76 mm, steps of 0.5 mm from 0.50 to 45 mm, steps of 0.1 mm from 0.10 to 10 mm, and 0.05 mm from 0.10 to 2.80 mm).

drilling equipment. Machines available for performing drilling operations are many and varied. The common floor drill press is widely used for drilling plain holes in simple, small, and medium-size parts, while for larger size work the heavy-duty radial drill is usually employed. Where mass production of holes is required, automatic machines are normally the only economical solution. These may be in the form of multiple-spindle units for producing many holes at once, or for small, intricate parts, an automatic-sequence-type machine capable of being retooled to suit numerous design requirements. Many additional operations such as tapping, light milling, reaming, and screw insertion as well as assembly by spinning over or staking can be performed with these machines. Complicated drilled parts or those requiring a large number of drilled holes accurately located are produced very often on "package unit" or transfer machines, which allow positioning of each component for automatic drilling as required. Similar machines are also used widely for large and intricate parts to produce all related holes at one loading. With these machines many additional operations such as reaming, counterboring, and light milling, can also be performed. Where abnormally long holes are needed, such as oil passage holes in camshafts, special deep-hole drilling machines are used. Such machines range in size up to those utilized for drilling gun barrels.

dry coloring. Method commonly used by fabricators for coloring plastics by tumble blending uncolored particles of the plastic material with selected dyes and pigments.

drying, microwave. By using microwave energy rather than steam to remove moisture, drying costs can be reduced along with the size of the machinery required. One of the unique features of microwave drying is that variations in moisture content can be automatically sensed to trigger an instant change in microwave-energy application, thereby ensuring more-consistent production quality. Typical applications of microwave-energy transfer are the drying of foundry sand core molds; processing potato chips; drying paper coatings; sterilization and pasteurization of wine, beer, and milk; freeze drying of food; the drying of printing inks; and in such advanced areas as the generation of plasmas for electronics manufacturing techniques and medical uses.

ductile iron (see cast iron, nodular)

ductility, metals. The ability of a material to be elongated plastically without fracturing is known as ductility. Materials usually do not elongate uniformly along their entire length when loaded, even though the length is originally of uniform cross section. It is common practice to express ductility in terms of the measured elongation over a specified gage length. One of the major problems in certain metalforming processes is to obtain full use of the total ductility that is available in a material by avoiding or preventing excessive localized deformation. Another indication of ductility is the reduction in area that occurs when necking takes place. The percentage decreases in area from that which originally existed are sometimes used as an indication of ductility. When a material fails with little or no ductility, it is said to be brittle. However, it should not be concluded that brittleness is associated with lack of strength. Some brittle materials are very strong. Likewise, brittleness should not be associated with a type of fracture. Some materials that actually are fairly ductile can be made to fail so that the fracture has much the same appearance as the fracture of some other materials which are very brittle. Another factor that has to do with the ductility of a material is malleability. This is the ability of a material to be deformed by mechanical working without rupture.

ductility test, Olsen. A cupping test in which a piece of sheet metal, restrained except at the center, is deformed by a standard steel ball until fracture occurs. The height of the cup at time of fracture is a measure of the ductility. **(See also Erichsen test.)**

duplex cold saw (see cut-off machines)

dyes, anthraquinone. The anthraquinone dyes have a *p*-quinoid structure in common with two other benzene rings in a fused-ring molecule. An example of an anthraquinone dye is alizarin.

dyes, azo. This class represents the largest and most important group of dyes. The characteristic feature of each dye in this class is the chromophoric azo group (–N = N–) which forms part of the conjugated system and joins two or more aromatic rings. Azo dyes are prepared by coupling a diazotized aromatic amine with another aromatic amine or with a phenol or a naphthol. This simple reaction can be carried out directly within the fabric, and results in a product complete with chromophore, auxochrome, and conjugated system. One of the first azo dyes to be prepared was para-red, obtained from diazotized *p*-nitroaniline and *β*-naphthol. More than one aromatic ring, when included in the coupling reaction, extends the conjugated system and deepens the color of a dye. For example, diazotized benzidine (4,4′-diaminobiphenyl) will couple with 1-naphthylamine-4-sulfonic acid to produce a dye called Congo red. Congo red is also used as a chemical indicator.

dyes, direct or substantive. Dyes of this class contain polar groupings, acidic or basic, which are able to combine with polar groups in the fiber. Such dyes color a fabric directly when the latter is immersed in a hot, aqueous solution of the dye. Direct dyes are particularly well adapted to the dyeing of silk and wool. Picric acid and Martius yellow are two examples of direct dyes. Both compounds are acids and can combine with the free amino groups found in protein fibers. Nylon, a polymide, also may be dyed by this method.

dyes, disperse. Dyes classified under this category are those that are soluble in the fiber but not in water. Disperse dyes are used in the coloring of many of the newer synthetic fibers. These are sometimes referred to as hydrophobic ("water-fearing") fibers and usually are lacking in polar groups. The dye in the form of a fine divided dispersion is dissolved in some organic compound (often a phenolic) and the absorption into the fiber is carried out in a surface-active bath (soap solution) at high temperatures and pressures. Examples of disperse dyes used for acetate rayons, Dacron, Nylon, and other synthetic fibers are Celliton Fast Pink B (1-amino-4-hydroxyanthraquinone) and Celliton Fast Blue B (1,4-*N*,*N*-dimethylaminoanthraquinone).

dyes, indigoid. The indigoid dyes have as their identifying feature two conjugated groups. Indigo is an example of a dye of this class. It does not occur naturally in a colored form but rather as a colorless plant glucoside, indican. The latter yields glucose and indoxyl when hydrolyzed. Indoxyl, when exposed to air, oxidizes to indigo.

dyes, ingrain. Sometimes called developed dyes, these are dyes that develop within the fabric itself. The azo dyes are good examples of ingrain dyes. The cloth to be dyed first is immersed in an alkaline solution of the compound to be coupled, usually a phenol or naphthol. A second immersion in a cold solution of a diazotized amine causes the couple reaction to take place within the fabric. Dyes produced in this manner were referred to at one time as ice colors because of the low-temperature requirements for stability on the part of the diazonium salt.

dyes, mordant. This class of dyes includes those that can form an insoluble, colored salt (called a lake) with certain metallic oxides. Mordant dyeing is one of the earliest methods practiced for anchoring color to a fabric. Mordant dyes can be used on silk and wool or on cotton fibers. The fabric appears colored because the colored precipitate is bound to and envelops each fiber. The oxides of aluminum, chromium, and iron usually are used for mordanting colors. Alizarin is an example of a mordant dye.

dyes, nitro and nitroso. Dyes of this class are the *o*- or *p*-nitroso, nitrophenols, and the nitroso and nitronaphthols. One of the oldest nitro dyes is the high explosive, trinitrophenol (picric acid).

dyes, organic pigments. A dye must be absorbed by the material to be colored; a pigment is applied to the surface. Pigments are not water soluble but, like the pigments in paints, are suspended in the vehicle. Pigments are used mainly in decorative coatings, in printing inks, and for imparting color to plastics. If the chemical structure of a pigment can be modified to confer upon it water solubility, then it may serve as a dye. The phthalocyanines comprise an important group of pigments. One of these, copper phthalocyanine, is prepared by heating phthalonitrile with copper at 200°F (93°C).

dyes, triphenylmethane. The triphenylmethane dyes can be identified from a common structural feature in which a central carbon atom is joined to two benzene rings and to a *p*-quinoid group. Malachite green is an example of a dye in this class and is one that has been prepared as a laboratory exercise by students for many years. Malachite green is prepared by condensing benzaldehyde with dimethylaniline to give a colorless leuco base. Oxidation of the leuco base followed by acid treatment yields the dye. The familiar phenolphthalein, an acid–base indicator rather than a dye, also has a triphenylmethane structure. It is prepared by condensing phthalic anhydride with phenol. Indicators and some dyes are able to change color with a change in the hydrogen ion concentration of their solutions. Such changes in color are possible because an indicator is itself a weak acid or a weak base and enters into a salt-producing reaction. Phenolphthalein in acid solutions is colorless but turns red in alkaline solutions. A strong alkaline solution converts the indicator into a triphenylcarbinol derivative. Without an extended conjugated system, the compound again is colorless.

dyes, vat. A vat dye is a substance that in a reduced form is water soluble and may be colorless. In this form it is introduced into the fabric. The fabric, after the dye is absorbed, is removed from the "vat" and exposed to air or treated with a chemical oxidizing agent. This step oxidizes the dye to a colored, insoluble form. The ancient dyes indigo and Tyrian purple are classic examples of vat dyes.

dynamic balancing. The improvement by machine of the mass distribution of a part that rotates about a predetermined axis in such a way that the uncompensated centrifugal forces resulting from unbalanced masses remain within permissible tolerances. These centrifugal forces naturally cause cyclic radial loads that impair efficiency and may cause noise or even damage. Imbalances of this type may be caused by production tolerances in centering the workpiece, eccentricity of ball bearing races, lack of homogeneity of the material or nonuniformities in machining, lack of symmetry of the electric winding on a rotor, etc. Displacement of the center of gravity and of principal inertia axis from the rotational axis is a direct consequence of this imbalance. A parallel displacement of these axes is called "static imbalance." This can be balanced by a single compensating weight. An inclination of these axes, that is to say, wobbling of the mass axis about the rotational axis, is the more general condition described as "dynamic imbalance." Both forms of displacement can be compensated simultaneously by correcting the mass distribution in two planes of the workpiece. Balancing machines, therefore, measure the magnitude and angular position of the imbalance in the workpiece in two planes while the workpiece rotates; compensate the imbalance with suitable mechanical means (addition of weight or removal of metal (weight) by suitable machining either within or outside the balancing machine); and check the residual imbalance on the balancing machine.

Dynel fibers. Dynel, a modacrylic fiber, is made in staple fiber and in tow by Union Carbide Chemicals Co. Much of the acrylonitrile is made from acetylene. Acetylene has been traditionally made by the action of water on calcium carbide, but more is coming from petroleum distillation. In the Sohio process the starting materials are propylene and ammonia. The propylene is oxidized to acrolein, ammonia adds on to it, and the complex is dehydrated and dehydrogenated to acrylonitrile. In manufacturing the acetone solution is deaerated, filtered, and then extruded by the usual wet spinning technique into a water bath where it coagulates to form continuous strands. After spinning it is stretched hot as much as 1300% and then annealed by heat treatment; it is crimped and cut to staple; alternatively it is supplied as tow; this tow is then cut into staple lengths and crimped. Dynel is a light-cream fiber; it can be bleached nearly white. Its filament cross section is irregular and ribbon-shaped. Tenacity ranges from 2.5 to 3.5 grams per denier, depending on filament denier, and the elongation varies correspondingly from 42 to 30%. The outstanding characteristic of Dynel is its extremely good chemical resistance; it has very good resistance on long exposure to weak or moderately strong solutions of inorganic alkalies, acids, and salts, and generally, outstanding resistance to organic chemicals. The chemical resistance of Dynel makes it very suitable for protective clothing. Because of its resistance to chemicals, it has been used for chemical filter cloths, acids, and acid dye liquors. It has also been used for diaphragms in copper-plating plants; for similar reasons it has found use in bags for collecting anode sludge, for holding water-softening chemicals, and in dye and laundry nets. Dynel is also used for paint roller covers. One of the most successful outlets has been in blankets which are easily washable, impervious to damage by disinfectants, and easy to clean. Its fire resistance is responsible for use in furnishings and drapery products for ships, hotels, aircrafts, etc. Its mothproof and nonallergenic properties have enabled Dynel to be used for filling cushions, pillows, and eiderdowns.

E

earing. A characteristic of sheet to form ears when deep drawn or spun.

ears. Wavy symmetrical projections formed in the course of deep drawing or spinning as a result of directional properties or anisotropy in sheet. Ears occur in groups of four or eight with the peaks of the projections located at 45° and/or at 0° and 90° to the rolling direction. Degree of earing is the difference between average height at the peaks and average height at the valleys, divided by average height at the valleys, multiplied by 100, and expressed as a percentage.

eddy-current testing. An electromagnetic nondestructive testing method in which eddy-current flow is induced in the test object. Changes in the flow caused by variations in the object are reflected into a nearby coil or coils where they are detected and measured by suitable instrumentation.

elasticity. A material is called elastic when the deformation produced in the body is wholly recovered after removal of the force. The relation between stress and the corresponding strain in the elastic range of the material is governed by Hooke's law, which states that stress is proportional to strain and independent of time. The law generally applies to most elastic materials for very small strains. For an isotropic material each stress will induce a corresponding strain, but for an anisotropic material a single-stress component may produce more than one type of strain in the material. It follows from Hooke's law that the ratio of stress to strain is a constant characteristic of a material, and this proportionality constant is called modulus of elasticity. Since there are three main types of stress—tension, compression and shear—there will be three corresponding moduli of elasticity.

elastic limit, material. This limit is the greatest stress a material is capable of developing without incurring permanent deformation on removal of the stress. The proportional limit is the greatest stress a material is capable of developing before it begins to deviate from Hooke's law. Hooke's law states that the strain is directly proportional to the stress. For many materials the difference in magnitude between the proportional and the elastic limit is very small.

elastic moduli for various materials. The elastic moduli for various materials are listed in the accompanying table.

Material	E^a		G		Poisson's
	GN/m²	10^6 psi	GN/m²	10^6 psi	Ratio, μ
Cast iron	110.3	16.0	51.0	7.4	0.17
Steel (mild)	206.8	30.0	81.4	11.8	0.26
Aluminum	68.9	10.0	24.8	3.6	0.33
Copper	110.3	16.0	44.1	6.4	0.36
Brass 70/30	100.0	14.5	36.5	5.3	
Nickel (cold drawn)	213.7	31.0	79.3	11.5	0.30
Titanium	106.9	15.5			
Zirconium	93.8	13.6	35.8	5.20	
Lead	17.9	2.6	6.2	0.9	0.40
Granite	46.2	6.7	19.3	2.8	0.20
Glass (soda lime)	68.9	10.0	22.1	3.2	0.23
Alumina sintered	324.1	47.0			0.16
Concrete	10.3–37.9	1.5–5.5			0.15[b]
Wood, oak (longitudinal)	89.6	13.0	3.2	0.47	
Wood, oak (tangential)	6.7	0.97	1.0	0.15	
Nylon	2.8	0.4			0.4
Phenolic resin	5.2–6.9	0.75–1.0			
Rubber, hard	2.8	0.40			0.43
PVC (polyvinyl chloride—rigid)	3.5	0.5			0.4
Glass-laminated plastic	20.7	3.0			

[a]The SI basic units are 1 psi = 6.895×10^3 N/m² (Pa);
GN = 10^9 N. E is Young's modulus and G is the shear modulus.
[b]Average value; μ may vary from 0.11 to 0.21 depending on the strength of concrete.

elastomers (polymers). Important group of polymeric materials that are subject to many cross-linking processes to impart desired properties to rubber. Any linear polymer can be made a good rubber if it meets the following characteristics. (1) The polymer chain should be very long and geometrically irregular so that thermal agitation will result in strongly entangled and coiled-type arrangement. (2) The intermolecular forces between the polymer chains should be such that at room temperatures thermal energy is sufficient to maintain them in a state of constant mobility. (3) There must be a possibility for introducing cross-links between the chains so that the required degree of rigidity can be obtained. To achieve these desired characteristics, synthetic rubbers are usually produced by copolymerization processes, which have a tendency to lower the symmetry and regularity of the chain and give a long chain of relatively weak intermolecular attraction. The presence of polar groups is usually avoided unless special characteristics such as oil resistance and improved heat resistance are required at the expense of flexibility of the polymer chain. **(See also rubber.)**

electrical conductivity measurement. pH measurements are concerned with the hydrogen concentration in a solution. All ions in a solution, however, affect its ability to pass an electrical current. Measurements of this ability (conductivity) are useful in many industrial processes. Such a measurement can be made by immersing a pair of electrodes of known area of a certain distant apart, and then measuring the resistance between them. Conductivity is expressed in ohms measured between two electrodes, each having an area of 1 cm² and placed 1 cm apart. A mho is equal to $^1/_{ohm}$ and represents the ability of 1 cm³ of a substance to pass 1 A of current at 1 V of potential.

electrical conductivity (metallurgy). A function of the ease with which electrons move about among atoms. Electrons circle the nucleus of the atom, which is composed of protons and neutrons, in orbits that allow the energy of the electrons to be constant within their orbit. Not more than two electrons can fit into any one orbit, and in order to fit together, they must spin in opposite directions; this is known as the Pauli exclusion principle. When an atom is in a higher-energy state, it is in an excited state. Each orbit differs from other orbits by one or more energy quanta, the energy level becoming greater with the orbit's distance from the nucleus. The orbits are grouped in shells, and inside the shells are subdivisions known as energy levels. The greatest number of electrons that can fit into each shell is determined by the expression $2N^2$, where N is the number of the shell.

electrical resistivity. The electrical resistance of a body of unit length and unit cross-sectional area or unit weight. The value of $^1/_{58}$ ohm-mm²/m at 68°F (20°C) is the resistivity equivalent to the International Annealed Copper Standard (IACS) for 100% conductivity. This means that a wire of 100% conductivity, 1 m in length and 1 mm² in cross-sectional area would have a resistance of 0.017241 ohms at 68°F (20°C).

electrical resistivity of some materials. The electrical resistivity of various materials is listed in the accompanying table.

Material	Resistivity (10^{-8} Ω·m) at 20°C	Temperature coefficient of resistivity (× 10^3) (0–10°C)
Aluminum	2.69	4.2
Chromium	12.9	2.1
Copper	1.67	4.3
Gold	2.3	3.9
Germanium	46 × 10^6	—
Iron	9.71	6.5
Lead	20.6	3.4
Magnesium	3.9	4.2
Mercury	95.8	0.9
Nickel	6.84	6.8
Platinum	10.6	3.9
Silicon	23 × 10^{10}	—
Silver	1.6	4.1
Tin	12.8	4.2
Zinc	5.92	4.2
80 Ni–20 Cr	109	—
18–8 stainless steel	70	—
55 Cu–45 Ni	45	—

electric arc welding. The basis for electric-arc welding is an electric arc between an electrode and the workpiece or between two electrodes. The arc is a sustained electrical discharge through a path of ionized particles called a plasma. The temperature may be over 30,000°F (16,650°C) inside and is almost 20,000°F (11,095°C) at the surface of the arc. Applications are classified as to whether the electrode is nonconsumable, for carbon-arc and tungsten-arc welding, or consumable, for metal-arc welding. The metal electrode in metal-arc welding is melted progressively by the arc and is advanced to maintain the arc length. Coated electrodes furnish a protective gaseous shield and slag that floats on top of the melt. Arc welding is quite versatile and able to make welds under many conditions, of producing high-quality welds, and of depositing metal rapidly. Examples of structures arc welded are tanks, bridges, boilers, buildings, piping, machinery, furniture, and ships. Most metals can be welded by one or more of the forms of arc welding. **(See also welding and type of welding.)**

electric furnace, induction process. The melting procedure in steel plants using the induction furnace is substantially that of a crucible of "dead-melt" process. The charge is carefully selected to produce the composition desired in the finished steel with a minimum of further additions except, perhaps, small amounts of ferroalloys, used as final dioxidizers. The charge may consist of a single lump of metal (ingot), a number of small pieces of selected steel scrap, or even turnings or other light scrap, mixed with a moderate amount of larger pieces to provide initial conditions favorable to the generation of heat. As a rule, no refining is attempted in acid-lined furnaces, and it is seldom tried in basic-lined furnaces. Very little oxidation actually occurs if the furnace is fitted with a tight cover over the crucible during the melting of heat from the surface of the molten metal. Because of this, from the standpoint of heat loss, the use of slag covering to protect the metal is unnecessary.

electric furnaces. The two types of electric furnaces most used for steel making are the direct-arc furnace (series arc) and the high-frequency coreless induction furnace. The direct-arc electric furnace widely used by steel mills and foundaries today receives the cold charge of scrap steel, etc., then an electric arc is drawn between the electrodes and the surface of the charge to be melted. Electrodes are made either from graphite or amorphous carbon. When the arc is struck and is sprung from the electrodes to the metal bath, the metal is heated both by direct conduction from the arc and by radiation from the roof and walls of the furnace. The height of the electrodes above the bath, and, consequently, the heat input, is automatically controlled by winch motors in the system used to raise and lower the electrodes. The indirect-arc furnace is of the rocking arc type, so-called because the arc is separate from the charge of metal. In this furnace heat is supplied to the metal by conduction and radiation from the hot refractory furnace walls, and conduction from the same hot refractory walls when the furnace is rocked and the molten metal washed over them. Although these furnaces are used more for melting copper and copper-base alloys, they are also used, at times, for melting cast iron and steel; however, with the high-temperature conditions that prevail the efficiency is comparatively low. **(See figure p. 112.)**

electric furnace, steel making. In most electric furnaces the heat is produced by means of an arc either above the bath, as in the Stassano furnace, or by means of arcs between the slag and electrodes suspended above the bath, the latter being the more satisfactory method. In a typical electric furnace there are three carbon or graphite electrodes suspended from the roof. Each of these electrodes is connected to one phase of a three-phase current and is lowered into the bath in such a manner that an arc is formed between the slag and each electrode. The modern arc-type electric furnace is cylindrical in shape and of the standard three-phase, three-electrode service-arc type. Basic differences of construction lie in the mechanical design. Some of these include different methods of filling, various methods of applying removable roofs to medium-size furnaces, and arrangement and application of electrical equipment. The advantages of the electric furnace include the nonoxidizing condition of the carbon arc, which is pure heat, making possible a tightly closed furnace and permitting maintenance of reducing atmospheres; the temperature attainable is limited only by the refractory nature of the furnace lining; close limits of regulation and control are obtainable; the efficiency of the unit is extremely high; and refining and alloying are readily accomplished and controlled.

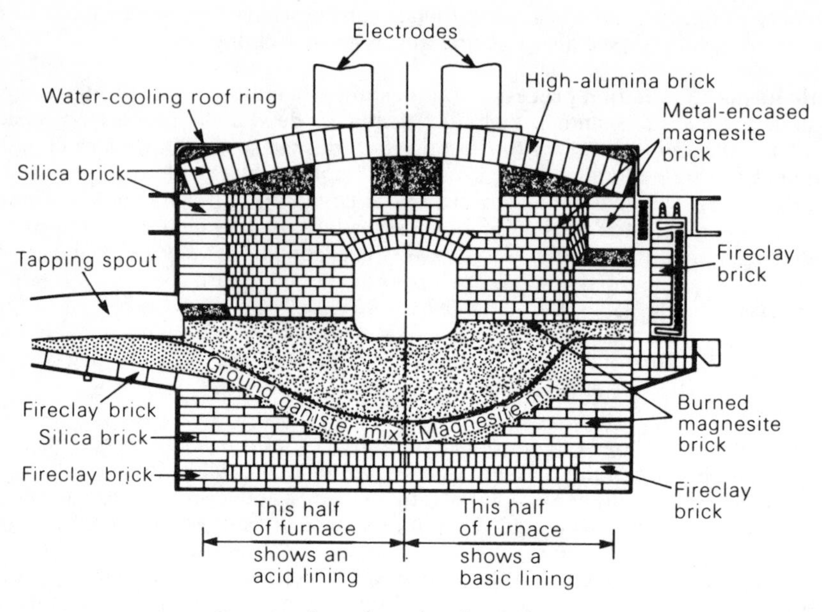

Cross section of an electric arc furnace.

electric truck. An industrial truck in which the principal energy is transmitted from power sources to motor(s) in the form of electricity.

electrochemical machining (see machining, electrochemical)

electroforming. A method of manufacturing a product that involves the electrodepositing of metal on a reversible or expendable mandrel or master form of a predetermined size, shape, accuracy, and finish. The metal being electrodeposited builds up to the required thickness after which the mandrel is removed, thus leaving the desired product, often without the necessity of further finishing of any kind. Given the correct electrolytic bath, mandrel material, and operating conditions, the metallurgical properties of an electrodeposited metal can be controlled and selected over a considerable range. Parts of intricate design with accurate dimensions and smooth surface finishes up to 1–2 μin. (0.0254–0.0508 μm) can be produced by the electroforming method. Electroforming is used in the manufacture of electrotype; phonograph master record disks, copper foil and tubing; leather-embossing dies; high-frequency wave guides for radar and television; searchlight reflectors; molds for forming plastics; copper floats; Pitot tubes; fine screening; seamless tanks; and sheet and strip.

electrogel machining (EGM) (chemical machining). EGM is an electrolytic process for removing metal in definite shapes by means of a formed tool made of a stiff gel consisting of cellulose acetate and acid. The molded gel is held against the workpiece, and as a current (about 8 A at 2.5 V) is passed through the stack, the pattern on the gel is etched into the workpiece at a rate of about 0.001 in./min (0.0254 mm/min). Forces are negligible, and there is no effect on the structure of the metal; the mild operation is suited for frail pieces like those made from thin metal honeycomb structures. Steel and nickel alloys, titanium, and high-pressure heat-resistant alloys have been found amenable to the process.

electrolyte. If two plates of dissimilar metals are placed in a chemical solution called an electrolyte, opposite electric charges will be established on the two plates. An electrolyte is technically defined as a compound that, when molten or in solution, conducts electric current and is decomposed by it. In simple terms, an electrolyte is a solution of water and a chemical compound that will conduct an electric current. The electrolyte in a standard storage battery consists of sulfuric acid and water. Various salts dissolved in water will also form electrolytes. An electrolyte will conduct an electric current because it consists of positive and negative ions. When a chemical compound is dissolved in water, it separates into its component parts. Some of these parts carry a positive charge, and others carry a negative charge.

electromagnet. When a soft-iron core is placed in a coil, an electromagnet is produced. The wire in the coil must be insulated so that there can be no short circuit between the turns of the coil. A typical electromagnet is made by winding many turns of insulated wire on a soft-iron core, which has been wrapped with an insulating material. The turns of wire are placed as close together as possible to help prevent magnetic lines of force from passing between the turns. The strength of an electromagnet is proportional to the product of the current passing through the coil and the number of turns in the coil. The value is usually expressed in ampere-turns. If a current of 5 amp is flowing in a coil of an electromagnet and there are 300 turns of wire in the coil, the coil will have an mmf of 1500 amp-turns. Since the gilbert is also a measure of mmf and 1 amp-turn is equal to 1.26 gilberts, the mmf may also be given as 1890 gilberts. The ultimate strength of the magnet depends on the permeability of the core material. The force exerted upon a magnetic material by an electromagnet is inversely proportional to the square of the distance between the pole of the magnet and the material.

electromagnetic induction. When a conductor is moved through a magnetic field, a difference of potential is set up between the ends of the conductor and an electromotive force is induced. This voltage exists only during the time when the conductor is in motion through the magnetic field. Thus, the current flow caused by the voltage is also present only during the time when the conductor cuts the lines of force by movement. The current that is caused to flow is known as induced current. The conductor can remain stationary and the lines of force can be shifted so that they will cut across the conductor. When this is done, an electromotive force is again induced. Thus, it is evident that the voltage may be induced in a conductor by moving the conductor through magnetic lines of force or by moving the source of the magnetic lines of force so that they cut across the conductor.

electrometallurgy. Essentially the electrorefining of metals after they have been produced by the many techniques of pyro- and hydrometallurgy. The principal use of electrorefining is to produce an extremely high purity metal from anodes that have been produced prior to going to the electrorefinery. In this case, the anode, such as copper, is placed in an electrolyte of copper sulfate in large cells, into which are also placed what are known as "starting sheets" of pure copper. An electric current is then passed through the cell, which causes a galvanic action, a dissolving of the anode metal, and a redesposition of the same onto the extremely high purity starting sheets. During this process, the impurities that are present in the anodes will either dissolve in the electrolyte or be deposited at the bottom of the cell, and in either case some of these "impurities" are valuable and are extracted profitably, such as gold and silver in the case of the electrorefining of copper, which cannot be recovered by pyrometallurgical methods. Electrorefining is based on Faraday's law that states, essentially, that the quantity of a substance that is liberated at an electrode is directly proportional to the quantity of the current that is flowing through the electrolyte. In addition to copper, nickel, lead, and tin are also refined by the electrorefining method.

electromotive force. The force that causes electrons to flow through a conductor is called electromotive force, abbreviated emf. The practical unit for the measurement of emf or potential difference is the volt (V). Electromotive force and potential difference may be considered the same for all practical purposes. When there is a potential difference, or difference of electrical pressure, between two points, it means that a field of force exists which tends to move electrons from one point to the other. If the points are connected by a conductor, electrons will flow as long as the potential exists.

electron compounds (Hume–Rothery compounds). Compounds in which the normal-valency laws are not observed but in which there is often a fixed ratio between the total number of valency electrons of all the atoms involved and the total number of atoms in the "molecular formula" of the intermetallic compound in question. There are three such ratios:

Ratio	Type of structure	Representative phases
(1) 3/2 (21/14)	bcc (β-brass type)	$CuZn$; Cu_3Sn; Cu_3Al; $CuBe$; Ag_3Al
(2) 21/13	complex cubic (γ-brass type)	Cu_5Zn_4; $Cu_{31}Sn_8$; Cu_9Al_4; Ag_3Zn_8
(3) 7/4 (21/12)	cph (ε-brass type)	$CuZn_3$; Cu_3Sn; $CuBe_3$; Ag_5Al_3

These Hume–Rothery ratios have been useful in relation structures which apparently had little else in common. In general, intermetallic compounds are less useful as a basis for engineering alloys than are solid solutions. In particular, intermetallic compounds are usually very hard and brittle having negligible strength or ductility. Often the compound has none of the physical characteristics of its parent metals. Since intermetallic compounds are of fixed compositions, and will generally dissolve only small amounts of their parent metals, they do not exhibit coring when viewed under the microscope. Very often, small amounts of these intermetallic compounds are present as essential phases in engineering alloys.

electronic computer-aided manufacturing (ECAM). A Tri-Service (U.S. Army-managed) program that uses ICAM methodologies for improving manufacturing capability in electronics.

electronic data processing (EDP). Here the computer is used to store, process, and produce vital manufacturing data on command. This area of computer use contrasts sharply with that where its capabilities are used to control equipment in the plant. Thus it must be considered as an automation tool in a different context. The emergence of EDP techniques was a prerequisite for truly advanced integration of mechanized machines and operations which automation entails. Such integration requires data concerning all the various aspects and interrelationships of a process as a whole. Generating, collecting, and handling large and varied amounts of data can be a key asset in the final accomplishment of plant automation. There are two basic kinds of information systems: tactical and strategic. A tactical information system deals with the day-to-day supervision and control of line activities. These activities include order processing, production scheduling and supervision, and inventory control. A strategic information system presupposes the existence of a tactical system. Output from the tactical system is accumulated and combined with financial data, market trends, business conditions, and monetary factors to provide a comprehensive input for long-range managerial decision-making. In effect, then, the strategic or management information system (MIS) is a total information system of which the tactical system is one part. Today, the tactical system is growing in use, but few as yet are going the total-systems route. Ultimately, efficient and effective strategic information systems will yield the most comprehensive and, hence, the most profitable results. **(See figure p. 115.)**

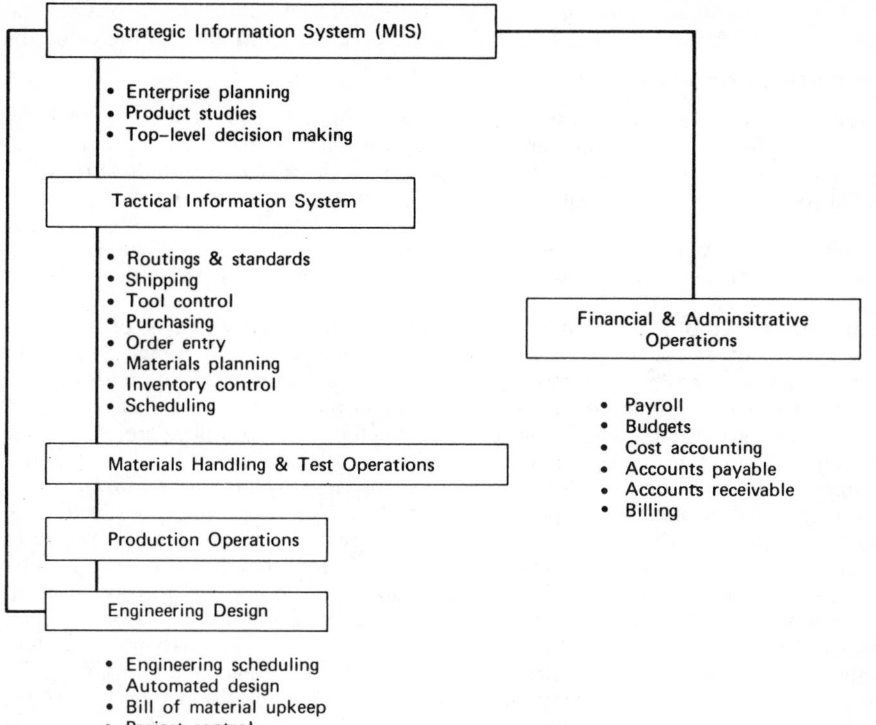

How information systems interface production and business operations.

electronic manufacturing processes. See ATE, CAD/D, dip plating, dip wiring, electroplating, fluxes, ion implantation, laminating, metallizing, photoengraving, printed-circuit assembly, printed-circuit boards, printed wiring, screen printing, soldering, sputtering, and vacuum plating.

electroplating. The deposit of metal coatings from electrolytic baths, which may be performed on any of the common metals. The part to be plated is made the cathode in the bath, and the anode is a bar of the metal to be plated. The anode may be soluble, in which case it consists of the metal to be deposited. When insoluble, salts of the plating metal must be added to the bath to maintain its concentration. The bath is conductive, containing either a solution of salts or molten salts of the depositing metal. A direct current applied across the terminals causes the metal to leave the solution and plate the cathode. Plating is used for protecting metals and for improving their appearance, nickel and chromium being most suitable for the latter purpose. The thickness of plate is controlled by the time of plating and the intensity of the current. Almost any metal may be used for plating, but those most used are zinc, copper, nickel, chromium, cadmium, and tin. Cadmium plating is common on small hardware items; it provides corrosion protection inferior to that of zinc and may result in hydrogen embrittlement. Cadmium plating is not permitted on food processing equipment because its compounds are toxic. Nickel tarnishes on exposure to atmosphere. Chromium does not tarnish, but electroplated chromium is porous.

elevator bucket. A conveyor for carrying bulk material in a vertical or inclined path, consisting of endless belt, chain, or chains, to which elevator buckets are attached, the necessary head and boot terminal machinery, and supporting frame or casing.

embossing (see cold working)

emery. A fine-grained impure corundum used for grinding and polishing, which was originally found on Naxos, an island in the Aegean Sea, Smyrna, in Turkey, and Chester, Massachusetts. Today, manufactured aluminum oxide or silicon carbide abrasives are used owing to superior quality. **(See also abrasives.)**

emery cloth. An outmoded term that applied to heavy cloth coated with emery used as a fine abrasive. It has been superceded by the term "abrasive cloth."

emulsifiers. Emulsifiers or emulsifying agents belong to a general class of compounds known as surface-active agents. They have the ability to adsorb at high-energy interfaces and transform them into low-energy surfaces. Emulsifiers are compounds exhibiting both polar and nonpolar character. The emulsifier molecule contains a lipophilic and hydrophilic group at its ends. The lipophilic end of the molecule will be preferably wetted by oil, whereas the hydrophilic end will be wetted by water. Thus, the oriented adsorption will occur at the interface of the two phases, resulting in the formation of a protective film around the dispersed droplet. A sodium soap molecule illustrates the functioning of an emulsifier. Emulsifiers can be divided into ionic and nonionic, into naturally occurring materials and finely divided solids. Ionic emulsifiers belong to a broad group of colloidal electrolytes such as soaps and certain detergents. A proper hydrophilic–lipophilic balance of the emulsifier molecule determines the type of emulsion that tends to be formed. Strongly lipophilic emulsifiers will tend to make a water/oil emulsion, whereas those of hydrophilic nature will favor an oil/water emulsion. Droplet size and distribution are important factors determining the emulsion's stability. The rate of sedimentation and creaming is proportional to the value p_1-p_2 as shown by Stoke's equation. Creaming is the rise of dispersed droplets under the action of gravity, the droplets remaining separable when they touch. The breaking of an emulsion occurs when many droplets coalesce and finally give rise to two separate bulk phases. This occurs when the electric charge on the droplets in the case of ionic emulsifiers has been neutralized. Almost infinite stability against coalescence is obtained with thick macromolecular or particulate interfacial film, which exhibit mechanical cohesion and stability.

emulsions. An emulsion is a two-phase system consisting of a fairly coarse dispersion of two immiscible liquids, the one being dispersed as finite droplets in the other. Any pair of immiscible liquids can be used, but the combination of water and an organic liquid (oil) is of great practical importance. The disperse, or internal, phase is the liquid that is broken up into droplets. The surrounding liquid is known as the continuous or external or dispersing phase. Emulsions are much coarser than colloidal dispersions and may be examined with a microscope or sometimes even with a lens. Usually the size of disperse droplets varies from 1 to 6 microns. Such a coarse system is inherently unstable. To increase its stability a third agent, called an emulsifying agent or simply an emulsifier, is added. Because of the presence of the emulsifier, which is a colloid, many properties of emulsions are those of colloidal dispersions.

emulsion types. Two types of emulsions can be obtained from such immiscible liquids as water and oil: oil-in-water and water-in-oil emulsions. These are usually designated in abbreviated form as O/W and W/O emulsions. An O/W emulsion has a creamy texture, whereas a W/O emulsion feels greasy because its external phase is an oil. Geometrical calculations of closely packed spheres indicate that the maximum amount of dispersion of one liquid dispersed in another can be 74.0% of the total available volume, provided that the disperse liquid is in the form of uniform and undistorted spheres. The order with which ingredients are added during mixing, whether oil to water or water to oil, may also influence the resultant emulsion type. The most important factor determining the type of emulsion is the emulsifier itself.

enamels and enameled metal. Porcelain, or vitreous enamel, is a ceramic mixture containing a large proportion of fluxes, applied cold and fused to the metal at moderate red heat. Complete vitrification takes place. The application of enamel to gold, silver, and copper is one that dates back to the ancients. Long valued as a material of great beauty in the field of decorative art, it has become into general commercial use because it provides a product of great durability and wide application; it is easy to clean and resists corrosion. Current uses are in plumbing fixtures, cooking utensils, industrial equipment, and glass-enameled steel for chemical use.

epoxies. An epoxy is a polymer with oxygen atoms in the polymer chain, though other polymers, such as silicones, may have oxygen in the main chain also. The epoxies are strong and brittle thermosetting plastics produced by a condensation reaction between bisphenol A and epichlorohydrin. The polymer chain terminates in a C—O—C group. A thermoplastic variant of the epoxies is the phenoxy plastics, which are ductile. The epoxies are principally used as potting, encapsulating, and casting resins, which are cast and cured, rather than molded with heat and pressure. They are equally important in the form of surface coatings for metals, producing hard, adherent, abrasion- and corrosion-resistant surfaces. Most uses of epoxies are specialized and have many applications such as in (1) civil engineering and construction, for bonding of concrete, repairs to concrete, durable epoxy surfaces for industrial floors, bonding of insulation board to concrete; (2) electrical engineering, in potting and encapsulating of electrical hardware, epoxy moldings, epoxy insulation; (3) aerospace engineering, in reinforced fiberglass plastics, adhesive for aircraft bulkheads, floors, wing flaps, and sandwich panels; (4) mechanical engineering, such as in epoxy tooling, jigs, dies, and foundry patterns, epoxy paints.

epoxy (plastics). Epoxy resins are characterized by the epoxide group (oxirane rings). The most widely used resins are diglycidyl ethers of bisphenol. These are made by reacting epichlorohydrin with bisphenol A in the presence of an alkaline catalyst. By controlling operating conditions and varying the ratio of epichlorohydrin to bisphenol A, products of different molecular weight can be made. Another class of epoxy resins are the novolacs, particularly the epoxy cresols and the epoxy phenol novolacs. These materials are recommended for transfer molding powders, electrical laminates and parts where superior thermal properties, high resistance to solvents and chemicals, and high reactivity with hardeners are needed. The cycloaliphatics comprise another group of epoxy resins that are particularly important when superior arc track and weathering resistance are necessary requirements. A distinguishing feature of these resins is the location of the epoxy group(s) on a ring structure rather than on the aliphatic chain. Cycloaliphatics can be produced by the peracetic epoxidation of cyclic olefins and by the condensation of an acid such as tetrahydrophthalic with epichlorohydrin, followed by dehydrahalogenation. Some of the epoxy resin products are: *Novolak Epoxy Resins:* These are prepared by reacting epichlorohydrin with a low-molecular-weight condensate of phenol and formaldehyde. Since they have a more-branched chain structure, giving a denser molecule, these resins have been used for high-temperature applications and are also said to have slightly better chemical resistance. *Phenol and Cresol Glycidyl Ethers*: These products are much less viscous than bisphenol glycidyl ethers and are often added to the latter to reduce viscosity. They can only be used in low percentages since they act as chain terminators and cannot form high-molecular-weight resins on curing. *Aliphatic Glycidyl Ethers:* Simple alcohols, such as butanol, yield low viscosity glycidyl ethers, which are used as viscosity reducers. Polyols, such as gylcerine and lower-molecular-weight poly glycols, give digylcidyl ethers that are partly water soluble and have uses in textile finishing. The cured products are no longer water soluble, but still show both lower water and temperature resistance when compared to the aromatic glycidyl ethers. *Glycidyl Esters:* Can be prepared by reacting epichlorohydrin with phthalic anhydride or stearic acid. They have not been found to have any significant commercial application. *Glycidyl Amines:* A number of glycidyl amines have been prepared and offered commercially. However, as a rule these products do not have equally good properties to the correspondingly glycidyl ethers. *Epoxidized Olefinics:* Considerable quantities of natural fatty acid esters that contain double bonds are epoxidized with hydrogen peroxide. These products find markets as plasticizers and vinyl stabilizers, but they cannot properly be termed resins. *Epoxidized Cycloolefinics:* A new class of epoxy resins, in which good stability is achieved, especially at high temperatures. The absence of aromatic and especially phenolic groups contributes to the color stability. For the most part these products are based on tetrahydrobenzaldehyde (THB), which is formed by the Diels Alder addition of butadiene to acrolein. THB can be made to undergo the Tischenko reaction to the ester of benzoic acid and tetrahydrobenzyl alcohol. Epoxidation of this ester with peracetic acid yields EP 201 (Union Carbide Chemical Co.). *Aliphatic Amines (epoxy curing agent):* Triethylene tetramine, diethylene triamine. The primary amine group can react with two epoxy groups thereby promoting chain formation. The secondary amine group reacts only once, and tertiary amine groups do not condense, but act as an accelerator. Aliphatic amines are used for room-temperature curing applications. *Aromatic Amines:* Metaphenylene diamine, methylene dianiline. Aromatic amines react slower and generally require heat to cure completely. The cured resins have somewhat higher heat distortion temperatures than those obtained with aliphatic amines. *Carboxylic Acid Anhydrides:* Phthalic anhydride, hexahydrophthalic anhydride, dodecenyl succinic anhydride. This is a large group that offers good properties but requires prolonged heating to achieve full cure. *Phenolic Melamine and Urea Resins:* These are used in coating applications, often in conjunction with other hardeners. Baking at relatively high temperatures is required.

Erichsen test. A cupping test in which a piece of sheet metal, restrained except at the center, is deformed by a cone-shaped spherical-end plunger until fracture occurs. The height of the cup in millimeters at fracture is a measure of the ductility. **(See also ductility test.)**

ES requirements. These are instructions and/or tests contained in Engineering Specifications (ES), which are extensions of released drawings and provide additional information necessary to produce parts/assemblies that meet engineering specifications. They are of two general types, those that define component performance requirements and those that define materials requirements.

etching. The art of preparing metal plates and other materials with relief designs, printing, or pictures by acid immersion. Ordinarily, a resistant coating is used that omits areas of the design or picture so that acid eats away only these areas which, after removal of the coating, leaves the desired result: printing, wiring, etc. **(See also photoengraving.)**

ETFE fluoropolymer. ETFE, a high-temperature thermoplastic, is readily processed by conventional methods, including extrusion and injection molding. As a copolymer of ethylene and tetrafluoroethylene, it is closely related to TFE. Manufactured by Du Pont Co. and marketed under the trademark Tefzel, it has good thermal properties and abrasion resistance, and good impact strength, resistance to chemicals, and electrical insulation characteristics. In addition, it will tolerate relatively large doses of radiation and is water resistant. Applications for injection molded items include gears, fasteners, pump components, column packings, automotive parts, seal glands, labware, valve liners, electrical connectors, and coil bobbings.

ethers. Compound of the general formula (*R*—O—*R* or *R*—O—Ar) in which an oxygen bridge joins two hydrocarbon groups. Ethers may be named in one of two ways. According to the IUPAC system of nomenclature one of the alkyl groups bonded to the oxygen is considered to be a substituted hydrocarbon. The smaller alkyl group with the oxygen is called an alkoyx substitute. In common nomenclature, both groups bridged by the oxygen are named and followed by the word ether. Ethers boil at much lower temperatures than do the alcohols from which they are derived because the oxygen is attached only to carbon. The boiling points of the ethers closely parallel those of the alkanes of the same molecular weight. The lower members of the aliphatic ethers are highly volatile and very flammable. Ethyl ether, the most important member of the family, is both an excellent organic solvent and a general anesthetic. Chemically, ethers are relatively inert, and in this respect they also are very much like the alkanes.

ether, ethyl. Most familiar as a general inhalation anesthetic. In organic chemistry, ethers are excellent solvents for facts, waxes, oils, plastics, and lacquers. Ethyl ether is the solvent used in the Wurtz reaction and in the preparation of the Gringnard reagent. Ethylene oxide is an important intermediate in the manufacture of ethylene glycol. Ethylene oxide represents a highly strained ring and, unlike the dialkyl ethers, is a very reactive compound. When treated with alcohols, ethylene oxide converts them into monoalkyl ethers of ethylene glycol. The products are called cellosolves and combine the solvent properties of both alcohols and ethers. Cellosolves are used as solvents for plastics and lacquers. A cellosolve, when combined with a second molecule of ethylene oxide, produces monoalkyl ethers of diethylene glycol. These products are called carbitols and also are good solvents.

Classes of Compounds And Their Functional Groups

Class Name	Functional Group	General Formula	Common Example
Alcohols	—O—H	R—O—H	$H{-}\overset{H}{\underset{H}{C}}{-}\overset{H}{\underset{H}{C}}{-}(O{-}H)$ Ethyl alcohol
Ethers	C—O—C	R—O—R	$H{-}\overset{H}{\underset{H}{C}}{-}\overset{H}{\underset{H}{C}}{-}(O){-}\overset{H}{\underset{H}{C}}{-}\overset{H}{\underset{H}{C}}{-}H$ Diethyl ether

ethyl alcohol. Sometimes called grain alcohol because starch from grain, when hydrolyzed to sugars and fermented by enzymes, produces ethyl alcohol and carbon dioxide. Starch from any source is a suitable material. The fermentation of sugar by yeast is a reaction practiced for ages and is the basis for the production of alcoholic spirits, and for the leavening action required in the backing process. **(See also alcohols.)**

ethylene polymers (polyethylenes). Made by the polymerization of ethylene, a hydrocarbon produced from ethyl alcohol or natural gas. They are obtainable in liquid resinous forms, as gums, or in semirigid solid forms. Because of their crystalline nature, cold working produces orientation that improves tensile strength and transparency. By varying tensile strength, brittleness, elongation, and break–tear resistance can be controlled.

explosive forming. The forming of metals in which the energy for forming is released over a very short time interval; loosely called high-energy-rate forming, or HERF processes. In these HERF processes energy is released at such a rate that local deformation occurs before the energy can be dissipated by "plastic waves" through the material. Mechanical efficiency is thus improved. Energy can be released quickly either by a sudden change in pneumatic pressure, by the use of high-voltage electrical discharge, or by the use of high explosives. Thus by detonating no more than 0.22 lb (0.1 kg) of a suitable high explosive, energy can be dissipated at a rate equivalent to 8×10^6 hp (6×10^6 kW). A method of explosion forming in which water is used to transmit the shock wave is illustrated. This is generally termed the "stand-off" method, the charge being suspended above the blank–die assembly. Dies can be of clay or plaster for sample jobs or of concrete, wood, resin, or metal for longer runs. The process is used at present for prototype production or for very short runs.

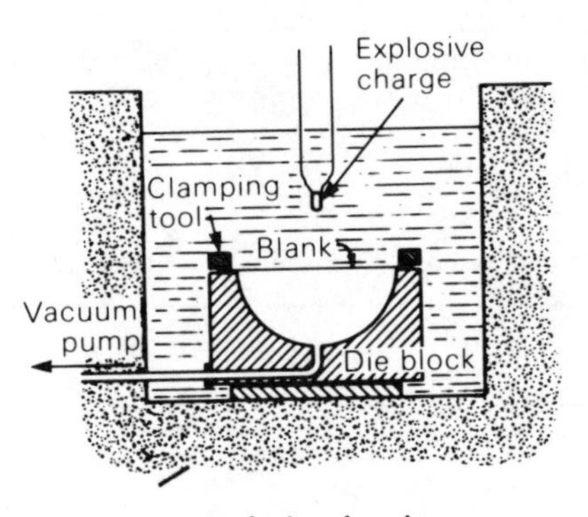

Explosion forming.

extrusion, cold metal. The names of impact extrusion, cold pressing, cold forging, and extrusion pressing have been given to and are descriptive of various forms of cold extrusion. Although nominally done at room temperature, cold extrusion is quick, and the heat liberated can raise the temperature of the metal several hundred degrees for an instant. Cold extruding takes less force if done quickly, particularly with metals that work harden. Cold extrusion pressure ranges from one to three times the finished yield strength of the metal; from 25 to 50 tons/in.^{2} (3.4 to 6.9×10^5 kPa) for softer metals to 200 tons/in.^{2} (2.8×10^6 kPa) for some carbon steels and higher for alloy steel and titanium. Cold extrusion is done both to improve physical properties of a metal and to produce specific shapes. Examples of cold-extruded parts are cans, fire extinguisher cases, trimmer condensers, aircraft shear tie fittings and brackets, and automotive pistons from aluminum, rocket motors and heads, hydraulic and shock absorber cylinders, wrist pins, and gear blanks from steel. Parts are also extruded from brass, copper, lead, magnesium, tin, titanium, and zinc. **(See also cold extrusion.)**

extrusion, hot metal. Forward hot extrusion of solid or hollow shapes enables the metal to be easily supported, handled, and freed from the equipment. When a piece is forward extruded, it is cut off, and the butt end is removed from the chamber. Preservation of the equipment subjected to high temperatures, as it is, is the major problem of hot extruding. Temperatures at 650–800°F (343–427°C) for magnesium, 650–900°F (343–482°C) for aluminum, 1200–2000°C (649–1094°C) for copper alloys, and 2200–2400°F (1204–1316°C) for steel. Pressures range from as low as 5000 psi (3.4 × 10^4 kPa) for magnesium to over 100,000 psi (6.9 × 10^5 kPa) for steel. Most hot extrusion is done on horizontal hydraulic presses especially constructed for that purpose. Most hot extrusions are long pieces of uniform cross section, but tapered and stepped pieces are also producible. Examples of commercial extruded products are trim and molding strips of aluminum and brass and structure rod, bars, and tube of all forms of aluminum and steel.

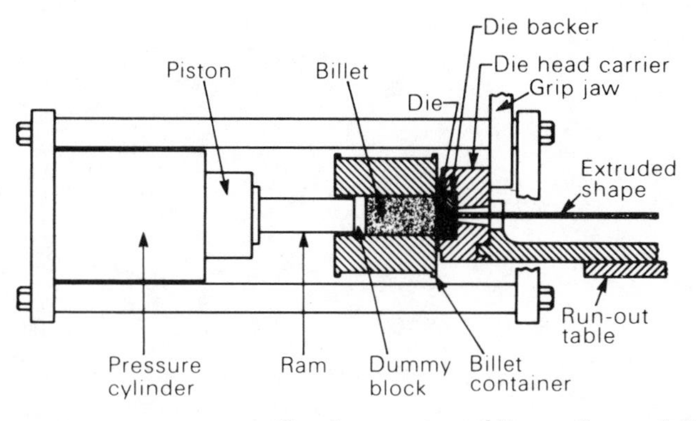

Drawing, courtesy of Revere Copper & Brass Inc.

Sectional view showing the general setup used for direct hot extrusion of metals.

extrusion, metal. Involves pressing a billet or slug of metal by movement of a ram through an orifice or nozzle. During extrusion, both compressive and shear forces are developed; however, there are no tensile forces present. It is therefore possible to deform the metal very heavily without fracture. Under the influence of the applied force, the metal is continuously deformed as it is pushed through a die and is shaped into a long bar of desired cross section. In the case of tubes, the extrusion is carried out through a round die, and a mandrel, integral with the ram, is pushed through the previously pierced billet. Two main extrusion processes are direct extrusion and inverted extrusion. In direct extrusion, the pressure ram is advanced toward the die assembly; in indirect extrusion, the die is moved down the container bore. This latter movement is characterized by a simple flow and lower extrusion pressures; however, the operation must use a hollow ram along which the extruded product can pass after being formed in the die.

extrusion ingot. A solid or hollow cast form, usually cylindrical, suitable for extruding. A drilled extrusion ingot is a solid extrusion ingot that has been drilled to make it hollow; a reamed extrusion ingot is a hollow extrusion into that has been machined to remove the original inside surface; and a scalped extrusion ingot is a solid or hollow extrusion ingot that has been machined to remove the original outside surface. **(See also fabricating ingot.)**

extrusion machines, plastic. Extrusion of plastics is a process of continuously forming rod, tubing, sheet, monofilament, and profile shapes in a continuous cross section. The extrusion method adaptable to almost every type of thermoplastic is called dry-melt extrusion. In this process, powders or pellets are fed from a hopper or silo to a conveying screw that melts, meters, and forces the hot metal through an orifice of the required shape. Extrusion molding ranks as one of the high-volume outlets for thermoplastic resins. Extrusion machines are used for pipe and tubing; wire and cable coating; profile shapes; film and sheet; monofilament; foam; coating substrates; compounding; forming parisons for blow molding; heating material prior to compression or transfer molding; and plasticating material in injection molding.

F

fabricating ingot. A cast form suitable for subsequent working by such methods as rolling, forging, or extruding **(See also extrusion ingot, forging ingot, and rolling ingot.)**

fabrics. Sheets of fibers or yarns woven, knitted, or otherwise physically bonded together.

fatigue (materials). Materials subjected to repeated loads may fracture at stresses considerably lower than the ultimate (maximum) strength of the material, and perhaps less than the yield strength. Such failures are referred to as fatigue failures. The failure is a gradual or progressive fracture due to the spreading of a small crack formed at a point of localized stress. Influencing factors are the range over which the stress extends, the number of cycles of stress, and, to a lesser extent, the speed at which the cycles are applied. The highest stress in a range of stress that can be applied an indefinitely large number of times without danger of fatigue failure is called the endurance limit.

fatigue testing. The fatigue strength of a material is indicated by the maximum stress that it can sustain for a specified number of cycles, without failure, the stress being completely reversed within each cycle unless otherwise stated. Fatigue resistance of a material, then, is its favorable reaction to loads applied more than once. Fatigue strength is usually proportional to tensile strength, but this generalization does not hold in many instances over wide ranges of tensile strengths. In making fatigue tests, the method commonly used consists of submitting specimens of or an entire structural or machined part to cycles of a known range of loads of various magnitudes, and noting the number of cycles required for fracture, or for arriving at a given specified "life" prior to fracture. The results of the test are plotted on what is called an *S-N* diagram, in which the maximum load or stress (*S*) is plotted as the ordinate and the number of cycles for fracture (*N*) as the absicissa.

fatigue testing, creep. Creep is the term given to a type of plastic flow, the continued change in dimensions that results from a particular condition of deformation, that is, under conditions of sustained stress (or load). In creep tests of metals the specimens are similar to those used in tension and compression tests. Where the total strain of metals is of the same order of magnitude as elastic strain, the magnitude of elastic strain is usually determined at each test temperature by applying load stepwise, and constructing a stress–strain diagram, or if first stage creep is too rapid for accurate measurements, by measuring elastic strain immediately upon release of the load at the end of the test.

fatigue testing, repeated bending or direct flexure. A type of fatigue test in which the specimen is bent back and forth but is not rotated. This kind of test is especially useful in the testing of flat rolled products. It has the added advantage that surface preparation of the specimen is not necessary, permitting the test to be made on specimens that have the actual surface exposed when in service.

fatigue testing, resonant frequency (direct flexure test). Consists of vibrating the specimen at its fundamental frequency by an oscillating applied force. Because of the characteristics of resonant vibrations, very small forces applied at or near the resonant frequency of the specimen are capable of producing large amplitudes of vibration and corresponding high stresses. Some testing machines of this type make use of an oscillating magnetic field tunable to the resonant frequency of the specimen. The specimen is supported at the nodes and it vibrates as a free–free beam. Specimens of relatively large cross sections can be tested in this manner, which would require very large machines if direct mechanical loading were used. This is possible with the resonant frequency type of machine by taking advantage of the resonance vibration in the specimen tested.

fatigue testing, torsion. In the torsion test the specimen is subjected to twisting or torsional loads much like those encountered in drive shafts, crank shafts, etc. Data obtained on the torsional strength, particularly the yield point or yield strength, are highly important to the designer of such structures. Data for torsion tests are usually obtained in the form of a torque–twist curve, in which the applied torque is plotted against the angle of twist. Torsion produces a state of stress known as pure shear, and the shear stress at yielding can be calculated from the torque at yielding and the dimensions of the specimen being tested. The stress varies from a maximum at the surface of the specimen, to zero at the axis. In the elastic range the variation is linear, and the maximum stress for a cylindrical specimen can be calculated from the following equation: $S = 16T/\pi d^3$, where S equals the maximum shear stress in lb/in.^{2}, T equals the torque in inch-pounds, and d equals the diameter of the specimen, in inches.

fats and oils. Fats and oils are glycerides, or esters of glycerol, a trihydroxyl alcohol. A simple glyceride is one in which all R groups in the general formula are identical. Natural fats and oils usually are not simple glycerides. The three acid residues produced when fats and oils are hydrolyzed usually vary not only in length but also in the degree of unsaturation. The principal structural difference between oils and fats lies in the degree of saturation of the acid residues, and this accounts for the differences in both the physical and chemical properties of these two classes of glycerides. The fats are glyceryl esters in which long-chain saturated acid components predominate. They are solids or semisolids and are principally animal products. Lauric, palmitic, and stearic are the most important acids obtained by the hydrolysis of fats. Such long-chain carboxylic acids are usually called fatty acids because they are obtained from fats. They are, for the most part, insoluble in water but soluble in organic solvents.

fats and oils, iodine value of. The extent of unsaturation in a fat or oil is expressed in terms of its iodine value. The iodine value is defined as the number of grams of iodine which will add to 100 g of fat or oil. The iodine value of tripalmitin with no unsaturation would be zero.

fats and oils, oxidation (rancidity). Edible unsaturated fats and oils, when exposed to air and light for long periods of time, not only are subject to slow hydrolysis, but the acid components produced are also subject to oxidative cleavage at the site of unsaturation. The lowest-molecular-weight and more volatile acids that are produced by this exposure impart an offensive odor to fats. This condition is known as rancidity. If the volatile acids are produced by hydrolysis, the resultant rancidity is known as hydrolytic rancidity. Butter, especially, when left uncovered and out of the refrigerator, easily becomes rancid through hydrolytic rancidity. A substantial portion of the fatty acid components of butterfat is made up of butyric, caproic, caprylic, and capric acids. These are liberated when butter is hydrolyzed and are responsible for the unpleasant odor of rancid butter. Oxidation leading to rancidity in fats and oils is catalyzed by the presence of certain metallic salts.

fats and oils, sponification value. The composition of fats and oils is variable and depends not only on the plant or animal species involved in their production, but also on climatic and dietetic factors as well. The approximate molecular weights of the acid components in fats and oils may be obtained from saponification values. Saponification is a term specifically applied to the hydrolysis of an ester when the action is carried out in alkaline solution. The saponification value of a fat or an oil is an arbitrary unit, defined as the number of milligrams of potassium hydroxide required to saponify 1 g of the fat or oil. Because there are three ester linkages in a glyceride to hydrolyze, three equivalents of potassium hydroxide are required to saponify one molecular weight of any fat or oil. A small saponification value for a fat or an oil indicates a high molecular weight.

feldspar. The term feldspar is used in a general way to refer to a group of aluminosilicates of potash, soda, and lime, and also with appropriate modifiers, to apply to subgroups or to individual minerals. Well-known characteristics of the feldspars include a white, gray, or pink color; two good cleavages at approximately right angles; and a hardness of 6.0–6.5. The feldspar of commercie falls within the microclineperthite–albite–oligoclase range. The mineral has two chief properties of value, and two major uses, dependent on these properties. When ground to powder and used as a constituent of ceramic mixes, feldspar acts as a flux, that is, it forms glass within the ceramic body at relatively low temperatures. It is a basic raw material in the manufacture of china, high-grade tile, and porcelain, as well as in whiteware glaze and sheet-iron enamel. The second property that makes feldspar of value is its alumina content, which is utilized in the manufacture of glass. Substitution of alumina for part of the silica in glass increases the resistance of the glass to impact, bending, and thermal shock.

felts. Nonwoven "entanglements" of fibers or yarn.

ferrite. A solid solution of alpha iron and carbon. The maximum amount of carbon present in the ferrite of plain carbon steels is 0.025% at 1333°F (723°C), and drops off, as the temperature decreases to 0.008% at room temperature. The characteristics of ferrite are nearly the same as of pure iron. It has low tensile strength of approximately 40,000 psi (2.8×10^5 kPa), and is relatively soft. After etching, it appears as white grains under the microscope. The ferrite in pure iron–carbon alloys consists of iron with a trace of carbon in solution; however, in steels it may also contain considerable amounts of alloying elements, including manganese, silicon, or nickel.

ferromagnetic materials. The most important materials used in magnetic applications in electricity and electronics are the ferromagnetic materials. These are characterized by a very high permeability and will become very strongly magnetized by a field relatively much weaker than that required for diamagnetic and paramagnetic materials. The direction of the induced magnetism is identical with that of the magnetizing field. Among the ferromagnetic materials are iron, steel, cobalt, magnetic lodestone, and alloys such as alnico and Permalloy. Alnico derives its name from the metals used in the alloy: aluminum, nickel, and cobalt (with some iron added). Permalloy (permanent alloy) contains iron and about four times as much nickel. Compared to soft iron, with a permeability of about 2000, these alloys have permeability ratings ranging, under certain conditions, as high as 50,000 or more.

fiber. A generic term for various types of matter, natural or man-made, which form the basic elements of textile fabrics and other textile structures. A single, continuous material whose length is at least 200 times its width or diameter.

fibers, acetate. Acetate fiber is spun from a solution of cellulose acetate in acetone, hardening as the solvent evaporates. It is weaker and lower in water absorbency than rayon, but more elastic and wrinkle resistant. It is used in a variety of fabrics, both alone and in blends. Triacetate can be heat set and thus used in "wash-and-wear" garments.

fibers, acrylic. Polyacrylonitrile alone and modified by a small amount of copolymer is made into textile fibers, mainly staple yarns. Acrylic fibers have moderate strength, low stretchability, low heat conductivity, and low softening point. Alone and blended they are used in light-weight warm, wrinkle-resistant, and good-wearing fabrics used in both men's and women's apparel. They are also used in industrial textiles and carpets. Similar fibers are made from a copolymer with vinyl chloride. Another copolymer is composed of equimolecular amounts of vinylidene dinitrile and vinyl acetate.

fibers, elastomeric. Natural and various synthetic rubbers are extruded into filaments used with other fibers in elastic fabrics. Spandex elastic fibers composed of not less than 85% polyurethane are superior to natural rubber in resistance to oxidation, chafing, and dry-cleaning damage. They have excellent resistance to abrasion, ultraviolet light, weathering, chemicals, and cosmetics, and are used in foundation garments, swim wear, surgical hose, and other elastic products.

fibers, nylon. Made from both nylon 6,6 and nylon 6 by melt spinning followed by cold drawing. They have high tenacity, wet strength, extensibility, and abrasion resistance. Nylon yarns in monofilaments, multifilaments, and staples are used in hosiery and in many fabrics both alone and in blends. Nylon is used extensively in parachutes, glider ropes, transmission and conveyor belting, cordage, fish lines, and tire cord. Coarse monofilament nylon is used in toothbrushes, in paint brushes, and in tennis rackets.

fibers, polyester. Products of a polyester reaction such as that between terephthalic acid and ethylene glycol. They are next to nylon in strength, have good resistance to wrinkling, and a high degree of stretch resistance. They are used in a variety of textile products such as "wash-and-wear" clothes, suits, dresses, sweaters, and underwear. There are also various industrial applications.

fibers, polyolefin. Both polyethylene and polypropylene are extruded into fibers in the molten form, followed by cooling in a water bath. Polypropylene apparently has better fiber-forming properties than polyethylene. While the polymers are readily colored before spinning, the fibers are difficult to dye. The polyolefins have a limited use in dress fibers and are used in protective clothing and in carpets. They are made into ropes for use on boats and in water skiing.

fibers, rayon. A regenerated cellulose fiber in which not more than 15% of the hydrogens in the hydroxyl groups have been replaced in a cross-linking operation. The two types, viscose and cupramonium, have similar properties, although the latter produces a fine yarn used in the sheerest of fabrics. Rayon yarns may be monofilament, multifilament, or staple, with a high luster or delustered or colorless pigments. Rayon is readily dyed and is resistant to mildew and moths. Textile rayon is used in a wide variety of fabrics, both alone and blended with other fibers. High tenacity rayon, made from purer cellulose with a resultant higher degree of polymerization, is used as a tire cord. Rayons can be cross-linked with urea-formaldehyde or melamine resins, glyoxal, diisocyanates, and alkyl titanates to improve strength and water resistance.

fibers, Saran. Saran is a copolymer of vinylidene chloride and vinyl chloride or acrylonitrile. The fibers are stronger than wool or rayon with low conductivity. Their low absorbency makes them unsuitable for most clothing use. They have excellent weather resistance, which makes them suitable for upholstery on outdoor furniture, automobile seatcovers, and window screens.

fibers, Vinyon. Composed of polyvinyl chloride or a copolymer such as 88% chloride and 12% acetate. They have excellent chemical resistance and low moisture absorption, making them suitable for protective clothing, fishing lines and nets, and filter fabrics. They are also used in felts and carpets.

filament. A variety of fiber having an extreme length not readily measured.

filament winding. Continuous strands of glass roving or other filaments are machine positioned and wound on a mandrel. The roving is impregnated during or after winding with resin that is cured to form a continuous bond. The mandrel is removed after the material has set. For enclosed shapes, a mandrel may be collapsible or made from rock salt which is dissolved to form metal alloys that melt in boiling water. Filament winding is done to produce tanks and pressure vessels, rocket tubes, motor cases, exit covers, radomes, and corrosion-resistant nonelectrical pipes. The process is best suited for surfaces of revolution, but can be applied whenever tension can be held as with square or triangular sections. Tolerances of ± 0.005–± 0.010 in. (± 0.127–0.254 mm) on various dimensions have been found practical. Many winding patterns are possible, such as circular, helical, variable, and lengthwise. Good design orients the filaments as nearly as possible in the direction of the tensile stresses in the structure.

fine blanking (see blanking, fine)

finish. The characteristics of the surface of a product.

finite element method (FEM). This analysis is roughly synonymous with computer-aided engineering (CAE), since it is by far the most frequently used tool for structural finite element analysis (FEA) and is also used for many other types of engineering analysis. This is an iterative process in which suggested or required design changes are made known to the designer. The redesigned product is then again subjected to analysis. Once a product is acceptable, it can be released to manufacturing.

firebrick, fireclay brick. Except when special conditions dictate the use of special materials, fireclay bricks are employed. Fireclay bricks are not generally suitable for metal melting, for slag and fume attack, nor for conditions of severe abrasion. Burned fireclay shapes are available in several grades of refractoriness; superduty, high heat duty, medium duty, and low duty. Such fireclay brick contains about 40% alumina and 60% silica. Both the stiff mud and the dry press process are used in their production.

firebrick, graphite. Because graphite oxidizes readily in the presence of air, graphite brick is used below the hearth line of metal-melting furnaces such as blast furnaces. Graphite is an excellent refractory since it cannot be spalled (because of high thermal conductivity) and its strength increases with temperature. If protected from oxidation, it may be used at 6000°F (3316°C). Few refractories are usable above 3000°F (1649°C).

firebrick, high alumina. For more-severe conditions such as in the manufacture of cement, high alumina brick is used in grades of 50%, 60%, 70%, up to 99% alumina, or mullite brick, 72% alumina. Alumina is resistant to the attack of many slags and fumes and has higher strength than fireclay.

firebrick, magnesite and chrome ore. Magnesite brick is magnesium oxide, the brick being a dark brown color. Chrome refractories, which are a rich green color and heavy, are largely chromium oxide with liberal amounts of alumina and iron oxide. A great amount of magnesite refractories are made from the magnesium chloride in sea water.

These are basic refractories and must be used to resist the attack of basic slags in metal melting. Most slags are basic, especially for steel, copper, and nickel; therefore, the linings of melting furnaces are largely magnesite brick. Chrome is usually employed as a plastic refractory, being then referred to as "chrome ore." Both materials have poor hot strength and poor resistance to spalling. The mortar for magnesite brick is rather strange. For mortar a thin steel plate is used, which makes it the strongest of all mortars. The magnesite bricks are laid with thin steel plates between them, then the furnace is heated and the steel oxidizes to ferric oxide, which combines chemically with the magnesia. After this chemical reaction it is hardly possible to break the brick joint with anything less than a pneumatic hammer or explosives.

firebrick, silica. Silica maintains its strength at high temperatures and is used in applications where great strength is necessary such as in the arched roofs of metal-melting furnaces. It is an acid brick, and therefore must be used in contact with acid slags in metal melting. Silica brick are rarely cooled below 1200°F (649°C). Below this temperature spalling is almost certain owing to a high coefficient of expansion. Silica brick, therefore, cannot be used in intermittently operated furnaces.

firebrick, silicon carbide. A versatile material, it is familiar as an abrasive, and is also a very fine refractory. Like graphite, it has a high thermal conductivity and great spall resistance. The thermal conductivity of firebrick is about 10 Btu/in. (4.15 kJ/cm), while that of silicon carbide is about 100 Btu/in. (41.5 kJ/cm). Silicon carbide is highly refractory, with a high strength and a high cost, and is an outstanding performer under conditions of abrasion.

firebrick, zircon and zirconia. Zircon is zirconium silicate; zirconia is zirconium oxide. The latter is probably the most photogenic of the refractories, being a beautiful yellow color. Zircon is resistant to sand, which is generally ruinous to refractories, and therefore finds its chief applications in glass melting. Both zircon and zirconia are suitable for the vacuum melting of metals.

flame cutting, metals. Involves heating metals either to a sufficiently high temperature so that it will oxidize in a stream of oxygen or to its melting temperature. In flame cutting a gas–oxygen flame is used for heating and a stream of oxygen then oxidizes the metal. Ferrous metals of almost any thickness can be cut by this process. Both ferrous and nonferrous metals can be cut by either the arc–air process, in which the metal is melted by an electric arc and then blown away by means of a high-pressure stream of air, or by means of a plasma-arc torch, which produces extremely hot plasma that impinges on the metal to melt it and blow it out of the kerf. The various heat-cutting processes can produce flat, internal and external cylindrical and conical, and irregular surfaces. While the accuracy is not as good as can be obtained by most of the chip-type or chipless machining processes, accuracies of $\pm 1/32$ in. (0.794 mm) are readily obtainable. Most heat cutting is done to rough-prepare metals for further machining or assembly operations, but in many instances parts are heat-cut to final form. **(See also torch cutting.)**

flame hardening, metal. A process whereby the surface temperature of the steel is raised above the critical temperature by means of an oxyacetylene torch and then quenched by a water spray. The depth of hardness can be varied from $1/2$ in. (1.27 cm) down to a thin skin. The core retains the softness and the toughness of the original condition of the steel. Flame hardening is used for very large pieces when it would be difficult to place the entire structure in a furnace or where warping might occur. **(See also steel, flame hardening.)**

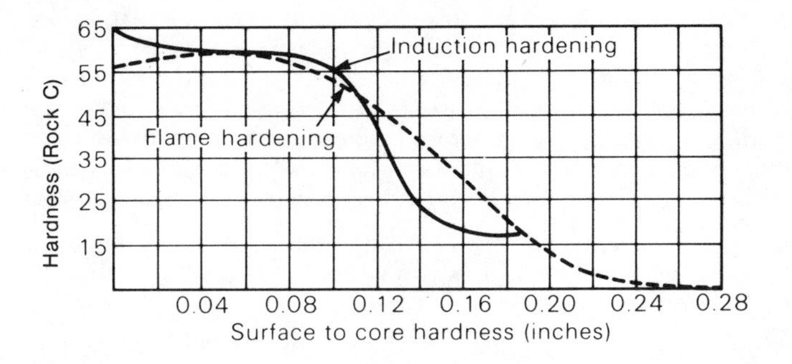

Graph showing typical hardness curves for flame and induction hardening.

flame retardants (antimony oxide). A white pigment that functions syneristically with organic chlorine and bromine compounds to reduce flammability. Antimony oxide (AO) is believed to interact with the halogen source to form a mixture of volatile chlorides and oxychlorides. The gaseous antimony halides provide some benefits by displacement of oxygen and also inhibit the combustion process by acting as free radical traps. The halogen source may be either the base polymer PVC, chlorendic anhydride-based polyesters or brominated epoxies, or various additives, e.g., chlorinated paraffins, brominated aromatics, mixed halogen cycloaliphatics or perchlorinated polycyclics.

flame retardants (borates). Zinc and barium borates are white powdery solids that are promoted as economical diluents and replacements for antimony oxide. The borates are believed to function by interacting with hydrogen halide to produce boric acid. This in turn inhibits the oxidation mechanism by promoting char formation and by forming a barrier layer to restrict access of oxygen to additional fuel. Most commonly the borates are used to extend the efficacy of the more expensive antimony oxide in halogenated and substrates, like certain polyesters, PVC, etc.

flame retardants (halogenated organics). A large number of products of widely diverse structure fall into this category: chlorinated paraffins, borminated aromatics, perchlorinated polycyclics, bromochlorocycloparaffins, brominated polyols, etc. In the absence of synergists these materials form hydrogen halide in the combustion process and the hydrogen halide interferes with the free-radical oxidation mechanism in the flame front. With synergists a dual mechanism operates with the synergists' activity competing with or complementing that of the halogen compound. Chlorinated paraffins are regarded as practically nontoxic, but many of the other halogenated organics are potentially more harmful. The combustion products, hydrogen halides, are well-known irritants and corrosive agents.

flame retardants (hydrated alumina). Hydrated aluminum oxides (alumina trihydrate, ATH) are white crystalline powders that function as flame retardants by absorbing heat to release water of hydration. This water is then available to dilute flammable vapors and cool the flame. ATH is introduced into phosphate-plasticized PVC for wall coverings and upholstery. Other useful applications include vinyl and polyurethane carpet backing, and potentially any resin system that permits simple replacement of some filler with ATH.

flame retardants (phosphate esters). These esters of orthophosphoric acid comprise a class of products, usually liquids, that provide dual functionality in plastic compositions. They are both plasticizers and flame retardants. The liquid phosphates are most commonly used with those resins that normally require external plasticizers, mainly vinyls, but also cellulosics. Commercial phosphate plasticizers may be categorized by structure into three types: triaryl, alkyl diaryl, and trialkyl. Flame retardance increases with the number of aromatic rings but varies little among members of the same class. Organophosphorus compounds are generally believed to act by pyrolysis to phosphoric acid and further condensation to exceptionally stable polyphosphoric acids (PPA). Char formation is promoted and with the PPA provides a wet protective barrier to the flame.

flame spraying. Method of applying a plastic coating in which finely powdered fragments of the plastic, together with suitable fluxes, are projected through a cone of flame onto a surface.

flash. A thin protrusion at the parting line of a forging or casting which forms when metal, in excess of that required to fill the impressions, is forced between the die or mold interfaces.

flash line. A line left on a forging or casting where flash has been trimmed off.

flash point and fire point, lubricants. The flash point is the temperature of the oil at which the vapors from the surface "flash" or form an inflammable mixture with air. The fire point is the temperature at which combustion continues. These characteristics are of importance in safety of shipping and storing; they are not indicative of lubricating ability.

flask, metal casting. A flask is a wood or metal frame in which a mold is made. It must be strong and rigid so as not to distort when it is handled or when sand is rammed into it. It must also resist the pressure of the molten metal during casting. Pins and fittings align the sections of a flask. They wear in service and must be watched to avoid mismatched or shifted molds. A flask is made of two principal parts, the cope (top section) and the drag (bottom section). When more than two sections of a flask are necessary to increase the depth of the cope and/or the drag, intermediate flask sections known as cheeks are used.

flexible manufacturing systems (FMS). The 1980 SME Machine Tool Task Force report defines FMS as a series of machining and associated workstations, linked by a hierarchical common control, and providing for automatic production of a family of workpieces. A transportation system, for both the workpiece and the tooling, is integral to an FMS as is computerized control. **(See figure p. 131.)**

flint (see abrasives)

floor-to-floor time. The total time elapsed for picking up a part, loading it into a machine, carrying out operations, and unloading it (back to the floor, bin, pallet, etc.); generally applied to batch production.

flow lines. Lines on the surface of painted sheet, brought about by incomplete leveling of the paint. Also, the line pattern revealed by etching which shows the direction of plastic flow on the surface or within a wrought structure.

flow process chart. This chart portrays the operations, inspection, delays, storages, and moves or transportations required to process material through a production process. **(See also product layout.)**

flow through. A forging defect caused when metal flows past the base of a rib resulting in rupture of the grain structure.

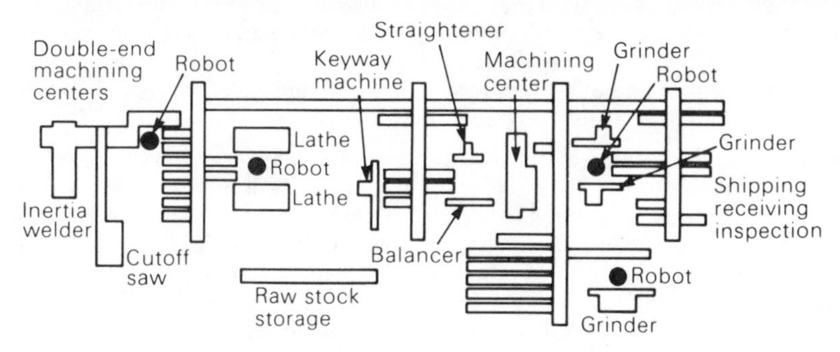

View of flexible manufacturing system shows an SI Cartrac® conveyor network, machine-loading robots and the welding, cutting, and grinding machines which produce rolls for printing presses. The automated line occupies an area about 250 ft (76.2 m) long by 40 ft (12.2 m) wide. The SI Cartrac conveyor system interfaces directly with 4 Cincinnati Milacron robots and 11 separate machining systems.

fluorescent penetrant inspection. Inspection using a fluorescent liquid that will penetrate any surface opening; after wiping the surface clean, the location of any surface flaws may be detected by the fluorescent, under ultraviolet light, of back-seepage of the fluid.

fluorocarbons. In the fluorocarbon polymers some or all of the hydrogen attached to the paraffinic carbon chain is replaced by fluorine. In order of decreasing fluorine content, the fluoroplastics include (1) polytetrafluoroethylene (PTFE or Teflon), which is a polyethylene formula with fluorine substituted for every hydrogen atom; (2) fluorinated ethylene-propylene (FEP), a copolymer, which has properties similar to PTFE, but is not as difficult to process; (3) polychlorotrifluoroethylene (CTFE), with one chlorine and three fluorine atoms in the monomer; (4) polyvinyl fluoride and polyvinylidene fluoride, which are fluorovinyls. Increasing fluorine content in these polymers is accompanied by improved electrical characteristics, higher service temperatures, lower coefficient of friction, and better corrosion resistance. PTFE is a crystalline thermoplastic, opaque, soft, waxy, and white in color. Though its monomer boils at – 106°F (– 76.7°C), the polymer has a service temperature for continuous exposure from absolute zero to 500°F (260°C). It has little strength, and creeps readily at room temperature. Specific electrical resistance is over 10^{18} ohm-cm, and dielectric strength is also high. PTFE is almost immune to chemical attack, except for molten alkali metals such as sodium and a few fluorine compounds. It has the lowest unlubricated coefficient of friction of any material, about 0.04. The electrical industry uses PTFE for a wide range of speciality applications, including insulation, coaxial cable components, tube bases, plugs, and sockets. The chemical and instrumentation industries use its chemical resistance for gaskets, tubing, valve packings, O-rings, valve linings, flexible couplings, and diaphragms. Its low friction and nonstick characteristics explain its numerous applications for rollers, bearings, conveyor coatings, etc. CTFE is somewhat less corrosion resistant and stronger than PTFE, and may be processed on standard equipment for thermoplastics.

fluoroplastics, chlorotrifluoroethylene (CTFEO). A thermoplastic resin produced in varying formulations from combinations of fluorine-containing monomers. It differs from other members of the fluoroplastic family in that its molecular structure contains chlorine. CTFE resins have inherent flexibility, radiation resistance, and excellent moldability. In addition their structures are such that crystallization is retarded during rapid cooling cycles after exposure to elevated temperatures thus tending to maintain resin toughness, flexibility, and clarity. CTFE is normally supplied as a homopolymer or as a copolymer with vinylidene fluoride and other monomers added to improve processability above 500°F (260°C) without excessive degradation.

fluoroplastics, perfluoroalkoxy (PFA). A class of melt-processible fluoroplastics introduced by Du Pont Co. in 1972 under the trademark Teflon, PFA resins resemble FEP fluorocarbon resins in having a branched polymer chain that provides good mechanical properties at melt viscosities much lower than those of TFE. However, the unique branch in PFA is longer and more flexible, leading to improvements in high-temperature properties, higher melting point, and greater thermal stability. Perfluoroakyl side chains are connected to the carbon–fluorine backbone of the polymer through flexible oxygen linkages. These perfluoroalkoxy branches lead to the general symbol, PFA.

fluoroplastics, poly-ethylene-chlorotrifluoroethylene (E-CTFE). A high-molecular-weight, 1:1 alternating copolymer of ethylene and chlorotrifluoroethylene, E-CTFE copolymer is a strong, highly impact resistant material that retains useful properties over a broad temperature range. Its tensile strength, impact resistance, hardness, creep properties, and abrasion resistance at ambient temperature are comparable to those of nylon 6. E-CTFE retains its strength and impact resistance down to cryogenic temperatures. In its radiation cross-linked form, it can withstand temperatures that range to as high as 400°F (204°C) for limited periods of time. E-CTFE is resistant at elevated as well as ambient temperatures to all corrosive chemicals commonly encountered in metal finishing, textile, paper, leather, mining, and other chemical-handling industries. These include mineral acids (sulfuric, hydrofluoric, hydrochloric), oxidizing acids (nitric, chromic, perchloric, aqua regia), inorganic bases (ammonia, sodium hydroxide), and inorganic etchants (ferric, zinc and cupric chlorides, and hydrogen peroxide). It is also resistant to essentially all organic solvents except hot amines. It becomes slightly plasticized in a few halogenated and oxygenated solvents. Barrier properties are excellent. Permeability is low to such gases as chlorine, hydrogen chloride, hydrogen sulfide, water vapor, oxygen, nitrogen, and carbon dioxide.

fluoroplastics, polytetrafluoroethylene (TFE) and fluorinated ethylene–propylene (FEP). Two fluorocarbon resins in the fluoroplastics class composed wholly of fluorine and carbon. The monomers, colorless low-boiling-point gases, are manufactured by pyrolysis of chlorodifluoromethane (Freon 22). They are then polymerized under pressure with suitable initiators to very-long, high-molecular-weight, chainlike molecules.

fluoroscopy. An inspection procedure in which the radiographic image of the subject is viewed on a fluorescent screen; this procedure is normally limited to low-density materials or thin sections of metals because of the low light output of the fluorescent screen at safe levels of radiation.

fluorosilicone (FSi). This compound combines the good high- and low-temperature properties of silicone with basic fuel and oil resistance. The primary uses of fluorosilicones are in fuel systems at temperatures up to 350°F (177°C), and in applications where the dry-heat resistance of silicone is required, but may have exposure to petroleum oils and/or hydrocarbon fuels. In some fuels and oils, however, the high-temperature limit is more conservative because temperatures approaching 350°F (177°C) may degrade the fluid, producing acids that attack fluorosilicone elastomers. High-strength-type fluorosilicones are available. Certain of these exhibit much improved resistance to compression set.

fluorspar. The only simple fluoride that occurs in nature is known as fluorite in mineralogy and as fluorspar in commerce. It is translucent to transparent and ranges from clear and colorless through shades of blue, purple, green, and yellow. It crystallizes in the isometric system and has perfect octahedral cleavage. The specific gravity is 3.18 and hardness 4. Fluorspar occurs as aggregates of cubic crystals, as granular masses, and in banded or crusted veins. If a small proportion of fluorspar is added to the charge in the basic open hearth method of making steel, it aids fusion, makes the slag fluid, and helps to transfer sulfur and phosphorus into the slag. About 6.5 lb (3 kg) of fluorspar is used per ton (907 kg) of steel produced. Fluorspar also aids in the smelting of gold, silver, copper, and lead. Another major use of fluorspar is in the manufacture of hydrofluoric acid (HF). Large amounts of this acid are used (1) by the aluminum industry, for the production of synthetic cryolite and aluminum fluoride, compounds that serve as electrolytes in the reduction of alumina to the metal; (2) in refineries, as a catalyst in the production of high-octane gasoline; and (3) in several minor applications such as the etching and polishing of glass. Hydrofluoric acid is also the chief source of fluorine for the chemical industry, including the field of fluorocarbon chemistry. Among the many products that contain fluorine are refrigerants, insecticides, and certain plastics. Fluorspar is used in the manufacture of opaque glass and colored or "cathedral" glass. It also goes into certain enamels.

fluxes (electronics). A solid or liquid that when heated exercises a cleaning action on surfaces to be soldered. Early soldering fluxes were rosin and acid (zinc chloride is an example of the latter). The acid fluxes were designed for soldering metal parts and were not considered for electrical applications because they are highly corrosive. Liquid rosin fluxes, on the other hand, were designed to be used in the manufacture of electronic circuitry. The flux composition based on a solution of rosin in alcohol constitutes an R-type flux. Further developments led to the addition of mild, noncorrosive activators yielding an RMA flux, and finally to the use of more aggressive and corrosive activators in the formulation resulting in an RA flux. These fluxes were distinguished one from another by a series of tests, which include a copper mirror test, a silver chromate paper test, and a flux extract resistivity test, still the major qualifying factors for fluxes to be used for military products. An RMA flux is distinguished from an RA flux according to its extract resistivity. The assumption is made that any activator that increases the conductivity of a water solution of the flux must also be potentially more corrosive.

Rosin fluxes have been developed that contain weak organic acids whose water solution is conductive. These weak organic acids in the flux, however, are no more corrosive than rosin itself, which is a natural mix of chemicals which include organic acids. One type of nonrosin flux is water soluble (WS) and provides high soldering efficiency but requires carefully controlled cleaning if later corrosion problems are to be minimized. Some nonrosin fluxes are not soluble in water. One flux family in this category is the solvent-soluble fluxes, which are as active as the WS fluxes, but which require solvent cleaning.

FMS (see flexible manufacturing system)

foam. Foams are dispersions of a large amount of gas in a small amount of liquid. When the volume of the liquid is much larger than that of a gas, the bubbles formed are widely separated from each other. Such a system is known as sphere-foam or gas emulsion. In a true foam, the bubbles are so crowded that they deform and assume nearly polyhedral shape with thin plane films of liquid between them. Such foams are called polyhedral foams. Foams, like emulsions and fogs, are fundamentally unstable systems because of very large surfaces and interfaces. Surface-active agents play an important part in implanting the stability to foams by lowering the interfacial surface tension and increasing the surface viscosity of the film. This, in turn, reduces drainage and prolongs the life of the foam. Antifoam agents appear to reduce bulk viscosity by displacing the foaming agent from the interface and producing a new interfacial film of lower viscosity. This lowers the film elasticity considerably. Solid foams with rigid strong wall films are indefinitely stable; they may be flexible and rigid depending on the nature of the film-forming materials. Foams may have open-cell structures or a combination of both. Closed-cell foams are produced when some pressure is maintained during the cell-formation sources. Open-cell foams are formed during a free expansion. Foaming can be accomplished by mechanical, physical, and chemical processes.

foam rubber, cellular polystyrene. Expandable polystyrene is marketed in the form of beads containing an expanding agent. When heated in a mold, they expand to fill the mold. The expanded material has closed-cell structures with densities as low as 1 lb/ft^3 (16 kg/m^3). Uses include thermal building insulation, flotation material, shipping containers, disposable drinking cups, and crash helmet liners. Expanded polystyrene is available in blocks and sheets, which are fabricated mechanically. In addition to use as a structural insulating material and in marine uses it is used in protective packaging, in display platforms, and in the production of many novelty items.

foam rubber, phenolics. The conventional type of phenolic foam utilizes the heat of reaction of the liquid resin polymerizing to vaporize the water and solvent. The product, which is a composite of 40% open cells and 60% closed cells, is used mainly as a building insulation in either preformed slabs or foamed-in-place material. Densities can be varied from 0.33 to 80 lb/ft^3 (5.3 to 1280 kg/m^3). Tiny phenolic spheres with densities from 10 to 40 lb/ft^3 (160 to 640 kg/m^3) are mixed with binder resin to form syntactic foam. This foam is used in sandwich construction as in boat hulls, and decks and airplane structures. Urea-formaldehyde open-cell foams are similar to foamed phenolics.

foam rubber, polyvinyl chloride foam. Three types of polyvinyl chloride cellular products are available: (1) open-cell flexible foam, (2) closed-cell flexible foam, and (3) closed-cell rigid foam. Uses are similar to those of the corresponding rubber products. Flexible vinyl foam can be laminated to fabric, producing a leatherlike material used in clothing and upholstery. It can be extruded into weather-stripping and gasket material. It is injection molded into such items as sandal soles and gaskets. It can be rotationally molded to produce doll bodies and heater ducts.

foam rubber, urethane. Urethane foams may be produced by vaporizing a blowing agent or by the carbon dioxide resulting from the reaction of water and isocyanate groups in the reaction mixture. Slabs may be formed continuously on conveyors, with curing in an oven. The foams may also be formed by pouring the foaming mixture into a cavity, such as in airplane members, and allowing them to cure in place. The foams are used in thermal insulation, cushioning, and electrical potting.

foil. A rolled product rectangular in cross section of thickness less than 0.006 in. (0.152 mm).

foil, annealed. Foil completely softened by thermal treatment.

foil, bright two sides (B2S). Foil having a uniform bright specular finish on both sides.

foil, chemically cleaned. Foil chemically washed to remove lubricant and foreign material.

foil, embossed. Foil on which a pattern has been impressed by means of an engraved roll or plate.

foil, etched. Foil roughened chemically or electrochemically to provide an increased surface area.

foil, extra bright two sides (EB2S). Foil having a uniform extra bright specular finish on both sides.

foil, hard. Foil fully work-hardened by rolling.

foil, intermediate temper. Foil intermediate in temper between annealed foil and hard foil.

foil, matte one side (M1S). Foil with a diffuse reflecting finish on one side and a bright specular finish on the other side.

foil, matte two sides (M2S). Foil having a diffuse reflecting finish on both sides.

foil, mechanically grained. Foil mechanically roughened for such applications as lithographing.

foil, mill finish. Foil having a non-uniform finish, which may vary from coil to coil and within a coil.

foil, roll. Foil in coiled form.

foil, scratch brushed. Foil abraded, usually with wire brushes, to produce a roughened surface.

foil, sheets of. Foil in rectangular form sheared to size.

forgeability. The term used to describe the relative workability of forging material.

forge welding. Modern forge welding is the diffusion of stacks of wrought sheets, which have been cut to shape, under heat and pressure. The sheets are stacked in a die cavity and pressure is applied through a punch. Titanium, for example, is welded at 2000 psi (13,800 kPa) and 1600°F (870°C). By using the proper combination, grain growth can be avoided and the wrought properties of the sheet retained. This process is a substitute for forging. The need for extremely large forging presses is avoided and better endurance limits can be obtained, because forgings made with marginal pressures tend to have fine grain structures only near the surface.

forging. The working of metal by localized compressive forces exerted by manual or power hammers, presses, or special forging machines. Forging may be done either hot or cold, but when done cold, various specialized names are usually given to the processes. Consequently, the term "forging" usually is associated with hot forging. In the forging process the metal may be either (1) drawn out, thus increasing its length and decreasing its cross section; (2) upset, resulting in an increase in section and a decrease in length, or (3) squeezed in closed impression dies, causing a multidirectional flow. The state of stress in the work primarily is either uniaxial or triaxial compression. The products of forging generally are discontinuous, treated, and turned out as discrete pieces rather than as flowing mass. Forging may be done in open or closed dies. The four principal types of forging operations are drop, machine, press, and smith.

forging, cold-coined. A forging that has been restruck cold in order to obtain closer dimensions, to sharpen corners or outlines, and, in non-heat-treatable alloys, to increase hardness.

forging, drop. A process used to shape metal in closed dies. Forms are machined into two halves of a closed die set. The metal is heated and under impact the metal becomes plastic. Under repeated blows the metal fills the die cavity. Excess metal (flash) is trimmed off. Most drop-forging dies have several cavities machined into their surfaces, to make it possible to preshape a forging before it is hammered into its final form. Bending, upsetting, and drawing may be incorporated in a set of drop-forging dies. Because the size and shape of the finished parts are controlled by the die cavity, parts may be duplicated readily. The grain structure obtained produces a workpiece of high tensile strength. The machine applying the blows is called a drop hammer; it may be actuated by gravity or by a combination of gravity and steam or air. There is practically no limit to the size or quantity of articles produced by drop forging, pieces ranging from less than 1 oz (28.4 g) to 800 lb (363 kg) being produced.

forging, hammer or smith forging. Normally done with a mechanical hammer. The heated metal is placed on an anvil and struck repeated blows with a power hammer. The metal has to be heated throughout to the proper temperature. Gas, oil, or electric furnaces usually are used, but induction heating is now used for some applications, mostly for upset or impact forging.

forging, hot press. A development of drop forging, which is generally used for producing simple shapes. In this method the hammer is replaced by a hydraulic-driven ram so that the work piece is shaped by steady pressure rather than by a succession of hammer blows. Deformation is then more uniform throughout the workpiece. Large forgings can be made direct from ingots of suitable size. Smaller forgings are generally produced from stock in the form of hot rolled bars. In such stock, impurities will already have been distributed directionally by the rolling process producing fiber. During forging this fiber will follow the contours of the die and so strength will be maintained.

forging, hot roll. Forging rolls are used to shape metal into long, thin sections; these rolls are semicylindrical, with several grooves machined into them through which the stock is passed progressively from one size groove to the next. By placing the metal stock between the rolls in a specific groove and permitting the same to roll toward the operator, the stock conforms to the size and shape of the groove, and with the proper contouring of the grooves, the heated stock may be formed into either tapered or straight sections as desired. The process of hot rolling consists of passing the hot metal between two rolls that are revolving at the same peripheral speed but in opposite directions, i.e., clockwise or counterclockwise, and so spaced that the distance between them is slightly less than the sections of stock entering the rolls.

forging ingot. A cast form suitable for forging. **(See also fabricating ingot.)**

forging, machine. Machine or upset forgings are made in a horizontal press. The hot metal in the form of a bar of uniform cross section is subjected to a squeezing action resulting from a rapid push. The section on the stock is more one of "gathering up" or "upsetting" than of flattening by hammering. The stock is held between gripping dies and then subjected to pressure from an upsetting or heading tool. The pieces produced by this operation vary in weight from several ounces to 500 lb (227 kg).

forging, mechanical. Either billet or bar stock is worked progressively into the shape required by the use of dies with the various cavities and contours previously cut into them. There are several steps used in the production of closed-die forgings, including fullering, edging, rolling, bending, blocking, finishing, and trimming.

forging, press. Press forging machines operate in a manner similar to upset forging machines but are vertical rather than horizontal. A heated slug is placed between dies and squeezed by a single blow to fill the die cavity. Comparatively simple and symmetrical shapes are produced in this way. The pieces produced are usually small and may be of nonferrous or ferrous metals. Either plain faces or dies are employed, and the forging is formed by slow steady pressures. In drop forging much of the energy of impact is dissipated near the surface of the metal and in the foundation of the press. As a result it is not possible to forge large sections on a forging hammer. The slow squeezing action of a forging press, on the other hand, penetrates throughout the metal. Forging presses are made in very large sizes, as large as 50,000 tons (45,360 tonnes) operating in the United States. Many press forgings are completed in a single closing of the dies. Press forgings usually require somewhat less draft than drop forgings and thus are more accurate dimensionally. **(See also deep drawing.)**

forging, smith. Smith hammers are steam operated and use flat dies with plain faces. By repeated blows, the cross section of a piece is reduced to that desired. The weight of finished pieces varies from under 1 lb (0.45 kg) to 20 tons (18 tonnes). Smith forging is used extensively to shape stock before it is forged in a drop forge.

forging stock. A wrought rod, bar, or other section suitable for forging.

fork lift truck. A self-loading industrial truck, equipped with load carriage and forks for transporting and tiering loads. **(See also industrial lift truck.)**

formaldehyde thermosetting polymers. A wide range of thermosetting plastics of considerable industrial importance is produced by reacting urea, melamine, phenol, and other compounds with formaldehyde. The first of the plastics was the thermoplastic cellulose nitrate. Phenol-formaldehyde was the first of the thermosetting plastics developed. It was named Bakelite, and the trade name is still applied to phenol-formaldehydes as a group. The phenol plastics are strong, elastic, and quite brittle. They are used in wall plates for electric toggle switches; mechanical parts of electric motor starters; miscellaneous hardware, such as lamp bases and receptacles; radio and appliance knobs; and toilet seats. Urea and melamine formaldehydes may also be used in the manufacture of these articles. Phenols are usually brown in color, and color limitations have restricted their range of uses. Phenolic, urea, and melamine resins are also used as impregnating, bonding, and laminating agents, especially in plywood and chipboard manufacture and in bonded grinding wheels. The general characteristics of melamine resins are familiar from their use in plastic tableware. Cellulose wood flour, asbestos, rag, and glass filler are often added to melamine and other molded formaldehydes to provide increased strength.

forming, contour. This process, as it is known today, covers the forming of metal sections of all types—sections, tubes, shapes, etc.—normally received in straight lengths, into various contours. Since the bending moment or forming action needed for the forming of sections can be applied in a great many ways, there is a wide variety of machines available to achieve these ends. Classified generally, these methods are as follows: The draw bending process for hollow sections is accomplished on a draw machine having a rotating die, to which the piece being bent is clamped, and a roller, against which it is drawn. The neutral axis lies between the center of the section and the inner fibers, with resultant thinning of the outer portions and slight thickening of the inner walls at bends. In general, the strength of such a bend is lower than that of an unbent portion. The flattening effect on these machines is severe where sections are not fully supported, consequently, tubular sections require an internal mandrel. Roll machines are composed of a pyramid of rolls, one of which is adjustable with respect to the other two. A section to be bent is passed through the rolls which, acting somewhat like a small rolling mill, produce the desired curve or radius on the piece. Both vertical and horizontal models in a wide range of sizes are available. Rings or continuous coils may be produced and a variety of radii may be had with only a simple adjustment of the rolls. A ram bender consists of two supports for the section to be bent and a central die operated by a hydraulic cylinder for applying the load. Simulating the action of the common stamping press, the ram machine is used for bends on extremely heavy sections such as pipe up to 14-in. (36-cm) diameter, railroad rails, reinforcing bars, structural sections, etc. Accuracy is poor and production low, making such machines more suitable for structural and maintenance work. The principle of ram bending, however, is utilized in other forming machines. The compression machines utilize a stationary die onto which the part to be formed is held while a rotating arm wipes the piece into the die contour. Similarly, the tangent bender utilizes a rocking die plate to wipe the shape onto the master die. Still another version utilizes a stationary arm and a moving die table. Shoes, rolls, or tangent dies mounted on the main ram of these machines are used to force the material into the die contour. Stretch machines overcome the internal strains of prior cold working by stretching a section of metal to or slightly beyond its elastic limit during forming. Held in special gripping jaws, the piece is first stretched to approximately its elastic limit and, with constant tension, is then laid onto the die form. Another type machine, without a rotating table and somewhat more limited in application, uses two stretching heads and a central die form, the stretch set being induced after forming. Many combinations of curves are possible—constant radii, varying radii, single-plane or multiplane, simple or compound—in springy materials such as tempered aluminum alloys and stainless steels with part-to-part uniformity and negligible springback. Complicated reverse curves, of course, make for slower production. Compression–stretch machines are used where the elongation characteristics of a specified material are insufficient to withstand the severity of stretch forming. A combination of stretch and compression can be resorted to for maintaining uniformity and obviating excessive springback.

free standing rack structure. This is a rack storage arrangement installed inside a manufacturing building of conventional construction, supported only by the floor and not supported or attached to any building structure.

friction saw (see cut-off machines)

friction welding (see welding, inertia)

fuel cells. Devices for the "direct" generation of electric energy. In a fuel cell a fuel such as hydrogen, natural gas, or propane can be converted directly without moving parts into twice the amount of electricity that would result from the burning of a corresponding amount of fuel indirectly through boilers, turbines, and generators to produce electricity. The fuel cell differs from the storage battery in that usually its gaseous or liquid fuel and its oxidizer are led in from outside, whereas the storage battery stores its solid fuel and oxidizer on plates, where they are consumed. A fuel cell operates electrochemically, the cell is actually a reactor, wherein hydrogen as the feed stream or fuel is conducted into the empty space paralleling the porous, electric conducting anode. The anode can be made of porous carbon with a metal catalyst like platinum, which chemically changes the hydrogen atoms to positively charged hydrogen ions and electrons. The electrons leave the anode, perform work, and enter the cathode. Meanwhile hydrogen ions migrate through the electrolyte, attracted by the oxygen from the cathode. To complete the reaction, the oxygen pulls in the electrons also, the water is generated and discharged from the cell. The cell is so arranged that the electrons must move up the anode, leave the cell through a wire, and enter the cell again at the cathode. While the electrons are outside the cell, they form the electric current and do work.

fuel oil. The term fuel oil may cover a wide range of petroleum products. It may be applied to crude petroleum, to a light petroleum fraction similar to kerosene or gas oil, or to a heavy residual left after distilling off the fixed gases, the gasoline, and more or less of the kerosene and gas oil. To provide standardization, specifications have been developed for five grades of fuel oil. Grades No. 1 and No. 2 are sometimes designated as light and medium domestic fuel oils and are specified mainly by the temperature of the distillation range. Grade No. 6, which is designated as heavy industrial fuel oil and some times known as Bunker C oil, is specified mainly by viscosity. The specific gravities of Grades, 4, 5, 6, are not specified because they will vary with the source of the crude petroleum and the extent of the refinery operation in cracking and distilling. In regard to specific gravity, this is the ratio between the weight of any volume of oil at 60°F (16°C) and the weight of an equal volume of water at 60°F (16°C). The common designation is Sp Gr 60/60 F and is expressed as a decimal carried to four places. Gravity determinations are made by immersing a hydrometer into the sample and reading the scale at the point to which the instrument sinks in the oil. The specific gravity is either read direct or the gravity is measured in degrees API. The heating value may be expressed in either Btu per gallon at 60°F (16°C) or Btu per pound. The viscosity is defined as the measure of the resistance to flow. The Saybolt Universal viscosity is expressed in seconds of time that it takes to run 60 cm^3 through a standard size orifice at any desired temperature. Viscosity is commonly measured at 100°F (38°C), 150°F (66°C), and 210°F (99°C). The oil is held at constant temperature within $\pm$0.25°F ($\pm$0.14°C) during the test period.

furans (plastics). Furan resins, derived principally from furfuraldehyde and furfuryl alcohol, offer advantages to specific industries such as plywood, utilizing furfuryl alcohol, urea formaldehyde adhesives; foundry trade, using acid-catalyzed furfuryl alcohol as a binder for sand cores; and construction, using chemical-resistant grouting compounds for chemical plants. In proceeding from the basic furfuraldehyde and furfuryl alcohol into other derivatives, the scope of furan product usage is considerably enlarged. For example, tetrahydrofurfuryl alcohol esters are excellent plasticizers for polyvinyl chloride and other thermoplastics. The furan molecule also serves as a starting point in the synthesis of raw materials for some types of nylon and urethane resins. Furfuraldehyde, as a coreactant with phenols and with ketones yields useful thermosetting resins. These forms of furan resins are noted for their chemical resistance and adhesive properties. Furfuraldehyde exhibits good wetting qualities. Small amounts in conjunction with thermosets such as phenolic produce better wetting of abrasive grains, as in the manufacture of grinding wheels.The low viscosity of furan resins permits high filler loading, valuable to manufacturers of chemically resistant material for flooring. Casting and curing in the vicinity of 77°F (25°C) are common. Furfuryl alcohol is an excellent solvent for many organic and some inorganic compounds.

furfuraldehyde (furfural). A very important industrial aldehyde obtained from agricultural wastes such as corn cobs and oat hulls. It is a colorless liquid when freshly distilled, but on exposure to air it becomes oxidized to a deep brown or black liquid. Treatment of furfural (a cyclic ether as well as a carbonyl compound) with a mineral acid results in a ring cleavage. In neutral or basic solutions furfural gives all the reactions of benzaldehyde. Furfural is used in the petroleum industry for the refining of lubricants, as starting material in the manufacture of some nylon, and in synthetic fabrics.

furnaces, arc (see arc furnaces)

furnaces, bath. Bath furnaces may be gas or oil fired or electrically heated. Gas- and oil-fired salt-bath furnaces have low first operating costs and are versatile. They can be restarted easily, and pots can be interchanged in one furnace to use a variety of salts. The bath may be heated externally or by immersed radiant tubes. Temperature control is not as uniform in externally heated furnaces as in others. There are several kinds of electrically salt-bath furnaces. All are surrounded by an insulated casing. Heating may be done by resistance elements around the pot in the externally heated type. Well-insulated heating elements are put directly in the bath in the immersion heating-element type of furnace. Temperatures are usually limited to 1100°F (593°C) for satisfactory resistor life. Higher temperatures can be held by passing electricity through the bath between electrodes. The immersed electrode salt-bath furnace has electrodes immersed in a metal pot. The submerged-electrode furnace has water-cooled electrodes extending through the sides into a ceramic brick pot. The molten salt penetrates the refractory material until it reaches a zone cool enough to freeze and thus seals the pot. Electrode furnaces use alternating current transformed to low voltages (5–15 V) because direct current decomposes the liquid salt. Temperatures are easy to control within 5°F (2.8°C). **(See also heat treatment.)**

furnaces, blast (steel industry). Blast furnaces are important where the substance and fuel are in contact. The function of the fuel is not only to heat the ore, but to produce gases that react chemically with it. The principal use of blast furnaces is for smelting iron. The shaft of an iron-smelting blast furnace may be from 40 to 80 ft (12.2 to 24.4 m) high, and its diameter at the top 10 ft (3.1 m), widening to 14 ft (4.3 m) at about 15 ft (4.6 m) from the bottom. It is filled to the top, and kept full during continuous operation with the correct mixture of iron ore (iron oxide or carbonate), metallurgical coke, and limestone. The purpose of the limestone is to form quicklime on heating, which removes sulfur from the iron; it also acts as a flux, collecting the slag-forming material from the ore and the ash from the coke to form a liquid slag, and so separating them from the iron. Air is forced through tuyeres into the bottom of the shaft, where the temperature is over 2732°F (1500°C), high enough to melt the iron and the flux so that they run down into the hearth from which they are tapped periodically. The air admitted is insufficient to burn the coke fully to carbon dioxide, and the gases which pass upward through the central and upper part of the shaft contain a high proportion of carbon monoxide and no oxygen. The carbon monoxide withdraws oxygen from the iron ore, reducing it to metallic iron; a relatively small proportion of the carbon monoxide is oxidized to carbon dioxide. Thus the blast furnace performs two functions; it reduces iron ore to iron and it separates the iron by melting, from the mixture of minerals in which it then lies. The gases which pass out of the top of a blast furnace still contain much carbon monoxide. This blast-furnace gas is usually dedusted and employed as a fuel, in hot-blast stoves for preheating the air of the blast furnace, or in coke ovens or reverberatory furnaces.

furnaces, car-bottom (hearth furnaces). Has a movable hearth like a flat car that is rolled out for unloading and loading. Commonly the load is stacked on heat-resistant alloy or refractory piers and spacers to facilitate circulation of gases and heat the materials uniformly.

furnaces, continuous. A continuous furnace is the type mostly used for in-line large-quantity production. It typically has a horizontal work chamber and a mechanical means of conveying the work from one end to the other. Heating may be by electricity or by burning fuel. Many furnaces have different temperatures precisely controlled in several zones for multistage heat treating. The name of cycling or semicontinuous furnace is given when the work is held for a different preset time in each section of the furnace. In either case, small pieces may be carried through in baskets or on trays. Common forms of semi- and full-continuous furnaces are the following: A motor-drive belt or chain-conveyor furnace transports the work on an endless hearth of heat-resistant chain or links. In a roller-hearth furnace to work rides on driven rollers, one set after another, through the length of the furnace. This type is suitable for uniformly sized parts on trays. The work sits on rollers or skids that are not driven in a pusher furnace. Instead the workpieces or trays are pushed, one against the other, by mechanical, pneumatic, or hydraulic means in a timed cycle. A screw-conveyor furnace has a coarse-pitch powered screw extending through the chamber. The hearth and work are gradually accelerated forward in a shuffle- or shaker-hearth or reciprocating furnace. Then the hearth is suddenly stopped, and the work slides forward. The work is shifted along by beams that rise as they move forward and fall as they move backward in a walking-beam furnace. **(See also heat treatment.)**

furnaces, indirect-fired (hearth furnaces). The indirect furnace has a heating chamber and a muffle, which separates the combustion space from the work space. The upper temperature limit for this furnace is approximately 2000°F (1094°C). Reduced scaling and contamination from the fuels are the advantages in its use.

furnaces, metals. Furnaces may be classified in numerous ways; however, the main classification is the batch type or continuous type. Batch-type furnaces are those into which the work is placed and remains stationary throughout the time it is in the furnace. Furnaces of this type may be either horizontal or vertical. Horizontal batch-type furnaces are often called box-type furnaces since they resemble a rectangular box. Such furnaces may be heated by either gas or electricity. In some box furnaces, particularly those for lower temperature, a circulating fan is provided to circulate the heated gas so as to obtain more-uniform temperatures. Vertical furnaces of the pit type are usually used for heating long, slender work. These are sunk into the floor, like a hole, with tops that can be swung off, so as to open the furnace. Long pieces can be suspended vertically in such furnaces to minimize warping. In addition, pit-type furnaces often are used to heat batches of small parts, which can be located conveniently into a basket and lowered into the furnace. The heating elements are contained within a bottomless bell, which is lowered over the work. This type is often used for annealing batches of steel or other metal sheets. After the work has been heated the bell furnace can be lifted off and transferred to another batch while the first lot is cooling. Sometimes an insulated cover is placed over the heated work after the furnace is lifted if slow cooling is desired. The elevator-type furnace is a modification of the bell furnace. This type of bell is stationary and the work is raised up into it by means of an elevator, the platform of which forms the bottom of the furnace.

furnaces, muffle and retort (hearth furnaces). Muffle and retort furnaces are indirect fired with the work in a protective or carburizing atmosphere and separated from the gases of combustion. The work space of a muffle furnace is surrounded by refractory material sufficiently tight to keep out contaminants. Semimuffle furnaces have a small amount of baffling in the work space to prevent direct impingement of the combustion flame upon the workpiece. A retort furnace takes a heat-resisting retort that is loaded to work outside the furnace capped and sealed, and then places in the furnace to be heated. A radiant-tube furnace has a tightly encased refractory-lined chamber with the hot gases forced through radiant tubes arranged around the work space. Electric furnaces look much like other kinds.

furnaces, recirculation (hearth furnaces). A recirculation furnace is indirect fired but the gases of combustion are circulated through the work space from the combustion area. The hot gases are channeled so the heating will be uniform. Application is mostly below 1300°F (704°C), commonly in ovens for tempering, toughening, and stress relieving.

furnaces, rotary. Rotary furnaces are used for both batch and continuous production. One kind is the rotary hearth furnace that is built in a wide range of sizes for from a few hundred pounds to 50 tons (45.4 tonnes) or so per hour. It has a round shell and horizontal hearth that turns slowly. Material is charged through a door and may be taken out of the same door or another one. The speed of the hearth is set so that the heating cycle is completed when the hearth makes one turn. A rotary-retort furnace consists of a revolving retort inside a heating chamber. Small parts are loaded at one end, tumbled, and pushed along by ribs or vanes, and emerge from the other end of the retort. Sizes are available to take loads from 100 to 1500 lbs (45.4 to 681 kg).

fusible alloys. Useful chiefly because they melt or soften at low temperatures. They most often contain lead, tin, and cadmium for the higher melting ranges [350–475°F (176–246°C)], and bismuth, lead, and tin in the lower melting ranges. Very-low-melting alloys contain indium and may melt as low as 117°F (47°C). Some of the fusible alloys are eutectic compositions, which melt sharply at one temperature, while others freeze over a range of temperatures. Those containing appreciable amounts of bismuth expand upon cooling below the solidification temperature. The fusible alloys have numerous applications, the most common of which is in the release mechanism of automatic sprinkler systems. They are also used as safety plugs in compressed gas cylinders. Actuating devices for fire-warning systems and automatically closed fire doors often depend upon a fusible operating link. They are also used as a substitute for wax or plastics in precision casting, and for cores or matrices over which other metals may be plated. **(See also solder, fusible alloy.)**

gage. (1) The thickness (or diameter) of sheet or wire. The various standards are arbitrary and differ, ferrous from nonferrous products and sheet from wire. (2) An instrument used to measure thickness or length. (3) An aid for visual inspection that enables the inspector to determine more reliably whether the size or contour of a formed part meets dimensional requirements.

galling. A condition whereby excessive friction between high spots results in localized welding with subsequent spalling and a further roughening of the rubbing surfaces of one or both of two mating parts.

garnet (see abrasives)

gas burners (torch). Devices for mixing gas and air in the proper proportions for combustion. The actual combustion process occurs in the visible flame after the fuel gas and the oxygen are mixed. Since the gases issue from the burner ports at a constant velocity, the size of the burner ports must produce a forward velocity of the same order as the flame propagation velocity, the speed at which the flame propagates through the gas mixture. Different gases have different flame propagation velocities and different air–gas ratios; this is the reason that the burner or torch must be changed if the fuel gas is changed. The flame velocity is influenced by many variables and cannot be readily stated for any fuel gas. Methane is a typical slow-burning gas; its flame velocity is only a few feet per second. Hydrogen burns at a high velocity, of the order of 20 ft/s (6.1 m/s). The heating value of the fuel gas is the amount of combustion heat released per cubic foot of fuel gas measured at standard temperature and pressure and the product of combustion cooled back to the initial temperature. The gross heating value is usually stipulated; this includes the latent heat of condensable water vapor.

gas cutting (see flame cutting)

gas-electric truck. An electric industrial truck in which the power source is a gasoline or LP-gas engine-driven generator.

gaseous fuels. Commercial gas fuels are classified as natural gas, manufactured gas, by-product gas, and liquefied petroleum gas (LPG). Natural gas is essentially methane, CH_4, plus small varying amounts of carbon dioxide and the lower gaseous hydrocarbons. The common manufactured gases are coal gas, from the destructive distillation of coal; oil gas, from the destructive distillation of petroleum oils; water gas, made by reacting steam with hot coke to produce a mixture of carbon monoxide and hydrogen carbureted water gas, a mixture of water gas and oil gas; and acetylene gas, made by severe cracking of gaseous hydrocarbons or by heating calcium carbide with water. By-product gases include various petroleum-refining gases that are produced in processing petroleum; coke oven gas, a by-product of the manufacture of coke and similar to coal gas; and blast furnace gas, principally carbon monoxide and nitrogen from iron-blast furnaces. Liquefied petroleum gases are a mixture of propane and butane recovered from natural gasoline or from refinery operations. The analysis of fuel gas is expressed in terms of volume percentages of the component gases. Determinations can be made by selective absorption in chemical solutions by separation of components through distillation, by infrared or mass spectrometry, or by means of gas chromatography. The heating value refers to the quantity of heat released during combustion of a unit amount of fuel gas. Determinations are made with a continuous-flow (constant pressure) gas calorimeter. The heating value as determined in calorimeters is termed higher heating value and is the quantity of heat evolved when the products of combustion are cooled to 60°F (16°C) and the water vapor produced is completely condensed to a liquid at that temperature. The lower heating value is the same as the gross heating value except that the water produced is not condensed but retained in the vapor form at 60°F (16°C). The heating value of manufactured gas is expressed as Btu/ft^3 when measured at 60°F (16°C) and 30 in. Hg, saturated with water vapor. The values for natural gas are commonly reported at a pressure of 14.7 psia or 30 in. Hg, at a temperature of 80°F (27°C) and generally on a dry basis. The heating values of gaseous mixtures can be calculated by multiplying the mole fractions of the component gases by their respective heating values; the sum of the products in the heating value of the mixture.

gas flame welding (see welding, gas)

gasoline, octane number. The ability of a gasoline to perform well in an internal combustion engine is given by its octane number. In order to standardize the performance of a gasoline, 2,2,4-trimethylpentane, $(CH_3)C–CH_2C–CH_2CH(CH_3)_2$, a very fine motor oil with no tendency to premature explosion under high compression is assigned an octane number of 100. Normal heptane, $CH_3CH_2CH_2CH_2CH_2CH_2CH_2CH_3$, an extremely bad "knocker" is assigned a value of zero. The octane number of any fuel is determined from a comparison of its performance to that of a blend of 2,2,4-trimethylpentane and *n*-heptane. Both the fuel in question and the prepared blend are tested in a specially instrumented engine. The percentage of 2,2,4-trimethylpentane in the blend that gives the same performance as the fuel in question establishes the octane number of the latter. Motor fuels can be prepared that have octane numbers well in excess of 100; that is, they will perform even better as motor fuels than 2,2,4-trimethylpentane.

gear grinding. Gear teeth can be produced entirely by grinding, cutting, or first cutting and then grinding to the required dimensions. Usually, gear grinding removes only a few thousandths of an inch of metal from precut gears to make accurate teeth for critical applications. Teeth made entirely by grinding are usually only those of fine pitch, for which the total amount of metal removed is small. Grinding of fine-pitch gear teeth from uncut blanks may be less costly than the two-step procedures if there is not much metal to be removed. The two basic methods for grinding of gear teeth are form grinding (nongenerating) and generation grinding. Either method can be used for spur or helical gears. Many varieties of machines have been built especially to grind gears and pinions.

gelatin. An organic nitrogeneous colloidal protein substance whose principal value depends on its coagulative, protective, and adhesive powers. Water containing only 1% high-test gelatin by weight forms a jelly when cold. Animal gelatin is obtained by hydrolysis from collagen, white fibers of the connective tissues of the animal body,

Typical Composition Ranges of Fuel Gases

	Volume %									
	Hydrogen, H_2	Methane, CH_4	Ethane, C_2H_6	Illuminants	Carbon Monoxide, CO	Carbon Dioxide, CO_2	Oxygen, O_2	Nitrogen, N_2	Specific Gravity, Air = 1.0	Gross Heating Value, Btu/ft^3
Natural gas	0	66–98	0–32	0	0	0–2.7	0–0.4	0.1–7.5	0.59–0.67	1047–1210
Mixed refinery gas[a]	2.0–20.9	21.6–53.3	13.4–19.3	10–16.2	0	1.5–8.0	1.5–8.0	1.5–8.0	0.83–1.15	1380–1828
Oil gas	24–55	26–60	0	3–15	0–10	1–4	0.3–0.5	2.5–4.0	0.37–0.50	540–700
Coal gas	37–50	27–40	0–3	3–4	6–8.5	1.6–2.4	0.4–1.0	3.2–11.0	0.45–0.55	540–700
Coke oven gas	48–53	30–35	0–1	3–4	4–6	1–3	0–1.1	4–6	0.4–0.5	550–650
Producer gas	10–19	0.3–6.3	0	0–0.4	17–30	2.5–7.3	0.3—0.7	49–58	0.82–0.87	135–170
Water gas	45–70	0–4.5	0	0–1.1	32–41	3.0–5.0	0.2–0.7	4–10	0.5–0.6	290–320
Carbureted water gas	32–37	11–14	0	8–10	26–30	4–6	0.4–0.9	4.5–12.0	0.55–0.6	528–545
Blast furnace gas	1–4	0	0	0	26–27.5	11.5–13.0	0	57.6–60	0.95–1.0	90–93
Acetylene (commercial)	0	0	0	95–96	0	0	0–0.8	3–4	0.91–0.96	1300–1400

[a]Contains 1.3–4.2% propane and 0.2–3.5% butane.

particularly the skin (corium), bones (ossein), and tendons. There are four different kinds of gelatin, edible, technical, photographic, and pharmaceutical. Gelatin is a widely consumed food, and a popular dessert. Gelatin has played an important part in the development of the motion-picture and photographic industries. It is coated on the film base, constituting the sensitized emulsion of the light-sensitive silver salts. Technical gelatin is quite an arbitrary name applied to small amounts used for miscellaneous purposes such as for sizing paper, textiles, and straw hats. Gelatin is used by pharmaceutical houses for making capsules and as an emulsifier.

gels. Formed from colloidal solutions (below 10% concentration), by reducing the solubility of the colloidal material enough to enable the particles to link together. This results in a coherent structure of solid character interpenetrated by the dispersing medium. Thus the gel can be regarded as an essentially two-component system exhibiting the mechanical properties of the solid state, within which both the dispersed constituent and the dispersing medium are continuous. Gels may be formed in various ways. One method is spontaneous swelling of a substance when brought into contact with a solvent, the swelling agent. The second method is somewhat similar to the coagulation process. This is brought about by a change of condition in a solution, by either changing temperature or composition. In a real coagulation process the dispersed particles form aggregates that settle down under the influence of gravity; however, in gel formation, the particles set to a coherent structure before large aggregates can be formed.

gel types. Gels can be distinguished as rigid elastic (xerogels), and thixotropic, according to their structural properties. Silica gel, as formed by the action of an acid on a sodium silicate solution, is an example of a rigid gel. The liberated monosilicic acid immediately begins to polymerize. First, two hydroxyl groups react forming an oxygen link between two molecules with the elimination of a water molecule. The remaining hydroxyl groups react in a similar way with other molecules, leading to the formation of colloidal particles, and finally to a three-dimensional gel structure. Elastic gels or xerogels include dry gelatin, cellophane, and many polymers in organic media, such as rubber, cellulose fibers, starch, leather, and various soaps. The constituent particles are connected into a coherent structure by secondary forces, such as the van der Waals forces. Thixotropic gels can pass into colloidal solutions as the result of mechanical agitation. Certain clays, such as bentonite, colloidal alumina, and ferric oxide, tend to form thixotropic gels. They are classified as nonthermal gels because the change of temperature does not affect the change from the sol to gel system.

gettering. Metals, like most substances, will readily adsorb gases on their surfaces, particularly water vapor. Getters are metals deliberately introduced into vacuum spaces for the purpose of adsorbing final traces of gases. In the manufacture of small and inexpensive vacuum tubes, the tube may be pumped down to a vacuum of 10^{-4} torr (10^{-4} mm Hg), and a getter metal then used to reduce the vacuum pressure to about 10^{-6} torr. The getter must have any adsorbed gases desorbed or baked out before it is inserted into the vacuum. The getter, in the form of a wire, a coating, or other shape, thus acts as an auxiliary pump or getter pump, though the gas thus ''pumped' is not actually removed from the vacuum space. Some of the materials used as getters are barium–aluminum alloys (for vacuum tube work), zirconium, titanium, tantalum, thorium, and molybdenum. Such metals can adsorb large amounts of residual gases. Getters must be selected for the residual gas to be removed.

glass. An inorganic product of fusion that has been cooled to a rigid condition without crystallization. Although silica is a perfect glass-forming material, it has a very high melting point and cannot be melted alone at reasonable cost. Basic metal oxides are added to lower the fusion point and viscosity of the melt and thus make easier the fabrication of glasswares. The addition of about 25% by weight of sodium oxide results in the formation of sodium disilicate (Na_2O–$2SiO_2$), giving a eutectic mixture with silica with a melting point of 1460°F (793°C). Such glass shows a little tendency to devitrification but, unfortunately, it is water soluble, making such a mixture of little use as a material of construction. The addition of suitable amounts of calcium oxides to the mixture gives a soda lime glass, which is insoluble in water. Commercial glass can be classified as soda lime or lime glasses, lead glasses, borosilicate glasses, and high-silica glasses. The ultimate tensile strength of glass is about 10,000 psi (6.9×10^3 kPa) for ½ in. (1.27 cm) diameter rods, while for fibers of 0.00005 in. (1.27×10^{-3} mm) diameter, very great strengths up to 3,500,000 have been reported. The modulus of elasticity of soda-lime glass varies from 8.2×10^6 to 12.0×10^6 psi (5.7×10^7 to 8.3×10^7 kPa). Compressive strengths up to 140,000 psi (9.7×10^5 kPa) have been obtained for glass rods. The creep resistance of glass is poor, since it is found to creep at low stresses and room temperatures.

Composition of Commercial Glasses

Component	Soda-Lime Glass	Lead Glass	Borosilicate Glass	Aluminosilicate Glass	High-Silica Glass
SiO_2	70–75[c]	53–68	73–82	57	96
Na_2O	12–18	5–10	3–10	1.0	—
K_2O	0–1	1–10	0.4–1	—	—
CaO[a]	5–14	0–6	0–1	5.5	—
PbO[b]	—	15–40	0–10	—	—
B_2O_3	—	—	5–20	4.0	3
Al_2O_3	0.5–2.5	0–2	2–3	20.5	—
MgO	0–4	—	—	12.0	—

[a]Maximum CaO up to 20%, because of devitrification.
[b]For extra dense optical glasses PbO may be as high as 80%.
[c]Values are in percent.

glass annealing. Applied to glass in order to relieve stress set up during the later stages of shaping as a result of rapid cooling. Failure to anneal the glass in a suitable manner would leave it weak and brittle. Annealing is usually carried out immediately after the shaping process. The glass is reheated in a special furnace to the necessary annealing temperature and then it is allowed to cool very slowly in the furnace. The small amount of viscous flow that takes place during annealing leads to stress relaxation within the glass, which is thus rendered relatively stress-free.

glass fibers. When in the form of fiber, glass is considerably stronger than it is in any other form and is used as an engineering material. The high strength of glass fiber is due partially to the fact that, unlike ordinary glass, these fibers are relatively free from such defects as are likely to lead to crack propagation and consequent failure at low stresses. In the absence of the possibility of deformation by slip, fairly strong chemical bonds operate along the length of the fiber. Glass fiber is widely used to produce composite materials in combination with various polymers.

glass, chemical strengthening. An intriguing method of chemical strengthening—of changing the composition of the glass near the surface—is by exchange of alkali ions. For example, if sodium ions, of larger atomic volume, replace some of the lithium ions near the surface of a lithium-silica glass, or similarly, if potassium ions replace those of sodium in a sodium-silica glass, the glass structure will be "crowded" by these larger ions. In this condition, the surface is in compression and the interior, correspondingly, in tension. A convenient ways to make these one-for-one exchanges is to immerse the glass part in a molten salt bath containing the desired alkali-metal ions. Magnitude of the compressive stress is much greater for the chemically strengthened glass; it may range from less than 50,000 (3.4×10^5 kPa) to considerably more than 100,000 psi (6.9×10^5 kPa), depending primarily on the glass composition. This provides a new range of available strengths for glass products, even those having complex shapes and thin walls.

glass, light-sensitive. Photochromic glass darkens when exposed to ultraviolet radiation and fades when the ultraviolet stimulus is removed or when the glass is heated. Some photochromic compositions remain darkened for a week or longer. Others fade within a few minutes after the ultraviolet is removed. A chief use for the faster-fading compositions is in eyeglass lenses that automatically darken and fade when exposed to or removed from sunlight. Photosensitive glass also responds to light, but in a different manner from photochromic glass. When exposed to ultraviolet energy and then heated, photosensitive glass changes from clear to opal. When the ultraviolet exposure is made through a mask, the pattern of the mask is reproduced in the glass. The image developed is permanent and will not fade, as would a similar image in a photochromic glass. The exposed, opalized photosensitive glass is much more soluble in hydrofluoric acid than the unexposed glass. Immersion in this acid produces shapes, depressions, or holes by etching away of those exposed and developed areas. Polychromatic glasses are the most recent addition to the family of glasses that are sensitive to light. They are full-color photosensitive glasses. Developed in 1978 by Corning Glass Works laboratories, their characteristics imply applications such as information storage, decorative objects, windows or other transparencies, and containers. Researchers believe the glasses are the first photographic medium having true color permanence.

glass, 96% silica. This is a highly heat-resistant glass made from borosilicate glass by a proprietary process. This glass can be formed more readily and in more shapes than vitreous silica. Its properties are so close to those of vitreous silica that it is frequently used as a substitute in optical components and spacecraft windows. It is also used as a heat-resisting coating such as on the exterior of NASA's Space Shuttle Orbiters. Other uses include laboratory ware and lighting components such as arc tubes in halogen lamps.

glass tempering. Glass may be toughened by a tempering process that seeks to reduce the formation of surface cracks by putting the surface in a state of compression. The principle involved is similar to that applied in the prestressing of concrete in order to prevent cracks from developing in its surface. Tempering of glass is achieved by heating it to much the same temperature as that used for annealing it, and then cooling the surface rapidly by means of air jets. In common with other substances, glass contracts as it cools. Initially the outside surface cools, contracts, and hardens more quickly than the inside. The latter, being still soft, will yield in the early stages of cooling. During the final stages of cooling, both inside and outside behave in an elastic manner, but their contractions are already out of step. Therefore, when the glass has finished cooling, the outer surface is in a state of compression while the inside is in tension. Since the surface layers are in compression, these compressive forces will balance any moderate tensile forces to which the glass may be subjected during service. Glass invariably fails as a result of tensile compnents of forces even when these forces are overall compressive.

glass, thermal strengthening. The traditional way to increase the strength of glass parts is by thermal tempering or strengthening. As a glass part is cooled, the surfaces cool faster than the interior. During cooling, the glass near the surface becomes rigid before the glass in the interior does. As cooling progresses to room temperature, the interior contracts over a larger temperature range and, in so doing, imposes a compressive stress on the glass near the surface. Any externally applied load on the glass must overcome the induced surface compression to produce a high enough tension at the surface to cause rupture; the effective tensile strength and the surface compression, usually about 20,000 psi (1.4×10^5 kPa)—perhaps up to 30,000 psi (2.1×10^5 kPa) for simple shapes with thick walls.

glue. An adhesive substance or solution used to join or bond various materials. Usually used in reference to animal-based collagen glue. **(See also adhesives.)**

grain flow. The directional characteristics of the metal structure after working, revealed by etching a polished section.

grain size. A measure of crystal size usually reported in terms of average diameter in millimeters, grain per square millimeter, or grains per cubic millimeter.

graphite. A soft, black, unctuous form of elemental carbon. It crystallizes in the hexagonal system and rarely is found in six-sided tabular crystals. It usually occurs as minute scales or flakes, as bladed or foliated masses, or as earthy cryptocrystalline lumps. Graphite has perfect basal cleavage, a dull to bright and metallic luster, and a gray streak. It is opaque in even the finest particles. In industry the mineral is called crystalline if it contains crystals large enough to be distinguished with the unaided eye and amorphous if it is of finer texture. The principal impurities are the other minerals of the enclosing schistose rocks, chiefly quartz, mica, feldspar, and clay. The softness and perfect cleavage of graphite make it highly slippery or unctous. It adheres readily to metal or other substances, filling the pores and making a slick surface. Graphite is extremely refractory being little affected by temperatures up to 5432°F (3000°C). It is immune to most acids and other reagents. It mixes readily with reagents, and with other materials, both solid and liquid, and is a good conductor of electricity. Much of that utilized in industry is produced from calcined petroleum coke. The largest use is in foundry facings, finely purverized material to give the surface of molds a smooth finish, so that casting may be easily removed after cooling, and for electric-furnace electrodes. Another metallurgical use is in crucibles and related equipment for melting nonferrous metals, mainly brass and aluminum. Some natural graphite is used for the manufacture of "lead" pencils, but batteries, carbon brushes for electric motors, protective paints, anodes, roofing compounds, and many other commodities use manufactured grades.

gray iron (see cast iron)

greases. The purpose of a grease is to provide a means of lubrication for an application where the fluidity of an oil would be disadvantageous. Greases are composed of a lubricating oil, a thickening soap, sometimes a filler, and sometimes additives to impart or enhance certain characteristics.

greases, aluminum-soap. Aluminum-base greases are characterized by brilliance and transparency. They are used for high operating temperatures and when nonspattering is a requirement. They are water resistant and do not separate with agitation.

greases, calcium-soap. Most greases are manufactured by cooking or boiling. A saponifiable animal or vegetable oil is mixed with a saponifying agent or metallic base, such as calcium hydroxide, until saponification occurs. The mineral oil is then added and worked into a homogeneous mixture. The calcium-base greases are also known as lime-base grease, cup grease, pressure gun grease, and water-pump grease. They vary in consistency from semifluid to semisolid, and have a smooth buttery texture. These greases are insoluble in water and are used in applications where water resistance may be a factor. Since they are "stabilized" by from 0.5 to 2% water, this water will start to evaporate at temperatures above 150°F (66°C), causing the grease to separate into soap and oil. High rotational speeds also cause separation.

greases, cold sett. Cold sett greases are not cooked or boiled but are made by mixing mineral oil, resin oil, and a "sett" or emulsion of light mineral oil, water, hydrated lime, and resin oil. The emulsion is stabilized by reaction between the lime and the resin oil. These greases are cheaper to manufacture and are of lower quality than cooked greases. They are used for the lubrication of rough machinery in mills and mines, and in such special applications as skids, launching ways, and track curves. They are water resistant but corrode brass bushings.

greases, lead-soap. Lead-base greases are intended for applications in which extreme pressure between mating parts is to be encountered as in hypoid gears and transmissions. They are not water soluble, but a settling out does occur when water is present.

greases, lithium-soap. Known as multipurpose greases because they can be used in a large number of applications under varied conditions. They can be made with oils of low pour points for low-temperature applications, being used as low as −60°F (−51°C). When high viscosity oils are used, lithium greases can be used for high-temperature applications. The same grease, however, cannot provide lubrication at both high and low temperatures. Lithium-soap greases are water resistant and have been found to operate satisfactorily in wet locations where other lubricants were quickly washed.

greases, mixed-base. Soda-lime soap greases have been developed for use in ball and roller bearings. They are less fibrous than soda greases and have an effective temperature range of −40–+250°F (−40–121°C) or higher. Mixed-base greases have good speed characteristics but are water susceptible. There are other greases available that have bases of lime-lead, lime-aluminum, etc., each having been developed for certain characteristics or applications.

greases, organo-clay. Organo-clay greases are prepared using organically modified smectite clays. Unmodified clays are hydrophilic (water attracting), and do not gel lubricating oils. To make the clay an effective gellant for lubricating oils, mixed chain quaternary amines (called quats) are reacted onto the surface of the clay. This organic modification makes the clay hydrophobic (water repellant) and, in turn, creates a product capable of swelling in lubricating oils and producing stable, efficient gels. Selection of the organic modification is quite specific to the nature of the fluid to be gelled. In a typical simple organo-clay composition, the base oil will comprise about 90–95% by weight of the grease; organo-clay rheological additives, 5–8%; and a polar dispersant, generally at 2%.

greases, residuum. May be blends of residues from the distillation of crude oils and heavy, dark oils to which fatty oils, resin oils, or tar may be added. Although not greases in a strict sense, they are applied where an adhesive lubricant is required, as in open gears, wire cable, and other heavy parts.

greases, sodium-soap. Sodium- or soda-soap greases are somewhat fibrous in texture. No water is used in their manufacture and they can be used at temperatures up to the melting point of 300–350°F (149–177°C). Melted soda greases resume their former appearance and characteristics when cooled. These greases are water soluble and cannot be used when there is danger of contamination by moisture. They have a much higher degree of cohesion than lime greases and do not channel or separate with centrifugal action.

greases, zinc-soap. Zinc-based greases find applications in hydraulic machinery and as rust-preventive slushing compounds. They are frequently used between press-fitted parts to inhibit frettage corrosion.

grinding, abrasive belt. Endless belts precoated with abrasive are widely used for both grinding and polishing. The contact wheel, over which the abrasive belt rides, provides an opposing pressure to the workpiece. Depending on its hardness, the contact wheel can provide either high unit pressure (hard wheel) or low unit pressure (soft wheel). Selection of contact wheel directly affects the rate of stock removal, the ability to bend, the surface finish obtained, and the cost of the grinding operation. The manufacturer of contact wheels is responsible for speed testing and balancing the wheels, as well as dressing the wheel, which effects the operating speed at which they will run. Although it is possible to operate a contact wheel satisfactorily at speeds from 2000 to 10,000 sfm (610 to 3048 smm), operating speeds usually range from 3500 to 7500 sfm (1067 to 2286 smm).

grinding, abrasive wheel. Used mainly for fine finishing of hardened surfaces and hard materials, grinding machines employ wheels of manufactured abrasives (silicon carbide and aluminum oxide) or, for very precise work, diamond wheels. Not unlike other machining methods, grinding is primarily a process of removing material in the form of chips. Abrasive grains of grinding wheels perform in much the same manner as single-point cutting tools, effecting by a multitude of contacts a virtual multiple-tool cutting action that produces extremely minute chips similar in character to those produced in turning. Since cutting edges of the grits are extremely thin, it is possible to remove much smaller chips and to refine surfaces to a much greater degree of accuracy with respect to both finish and dimension than with other methods of metal cutting presently available. **(See also abrasives.)**

grinding, electrical discharge (EDG). Much like electrical discharge machining except that the electrode (tool) is a rotating graphite wheel. The work is fed to the wheel by a servo-controlled worktable. The workpiece is cut by the action of a stream of electric sparks between a negatively charged wheel and a positively charged workpiece immersed in a dielectric fluid. Each spark discharge melts or vaporizes a small amount of metal from the workpiece surface producing a small crater at the discharge site, as in EDM. The stream of sparks is produced by high-frequency pulses of direct current. The power supply and the dielectric fluid are similar to those used in EDM, but lower amperage is used in most EDG applications because the cutting area is usually small and the method is used primarily to achieve accuracy and smooth finish.

grinding, electrochemical discharge (ECDG). A combination of the process of electrochemical grinding (ECG) and electrical discharge grinding (EDG), with some modification of each. The process resembles ECG in the electrochemical formation of oxides on a positively charged workpiece (anode); however, it employs alternating current or pulsing direct current and does not use an abrasive coated wheel. It resembles EDG in the use of a graphite wheel that does no mechanical grinding and in the use of intermittent spark discharges to remove material from the workpiece surface, but differs from EDG in using a highly conductive electrolyte instead of a dielectric fluid and in using low-voltage, high-frequency current. Like ECG, the process is most useful for grinding carbide tools, hardened tool steel, nickel-base alloys, and parts that are fragile or sensitive to heat.

group technology. (1) A system for coding parts based on similarities in geometrical shape or other characteristics of the parts. (2) The grouping of parts into families based on similarities in their production so that the parts of a particular family could then be processed together.

gypsum. A hydrous calcium sulfate. Mineral varieties include selenite, a transparent cleavable form; satin spar, a fibrous form with a silky luster; and alabaster, massive and finely crystalline, pure white or delicately tinted and translucent. Gypsum is a compact, massive, finely crystalline to granular rock. It is white when pure, but may be gray, bluish-gray, pink, or yellow, owing to such impurities as organic matter, clay, and iron oxide.

gypsum plaster. Obtained by removing the water of crystallization from gypsum rock. The rock is first ground, then heated or calcined, and finally reground. With incomplete calcining at lower temperatures, dehydration is not complete and plaster of Paris is obtained. Complete dehydration produces flooring plaster ($CaSO_4$). When alum, borax, or other similar materials are added to flooring plaster, hard-finish plaster is obtained. The setting time for gypsum plasters varies from about 5 min for plaster of Paris to several hours for hard-finish plaster. The time of setting may be decreased by the addition of various ingredients. The tensile strength of gypsum products is low, and to increase the cohesive strength, wood and hair fibers are added. The strength of plaster increases with time after setting and approximately one-half the maximum strength is reached in about 24 h. Plaster of Paris is useful in making casts and architectural adornments, as it expands on hardening and leaves a clear impression of the mold used. It is used for molds in the ceramic industry and also for casting metals. Wall plaster with sand is used for covering walls. Wall plaster mixed with sawdust and cinders is cast into blocks for construction of floors, wall, roofing, or fireproofing. Finely ground gypsum is used as a filler for toothpaste, plastics, cloth, paint, and paper.

hacksaw (see cut-off machines)

Hall process, aluminum refining. The main commercial method used in the production of metal aluminum. A rectangular steel cell about 8 ft (2.4 m) long, 5 ft (1.5 m) wide, and 2 ft (0.6 m) deep is the cell or pot used in the extraction (or smelting) of aluminum. Buried in the carbon lining is a collection plate to conduct the current, the entire carbon lining being the cathode of the cell. Carbon anodes are attached to metal rods, which hang from the superstructure. The electrolyte is molten crylite, Na_3AlF_6, a very stable compound that will dissolve about 20% Al_2O_3. As long as there is sufficient Al_2O_3 in solution, there is little or no liberation of fluorine or other decomposition products of the electrolyte. This operation is conducted at 1652°F (900°C). A crust of frozen electrolyte forms on top of the cell during the operation. There is a drop in voltage across a cell of from 6.5 to 7.5 V with a current density of 650 to 750 A/ft^2 of cathode surface, and a cell will draw from 8000 to 30,000 A. Aluminum is liberated at the cathode, where it collects in a pool. At the anode there is a liberation of oxygen which combines with the carbon to form CO, and $CO_2Al_2O_3$ is stored in hoppers above each cell, and it is usually dropped onto the electrolyte crust by gravity to be warmed prior to adding it to the cell. When the cell is operating normally, the electrolysis proceeds continuously. The aluminum produced from Hall cells is commercially pure aluminum and requires no further refining.

hardenability test, end-quench. An index of the depth to which a given steel will harden when quenched in a particular manner, or the ease with which martensite can be produced in the steel. Determined by a highly standardized test such as the Jominy end-quench test, but is best assessed by the diameter of the largest steel cylinder that can be hardened all the way to its center by water quenching. During quenching, a piece of steel will cool more rapidly at the surface than in the interior. This may cause the interior of the piece to be hardened to a lesser degree than the surface, since, in a less rapidly cooled portion, some pearlite may be formed before martensite. For a piece of steel to be hardenable throughout, the critical cooling rate should be such that martensite forms before any pearlite has a chance to appear. High ratios of surface to mass tend to produce greater depths of hardening because the rate of cooling depends on the speed with which heat leaves the specimen. Other factors are the surface conditions and the austenite grain size. Any scale formed on the quenched piece impairs hardenability, because it reduces the rate of heat removal from the interior. **(See also Jominy end-quench test.)**

hardening, strain (see strain hardening)

hardness, material. Hardness of a material can be measured by its resistance to scratching or to indentation. Most widely used hardness tests involve the determination of the material resistance to indentation under strictly standardized conditions. The hardness of materials depends on the type of binding forces between atoms, ions, or molecules and increases, like strength, with the magnitude of these forces. Thus molecular solids such as plastics are relatively soft, metallic and ionic solids are harder than molecular, and covalent solids are the hardest materials known. The hardness of metals is also increased by alloying, cold work, and precipitation hardening. There is also a close connection between the yield strength of metals and their hardness.

hardness test, Brinell. Made by pressing a hardened steel ball, usually 10 mm in diameter, into the test material by weight of a known load, 500 kg for soft materials such as copper and brass, and 3000 kg for materials such as iron and steel, and measuring the diameter of the resulting impression. The hardness is reported as the load divided by the area of the impression, and tables are available from which the hardness may be read, once the diameter of the impression is obtained. A small microscope is used for measuring these impressions. On a Brinell hardness tester, the specimen is placed upon the anvil, which is raised or lowered by means of the screw. The anvil is then raised until the specimen is in contact with the steel ball. The load is then applied by pumping oil. The load is applied for 30 s; then the diameter of the resulting impression is measured. The Brinell hardness of annealed copper is about 40, of annealed tool steel about 200, and of hardened tool steel about 650.

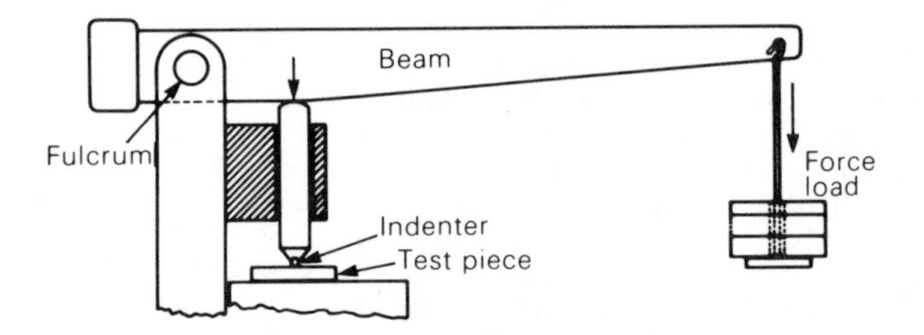

The principle of most hardness-measuring machines. The indenter may be a steel ball, as shown, a diamond pyramid or a diamond cone.

hardness test, Knoop. Microhardness determined from the resistance of metal to indentation by a pyramidal diamond indenter, having edge angles of 172° 30′ and 130°, making a rhombohedral impression with one long and one short diagonal.

hardness test, Moh's scale. A scratch hardness test for determining comparative hardness using 10 standard minerals from talc (the softest) to diamond (the hardest).

hardness test, Monotron. The Monotron measures the indentation made by pressing a hard penetrator into the material under test. The difference between it and other testing machines is that the depth of impression is made constant instead of the pressure. Two dials at the top of the machine enable the operator to read depth of impression on one and pressure on the other. Pressure is applied until the penetrator sinks to the standard depth, 9/500 in. (0.457 mm); the amount of pressure so needed is read on the upper dial. The penetrator used is a diamond having a spherical point about 0.025 in. (0.635 mm) in diameter. An advantage of this method, based on constant depth of impression, is that the same amount of cold work is performed regardless of whether the material tested is extremely hard or comparatively soft. Because of this, a wider range of materials can be tested without making any change in the penetrator of the methods of the tests.

hardness test, Rockwell. Accomplished with a Rockwell tester, an apparatus that automatically elevates the specimen very slowly against the indenter until a minor load is applied, as indicated by an index hand on the dial gage. Then a major load is applied by releasing a loaded lever system, the speed of descent of the lever is controlled by an adjustable oil dashpot. When the descent of the lever is completed, the major load is removed and with the minor load still acting, the Rockwell hardness number may be read on the dial gage. This number is based on the depth of indentation, less the elastic recover following the removal of the major load, less the penetration caused by the minor load. Because of the reverse order of the numbers on the dial gage, a high number indicates a shallow impression produced in a hard material, and a low number is the case of a deep impression in a soft material. In order to cover the range of hardnesses found in metals, it is necessary to have several Rockwell scales, each one associated with a specific combination of load and indenter.

hardness test, Shore scleroscope. This instrument, rather than measuring the indentation of penetration into a material, the height of rebound of a diamond-tipped weight which is permitted to fall on the specimen from a height of 10 in. (25.4 cm) is measured. Since hard metals have a higher resiliency, the weight will rebound to a higher distance than on softer metals. This scleroscope is one of the very useful and valuable testers used primarily because of its portability and because the test may be made without damaging or having to disassemble the component part being tested from the main portion of the machine. The diamond-tipped weight of the scleroscope operates within a vertical glass tube on which is inscribed a scale for measuring the height of rebound; then a dial gage can be used for measuring and recording this height. This test is particularly useful for measuring the hardness of extremely smooth surfaces not marked by an indentation.

hardness test, Tukon micro. Applies from 25 g to 50 kg and uses either the ''Knoop'' type diamond indenter or a conventional square-based diamond pyramid indenter, similar to that used in the Vickers hardness tester. The ''Knoop'' indenter is of pyramidal form, with an included longitudinal angle of 172°30′ and an included transverse angle of 130°0′. Entirely automatic, the Tukon tester is electronically controlled in a synchronous cycle. The specimen to be tested is placed on a special micrometer stage that has a two-way adjustment in a horizontal plane. The indentation is made by elevating the specimen against the indenter until it resists any further indentation, at which time the electronically operated contacts are opened, elevation of the specimen ceases, and the load remains on the specimen for a fixed period of time, after which the specimen is automatically lowered to clear the indenter and then the specimen removed to a micrometer measuring microscope. The Knoop impression appears rhombic under the microscope, with a ratio of the long diagonal to the short diagonal of 7.11 to 1.

hardness test, Vickers. Uses a square-based diamond pyramid indenter with an included angle of 136° between opposite faces, with loads usually ranging from 5 to 100 kg. With the Vickers machine, the specimen is elevated to within 1 mm of the indenter, and the lever-loading system is set in the starting position by means of a foot lever acting on a weight-and-cam mechanism that produces the motive power for the test. When the weight-and-cam mechanism is released, the indenter descends under the load, indents the specimen, and then returns to the starting position. Both the rate of descent of the indenter and the time the indenter is in contact with the specimen are controlled by an oil dashpot that is usually adjusted for a time cycle of from 10 to 20 s. The operator is warned by a buzzer when the specimen is not supporting the full load, because it is too far from the indenter or not held rigidly enough for the size of the load applied. In the Vickers tester, the hardness number is the quotient of the load and the area of surface of the indentation, the same as is true in the Brinell hardness test. Here, however, the diamond square-based penetrator produces an indentation having equal diagonals, which are measured by use of a microscope. In the Vickers machine, the hardness numbers are obtained from tables or charts after the diagonals have been measured, rather than from specific calculations of the area of the surface of the indentations produced.

heading (see cold forging)

heating processes (see furnaces, gas burners, gaseous fuels, induction heating, and infrared radiant heating)

heat sealing. All methods of sealing plastics or plastic-coated materials by heat are alike in that the interfaces to be sealed must be simultaneously in intimate contact and at the proper sealing temperature. Where both interfaces are made of the same material, the proper sealing temperature is the welding temperature, or the temperature at which the interface disappears. Pressure is used to ensure intimate contact. Sealing methods may be roughly classified into bar, impulse, band, electronic, ultrasonic, radiant, hot-melt, solvent and friction. The most important features of each are: *Bar.* Sealing with a hot bar or iron, wielded by hand or mounted in a machine, is the oldest and still most widely used method. Simple hand- or foot-operated electrically heated reciprocating bar or rotary type of heated bar sealer with optional preheaters, through which material may be fed continuously are typical. *Impulse.* Although usually more expensive and slower than the jaw-type bar sealer, it is superior in most instances for sealing unsupported thermoplastic films. Material is carried through preheaters, then between rollers, manually or on belts. One roller may be resilient; one or both may be serrated to produce crimped seals in finished package. *Band.* This sealer is, in effect, a continuous impulse sealer. Endless metal bands carry material between heated jaws, pressure rollers, and cooling jaws. Thin bands transmit heat rapidly, hold material in contact while heating to form seal and cooling to "set" seal. *Electronic.* High-frequency sealing equipment is relatively expensive but certainly it is most economical for sealing many heavier materials. Shaped electrodes perform both the sealing and the cutting function. Rotary electrodes, too, are used for continuous sealing, analogous to the rotary sealer. *Ultrasonic.* This method relies on the "hammering" of two films (or foils) together at high frequency, causing them to weld without generating a temperature that would deorient a material such as polyester. It is the only method known that will satisfactorily weld this film to itself without great loss of strength adjacent to the weld. *Radiant.* Since radiant energy supplies most of the heat used in this method, sealing by means of a hot wire, rod, or flame is called radiant sealing. Hot-wire severing has been used on thermoplastic films for many years. Materials like polyethylene, which stand high temperatures for a short time, are also commercially sealed by radiant heat in bag making and collapsible-tube closing machinery. *Hot melt.* When rapid heat sealing of poor thermal conductors like paperboard is required, a thin bead of molten adhesive is extruded directly onto one of the boards. The other board member is immediately pressed against the molten bead, chilling it sufficiently to effect a primary seal. Composite containers of paperboard and plastic can be quickly assembled. *Solvent.* With water-soluble packaging films, the technique of softening the seal area with a solvent can be used. The film is moistened lightly with a felt wick just prior to passing between heated pressure rollers where it meets its mating film, effecting a good seal simply. Also, polystyrene parts are often sealed with the aid of a solvent. *Friction.* Products can be assembled from injection molded or thermoformed "halves" by holding one half stationary and spinning the other half in contact with it. The resulting frictional heat melts the material and welds the halves. Nylon spin welds readily. Other thermoplastics may have to be specially tailored in order to be properly spin welded.

heat treating. The heating and cooling of metal for the specific purpose of altering its mechanical and metallurgical properties. Thus, metals can be hardened and softened, and their grain structure changed by heat treatment. Consequently, heat treatment is an important and widely used manufacturing process. Since various metals do not react in the same way to heat treatment, the designer must know the response of a selected material to heat treatment. Proper selection of metal and use of heat treatment often will permit an inexpensive metal to serve as well as a more costly material. There are five practical methods by which metals may be strengthened. (1) Strain hardening. Method involves physical distortion of the normal metal crystal lattice so that movement of dislocations is impeded, resulting in increased strength. (2) Phase transformation. This procedure requires that the metal be heated to dissolve all the alloying elements in a single high-temperature phase. Heating is followed by rapid cooling, bringing about transformation to various room-temperature phases which have desired strength characteristics. (3) Precipitation hardening. Metal is heated to put all the alloying elements into solid solution in the matrix. This solution treatment is followed by rapid cooling so as to retain the elevated-temperature structure. There follows an "aging" period, either at room or elevated temperature, during which solute atoms precipitate out of the supersaturated matrix and act as "keys" in the lattice to impede deformation. This process requires that the alloying elements must be more soluble in the base metal at higher temperatures than at lower temperatures. (4) Solid-solution hardening. Complex solid solutions are formed with the additive elements having an atomic size substantially different from those of the matrix so that the resulting lattice is distorted. Strengthening occurs only as long as a single-phase solid solution is maintained. This procedure is useful in strengthening the refractory metals, such as columbium. (5) Dispersion hardening. This procedure is used primarily with powdered metals. It involves the addition of an insoluble second phase, such as an oxide, fine wires, or whiskers.

heat treating, annealing. A heat treatment in which the steel is heated to above 100°F (55°C) above the upper-critical (Ac_3) temperature, then slowly cooled in the annealing furnace. Annealing is often applied to hot workpieces, which are air cooled after working, so that there has been little opportunity for complete precipitation of the proeutectoid constituent from austenite. Since sections of varying thickness cool at different speeds, they exhibit different mechanical properties. This is due to the difference in fineness of pearlite and varying amounts of proeutectoid. The objectives of annealing are: (1) Softening. In hypoeutectoid steels containing lamellar pearlite, maximum softness results if the soft ferrite network is permitted to reach a maximum size and if the pearlite laminations are allowed to become coarse. (2) Grain size refinement. The nonuniform and sometimes large grain sizes resulting from hot working are eliminated by reheating the steel to austinite range. The grain size in castings is also refined. (3) Residual stress relief. Any stresses that remain in the steel as a result of prior processing are relieved upon full annealing. (4) Homogenization. The heating period and slow cooling involved in annealing allow diffusion to occur, and thus tend to eliminate segregation and nonuniform distribution of component.

heat treating, austempering. An interrupted quenching process, in which in order for the process to be successful the piece must be cooled rapidly enough to prevent

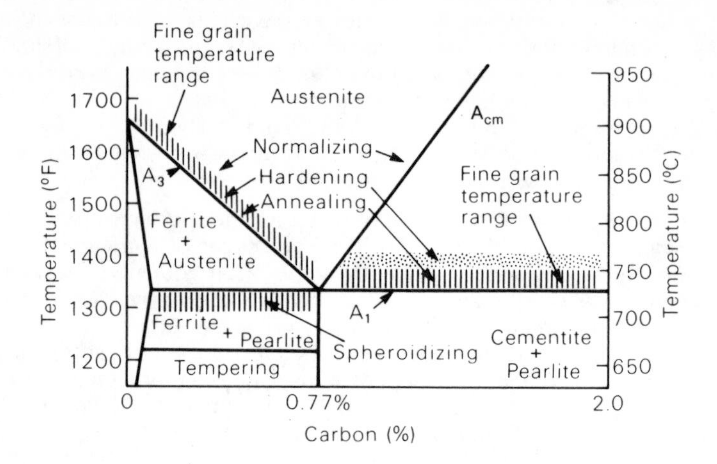

Heat-treating temperatures for carbon steels.

the formation of pearlite. The quench is interrupted at a transformation temperature, which produces acicular bainite. The piece is held at this temperature until the transformation of austenite to bainite is complete. The resulting transformation product may be as hard as the martensite produced by conventional quench and temper methods. Austempering is applied to small parts such as shoe shanks, or shapes of small cross sections such as wire, in which it is possible to prevent formation of pearlite during the initial stages of cooling.

heat-treating, cryogenic. A method of exposing tool steel to extremely dry cold temperatures in order to attain longer production life. Used in addition to normal heat treatment, the process consists in slowly cooling tool steel to temperatures less than −310°F (−190°C), changing austenite in the material to the more stable martensite.

heat treating, cyaniding (see cyaniding)

heat treating, martempering. Applied to steel sections or irregular shapes that are likely to crack during conventional quenching. The initial quench is the same as that for austempering, except that the steel is not held at the quenching temperature long enough to allow any bainite to form. Instead, it is held only long enough to allow for equalization of the temperature throughout the piece. When this is achieved, it is slowly cooled to room temperature, during which time transformation occurs slowly and uniformly to martensite. Thus the high stresses that accompany conventional quenching to martensite are avoided.

heat treating, normalizing. Steel that is heat treated well into the austenite range, then given a mild quench in still air. Normalizing accomplishes the following: (1) The amount of proeutectoid network is reduced. In hypereutectoid steels this is the result of dissolving the cementite at high temperatures and preventing its subsequent precipitation by a mild air quench. Hypereutectoid compositions are thus rendered more suitable for spheroidization because of elimination of the massive network of cementite, which does not readily spheroidize. In hypoeutectoid steels the size of the ferrite network is reduced and pearlite is rendered finer. (2) Homogenization and grain size refinement. (3) Pearlite is rendered finer than in the annealed condition. (4) Improvement of machinability of low carbon (0.3% carbon) steels by decrease in the amount of soft, ductile ferrite.

heat treating, nitriding (see nitriding)

heat treating, quenching (see quenching)

heat treating, spheroidizing. When steel is heated to temperatures below A_1, the lower critical temperature, for long enough periods of time, the lamellar cementite in pearlite becomes spheroidal. Spheroidizing places a steel in the softest possible condition. It is widely used for high-carbon steels, to render them more machinable. After machining to shape, the steel is in an ideal condition for heat treating, because the carbide dissolves in austinite most readily when in the finely dispersed spheroidal condition. Spheroidizing is also the end result of prolonged tempering of martensitic steels. As tempering proceeds, carbide particles at first distributed on a submicroscopic scale grow to larger sizes until they can be seen even at fairly low magnification. This amounts to a spheroidizing treatment.

heat treating, solution (see solution heat treating)

heat treating, tempering. Tempering is also known as drawing. A controlled heat treatment consisting of reheating martensite, which is extremely brittle when produced during quenching of carbon steel. The operation involves heating the hardened steel to some temperature below the critical temperature. The principal objectives of tempering are to achieve increased toughness, to relieve residual stresses, and to stabilize structure and dimensions. Tempering achieves structural stability by hastening the decomposition of martensite and by causing either decomposition or stabilization of retained austenite. Dimensional changes are thus minimized or eliminated. The three most important considerations in tempering are the temperature used, the time at temperature, and the effect of alloying element in resisting softening and causing brittleness under certain conditions. Three stages of tempering are distinguished. In the first stage the quenched steel is heated to a temperature of 176–320°F (80–160°C) during which martensite loses some epsilon-carbide. This latter separates into very small particles and becomes less tetragonal. During the second stage, 446–536°F (230–280°C), any retained austenite transforms to bainite, and large dimensional changes occur. Finally, in the third stage, 500–680°F (260–360°C), epsilon-carbide changes to cementite platelets, producing a structure of ferrite and cementite. This is accompanied with a marked softening of the steel. The epsilon-carbide particles are extremely small in size, but they can be identified with the aid of an electron microscope.

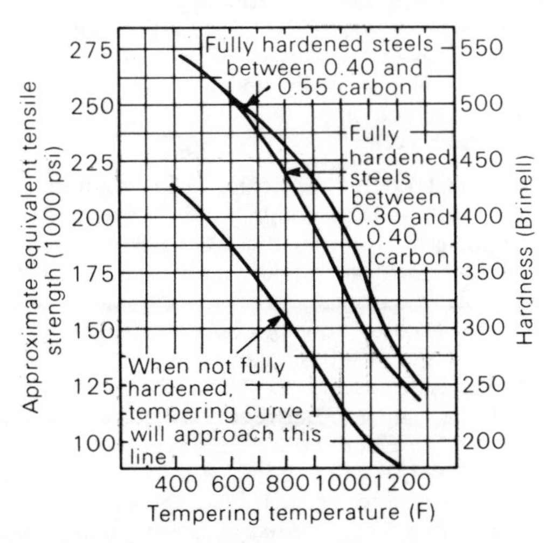

Typical curves showing relation between hardness and tempering temperatures (SAE Handbook).

helium (He). A monatomic inert gas, which does not dissolve in liquid metals. It competes with argon in its industrial use. Like argon, helium may be extracted from air, but is also available from natural-gas wells. Its specific weight at 70°F (21°C) and 14.7 psia (101.4 kPa) is exactly 10% that of argon. Its boiling point is very close to absolute zero. The specific heat of helium is very high, 1.25 Btu/lb (2.9 kJ/kg), second only to that of hydrogen. Because of its small molecule, helium serves as a most sensitive leak detector. It is also used as a refrigerant in the attainment of temperatures near absolute zero.

high-energy forming (see cold drawing and explosive forming)

hog fuel. In the manufacture of lumber the amount of material removed from the log to produce sound lumber is approximately 18% in the form of slabs, edgings, and trimmings; 10% as bark; and 20% as sawdust and shavings. While the total waste material will usually average 50%, distribution of different types of waste may vary widely from the approximations given above, owing to mill conditions as well as finished product. The mills frequently use either the sawdust or a mixture of sawdust and shavings for steam-production purposes because these can be burned without further processing. The remainder of the so-called waste products require further size reduction in a "hog" to facilitate feeding, rapid combustion, transportation, and storage. These newly sized products, together with varying percentages of sawdust and shavings present, constitute hog fuel. The percentage of sawdust and shavings present may be quite high if the fuel is to be burned at the sawmill. Hog fuel, normally delivered to a furnace, contains variable amounts of moisture, which may be taken as approximately 50% and most of which is in the cellular structure of the wood. Hog fuel, on drywood basis, contains approximately 81% volatile matter and somewhat less than 18% fixed carbon. The noncombustible residue, in the form of wood ash, is only a small fraction of 1%.

honing. A low-speed surface finishing process in which stock is removed by the shearing action of the bonded abrasive grains of a honing stone or "stick." The usual purpose of honing is to produce uniform high accuracy and fine finish, most often on inside cylindrical surfaces, and in the most common applications only a few thousandths of an inch of stock is removed. In honing a simultaneous rotating and reciprocating action of the stone results in a characteristic crosshatch lay pattern. For some applications, such as cylinder bores, angles between the cross-hatched lines are important and may be specified within a few degrees. Because honing is a low-speed operation, metal is removed without the increase in workpiece temperature that accompanies grinding, and thus surface damage caused by heat is avoided. Although the most frequent application of honing is for finishing inside cylindrical surfaces, numerous outside surfaces are also honed. Gear teeth, valve components, and races for ball bearings and roller bearings are typical applications of external honing.

honing, electrochemical (ECH). A modification of conventional honing, in which anodic dissolution of the work metal is combined with mechanical abrasion. It offers the same potential advantages over honing that electrochemical grinding (ECG): faster metal removal, improved deburring action, and increased life of bonded abrasive. The finished surfaces are virtually free from stress or heat damage. The operations that can be done by ECH are the same as those done by conventional honing, and the mechanical aspects of the two processes are closely similar. In its chemical and electrical aspects the ECH process most closely resembles the ECM. The workpiece is the anode and the metal tool is the cathode. The gap between workpiece and tool is usually about 0.002–0.005 in. (0.051–0.127 mm) at the start of the cycle and increases by the amount of stock removed per surface. The gap can increase to 0.020 in. (0.51 mm) or more. ECM electrolytes can also be used for ECH. The composition of the electrolytes is less critical than for ECM, because unreactive materials are removed continuously by the stones. A solution of sodium chloride is suitable for most work metals. Most of the metal removal is by electrochemical action.

horizontal bar code. A bar code or symbol presented in such a manner that its overall length dimension is parallel to the horizon. The bars are presented in an array which look like a "picket fence". **(See also bar code.)**

hot drawing, metals. A process for forming relatively thick-walled parts, having a generally cylindrical shape, by causing heated metal to be drawn through the opening between a die and a punch. The process starts with a flat, heated disk of metal, which is placed on the top of a female die. The difference between the diameters of the punch and the die opening is somewhat less than twice the thickness of the metal to be drawn. The stress condition is primarily a combination of biaxial compression and tension. As the punch pushes the metal through the die, the disk is formed into a cup. The cup can be further drawn out into a cylinder by drawing it through additional dies, which have reduced clearance between the die and the punch. It is also possible in some cases to draw a cup through several dies with a single punch. Hot drawing is used primarily for forming thick-walled cylindrical tanks, such as oxygen tanks and artillery shells. It also can be used for drawing other parts wherein the female die is closed.

hot rolling, metals. A method by which cast steel ingots are reduced to the size and shape desired. The ingots are heated in soaking pits at about 2200°F (1205°C) until white hot. They are then reduced in section and correspondingly lengthened in a rolling mill. The first mill through which the ingot passes is called a blooming mill, or bloomer, and the reduced ingot is called a bloom. The bloom is then further reduced in size and becomes a billet. Billets are about 2 in. (5.08 cm) square and are usually rolled further to make bars and rods of various cross sections including round, oval, hexagonal, octagonal, and flat. Ingots may be rolled directly into slabs, finished plates, and rails rather than into blooms. Blooms may also be rolled directly into plates, sheets, rails, and structural shapes instead of billets.

hot rolling mills. Hot rolling is done in stages on a series of rolling stands. Ingots are first rolled into blooms. Blooms are large bars having a minimum thickness greater than 6 in. (15.2 cm) and usually square in cross section. Blooms are rolled into billets or slabs. Billets are rolled bars having dimensions between 1.5 and 6 in. (3.8 and 15.2 cm), but are not in finished shape. Slabs are much wider than they are thick; usually more than 1.5 in. (3.8 cm) thick and at least 10 in. (25.4 cm) wide. Billets and slabs are the raw material from which finished hot-worked shapes are made. The rolls on rolling-mill stands may be arranged in six configurations. Blooms are rolled from ingots on two-high reversing or three-high mills. The three-high type is equipped with an elevator on each side of the stand for raising or lowering the bloom and mechanical manipulators for turning the bloom and shifting it to the various roll passes as it is rolled back and forth to final size. Four-high and cluster-roll arrangements are used in hot rolling very wide plates and sheets, and in cold rolling, where the deflection of the center of the roll would result in a variation in thickness. The small rolls, which are subject to wear, are less expensive, and the large back-up rolls provide the rquired rigidity. **(See also rolling mills.)**

hot working, metals. The properties of a metal are different when above and below its recrystallization temperature. The strength of a metal decreases as temperature rises, and its grains can be distorted more easily. If metal is held above the recrystallization temperature and below the melting point, its crystals grow larger. Small crystals tend to combine, and large ones tend to absorb the small ones. The higher the temperature, the faster the growth. The longer the time, the larger the grains become. The advantages of pressing or working hot metals are: (1) True hot working does not change the hardness or ductility of the metal. Grains distorted and strained during the process soon change into new undeformed grains. (2) The metal is made tougher because the grains are reformed into smaller and more numerous crystals. (3) The metal is made tougher because its pores are closed and impurities segregated. Slag and other inclusions are squeezed into fibers with definite orientations. (4) Less force is required, the process is faster, and smaller machines can be used for a given amount of hot working as compared with cold working because the metal is weaker. (5) A metal can be pushed into extreme shapes when hot without ruptures and tears because the crystals are more pliable and continually reform. Hot working is done well above the critical temperatures to gain most of the benefits of the process but not at a temperature high enough to promote extreme grain coarsening.

hubbing (see cold forging)

Hume–Rothery rules (metallurgy). The Hume–Rothery rules are helpful in making proper selection of alloying elements, and are, briefly, as follows: (1) Relative Size Effect. Upon the investigation of the sizes of two metallic atoms revealed in a measure by their lattice constants, if they are found to differ by less than 15%, the metals are concluded as having a favorable size factor for the formation of a solid solution. If the comparative size factor is greater than 15%, the formation of a solid solution is severely limited, being only a fraction of 1%. (2) Relative Valence Effect. In cases where the alloying element has a different valence from that of the base metal, the number of valence electrons per atom (the electron ratio) will be modified as a result of alloying. Inasmuch as crystal structures are more sensitive to a decrease in the electron ratio than to an increase, a metal with a high valence will dissolve only a small amount of a metal with a lower valence, although the lower valence metal may have satisfactory solubility for the higher valence metal. (3) Chemical Affinity Factor. This involves the electropositivity of the metals; the greater the chemical affinity of two metals, the more restricted will be their solid solubility. Those metals that differ greatly in electropositivity will usually form intermediate compounds rather than solid solutions. In cases where an unusually stable metallic–intermetallic compound is formed, the solid solubility is somewhat restricted. These Hume–Rothery empirical observations are extremely important in metallurgy in the consideration of binary alloys. All ferrous metallurgy depends essentially on the interstitial solid solutions of carbon in iron.

humidity. The measure of the amount of water vapor in a give volume of air, or other gases. The terms used to describe humidity are absolute humidity and relative humidity. Absolute humidity is the actual amount of water present in a specified volume of air. It is expressed in grains of water per cubic foot of air, or grams per cubic centimeter (1 grain equals 1/7000th of a pound). Relative humidity is the ratio of the actual amount of water vapor in the air to the maximum amount the air would contain at the same temperature, if saturated. Relative humidity is also defined as the ratio of the actual water vapor pressure in the air to the water vapor pressure in saturated air (containing the maximum amount) at the same temperature. Humidity expressed as 100% relative humidity indicates the air is saturated.

hydraulic lime (cements). Hydraulic lime is manufactured by burning limestone containing silica in kilns similar to those used in producing quicklime. The temperature used is somewhat higher than that required for quicklime manufacture. The "clinker" formed contains enough lime silicate for hydraulicity and enough quicklime for slaking. Slaking, which is done at the place of manufacture, causes the clinker to disintegrate and the fine powder which results is bagged as hydraulic lime. This cement is not suitable for underwater construction as it is too slow in setting, and it is too weak for general construction. It is primarily used for architectural decoration and for masonry, and together with Portland cement, for concrete.

hydrochloric or muriatic acid. A heavy chemical, hydrogen chloride (HCl) is a gas at ordinary temperature and pressure; aqueous solutions of it are known as hydrochloric acid or, if the HCl in solution is of the commercial grade, as muriatic acid. The largest users of hydrochloric acid are the metal, chemical, food, and petroleum industries. The major use of hydrochloric acid is in steel pickling (surface treatment to remove mill scale). Obtained from four major sources: as a by-product in the chlorination of both aromatic and aliphatic hydrocarbons, from reacting salt and sulfuric acid, from the combustion of hydrogen and chlorine, and from Hargreaves-type operation.

hydrogen. Has long been an important gaseous raw material for the chemical and petroleum industries. High-purity hydrogen is supplied in cylinders or trucks. It is frequently manufactured by electrolysis of alkalized water or from steam hydrocarbon reforming. About one-third of the production is used for ammonia. Other important chemical uses involve hydrogenating edible oils, electric machinery and electronics, missiles, and heat-treating ovens for metals and start-up of catalytic crackers. Hydrogen is derived almost exclusively from carbonaceous materials, primarily hydrocarbons and/or water. These materials are decomposed by the application of energy, which may be either electric, chemical, or thermal. Examples include electrolysis of water, steam reforming of hydrocarbons, and thermal dissociation of natural gas. Hydrogen is also produced by partial oxidation of hydrocarbons and by steam iron process, water gas and producer gas processes, and separation from coke-oven gas and refinery off-gas streams. Diffusion through a palladium–silver alloy furnishes very-high-purity hydrogen.

hydrogen bond (atomic structure). The hydrogen bond is formed by hydrogen between two strongly electronegative atoms or groups of atoms (oxygen, nitrogen, fluorine). A hydrogen atom lying on the periphery of such molecules as the OH, HF, and NH groups is capable of carrying an excess positive charge. This charge may interact strongly with the negative end of another molecule. The hydrogen bond appears to be essential in the association of molecules possessing high values of electric dipole moment. Water and hydrogen fluoride are examples, and their abnormally high boiling points are a consequence. The hydrogen bond, although slightly stronger than the pure van der Waals–London interaction, is generally of the same magnitude. The interaction energy for the hydrogen bond is given by $U_{CH} = c/r^2$, where c is the constant and r is the distance between atoms or groups of atoms forming the bond.

hydrogen peroxide. Hydrogen peroxide is manufactured by electrolytic and two organic oxidation processes. Applications include commercial bleaching-dye oxidation, the manufacture of organic and peroxide chemicals, and power generation. Bleaching outlets consume a great deal of the hydrogen peroxide produced. These include bleaching of wool, textile-mill bleaching of practically all wood and cellulose fibers, as well as of major quantities of synthetics, and paper- and pulp-mill bleaching of groundwood and chemical pulps. The advantage of hydrogen peroxide in bleaching is that it leaves no residue, with little or no deterioration of the organic matter that is bleached. Organic applications include manufacture of epoxides and glycols from unsaturated petroleum hydrocarbons, terpenes, and naturally fatty oils. The resultant products are valuable plasticizers, stabilizers, diluents, and solvents for vinyl plastics and protective coating formulations. It is also being used as a source of oxygen in municipal and industrial wastewater treatment systems.

hydrometallurgy. Basically used in large-scale operations, to separate soluble materials from those that are insoluble, by means of a solvent. An extraction process in which solvent used is either water or some aqueous solution. In every hydrometallurgical operation there are two essential steps required: first, leaching to dissolve the metal, and second, a precipitation to remove the dissolved metal from the leaching solution. Oxidized copper ores, ores of gold, oxidized uranium and vanadium, are all treated by hydrometallurgical methods. Many metals do not respond to leaching methods, consequently, hydrometallurgy is not as effective as pyrometallurgical techniques. This is due to the fact that it is used primarily on low-grade ores and such other materials as ore concentrates, artificial sulfides or metals (matte), the artificial arsenides and antimonides (speiss), as well as the alloy of some metals. All of this hydrometallurgical processing must be done on a large scale in order to prove profitable.

hysteresis. The ability of a magnetic material to withstand changes in its magnetic state. When a magnetomotive force (mmf) is applied to such a material, the magnetization lags the mmf because of a resistance to change in orientation of the particles involved.

hysteresis loop (electronics). The characteristics of a magnetic material within a coil may be plotted on a curve. In the figure below the x axis represents the field intensity measured in ampere-turns per inch. The Y axis represents flux density in kilolines per square inch. As H is increased the flux density increases to the point of saturation. This means that the magnetic core material is so full of flux lines that any further increase in force can produce no more. Now the force H is returned to zero, but the curve did not return to zero. This point marks the residual magnetism in the material after the magnetizing force is removed. A certain amount of force H in the opposite polarity is required to demagnetize the material. Continuation of the curve is shown as H is applied to the opposite polarity to saturation. The curve may be returned to saturation at opposite polarity by again reversing force H. The hysteresis loop graphically shows the characteristics of a magnetic material with a magnetizing force alternating in polarity.

I

ICAM (see integrated computer-aided manufacturing)

image enhancement, computer. A product inspection picture via computerized techniques may not always provide a satisfactory representation of the original object or scene; it may be geometrically distorted or it may be blurred or "noisy." There are many cases in which one can reduce the difference between a picture and its original by operating on the picture; this is the goal of image restoration. One can sharpen a picture (increase local contrasts in it, deblur it) by emphasizing its high spatial frequencies. This can be done by multiplying the Fourier transform of the picture by a weighting function from the original, and then taking the inverse Fourier transform to obtain the sharpened picture. Similar effects can be obtained by performing a differencing operation on the picture and combining the results with the original picture. If the picture is not only blurred but also noisy, these methods make it still noisier. It is often possible to achieve a useful compromise by emphasizing only a selected band of spatial frequencies (or, analogously, using a difference between average gray levels rather than between the gray levels of single pixels). Similarly, one can smooth a picture by deemphasizing its high spatial frequencies or simply by locally averaging it, but this is usually undesirable because it blurs the picture. If the noise is random, and several copies of the picture are available in which the samples of the noise are independent, smoothing without blurring can be achieved by averaging the copies. A known geometrical distortion in a picture can be corrected by resampling it at an irregularly spaced array of positions and outputting the samples as a regular array. Gray levels can be assigned to the new samples by interpolation from the levels of the nearby pixels. To correct an unknown relative distortion between two copies of a picture, one can find matches between pairs of distinctive local patterns, measure the relative displacement of each pair, and construct a geometrical distortion function by interpolation from these displacements.

image processing. A wide variety of techniques exist for processing pictorial inspection information by computer. The information to be processed is usually input to the computer by sampling and analog-to-digital conversion of video signals obtained from some type of two-dimensional scanner device. Initially, the information is in the form of a large array in which each element is a number representing the brightness (and perhaps color) of a small region in the scanned image. A digitized image array is sometimes called a digital picture, its elements are called "points," "picture elements," or "pixels." The values of these elements are typically six-bit or eight-bit integers. They usually represent brightness (or gray color). A digital picture may contain millions of bits, but most of the classes of pictures encountered in practice are redundant and can be compressed without loss of information. One can take advantage of picture redundance by using efficient encoding techniques in which frequently occurring gray levels or blocks or gray levels are represented by short codes and infrequent ones by longer codes. The pixels are encoded in a fixed succession. One can capitalize on the dependency of each gray level on the preceding ones by encoding differences between successive levels rather than by the levels themselves. If the dependency is very great, it may be even economical to represent the picture by the positions (or lengths) of runs of constant gray level or, more generally, to specify the positions and shapes of regions of constant gray level. Except for very simple classes of pictures, only a limited degree of compression can be achieved using efficient encoding. The digitization process itself, based on spatial sampling and gray-level quantization of a given real image, is a process of approximation. In designing approximation schemes, one can take advantage of the limitations of the human visual system, for example, quantization can be coarse in the vicinity of abrupt changes in gray level.

immersion plating, dip plating. The deposition, without application of an external electromotive force, of a thin metal coating upon a less noble metal by immersing the latter in a solution containing a compound of the metal to be deposited. **(See also electroplating.)**

impact extrusion. A part formed in a confining die from a metal slug, usually cold, by rapid single stroke application of force through a punch causing the metal to flow around the punch and/or through an opening in the punch or die.

impact mills (see coal)

impact test. A test to determine the behavior of materials when subjected to high rates of loading, usually in bending, tension, or torsion. The quantity measured is the energy absorbed in breaking the specimen by a single blow, as in the Charpy or Izod tests.

induction furnace. There are two types, one of which is the high-frequency type consisting of a crucible surrounded by a coil of copper tubing. When a high-frequency electrical current is passed through the copper coil, a rapidly alternating magnetic field is established, which induces secondary currents in the metal contained in the crucible. These secondary currents heat the metal very rapidly. Cooling water is passed through the copper tubing. The second type is the low-frequency induction furnace, used particularly in connection with die casting. The principle of a low-frequency induction furnace consists of a primary coil with a loop of molten metal acting as the secondary coil. Because the secondary has only one turn, a low-voltage, high-amperage current is induced in it. This high-amperage current provides the desired heating. Furnaces of this type require molten metal in order to start them operating, but the temperature can be controlled rapidly, so that they are useful in holding furnaces where it is desired to maintain molten metal at a constant temperature for an extended period of time, as in die-casting machines. Since there is no contamination from the heat source, very pure molten metal can be melted at one time, usually from a few pounds up to 4 tons (3.6 tonnes). **(See also coreless electric induction furnace.)**

induction hardening, metals. The procedure and results for induction hardening are the same as for flame hardening except that the high temperature is produced by high-frequency alternating currents. An induction coil is placed at the location to be hardened and the passage of current through the coil causes induced currents to heat the steel to the upper critical range. This is done very rapidly and is immediately followed by spray quenching. Induction hardening produces higher hardness for a given carbon content than normal heating. It is used extensively in producing objects in mass production, particularly in the automotive industry.

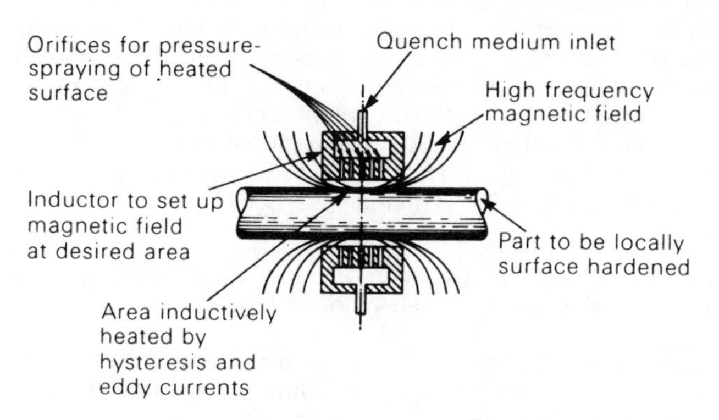

The scheme of induction hardening.

induction heater, interference terminology. A device for causing electric current to flow in a charge of material to be heated. Types of induction heaters can be classified on the basis of the frequency of the induced current, for example, a low-frequency induction heater usually induces power-frequency current in the charge; a medium-frequency induction heater induces currents of frequencies between 180 and 540 Hz; a high-frequency induction heater induces currents having frequencies from 1000 Hz upward.

induction heating. The heating of a nominally conducting material in a varying electromagnetic field due to its internal losses.

industrial process supervisory system. A supervisory control system that initiates signal transmission automatically upon the occurrence of an abnormal or hazardous condition in the elements supervised, which include heating, air-conditioning, and ventilating systems, and machinery associated with industrial processes.

industrial tow tractor. A powered industrial truck designed primarily to draw one or more nonpowered trucks, trailers, or other mobile loads.

inertia welders (see welding, inertia)

infrared radiant heating. Infrared energy is radiated by every object with a temperature above absolute zero. The amount of infrared radiation is directly related to the object's temperature. Thus, the operating temperature of the source is an important consideration. Among the various sources used by industry, the tubular quartz lamp with tungsten filament is noteworthy for its high concentration of radiant energy. Standard argon-filled lamps operate at a filament temperature of 4000°F (2205°C); halogen types are available that operate at temperatures up to 5600°F (3093°C). These lamps are efficient transducers of electrical power into radiant energy. Tests have shown that as much as 86% of the electrical power dissipated by the lamp is converted to usable infrared energy in air (radiation energy is not measured in degrees of temperature but is described as a field of flux with a density measured in terms of watts per unit area). Radiant energy striking an object is reflected, transmitted, or absorbed in varying proportions according to the color, composition, and temperature of the object. Energy that passes through the object, or is reflected, or reradiated by it, is lost. Some of these losses may be minimized by suitable reflectors. Only the portion that is absorbed is converted to heat and is effective in raising the object's temperature. Infrared energy is a practical and economical source for heating in a variety of production processes in a variety of industries. Examples of processes include drying, curing, softening, soldering, fusing, shrinking, annealing, and stress relieving. Examples of the industries include such areas as metalworking, metalforming, electronics, plastics, chemicals, food processing, and paper and foil conversion. The main advantages of using infrared radiant heating are: energy transfer by radiation is very rapid; heat losses are minimal; equipment is compact and easily installed; there are no products of combustion to contaminate the air or product (unless they come from the product heated); energy can be focused to concentrate the heat; and the energy output can be precisely adjusted and maintained—with almost instantaneous response to control.

ingot. A cast form suitable for remelting or fabricating. **(See also extrusion ingot, fabricating ingot, forging ingot, remelt ingot, rolling ingot.)**

ingot production. Regardless of the method by which steel is made, it must be either poured into some kind of mold in which it solidifies into the final desired form or into a form suitable for further processing. In most cases, it is made into a form suitable for further processing, which will be some type of ingot. These are the raw materials for forging and rolling operations. In casting ingots the molten steel is tapped from the furnace into pouring ladles. Nearly all of these are of the bottom pouring type. The metal is poured into the ingot molds through either the top or bottom. The majority of steel ingots are poured from the top. Ingots sometimes are poured from the bottom to avoid the hot metal that may be splashed onto the side walls of the mold and solidifies, and then becomes a part of the ingot when poured from the top.

injection molding machines, fixed screw and ram. These machines are similar to the two-stage plunger, except that the first plunger is replaced by a screw that carries the material through the heating chamber. The material is forced through the shut-off valve and into the injection chamber. When the injection chamber has been filled with a predetermined amount of plasticized material, the screw stops rotating and the shut-off valve turns so that the injection chamber is connected to the nozzle. After the clamp has developed the required tonnage, the material is injected into the mold.

injection molding machines, plastics. The process of injection molding consists of using a specially designed machine for melting solid plastic particles into a viscous liquid and forcing the liquid into a mold. The liquid plastic assumes the shape of the mold cavity and is held under pressure until it cools sufficiently to become solid. The mold halves are then opened and the solid part is removed. The solid part in many cases may be ready for use without any further operations. The viscous liquid is normally injected into the mold at pressures up to 20,000 psi (1.4×10^5 kPa) and the force required to clamp the mold halves together will normally range from 1 to 5 tons/in.^{2} (1.5 to 7.5 hectobars) of projected area of the part being molded.

injection molding machines, reciprocating screw. The in-line reciprocating screw is the most popular injection system. Plastic pellets are placed in the material hopper above the screw. As the screw rotates, the material feeds by gravity from the hopper down into the barrel. The flights on the screw carry the material toward the left. The barrel is wrapped with heating bands that supply heat to the barrel and the plastic material. Rotation of the screw imparts a churning action to the material and is constantly forcing unmelted material against the hot barrel. The screw design is such that the material is compressed as it moves along the screw. The screw is divided into three sections: the feed section, the transition section, and the metering section. The root diameter of the screw is constant in the feed and metering sections, but increases in the transition section. As the screw rotates the plasticized material is deposited in front of the screw, causing the screw to move to the right. By regulating the hydraulic pressure in the injection cylinder that opposes the screw movement to the right, the quality of the plasticized material can be controlled. This hydraulic pressure is called "back pressure." The screw continues to rotate and plasticize material until it is stopped either by a preset limit switch or timer.

injection molding machines, single-stage plunger. A molding machine that uses a plunger to pack plastic pellets into a heat chamber. Each time a quantity of solid material is pushed into the chamber a quantity of melted plastic is displaced from the other end of the heating chamber into the mold.

injection molding machines, two-stage plunger. This machine uses one plunger to force solid materials into the heating chamber, which in turn forces melted material from the chamber. But the melted material is deposited in front of a second plunger rather than directly into the mold. When the second plunger is charged with a predetermined amount of material, it forces the material into the mold.

injection molding, powder metals. Using conventional thermoplastic injection molding equipment commercial production of powder metal precision parts is possible. Metal powders of less than 394 μin. (10 μm) in size as compared with the 1.2×10^{-3}–7.9×10^{-3} in. (30–200-μm) size used in powder metal pressing are combined with a plastic binder and injection molded. Solvent extraction sintering removes the binder and complex, close-tolerance parts are possible with 94–98% of theoretical density.

input signals. Input signals, determined by the presence or absence of a specific voltage at the input signal terminals, tell a robot when or where not to do something. A typical condition is "if input signal three is present, put the part in chute one." The robot makes a programmed decision to perform the routine of placing the part in the chute based on the state of input signal three. More complex applications may require decisions to be based on more than one condition. For example, "if a part is ready, and if the oven is up to temperature, and if the oven is empty, then put the part in the oven." Three conditions are required to make this decision, which normally means wiring three switches in series to an input signal terminal, or first wiring them to a relay panel or programmable controller if additional decisions are to be made with those same switches. It is this type of application that can bring out the cost effectiveness of a computer-controlled robot, because additional switching logic hardware is not required.

inspection. The evaluation of supplies and/or products being manufactured either by visual means or by use of measuring devices to determine conformance to engineering specifications covering appearance, dimensional, or functional characteristics.

insulation (see ceramic fiber)

insulators. Materials that have relatively few free electrons. There are no perfect insulators, but many substances have such high resistance that for practical purposes they may be said to prevent the flow of current. Substances having good insulating qualities are dry air, glass, mica, porcelain, rubber, plastic, asbestos, and fiber compositions. The resistance of a wire varies inversely with the area of the cross section. Temperature is another factor that affects the resistance of a wire. Usually, the resistance of a wire increases with an increase in temperature. However, some substances such as carbon, decrease in resistance as the temperature increases. The degree of resistance change due to temperature variation is not constant but depends on the material. The general rule for the resistance of a conductor is as follows: The resistance of a given conductor varies directly as its length, and inversely as the area of its cross section, when the temperature remains constant.

integrated circuits (electronics). Tiny circuits called integrated circuits formed by masking, etching, and diffusion on a monolithic substrate of silicon. The integrated circuits are made on a thin slice of silicon from 1 to 2 inches (2.54 to 5.08 cm) in diameter. A normal slice may contain from 100 to 600 circuits side by side, all processed at the same time. After processing, the slice is divided into separate circuits called "dies". The number of complicated ICs will depend on the "die" size. These could be in squares of 60 × 60 mils. The following are steps to make a simple integrated circuit (expitaxial-diffused process), starting with a wafer of *p*-type silicon as a substrate. (1) On the wafer is grown an epitaxial layer of *n*-type silicon about 0.25 microns thick. (2) A thin coat of silicon dioxide is now grown over the *n*-type material by exposing the wafer to an oxygen atmosphere at about 1832°F (1000°C). (3) To prepare the wafer for isolation between various components, the wafer is covered with photoresist and exposed under ultraviolet light through a specific photographic mask. The nature of the mask depends on the circuits to be made. (4) The wafer is then etched with hydrofluoric acid and unexposed areas of silicon dioxide are etched away. (5) The wafer is now subjected to a diffusion process using boron. The boron diffuses into and forms a *p*-type material on all areas not protected by the silicon dioxide. Sufficient time is allowed for diffusion completely through the epitaxial layer to the *p*-type substrate. The wafer now appears with isolated islands of *n*-type materials. Isolation is realized by the formation of the *npr* junctions around each island, and there are back-to-back diodes between each *n*-type island. (6) During diffusion, a new layer of silicon dioxide forms over the diffused *p*-type areas as well as on the top of the islands. (7) Using the photoresist coating again and exposure under a specified mask, areas in the *n*-type islands are etched away. Once again the wafer is subjected to a *p*-type diffusant and areas are formed for transistor-base regions, resistors, or elements of diodes or capacitors. (8) Again the wafer is given an oxide coating. (9) The wafer is again masked, exposed, and etched to open windows in the *p*-type regions for transistor emitters and regions for diodes and capacitors. Small windows are also etched through to the *n*-layer for electrical connections. The total wafer is again given the oxide coating. (10) For this particular IC, the monolithic circuit is complete except for the interconnections. A thin coating of aluminum is vacuum deposited over the entire circuit. The aluminum coating is then sensitized and exposed through another special mask. After etching, only the interconnecting aluminum forms a pattern between transistors diodes and resistors and pads for wires to connect the wafer to an external circuit. (11) The wafers are now scribed with a diamond-tipped tool and separated into individual circuits.

integrated computer-aided manufacturing (ICAM). A U.S. Air Force project established to automate production in the aircraft industry.

integrated quality-management systems (IQMS) (electronics). In electronic chip, circuit board, and total systems manufacture, this comprises a local-area network and database management software system to process the data gathered from all the automatic-test-equipment (ATE) stations in the plant.

interchangeability. This means that any product or element in a manufactured lot can be substituted for any other at random. The technique of mass production permits production of piece-parts at diverse locations to specified tolerances so that all are sufficiently identical to permit random use or random assembly with other products. Its basic feature is the elimination of separate individual fitting up in manufacture and assembly.

interfacing. A shared boundary. Except in a few applications, most robots need to communicate and interact with the outside world. This can take the form of simple on/off signals by means of electrical, or pneumatic, contacts, or of more complex electronic signals. Inputs are the number of lines over which the robot will accept signals from the outside world, and outputs are the lines over which it will send signals to external equipment.

intermetallic compounds. Defined as compounds of two or more metals in an apparently stoichiometric proportion, but they generally do not follow ordinary valency rules. Both ionic and covalent bonding may occur in metallic compounds, depending on the difference in their electronegativities. The extent to which the bonding deviates from the covalent types determines properties such as electrical conductivity and melting points. Generally, intermetallic compounds with the covalent bond do not exhibit to a marked degree the properties characteristic of metals. They form, however, an important group of intermetallic compounds exhibiting semiconducting properties. The compounds are formed between the elements of the III and V groups of the periodic table, and examples of them are InAs, GaSb, and InSb. Other types of compounds showing a more ionic character are represented by Mg_2Sn and Mg_2Pb.

internal combustion engine truck. An industrial truck in which the power source is a gas or diesel engine.

Invar (see nickel-iron alloys)

investment casting (see casting, investment)

iodine. Most iodine used is converted directly to chemical compounds. The largest use is for organic compounds; potassium iodide, sodium iodide, and other inorganic compounds. Iodine is used as a catalyst in the chlorination of organic compounds and in analytical chemistry for determination of the so-called iodine numbers of oils. Iodine for medicinal, photographic, and pharmaceutical purposes is usually in the form of alkali iodides, prepared through the agency of ferrous iodide. In addition the element is also employed in the manufacture of certain dyes and as a germicide. Simple iodine derivatives of hydrogen, such as iodoform have antiseptic action. Organic compounds containing iodine have been used as rubber emulsifiers, chemical antioxidants, and dyes and pigments.

ion implantation. This is a method of doping semiconductors. All semiconductors, both discrete devices and integrated circuits, require that their electrical characteristics be adjusted in certain locations. It may be necessary to make some regions "positive," or *p*-type, and other regions "negative," or *n*-type. This alteration of electronic properties of the starting silicon material is usually accomplished by introducing minute quantities of impurities, or dopants, into the silicon crystal lattice. Customarily, boron is used as the *p* dopant, and phosphorus, arsenic, and antimony are used as *n* dopants. In the ion implantation process the chips are placed under vacuum. A beam of ions of the species of interest is generated and accelerated electrically to a final energy. They then impinge on the surface of the water at high velocity and are driven up to 39.4 μin. (1 μm) into the crystal lattice by their high kinetic energy. An ion is an atom or molecule that has a net electric charge by virtue of having a deficiency or excess of electrons. In ion implantation, the ions are usually single atoms that have had one negative electron removed, and so are positively charged.

ionomers (plastics). The term ionomer was coined by the Du Pont Company to describe a polymer which contains both organic and inorganic materials linked by both covalent bonds and ionic bonds. The major constituent of the product is ethylene. The ionic bonds involve anions from the polyethylene chain and metallic cations such as sodium, potassium, magnesium, or zinc. The properties are given as high transparency, toughness, unusual resilience, and resistance to oils and greases. The ionomers can be processed by the usual thermoplastic techniques. Uses may include vacuum packaging, bottles, houseware, goggles, shields, refrigerator trays, automotive steering wheels and trim, toys, novelties, and electrical parts.

IQMS (see integrated quality-management systems)

iron. Pure iron does not occur in nature nor is iron produced in pure form for engineering use. The term iron means, in commercial usage, pig iron, cast iron, or wrought iron. In any of these forms iron is combined with other elements which influence the physical and mechanical characteristics. Steel is iron alloyed with carbon and various other elements. Special types of steel are sometimes identified by one or more of these elements, such as chrome steel, nickel steel, chrome–molybdenum steel. At other times the ultimate use of the steel is employed to designate the type, such as structural steel, tool steel, and stainless steel.

iron–carbon alloys (metallurgy). A small amount of carbon added to iron and subsequently heat-treated properly can raise the tensile strength of otherwise pure iron from 40,000 to 150,000 psi (2.8×10^5 to 1.03×10^6 kPa). The most common form of commercially pure iron, known as ingot iron, is manufactured in the open hearth furnace. Such ingot iron is useful when fabricated into culverts and roofing, and as a base for porcelain enameled refrigerators, stoves, etc. Iron and carbon alloys, with other elements added for specific purposes to improve the mechanical properties, constitute the important ferrous alloys referred to as steels and cast irons. Pure ingot iron is soft but can readily be toughened by cold working. Ingot iron is used extensively in direct-current magnetic circuits because of its high magnetic permeability and low remanence.

iron–chromium alloys (metals). There are four major groups of chromium steels: (1) pearlitic, 0–2% chromium; (2) martensitic, 2–17% chromium; (3) ferritic, 17–35% chromium; and (4) austenitic, 10–35% chromium. Pearlitic chromium steels are used when carburizing for case hardening is required; they find application in gears, bearings, springs, files, drills, chisels, and other small tools. The effect of the chromium is to increase hardenability and toughness; in amounts of 1–2% it enhances wear resistance. Martensitic steels containing from 2 to 4% chromium have high ductility, toughness, and wear resistance, and are used for railroad rails and for moderately high-temperature service. The 4–6% chromium steels are used in oil stills because of their resistance to oxidation and corrosion. Steels containing more than 12% chromium are very resistant to oxidation, and therefore are suitable for turbine blades and cutlery. With up to 2% carbon they are used for tools, rolls, and dies when wear might otherwise be excessive. Ferritic chromium steels have high corrosion resistance. Those that resist corrosion at high temperatures contain greater amounts of chromium. With a low carbon content, they are easily cold worked and do not harden readily. They are not easily machined unless they contain additional elements. Austenitic chromium steels contain from 10 to 30% chromium, amounts that would normally bring them into the martensitic or ferritic range. However, the austenitic condition is obtained at room temperature by the addition of manganese, nickel, or copper. They are used in corrosion- and oxidation-resistant applications, as in chemical and food plants and for furnace parts and architectural trim. They do not machine well, but are easily cold rolled or drawn into shape. Their great strength and stiffness make them useful for lightweight structures, such as aircraft. They are also used in jet engines because of their relatively high strength at elevated temperatures.

isoprene rubber–synthetic (IR). Polyisoprene has the distinction of being a synthetic elastomer which has the same chemical composition as natural rubber. **(See also natural rubber.)**

isopropanol process (petrochemical). Process for the manufacture of 99.9% min purity isopropanol from 90 to 99% or lower concentration propylene. The direct hydration of propylene to isopropanol is carried out using a trickle process with an ion-exchange catalyst. Liquid propylene at elevated pressure is mixed with preheated water, the heat capacity of the water being used for evaporation of the propylene. The mixture of water and propylene in a supercritical state is charged to the top of a fixed-bed reactor and allowed to trickle downward concurrently over a bed of an ion-exchange resin. An intensive exchange between the liquid and gas phases occurs at a temperature between 266 and 320°F (130 and 160°C) and a pressure in the range of 1190.7–1484.7 psig (8210–10,237 kPa). Aqueous alcohol and nonconverted propylene are drawn off from the bottom of the reactor and passed to a high-pressure separator where the alcohol-containing aqueous phase is separated from a propylene containing gas phase. The liquid phase is then passed to a low-pressure separator. The crude alcohol from the low-pressure separator is charged to a distillation column where diisopropyl ether is removed overhead. The bottoms are charged to a second column where isopropanol is taken overhead as an aqueous azeotropic mixture. Water from the bottom of this column is desalted by ion exchange and recycled to the reactor. Dehydration of the azeotropic mixture of isopropanol and water is carried out using benzene and an entrainer.

isostatic or hydrostatic pressing. This is a method of compacting powdered materials by applying pressure simultaneously and equally in all directions. Since the behavior of powders under pressure is dependent on the number of directions from which pressure is applied, isostatic is the ideal way of forming powdered materials. The procedure is as follows: powdered material is placed in a flexible container, which serves as the mold, and is tightly sealed. This container or mold is then placed inside a pressure chamber. The pressure vessel is sealed, and hydrostatic pressure is applied equally to all surfaces and released. The result is uniform compression and density in the finished compact.

isostatic pressing, cold. This is a method mainly intended for the compacting of powdered parts that owing to their shape and size cannot be produced in conventional compacting presses. The powder is enclosed in a container (mold) of yielding material having the same shape as the finished product. After air evacuation and sealing, this container is lowered into a pressure chamber and subjected to a high, uniform hydrostatic pressure. The powder is compacted uniformly in all directions so that the compact becomes an accurate scale-down of the mold. Normally, the compact has such a strength that it can withstand handling up to subsequent sintering. The uniform density makes it possible to avoid distortions during the sintering process.

isostatic pressing, hot. Often termed gas-pressure bonding, this is a method where powder, mostly in the form of a cold compact, is enclosed in a pressure-tight metal container of appropriate geometry, inserted into a pressure vessel, and subjected to simultaneous heat and pressure. The pressure medium in the vessel is an inert gas. The combined action of high pressure and high temperature promotes the sintering. Powder parts of full density can be produced at much lower temperature than with conventional sintering at atmospheric pressure or under vacuum. For refractory metals, this high density may occur at temperatures only 0.5 to 0.6 times the melting point. The lower temperature required for this process allows for much greater control of the crystal or grain growth. The process can also be used for the joining of metals, ceramics, and cermets.

Izod test. A pendulum-type single-blow impact test in which the specimen, usually notched, is fixed at one end and broken by a falling pendulum. The energy absorbed, as measured by the subsequent rise of the pendulum, is a measure of impact strength or notch toughness. Contrast with the **Charpy test.**

J

jig. Tool for holding component parts of an assembly during the manufacturing process, or for holding other tools. Also called a fixture.

jig borers and special boring machines. Entirely special machines for extremely large-production precision boring are designed primarily for maintaining exact hole spacings and alignment. Inasmuch as many parts, by both design and quantity, fall outside the scope of economy considering totally special machines, the jig borer is normally resorted to for precision results. This machine is primarily designed for precision work on jigs and fixtures in the toolroom, but where accuracy is such as to be beyond the scope of drills and drill jigs, the part too fragile to withstand heavy cutting loads, or the material subject to poor finish by ordinary methods, the jig borer is especially well suited. Placed in a fixture, the part can be finish bored in the sequence required, by means of the gaging features on the table positioning axes of these machines, without relocation of the part during the process.

job shop. A discrete parts manufacturing facility characterized by a mix of products of relatively low-volume production in batch lots.

Jominy end-quench test (steel). The Jominy end-quench test holds all factors constant except composition to measure the hardenability of steels. A 1-in. (2.54 cm) diameter by 3- or 4- (7.6- or 10.2-cm) long bar is properly austenitized and quenched on the end in a standardized way. Heat is removed substantially from the quenched-end surface and is thus withdrawn at different rates along any one bar but in the same way along any bars of steel tested. The result is a gradient of hardness along the bar that depends only on the composition of the material. After the piece has cooled to room temperature, two flats are ground lengthwise on diametrically opposite sides. Rockwell hardness readings are taken at 1/16-in. (1.6-mm) intervals along the bar and plotted. The hardness for a plain carbon steel drops off rapidly a short distance from the end. An alloy steel would probably show a much smaller rate of decline.

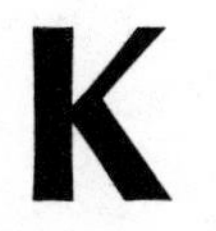

kaolin (see clay)

ketones. Often referred to as carbonyl compounds, the carbonyl carbon of ketones is bonded to two organic groups. Such groups may be identical or they may be different. Common names for the simple ketones are formed by naming both groups attached to the carbonyl carbon atom, then adding the word ketone. With the exception of formaldehyde, a gas, the lower-molecular-weight aldehydes and ketones are liquids that have lower boiling points than alcohols of the same carbon content. Ketones except for the lower members, which contain up to four carbon atoms, are practically insoluble in water.

kirksite. A low-melting-point alloy of aluminum and zinc used for the construction of molds; it imparts a high degree of heat conductivity to the mold.

L

laminated wood. A high-pressure bonded wood product composed of layers of wood with resin as the laminating agent. Plywood is a form of laminated wood in which successive layers of veneer are ordinarily cross laminated, the core of which may be veneer or sawed lumber in one piece or in a number of pieces.

laminated metals. A strong but more-economical base substrate is combined with a thinner, less-expensive facing sheet. The composite metal provides: exceptional sound and vibration damping qualities; hostile environmental resistance at lower cost; lighter weight; lower cost radiation shielding; EMI/RFI shielding; and insulation properties. Just about any two or more materials can be combined in a suitable laminate—aluminum, copper, brass, bronze, carbon steel, galvanized steel, stainless steel, lead, magnesium, nickel alloys, titanium, honeycombs, rigid plastics, foamed plastics, hardboard, and wood.

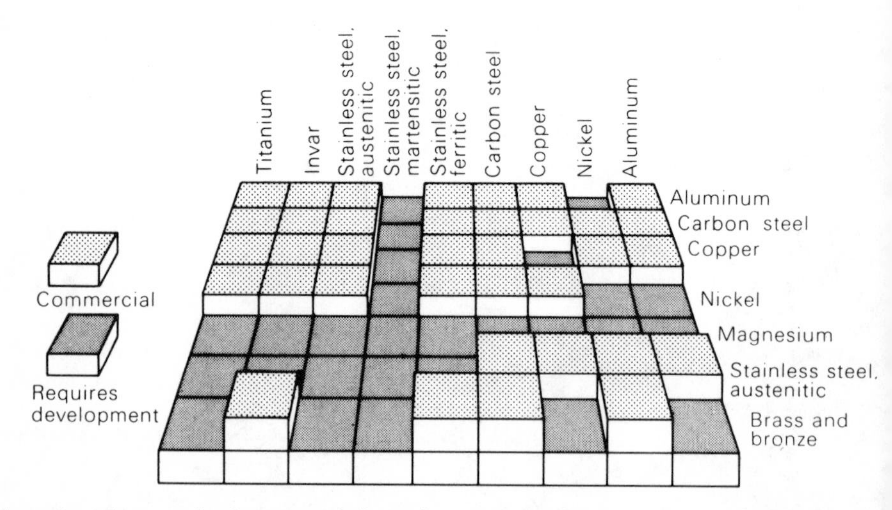

Illustration shows 36 laminated or clad metal systems that have found commercial use. The other 27 systems are considered feasible.

laminate, textiles. A fabric made by fusing a woven or knitted fabric with another fabric, usually a plastic film.

laminating, thermoplastics. A process wherein sheets of cloth or paper are impregnated with thermosetting liquid and then placed together to build up a desired thickness. The resulting "sandwich" is cured, usually under considerable pressure and at elevated temperature. In such laminated products the cloth or paper filler is not ground, shreded, or macerated. Such products may be produced so as to have unusual strength properties. Since the surface is a thin layer of pure resin, they usually possess a smooth attractive appearance. By using transparent resins the sheet filler may be made visible. Laminated plastics are produced as sheets, tubes, and rod. Because of their excellent strength qualities, plastic laminates find a wide variety of uses. Laminated sheets can be blanked and punched readily. Thick laminated sheets are often used for gear blanks. Many laminated plastic products, which are not flat and which contain relatively simple curved shapes, such as safety helmets, boats, and automobile bodies, are made using moderate pressure and temperature. Usually a simple female mold is used, made from metal, hard wood, or hardboard of some type. The laminated material is placed in the mold and both are put in a bag from which air is evacuated. The external pressure then holds the laminate against the mold during curing in live steam and under heat lamps. When plastics that cure at room temperature are used, the vacuum bag may be eliminated, the pliable resin-dipped material merely being placed in the mold or over a form.

lamination. An internal crack aligned parallel to the principal surfaces of a sheet or plate. Also, a composite product consisting of two or more sheets or films joined together, e.g., paper to foil or plastic films to foil.

lampblack. A product made by the restricted combustion of resins, oils, or other hydrocarbons. It has gradually been replaced in the pigment trade by carbon black, which has superior tinting strength and coloring qualities. In the manufacture of carbon brushes for electrical equipment and lighting carbons, lampblack is still a very important constituent. Its color is bluish-gray black. In the United States either tar oils or petroleum oils are burned with restricted air to form soot, or lampblack. The soot is collected in large chambers from which the raw lampblack is removed, mixed with tar, molded into bricks, or pugs and calcined up to about 1832°F (1000°C) to destroy bulkiness. The calcined pugs are ground to a fine powder. Carbon brushes for use in electric machinery are made by mixing lampblack with pitch to form a plastic mass. Petroleum coke or graphite may be added to this mix to impart special properties. Plates or blocks from which the brushes are later machined are formed by extrusion or high-pressure molding and these green plates are then baked at a high temperature for several days to drive off volatile matter. When carbons for the production of special arc lights are desired, a mixture of petroleum coke and thermally decomposed carbon is extruded in the form of a tube. This tube is baked at 2642°F (1450°C), and a core of selected material is forced into its center and calcined again.

lapping. A low-speed low-pressure abrading operation that accomplishes one or more of the following: (a) extreme accuracy, (b) correction of minor imperfections in shape, (c) refinement of surface finish, and (d) extremely close fit between mating surfaces. In general, the quality that can be obtained by lapping is not easily or economically obtained by other processes. Either loose or bonded abrasives may be used for lapping. When bonded abrasives are used, the lapping process resembles conventional grinding, except that lapping is done at low speeds and is nonsparking. Lapping with bonded abrasives differs from honing mainly in the type of tooling employed. Also, unlike honing, lapping is not necessarily a two-motion process. Lapping operations usually fall into one of two categories; individual-piece lapping or matched-piece lapping.

lapping machines. For production lapping a variety of machines are available, which use abrasive-charged laps, bonded-abrasive laps, and abrasive-coated cloth or papers. When desirable, cast iron or other charged laps can be substituted for the honing tool on various standard honing machines to obtain a velvet lapped finish. Vertical-spindle machines with parallel flat charged cast-iron laps will handle flat parts up to 4 in. (10.2 cm) thick and 9 in. (22.9 cm) in longest dimension or cylindrical work up to 4 in. (10.2 cm) in diameter by 9 in. (22.9 cm) long. Centerless lapping machines are available which employ a charged lapping roll and a control roll both of which are driven, the charged roll normally being the faster. Where desirable, dual or triple-roll machines can be used to rough and finish lap parts with one handling. Either manual or automatic feeds can be used, depending on production requirements, for lapping cylindrical, tapered, or shouldered parts from 0.030 to 10 in. (7.6 mm to 25.4 cm) in diameter. Contoured parts require special rolls.

lap-welded tube (see tube)

laser hardening. Surface hardening via a laser is a process used to increase the strength and wear resistance of metal parts. This surface modification technique requires no addition of material to the workpiece. After laser surface hardening, a part can be used without posthardening machining. **(See also case hardening.)**

laser machining and fastening. Of the many uses for power laser beams in industry the main applications are cutting, welding, soldering, drilling, and heat treating. Most are YAG, CO_2, or solid-state lasers built into appropriate machine-tool systems. Holes can be drilled from 0.008 to 0.03 in. (0.2 to 0.8 mm) diameter in metals. Larger holes can be done with a beam-rotating device or on an NC machine. Thickness of the metal

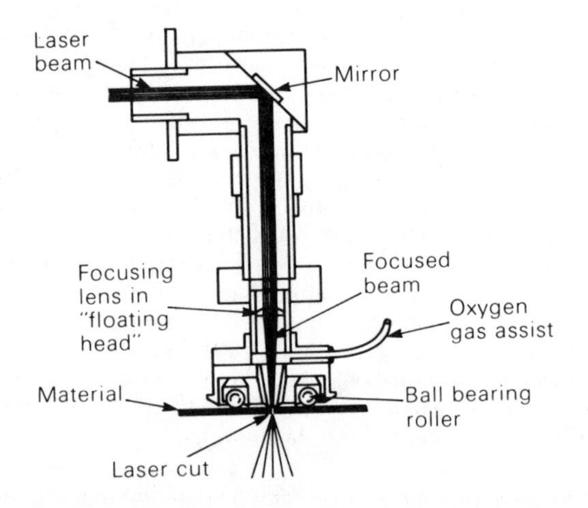

Cutting with a laser is achieved by heating the metal to its melting point while at the same time applying a jet of gas through a nozzle.

drilled or cut is limited to less than 0.2 in. (5.1 mm) although up to 0.5 in. (12.7 mm) has been done in Hastelloy-X. Depending on speed, mild steel from 0.04 to 0.13 in. (1 to 3.3 mm) thickness can be cut, wood from 0.15 to 1.2 in. (3.8 to 31 mm), particle board from 0.125 to 1.1 in. (3.2 to 28 mm) and acrylic plastic from 0.11 to 1.0 in. (2.8 to 25.4 mm). Spot welds at rates over 20 spots per second can be done. Continuous seam welds are done with overlapping spots. Welding is excellent in thin-gage materials, because minimum heat is introduced so warping and cracking is eliminated. Although welding is mainly autogenous, in some cases filler metal can be used. Ceramic substrates as used in electronics can be cut or drilled, diamond dies can be drilled, silicon wafers can be drilled, resistors can be trimmed, etc., all without direct mechanical contact. Beams can be focused down to a spot small enough to machine lines thinner than one-tenth the diameter of a human hair. Unusual applications range from perforating holes around the filter of cigarettes to admit air to cutting cloth or fabrics in the production of clothing and the slicing of glass tubes for fluorescent bulbs.

laser printing. In production of electronic components lasers are used to print identification numbers on each part manufactured. Used in lieu of conventional rubber stamping, the laser is beamed through a stencil onto the component where it changes the cured colored powder coating to create the proper identification marks. This method is faster by 10 times and can be used to clearly mark two sides at the same time with no pressure damage to fragile products.

laser scanner. An optical bar code reading device using a low-energy laser light beam as its source of illumination. **(See also bar code and scanner.)**

lasers (electronics). Quantum electronic devices that generate intense monochromatic and coherent electromagnetic radiation. The words laser and maser stand for acronyms of the phrases "light (microwave) amplification by the stimulated emission of radiation." Amplification by stimulated emission has been obtained at many wavelengths in the infrared, visible, and ultraviolet portions of the spectrum. Gaseous, liquid, and crystalline materials, both inorganic and organic, have successfully been used as active media for light amplification. Thermodynamic equilibrium between two states of a group of atoms requires that the distribution of these atoms follow Boltzmann's law. Stimulated emission takes place if the atom is struck by a wave packet with exactly the same frequency as previously emitted. Only a medium in which more atoms are excited to the upper-energy state than there are in the lower-energy state is capable of amplification by stimulated emission. To produce a population inversion, the energy must be pumped in the system very quickly because the excited atoms will stay at the upper level for only a few seconds. This can be obtained by optical pumping (flash tube) as used in solid-state lasers, by electric discharges through gases, and by electron bombardment. The condition for the start of oscillation necessitates that the gain produced by the stimulated emission be equal to that lost by the resonator. To make a coherent oscillator from the excited atoms, they are enclosed in a cavity resonator. The most convenient way to form the cavity resonator is to contain the active or laser medium between two mirrors facing each other at a distance that is large compared to their diameter. One of the mirrors is partly transparent or has a small hole so that a part of radiation escapes through it. This radiation emerges as an extremely directional and coherent beam of light. Since a major portion of the wave in the resonator is always retained, stimulated emission continues to follow its path.

lasers, gaseous. Gases may be used as CW (continuous-wave) lasers. To produce population inversion, an electric-arc-discharge or RF (radio-frequency) generator of suitable frequency is used. Collision by electron impact in a suitable discharge causes a population inversion in certain gases such as pure neon, helium, argon, or krypton. The first gas-discharge laser used a He–Ne mixture under low pressure and produced a coherent output at a few wavelengths, including several in the visible spectrum, with the strongest visible line at 6328 Å. The laser action occurs by the transfer of excitation from helium to neon atoms, which in turn, fall to lower staees; emitting radiation. Transitions between the electron levels within the excited atom give rise to emission spectra in the visible or ultraviolet region with some superimposed near-infrared radiation. When the molecular gases are used as a lasing medium, then, in addition to the electronic transitions, vibrational and rotational transitions within the excited molecules occur, producing infrared emission. The energy level spectra of molecular gases are considerably more complicated than those of atomic gases. There are possible transitions between different vibrational and rotational energy levels, causing emissions at quite a few wavelengths. The type of emission can be modified by the structure of the lasers, such as the length of the lasing tube, the position of the mirror, the composition and the pressure of the gases (lasing medium), the type of energy input, and other specific physical characteristics of the lasing system.

lasers, optically pumped. The first commercially operated laser was the ruby laser (chromium ions in alumina), which yields red light. The ruby laser is operated normally only in pulses because of the high power required to reach a threshold for lasing. The chromium ions in ruby have broad absorption bands, so that light over a broad range of wavelengths can be absorbed to produce excited ions. A crystal ruby of certain dimensions having dielectric or metallic mirrors at both ends forms its own resonator cavity. The crystal is irradiated on its parent side by light from a flash lamp. Solid-state lasers generally operate intermittently. Ordinary ruby lasers are excited for periods of a few milliseconds, the length of the period being determined by the duration of the exciting flash. Other solid lasers and certain liquid lasers have various rare-earth ions for the same purpose.

lasers, semiconductor. In semiconductor lasers a most convenient way to provide for the light emission is through a *pn* junction. The light emission, which results from a recombination of holes and electrons, may occur by various processes dependent on the type of the energy band of the semiconductor. The most promising for light emission are semiconductors that have the direct energy gaps corresponding to the visible region of wavelength. These are GaAs, GaSb, and CaP. An inverse population arises in a very narrow region near the *pn* junction at a distance of several microns. Laser action takes place along the line of the junction, which is extremely narrow and which must be perfectly straight for a to-and-from reflection to occur. Three methods of excitation are used: optical pumping, a beam of fast electrons, and the injection of electrons and holes through a *pn* junction. The small dimensions of semiconductor lasers make it possible to construct quantum amplifiers with extremely high sensitivity. The output wavelength depends on the semiconductor energy gap, which can be modified by alloying as required.

lathe, automatic tracer. Similar to the automatic-cycle lathes, automatic-tracer lathes employ single-point tooling but, ordinarily, only one tool. Primary limitation to design is that all details must be possible to reach with a sharp-point tool. In general, parts which are considered suitable for obtaining the advantages of the automatic tracer usually contain a complicated form, taper, or other details. Characteristic parts are valve plugs, valve bodies, nozzles, orifices, stepped shafts, contoured rolls, etc. Employing low-cost templates for reproduction, automatic tracers are ideally adapted to production quantities ranging from only a few pieces to large-quantity output. A compressor crankshaft with five stepped diameters, two tapers, radii, and shoulders is turned in 24 s from the rough castings.

lathe, engine. A lathe, fully powered, in which the spindle speed may be changed through the use of belts or speed selectors. Many lathes are equipped with rheostats for a wider selection of spindle speeds. The longitudinal and cross feeds may be engaged to feed automatically. The swing of the lathes ranges approximately from 8 to 40 in. (20.3 to 102 cm).

lathe, turret. May be of the hand-screw machine variety, the power feed (common turret lathe), semiautomatic, fully automated, or vertical type. The distinguishing characteristic of the turret lathe is the hexagon turret used for longitudinal feed, which is mounted on the bed of the machine, and the square turret, which is mounted on the cross slide. The hand-screw machine requires the turret to be hand fed into the work. In the common variety of turret lathe, the indexing of the turret is done by hand. Once indexing has been accomplished the tool may be fed into the work automatically. The major components of a lathe are the headstock, the tailstock, the bed, the saddle, the apron, and the quick change box. The headstock is mounted at the left end of the machine and houses the spindle which goes through a cone pulley (for belt driven machines) or gears (for gear driven machines). The tailstock is mounted on the bed at the right end of the machine. It may be slid along the bed to hold different length workpieces between centers. The saddle is mounted on the ways and houses the cross-feed spindle and compound rest. It makes possible longitudinal, cross, and angular feeding of the tool bit. The assembly of the saddle, apron, and compound is called the carriage. The apron is mounted on and below the saddle.

lead. Lead ores are widely distributed, usually with zinc ores present in greater or lesser degree. The principal ore is galena, PbS, with cerussite, $PbCO_3$, of secondary importance. Treatment consists of concentration and roasting, followed by blast-furnace smelting. The furnace charge consists of ore, coke, flux, and some scrap iron. Some lead is extracted by electrolytic deposition. Lead from the blast furnace contains impurities in the form of other metallic elements such as zinc, silver, antimony, arsenic, bismuth, cadmium, iron, nickel, gold, cobalt, and copper. Many of these impurities can be removed by melting the lead in kettles or on hearths. Since they are more easily oxidized than lead, they form a dross that can be skimmed from the surface. This is referred to as a softening process, because these impurities tend to increase the hardness of lead. The precious metals and bismuth are removed in one of several types of "desilvering" processes. Lead has the highest density of the common metals, with a specific gravity of approximately 11. The short-time tensile yield strength of lead is about 1600 psi (1.1×10^4 kPa), and the ultimate strength is about 2400 psi (1.7×10^4 kPa). Lead is soft, and when cut, it has a bright silvery luster that quickly oxidizes to dull gray. Lead has a low melting point, 621°F (327°C), and a high vaporization temperature, 3171°F (1744°C). The creep rate is high, creeping at room temperature and at stresses as low as 150 psi (1034 kPa). When exposed to air, it forms a thin protective coating of lead oxide, which prevents further corrosion. Lead cannot be work hardened at ordinary temperatures, but the addition of small quantities of tin, antimony, and cadmium increases hardness considerably. It is malleable but not ductile. Water, air, and sulfuric acid do not affect lead, and its resistance to penetration by x-rays and nuclear particles increases its importance in medicine and physical research. Lead is used in large quantities in the manufacture of storage batteries and electric cable coverings. Because of its softness it is easily extruded into pipes for plumbing and for chemical plants, and rolled into sheets for tank linings. Other uses are for roofing and foil. Many alloys are produced from lead, particularly bearing metals, solders, and type metals.

lead, commercial forms. Lead is cast into pigs weighing from 80 to 100 lb (36.3 to 45.4 kg). The types of lead available are specified in ASTM standard specification B29049. These types are: (1) Corroding lead. Lead refined to a high degree of purity. (2) Chemical lead. Undesilvered lead produced from ores of southeastern Missouri. (3) Acid lead. Fully refined lead with enough copper added to produce a minimum percentage of 0.040% and a maximum of 0.080%. (4) Copper lead. Similar to acid lead. (5) Common desilverized lead A and lead B. Fully refined desilvered leads. (6) Soft undersilverized lead. Lead produced from ores of the Joplin, Missouri, district. Minimum purities of from 99.73 to 99.93% are covered in this specification but a purity of 99.99% is common for commercial lead. Besides pigs, commercial lead is available in pipes in a large range of diameters and sheets up to 11¾ ft (3.6 m) wide. The sheets may be of almost any desired thickness.

leaded bronze (see bronzes)

ledeburite (metals). Ledeburite is a mechanical mixture of cementite and austenite. It is the eutectic of the iron–carbon system. Ledeburite appears in all compositions of more than 2% carbon, and for this reason the figure 2% is used to separate steels from cast irons.

level of automation. The degree to which a process has been made automatic. Relevant to the level of automation are questions of automatic failure recovery, the variety of situations that will be automatically handled, and the situation under which manual intervention or action by humans is required.

life test. The test included as part of ES requirements which is used to determine the value of a specific functional or appearance characteristic that is inherent in the design of an item being tested and to compare it to a specified objective value.

lift truck (see fork lift truck)

lift truck, counterbalanced front/side loader. A self-loading high lift counterbalanced industrial truck, equipped with a fixed or tiltable elevating mechanism, capable of transporting and tiering a load in both the counterbalanced forward position and any location up to and including 90° from the longitudinal centerline of the truck, while possessing the capability of transversing the load laterally.

lime. Referred to as a carbonate cementing material because it is manufactured by driving off carbon dioxide from calcium carbonate, and hardens by reabsorbing this gas. Lime comes from limestone, which is principally calcium carbonate with silicates, aluminates, magnesia, and magnesium carbonate present as impurities. Limestone is converted into lime by heating in vertical kilns, similar to small blast furnaces or in horizontal rotary kilns. The fuels used are coal, coke, oil, or gas. The calcium oxide, resulting from calcining limestone, is called quicklime. Sometimes magnesium limestone is used, in which case the lime produced is a combination of calcium oxide and magnesium oxide. If the amount of magnesia in the combination exceeds 25%, the lime is referred to as dolomitic; it is called high-calcium lime when the calcium oxide is 90% or more. Limestone converts to calcium oxide and CO_2 at a temperature of about 1650°F (899°C). Higher temperatures may be used to hasten the process, but there is a danger of burning, particularly in the case of high-magnesium limes. Lime is used in agricultural, chemical, and metallurgical applications, but its greatest use is in the building industry. As a construction material, quicklime is first slaked by the addition of water to form calcium hydroxide. In addition to use in masonry mortars, limes are widely used as plastering materials. Lime is used in the manufacture of glass, paper, iron and steel, Portland cement, and so-called sand–lime brick. It is used to correct agricultural soil deficiencies, and it has other chemical and metallurgical applications.

lime, hydrated. Actually quicklime that has been treated with water (slaked), dried, repulverized, and thus made ready for immediate use. It uses duplicate those of quicklime and it has largely superseded quicklime as a building material. In mixing hydrated lime with other materials to form mortars, one volume of hydrated lime is considered equivalent to one volume of lime putty.

lime, hydraulic. Manufactured by burning limestone containing silica in kilns similar to those used in producing quicklime. The temperature used is somewhat higher than that required for quicklime manufacture. The "clinker" formed contains enough lime silicate for hydraulicity and enough quicklime for slaking. Slaking, which is done at the place of manufacture, causes the clinker to disintegrate, and the fine powder that results is bagged as hydraulic lime. This cement is not suitable for underwater construction as it is too slow in setting, and it is too weak for general construction. It is used primarily for architectural decoration but is also used for masonry, and together with Portland cement for concrete. The coarse particles or hard lumps that remain after hydraulic lime is produced are ground finely and either added to the prepared lime or sold separately as grappier cement.

lime, quicklime. The ASTM Designation C51-47 classifies quicklime according to size as follows: Large lump 8 in. (20.3 cm) and smaller. Pebble or crushed, 2$^1/_2$ in. (6.4 cm) and smaller. Ground, screened, or granular, $^1/_4$ in. (0.64 cm) and smaller. Pulverized, substantially all passing through a No. 20 (840 micron) sieve. Quicklime must always be slaked in order to be used for structural purposes, and the manner of slaking has an important bearing on the properties of the finished product. Much testing and experience are usually necessary to ensure the best possible quality. In utilizing quicklime for mortar, lime putty is first made by slaking. This must be done carefully, with enough water to ensure complete hydration and to avoid overheating, the reaction being exothermic. An excess of water may reduce the temperature to a point where a putty is formed that is not sufficiently plastic; it will also cause a decrease in tensile strength. The putty is cooled and aged for 20–24 h, when pulverized lime is used, and from 7 to 10 days, when lump lime is used. Limes high in magnesia require longer periods of slaking than purer limes. The lime putty or paste is not used in this form as a mortar because it contracts greatly and cracks. Instead, sand in varying amounts is added, thereby preventing almost all shrinkage and increasing the porosity so that more surface is exposed to carbon dioxide. Sand, however, tends to decrease tensile strength. A mixture of one volume of lime putty with two to three volumes of sand is generally used.

liquefied petroleum gas (LPG). The term liquefied petroleum gas, LPG, is applied to certain hydrocarbons, which are gaseous under normal atmospheric conditions, but can be liquefied under moderate pressure at normal temperatures. LPG is derived from natural gas and from various petroleum refinery sources such as crude distillation and cracking. The hydrocarbons in LPG are mainly of the paraffinic (saturated) series, principally propane, isobutane, and normal butane. The greatest use of LPG is as a domestic fuel with an appreciable amount being consumed in synthetic rubber production and chemical industries.

liquid penetrant inspection. A type of nondestructive inspection that locates discontinuities that are open to the surface of a metal by first allowing a penetrating dye or fluorescent liquid to infiltrate the discontinuity, removing the excess penetrant, and then applying a developing agent that causes the penetrant to seep back out of the discontinuity and register as an indication. Liquid penetrant inspection is suitable for both ferrous and nonferrous materials, but is limited to the detection of open surface discontinuities in nonporous solids.

lithium. The lightest of all metals, and the third lightest among the elements. Its chemical properties place it in the alkali group, along with sodium and potassium. Lithium is highly reactive, does not occur in the native form, and is difficult to produce from its ores. The major usefulness of lithium is in its compounds, nearly 20 of which have industrial applications. Lithium and its compounds are derived chiefly from certain pegmatite minerals. The chief lithium-bearing pegmatite minerals are spodumene, lepidolite, amblygonite, and petalite. Spodumene is a member of the pyroxene group and occurs in prismatic or elongate bladed crystals of a dull gray-white color. Lithium metal is combined with chromium, copper, and other nonferrous metals to produce special-purpose alloys, and is utilized as a reagent in the production of synthetic vitamins. Its most important uses lie in the field of nuclear energy. A naturally occurring isotope, ^{6}Li, is said to be capable of fusioning explosively. ^{6}Li is also an effective shield against nuclear radiation. Lithium minerals are used directly, to a minor extent, in glassmaking and ceramics. Lepidolite reduces the viscosity of glass melts and improves their workability. Spodumene and petalite are introduced directly into certain ceramic mixes. Lithium compounds are widely applied in industry, the two most important are the carbonate and the hydroxide. Lithium carbonate is used chiefly in glass and in ceramic enamels. Lithium hydroxide monohydrate is reacted with stearic acid to produce lithium stearate greases. Lithium chloride is used in air conditioning and industrial drying, in dry-cell batteries, and in welding and brazing. Lithium hydroxide is utilized in alkaline storage batteries.

lithium salts. Lithium carbonate, the most widely used of the compounds, is employed in the production of lithium metal and frits and enamels. Together with lithium fluoride, it serves as an additive for cryolite in the electrolytic pot-line production of primary aluminum. Lithium-base greases, often the searate, are a large outlet. These lubricants are efficient over the extremely wide temperature range −60–320°F (−51–160°C). Lithium hydroxide is a component of the electrolyte in alkaline storage batteries and is employed in the removal of carbon dioxide in submarines and space capsules. Lithium bromide is used for aid conditioning and dehumidification. The hypochlorite is a dry bleach used in commercial and home laundries.

load configuration. Pallet loads of goods are defined as load length, the maximum overall dimension of a pallet or load module and load in the direction perpendicular to the length of the aisle; load width, the maximum overall dimension of pallet or load module and load in the direction parallel to the aisle; and load height, the maximum overall dimension from bottom of pallet or load module to top of load.

lock-seam tube (see tube)

loom. A machine or device from which cloth is produced by interweaving thread or yarn at right angles.

loom, needle punch. A machine or device from which cloth is produced by interweaving and tangling a layer of short fibers, natural or synthetic, into a soft, medium, or hard fabric, rug materials, or even hard fiber material.

loom, triaxial. A machine or device from which cloth is produced by interweaving thread or yarn in three equiangular directions.

lot inspection. The inspection performed on random samples from successive isolated production lots that were produced from a continuous process.

lox. abbreviation for liquid oxygen.

lubricant residue. The carbonaceous residue resulting from lubricant burned on the surface of a forged part.

lubricants (see ceramic)

lubricating oils. Most lubricating oils are derived from petroleum. The crude oil is distilled and certain distillates and residues are used as lubricants, usually after other processing, refinement, and blending. Lubricating oils are made from paraffinic, naphthenic, or mixed-base crude oils. Paraffin-base oils, the so-called "Pennsylvania" oils, exhibit the least change in viscosity with temperature change, whereas naphthenic-base oils "thin out" more rapidly with temperature rise. Modern methods of refinement and treatment enable producers to impart characteristics of one type of oil to other types. Thus the viscosity index, or rate of change of viscosity with temperature, of naphthenic oils can be made to approach that of paraffin oils. Lubricating oils may be distillates or residues or blends of both. Distillates are fractions that have been vaporized from the crude and then condensed. Residues come from the portion of the crude that did not volatilize. The distillates contain crystalline and amorphous wax (petrolatum), which must be removed. Naphthenic oils contain little wax but do have undesirable constituents that tend to produce sludge in service. These are removed by solvent refining and treatment with sulfuric acid. Acid refining is more expensive and will produce, from the same stock, lubricating oils of greater "oiliness," oxidation resistance, and less tendency to produce "lacquer" and corrosive oxidation products. Products of solvent refining have a lower carbon residue and a higher viscosity index. The chief advantage of solvent refining is the cheap production of lubricating oils from sources not possible with the older acid refining. Mixed-base crudes are treated for the removal of waxes and sludge-forming compounds. Some oils can be used for a great number of applications. However, certain applications require special lubricants that conform to rigidly held specifications. Turbine oils, transformer oils, and diesel lubricating oils are examples.

lubrication. Lubricating aims to interpose films on moving material surfaces, thereby reducing friction and reducing wear and damage to the rubbing surfaces. Lubrication is effected by two main mechanisms, fluid lubrication and boundary lubrication. Fluid lubrication, also called hydrodynamic lubrication, involves a complete separation of the sliding surfaces by a thick layer of the liquid lubricant sot that there is no direct contact between the material surfaces. The coefficient of friction becomes very low, 0.001–0.01, and is determined by the viscosity of the lubricant. Boundary lubrication occurs when the layer of lubricant is only a few molecules thick, comparable in dimension to the pits and asperities on the moving surfaces. Between these two extreme types of lubrication, there is an intermediate region in which the bearing surfaces are partly in direct contact and partly separated by a liquid film of the lubricant. Boundary lubricants are generally produced by adding a relatively small quantity of polar organic compounds such as fatty acids to a lubricating oil. The reaction of the polar groups, such as carboxyl, with the metal surface results in the formation of a monomolecular layer strongly adherent to the surface, whereas a long hydrocarbon chain is oriented outward in an approximately perpendicular direction.

Lüder lines. Surface markings resulting from localized flow, which appear on some alloys after light straining. They lie approximately parallel to the direction of maximum shear stress (about 45° to the direction of the applied stress), and appear as depressions when forming is in tension and as elevations when in compression.

M

machine cycle. A set period of time in which the computer can perform a specific machine operation.

machine vision. A robot or "smart machine" must be able to sense its environment and react to it, to perform a task—whether that task is part recognition, monitoring thicknesses, measuring levels, checking quality, or range finding in the case of mobile robots. There are many vision and sensing systems in use and in development today. The most common machine vision systems use: linear diode arrays, to detect a light source such as a LED, laser, or visible light; binary black-and-white digitizing cameras, either charge-coupled devices (CCD) or charge-injected image devices (CID); gray-scale vision systems solve the same problems as binary vision devices but have greater computer power and additional processors to detect color, shades of gray or color, and texture; x-ray system using image enhancement for industrial inspection; ultrasonic imaging through transmitting ultrasonic sound from transducers focused at the surface of material; eddy current sensors use a magnetic probe to scan an object using a magnetic coil; a pulsed laser can produce a hologram of an object under stress revealing internal defects; a low-power laser beam for inspection. **(See also image processing, and image enhancement, sensors.)**

machining, abrasive jet (AJM). The removal of material from a workpiece by a high-speed stream of abrasive particles carried by gas from a nozzle. The process is used chiefly to cut materials that are sensitive to heat damage and thin sections of hard materials that chip easily, and to cut intricate holes that would be more difficult to produce by other methods.

machining, chemical (CHM). The production of desired shapes and dimensions through selective or overall removal of metal by controlled chemical attack or etching. Areas from which metal is not to be removed are protected from attack by masking. The process ordinarily is not electrically assisted. Chemical machining is an extension of the metal finishing processes of etching, chemical brightening, and chemical polishing, in which controlled chemical attack is used to alter the surface condition of metals. The same types of chemical solutions are used for chemical machining, with appropriate modification of concentration of composition, and of operating conditions, to increase the rate of metal removal. The process is also closely related to the etching of metal printing and engraving plates and of nameplates. Nearly all metals can be chemically machined. There are two types of chemical machining: chemical blanking, for cutting or "stamping out" parts from thin sheet metal; and chemical contour machining, or chemical milling for the selective or overall removal of metal from thicker material. These processes are employed chiefly when the desired blanking or removal of metal by conventional methods is difficult or impractical because of hardness, toughness, or brittleness of the work metal, or because of the size of part, complexity of shape, or thinness of final section.

machining, electrical discharge (EDM). A method for producing holes, slots, or other cavities in electrically conductive material by means of the controlled removal of material through melting or vaporization by high-frequency electrical sparks. The spark discharge is produced by controlled pulsing of direct current between the workpiece (which is usually anodic or positively charged) and the tool, or electrode (which is usually the cathode or negative electrode). The end of the electrode and the workpiece are separated by a spark gap of 0.0005–0.020 in. (0.0127–0.508 mm) and are immersed in or flooded by a dielectric fluid. The dielectric in the gap is partially ionized under the pulsed application of a high voltage, thus enabling a spark discharge to pass between tool and workpiece. Each spark produces enough heat to melt or vaporize a small quantity of the workpiece, leaving a tiny pit or crater in the surface of the work.

machining, electrochemical. This process is often confused with chemical milling and electrochemical milling. It is a distinctly different method and is used for much different component production. Unlike the former two processes, this one uses fairly expensive equipment with costly tooling. No photoresist masks are used. Instead, a shaped cathode tool is used, which is connected to the negative terminal of a high-amperage dc power source. The workpiece is connected to the positive terminal. In operation, the tool is advanced into the workpiece while immersed in an electrolyte (usually a sodium chloride solution) that completes an electrical circuit between the two. Metal is removed through a reverse plating action or electrolysis. As the current passes from the work to the tool, metal particles or ions are caused to go into solution due to electrochemical reaction. With this system there is no tool wear and nearly all metals can be machined with ease regardless of hardness. Speed of metal removal is roughly 1 in.^{3}/min (16.4 cm^3/min) for every 10,000 A of electric current. Machines using up to 40,000 A are not uncommon. Electrochemical machining can be used to do work that would be difficult or impossible by mechanical machining. The work includes hard materials (such as hardened steel and heat-resisting alloys) and odd-shaped, small, deep holes. Electrochemical machining is used for operations as widely different as face milling, deburring, etching, and marking. Although electrochemical machining is sometimes applicable to small lot production, the process is best suited to mass production applications because of high tooling and setup costs and high capital equipment costs. The most frequent application of ECM is in the production of jet engine parts and for other aerospace applications; it is also used in automotive and general manufacturing applications.

machining, electron beam (EBM). A method of cutting material under vacuum using a focused beam of high-velocity electrons. On impact of electrons with the workpiece, the kinetic energy of the electron changes into heat, which vaporizes a small amount of the workpiece. The vacuum is necessary to prevent scattering of the electrons by collision with gas molecules. By controlling beam energy at a lower level, the process is used for welding instead of machining. In machining, electrons are accelerated in an electrostatic field to velocities of more than half the speed of light. The electron beam, and the laser beam, exceed ordinary heat or light sources in energy density, precision, and mobility. By focusing the beam with optical precision on a 0.0005–0.001 in.^{2} (3.25×10^{-3}–6.5×10^{-3} cm^2) area of the workpiece, energy is delivered at a power density of 10^7 W/in.^{2} and can vaporize any material instantly. Electron beam machining is applicable to parts 0.010–0.250 in. (0.254–6.4 cm) thick and can drill holes as small as 0.0005 in. (0.0127 mm) in diameter in all materials, including ceramics, at a penetration rate of 0.010 in./s (0.254 mm/s) or faster. It cuts slots as narrow as 0.001 in. (0.0254 mm) at a spacing as close as 0.005 in. (0.127 mm). The process is used also to scribe thin films and to remove small, broken taps from holes.

machining, laser beam (LBM) **(see laser machining)**

machining, plasma arc (PAM). Done with a high velocity jet of high-temperature ionized gas. The relatively narrow plasma jet melts and displaces the workpiece material in its path. Because plasma machining does not depend on a chemical reaction between the gas and the work material, and because plasma temperatures are extremely high, the process can be used on almost any metal, including those that are resistant to oxy-fuel gas cutting. The method is of commercial importance in the United States mainly for profile cutting of stainless steel and aluminum alloys.

machining, ultrasonic (USM). The removal of material by particles of abrasive that vibrate in a water slurry circulating through a narrow gap between the workpiece and a tool that oscillates at about 20,000 cycles per second. The tool reproduces its shape in the workpiece, generally to an accuracy of ±0.001 in. (±0.0254 mm), and sometimes to a tolerance of 0.0005 in. (0.0127 mm) or less, without burrs. Accuracy depends on the size of the tool, rigidity of the machine and the tool, temperature of the slurry, grit size, and the procedure for roughing and finishing. Ultrasonic machining is used chiefly on hard, brittle materials that do not conduct electricity; however, it is used on both metals and nonmetals, and on ductile as well as brittle materials. It is particularly well suited to the production of relatively shallow, irregular cavities and is one of the few processes suitable for machining extremely fragile material, such as honeycomb. The main disadvantages of the process are low metal removal and high cost.

magnesite. Magnesium carbonate, a member of the isomorphous group of minerals that includes calcite and dolomite, crystallizes in the hexagonal system and has rhombohedral cleavage. It occurs commercially as crystalline masses that resemble marble or coarse-grained dolomite, and a cryptocrystalline (amorphous) masses, which are a dense porcelainlike texture and conchoidal fracture. There is some variation in color, but the mineral is generally white or grayish. Specific gravity is 2.9–3.1; the hardness is 3.5–4.5. Magnesite loses carbon dioxide on heating; calcining at a temperature of 1310–1832°F (700–1000°C) removes all but 2–10% of the carbon dioxide, and give a product known as caustic-calcined magnesia. Calcining at 2642–3182°F (1450–1750°C) drives off all the carbon dioxide except about 0.5% yielding a dense, sintered, inert product called refractory magnesia. More than 90% of the magnesite output is dead-burned to produce refractory magnesia, either in the form of loose grains or as shaped bricks. These find their major applications in the steel industry where they are used to line basic open-hearth furnaces and converters. Magnesite refractories are also utilized in copper smelters, cement kilns, and other high temperature installations. Refractory magnesia contains 4–5% of iron oxide and a comparable amount of silica, with minor proportions of alumina and lime. Caustic-calcined magnesia has several uses, the most important being in the manufacture of oxychloride cement. Finely ground magnesia is mixed with a solution of magnesium chloride to form a tough, dense cement suited for heavy-duty interior flooring. Caustic calcined magnesia is also utilized in refractories and insulation, and in the chemical, paper, and rayon industries.

magnesium. Found as chloride in sea water and in brine wells, and in the following ores: magnesite, a carbonate; dolomite, a carbonate of magnesium and calcium; and brucite, a hydroxide. The metal is recovered by either electrolysis or chemical reduction, each method having several major variations. Magnesium is the lightest of the commercially used metals, having a specific gravity of 1.74 compared to 2.70 for aluminum. It is silvery white in appearance. ASTM Specification B92-45 for ingots and sticks intended for remelting, calls for a minimum purity of 99.8%. Cast magnesium has approximately the same tensile strength as aluminum, 14,000 psi (9.7×10^4 kPa), but is less ductile. Rolled, it has a tensile strength of 25,000 psi (1.7×10^5 kPa), but cold working hardens it rapidly. It is difficult to cast magnesium and its alloys because of the tendency to oxidize and to retain gases. Magnesium forms a film of oxide and carbonate on exposure to air. It is subject to corrosion in salt air, hence it usually is protected by paint. It is dissolved by dilute hydrochloric, sulfuric, and nitric acids, but is not affected by alkalis. It is, however, attacked by salts; when salts are contained as impurities, magnesium disintegrates rapidly. Magnesium does not have much use in pure form but is used extensively when alloyed with aluminum, zinc, and other metals. The pure metal does find use in powder or ribbon form, for flares and flash bulbs. It is also used as a deoxidizer in the production of nonferrous metals and their alloys. It is used for the cathodic protection of underground pipe lines and other structures. Pure magnesium is furnished in ingots and sticks, and as powder, wire, ribbon, and extruded strips. The ingots usually weigh 17 lb (7.7 kg) and the cylindrical stick ranges from 0.25 to 2 lb (0.11 to 0.91 kg).

magnesium alloys. Aluminum is the most important alloying element for magnesium. Among other things, it can triple the tensile strength and the hardness of cast magnesium, and increase the yield point seven times. For the wrought material, the effects are somewhat less spectacular, although the tensile strength and hardness may both be doubled. Cold working causes magnesium alloys to harden rapidly, and for this reason

Magnesium-Base Alloys

Composition (%) (balance Mg)	Condition	Typical mechanical properties: 0.1% Proof stress (N/mm²)	Tensile strength (N/mm²)	Elongation (%)
Cast alloys				
10.0 Al, 0.3 Mn, 0.7 Zn	Chill-cast	115	200	2
4.0 Zn, 0.7 Zr, 1.2 Rare earths	As-cast	95	170	5
	Heat-treated	130	215	4
0.7 Zr, 3.0 Th	Heat-treated	100	210	8
Wrought alloys				
1.5 Mn	Rolled	95	200	5
6.0 Al, 0.3 Mn, 1.0 Zn	Forged	155	280	8
	Extruded	140	215	8
3.0 Zn, 0.7 Zr	Rolled	170	265	8
	Extruded	215	310	8
1.0 Mn, 3.0 Th	Rolled	215	280	10

many shapes are extruded rather than rolled. At elevated temperature they can be worked without trouble. Extruding and rolling alloys usually contain from 2.5 to 10.5% aluminum, forging alloys 8.5%, casting alloys from 6 to 8.5%, and die casting from 9 to 11%. Above 12%, the alloys are brittle and of no commercial value. Magnesium alloys are subject to solution heat treatment and age hardening when the aluminum content is above 6–7% for castings and 9% for wrought alloys. The low weight-for-strength ratio of these alloys makes them useful for aircraft parts, and are used in household appliances. These alloys are easily machined. The names "Dowmetal" and "Mazlo" are used in commercial practice. Manganese is added to all magnesium–aluminum alloys to improve corrosion resistance and impart weldability. Zinc also improves the corrosion resistance of "Dowmetal." "Electron metal" is the tradename applied to alloys of magnesium in which zinc is the major alloying element. The strength and utility of these alloys are about the same as for the magnesium–aluminum series, but zinc by itself, has a bad effect on corrosion resistance. Cadmium and tin are included in alloys especially suited for hammer forging because of their effect in improving ductility. Silicon reduces hot shortness in magnesium castings.

magnesium processing. The Dow process is one of the most important means for the production of magnesium. Hydrated $MgCl_2$ is obtained from natural brines by recrystallization. This is dehydrated and mixed with NaCl and KCl to form a mixture which will melt at about 1300°F (704°C). Current is passed through the melt in an electrolytic cell. Magnesium forms at the cathode and floats to the surface, from which it is periodically ladled. Chlorine forms at the anode and is collected as a valuable by-product. Magnesium chloride for electrolysis is also obtained from sea water and magnesite ores. Aluminum is relatively weak in the pure state and for engineering purposes is almost always used as an alloy. Its modulus of elasticity is even less than that of aluminum, being only about 6,000,000 psi (4.14×10^7 kPa). It is, therefore, usually necessary to use it in considerable thickness or utilize deep sections so as to obtain adequate stiffness. Because of its high cost, magnesium is economical only for applications where light weight is very important. Magnesium does not possess good corrosion characteristics of aluminum and under many conditions corrodes badly, and must be used with some care.

magnetic-analysis inspection. A nondestructive method of inspection to determine the existence of variations in magnetic flux in ferromagnetic materials of constant cross section, such as might be caused by discontinuities and variations in hardness. The variations are usually indicated by a change in a pattern on an oscilloscope screen.

magnetic-particle inspection. A nondestructive method of inspection for determining the existence and extent of surface cracks and similar imperfections in ferromagnetic materials. Finely divided magnetic particles, applied to the magnetized part, are attracted to and outline pattern of any magnetic-leakage fields created by discontinuities.

magnetic disk storage (computer). The magnetic disk is a thin metal disk resembling a phonograph record; it is coated on both sides with a magnetic recording material. Data are stored in magnetized spots arranged in binary form in concentric tracks on each face of the disk. A characteristic of disk storage is that data are recorded serially bit-by-bit, eight bits per byte along a track rather than by columns of characters. Disks are normally mounted on a stack on a rotating vertical shaft. Enough space is left between each disk to allow access arms to move in and read or record data. A single magnetic disk unit is capable of storing several million characters. Usually more than one disk unit can be attached to a computer.

magnetic drum storage (computer). The magnetic drum is a cylinder on which data are recorded serially in a series of bands around the drum in a manner similar to that utilized on disk storage. As the drum rotates at a constant speed, data are recorded or sensed by a set of read–write heads. The heads are positioned close enough to the surface of the drum to be able to magnetize the surface and to sense the magnetization on it. The heads contain coils of fine wire wound around tiny magnetic cores. There may be one or more heads for each drum track or one or more heads that can be moved to the various tracks. The drum may rotate up to 3500 revolutions per minute, and the data transfer rate to or from the processing unit may be up to 1,200,000 bytes per second. The rotational delay to a specific part of the track ranges from 0 to 17.5 ms and averages 8.6 ms. Data are stored in the form of minute magnetized spots, arranged in binary form on the individual recording track. Spots are magnetized by sending pulses of current through the wire coil. The polarity of a spot is determined by the direction of the current flow. Depending on their polarity, spots can represent either 1's or 0's.

magnetic film storage (computer). Magnetic film storage functions similarly to core storage; however, instead of individual cores strung on wires, magnetic film is made of much smaller elements in a form. One type of magnetic film known as planar film (thin film) consists of very thin, flat wafers made of nickel–iron alloy. These metallic spots are connected by ultrathin wires and are mounted on an insulating base such as glass or plastic. Magnetic film may also be in the form of plated wire. This is a type of cylindrical film, essentially the same as planar film except that the film is wrapped around a wire. Another type of cylindrical film is the plated rod. This technique, known as thin film rod memory, consists of an array of tiny metal rods only 0.1 in. (2.54 mm) long. The operation of a magnetic film memory unit is similar in principle to that of a magnetic card unit as the storage elements in both are formed into planes that may be stacked.

magnetic ink character readers. Magnetic ink character recognition (MCR) is a high-speed data input technique that reduces manual keystroke operations and allows source documents to be sorted automatically. Magnetic ink reading heads produce electrical signals when magnetic characters are passed beneath them. These signs are analyzed by special circuits and are compared with stored tables to determine what character has been sensed. The data are then transmitted to the memory of the computer for processing.

magnetic reluctance testing. This test is used in continuous measurement of thickness of material in processes such as boxboard, paperboard, sheet plastic, etc. The signal, which is proportional to the space between the sensing elements, can be processed electrically for measurement, automatic control, or management information purposes.

magnetic tape (computer). A tape or ribbon of material impregnated or coated with magnetic material on which information may be placed in the form of magnetically polarized spots. The use of magnetic tape as a means of storing information is based on the same principles of magnetization and induction as are utilized in the magnetic drum. The tape is surfaced with a magnetizable material, and is moved past a reading and writing head that obtains or stores binary digits. The tape is wound around a reel. As the reel unwinds the tape passes the read–write station and is taken up by a second reel, like a roll of film in a camera. The physical end of the tape is sensed by a photoelectric cell near the read–write head. Special procedures are incorporated into programs to take care of the end-of-tape condition. An erase head is energized to remove previous information from the tape whenever the machine begins to "write" on the tape. Information on magnetic tape usually appears as a succession of characters written vertically. Records are demarcated by empty spaces called interrecord gaps. When writing on tape, the machine creates an interrecord gap at the end of each new record; information previously on the tape is erased as writing occurs. Additional instructions are available to backspace and to rewind the tape.

magnets (see ceramic magnets)

magnet steels. Carbon, chromium, tungsten, and cobalt steels are used for permanent magnets. These steels are used in the hardened condition. The carbon steels contain from 0.80 to 1.20% carbon and are, in fact, carbon tool steels. The chromium magnet steels contain from 0.70 to 1.00% carbon and 2 or 3% chromium. The tungsten steels contain about 0.70% carbon and 5% tungsten. The best and most expensive cobalt steels contain 35% cobalt, together with several percent of both chromium and tungsten. A permanent magnet alloy (Alnico) contains approximately 60% iron, 20% nickel, 8% cobalt, and 12% aluminum. This alloy cannot be forged and is used as a casting hardened by precipitation heat treatment. Alloys of this type have great magnetic strength and permanence.

main control unit (computer). In a computer with more than one instruction control unit, that instruction control unit to which, for a given interval of time, the other instruction control units are subordinated. An instruction control unit may be designated as the main control unit by hardware or by hardware and software. A main control unit at one time may be a subordinate unit at another time.

malleability (materials). The ability of a material to be deformed plastically by hammering or rolling without rupture. This property is similar to ductility but does not bear a direct relationship to it.

malleable iron (see cast iron).

management information systems (MIS). There are two basic kinds of information systems in manufacturing: tactical and strategic. A tactical information system deals with the day-to-day supervision and control of line activities. These activities include order processing, production scheduling and supervision, and inventory control. A strategic information system presupposes the existence of a tactical system. Output from the tactical system is accumulated and combined with financial data, market trends, business conditions, and monetary factors to provide a comprehensive input for long-range managerial decision-making. In effect, then, the strategic or management information system (MIS) is a total information system of which the tactical system is one part. Today, the tactical system is growing in use, but few as yet are going to total-systems route. Ultimately, efficient and effective strategic information systems will yield the most comprehensive and, hence, the most profitable results. Much effort is still required in order to reduce the cost of the ''software'' programming.

manganese steels. Manganese is always used in steel, having a beneficial effect in the steel both directly and indirectly, and is always added to steel during its manufacture. Manganese combines with sulfur, iron, and carbon. With no carbon present it forms a solid solution with both gamma and alpha iron. Manganese has only a mild tendency to combine with the carbon and therefore to form a carbide (Mn_3C), which accounts for one of the principal advantages obtained from alloying manganese with steels. All steels contain manganese from a few hundredths of 1% in many of the low-carbon structural and machine steels to about 2% in steels that remain pearlitic in structure. As the manganese content of the steel is increased, the steel changes from a pearlitic to a martensitic, to an austenitic steel. The pearlitic steels are the most important. Manganese structure steel contains from 1 to 2% manganese and from 0.08 to 0.55% carbon. In tool steels, increasing the manganese content from 0.30 to 1.0% changes the steel from water hardening to oil hardening, owing to the greater hardenability of the steel and with the higher manganese content.

man-on-board. A storage and retrieval concept whereby materials are accessed by taking the operator to the materials on board an S/R machine.

manufacturing cell. A collection of machines, grouped together for processing a family of parts.

manufacturing cycle efficiency (MCE). This compares the time materials spend being processed to the total time spent in the manufacturing department. MCE = Total time spent on machining ÷ total time spent in production system.

martempering (see heat treating, martempering)

mass production. The large-scale production of parts in a continuous process uninterrupted by the production of other parts.

master alloy. An aluminum-base alloy in remelt ingot form containing at least 50% aluminum and one or more added elements for use in making alloying additions. Also referred to as rich alloy or hardener.

material requirements planning (MRP). Materials or inventory control is an important ingredient in the overall productivity of a company. Doing a better job of getting the right amount of material to the right place at the right time means productivity improvement. The materials management function is concerned with purchasing, inventory control, quality control, materials handling, and distribution; and each subfunction of materials management should be included in the overall productivity program. Great importance is placed on increased real-time control of materials from receiving through shipping. The need for increased control of materials is especially great in the areas of work-in-process—good visibility of the status of material once it has been released to manufacturing. Raw material, work-in-process, and finished goods inventories need to be monitored and controlled. Turnover ratios, shortage or fill ratios, stock replenishment ratios, error ratios, and obsolete material ratios are used to improve materials control and increase productivity.

matte one side foil (M1S) (see foil)

matte two sides foil (M2S) (see foil)

MCE (see monitoring cycle efficiency)

mean-time-between-failures (MTBF). The average time that a device will operate before failure.

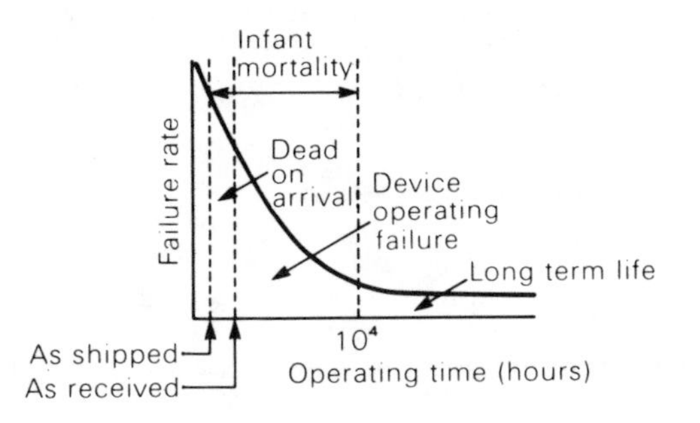

Typical failure rate curve for electronic apparatus, industry-wide.

mean-time-to-repair (MTTR). The average time that a device is expected to be out of service after failure.

mechanical plating. Any plating operation in which the cathodes are moved mechanically during the deposition. **(See also electroplating.)**

mechanical properties. Those properties of a material that are associated with elastic and inelastic reaction when force is applied, or that involve the relationship between stress and strain, for example, modulus of elasticity, tensile strength, endurance limit. These properties are often incorrectly referred to as "physical" properties.

melamine–formaldehyde resins (plastics). Melamine, $(CN)_3(NH_2)_3$, reacts similarly to urea under neutral or slightly alkaline conditions to form methyol melamines, the monomers of the resin. The reaction is carried out in much the same manner as for urea resins. Melamine resins are transparent and can readily be colored in light, pastel shades. They can be filled with alpha cellulose, like ureas, to give translucent products. Other fillers, such as asbestos, cotton flock, macerated fabric, and glass fiber result in products that are opaque but that have properties suitable for many applications. Melamine resins are stronger and have a much lower water absorption than the urea resins. Melamine molded products can be used above the boiling point of water continuously. A considerable part of the malamine resin production goes into the manufacture of tableware. Melamine moldings can be used for electrical parts because they operate at higher temperature than ureas. Melamine resins are used in light-colored laminated products and to add wet strength to paper.

mercury cells. Commonly used for many applications and are made in sizes to fit the device in which they are to operate. A mercury cell consists of a positive electrode of mercuric oxide mixed with a conductive material and a negative electrode of finely divided zinc. The electrodes and the caustic electrolyte are assembled into a flat circular shape and others are formed into hollow cylindrical shapes depending on the type of cell for which they are made. The electrolyte is immobilized in an absorbent material between the electrodes.

metal cladding. Consists of fastening two different metals together and then rolling, drawing, or otherwise shaping the composite material as if it were homogeneous. The metal may be joined by welding, casting, brazing, or any other method that ensures a good bond. The two materials must have rolling characteristics that are nearly the same. The final product may be two-layered or one metal may be completely coated by the other. Thin sheets, plates, bars, and wire are produced in this fashion.

metal cold working (see cold working, metals)

metal decorating. In manufacturing, includes techniques of coating, coloring, printing, and like modification, usually of the plane surface of basic metal sheet or cylinder, to reproduce on the surface designs, pictures, art, logos, and trademarks called for by the product design.

metal hot working (see hot working, metals)

metallic alloy composition. The weight percentage of each chemical element of which the alloy is composed. Composition does not change with the temperature or heat treatment to which the alloy is subjected unless a temperature high enough to cause chemical changes such as combustion is reached. The chemical elements of which the alloy is composed may be combined in each of several manners or phases. The usual phases found in alloys are (1) pure metal, (2) intermediate compound, (3) solid solution. An alloy may contain all three phases at the same time. Phases have several distinguishing characteristics. They are physically and chemically homogeneous and can be distinguished visually and separated mechanically from one another. Mechanical mixtures of phases also exist in some alloys; although they may be phaselike in appearance, they are not chemically homogeneous and therefore do not qualify as phases. They may, however, be classified along with phases as constituents. Constituents are visible under the microscope as definite units of the total structure.

metallic alloys intermetallic compounds. The chemical combinations formed in metallic alloys are known as intermetallic compounds. They are similar to chemical compounds in having definite compositions, properties, and characteristics but do not adhere to the laws of chemical valence. Each lattice contains the same number of atoms of each type as every other lattice of the same compound. The appearance of an intermetallic compound under the microscope is similar to that of a pure metal.

metallic alloys, solid solution. A solid solution is similar in many ways to a liquid solution. It consists of one element in solution with another element, although both are in the solid state. The element present in greater quantity is frequently called the solvent, the other being called the solute. In the case of an alloy, the solute (metal or nonmetal) is dissolved in the solvent, which is usually a metal. The proportions of solute to solvent may vary between fixed limits, without any abrupt change in physical properties. Under the proper temperature conditions, the dissolved elements diffuse throughout the solvent so that complete physical and chemical homogeneity will result if enough time for complete diffusion is allowed.

metallic bond (atomic structure). A chemical bond which is confined to metals. In the ideal metallic bond the valency electrons are not bound to any particular pairs of atoms but move freely throughout the metal. This is because the valency electrons in a metal lie much farther out from the core of the atom than those in nonmetals do. Being subjected to the attraction of the positive charges of neighboring nuclei, the valency electrons pass easily to regions remote from their parent atoms. Consequently, the valency electrons in metals are never permanently associated with any particular atom but flow freely in a random arrangement as free "electron gas". The typical metallic bond can, therefore, be regarded as an assemblage of positive ions, each consisting of the core of an atom (nucleus plus nonvalency electrons) immersed in a "gas" of free electrons. The free mobility of the electron "gas" accounts for the high electrical and thermal conductivity of metal, as well as for their luster. The metallic bond is nonspecific and nondirectional, acting equally strongly in all directions. This leads to highly coordinated close-packed structures accounting for the unique plastic properties of metals and for their ability to form alloys.

metallic yarns. Basically consists of filaments of aluminum covered with plastics; two kinds of plastics are mainly used for covering. The first is cellulose acetate–butyrate and the second is Mylar, du Pont's polyester film which is similar chemically to Dacron and Terylene. The mixed ester of cellulose with acetic and butyric acids is more popular for use in plastics than in straight cellulose acetate, because it has a lower melting point and is more easily worked. The raw material is a roll of aluminum foil of 0.00045 in. (0.01143 mm) thickness, and about 20 in. (50.8 cm) wide. To both sides of the sheet there is applied a thermoplastic adhesive to which has already been added the required coloring matters. The adhesive-coated foil is heated to about 194–203°F (90–95°C) and a sheet of cellulose acetate–butyrate transparent film is laminated to each side of the foil by passing through squeeze rollers at a pressure of 2000 psi (1.4×10^4 kPa). The laminated material is then slit into filaments of the required width, the most popular width being $^1/_{64}$ in. (0.4 mm), although other sizes from $^1/_8$ to $^1/_{120}$ in. (0.2 to 3.2 mm) are also made. Because the metal basis is aluminum, there is no tarnishing and because it is sealed in, it cannot be affected by salt water, chlorinated swimming pools, or climatic conditions. Because the covering film is plastic the yarns are flexible and have some extensibility. The acetate–butyrate-coated metallic yarns can be washed at temperatures as high as 158°F (70°C), but not higher, otherwise delamination occurs. Metallic yarns are used wholly for decorative purposes. A major application is for car upholstery. It is used in clothing, upholstery, shoe-laces, bathing dresses, table linens, excelsiorlike packing material for such gifts as perfume and liqueurs. They are also used in women's sandals, and have been chopped into very short pieces and added to Vinylite floor-coverings to give bright sparkling flecks.

metallized plastic. Identified as formable metallized plastics (FMP). The developers, Sheldahl Inc. and Dow Chemical, describe FMP as a composite plastic–metal–plastic structure. FMP can be thermoformed into complex shapes on conventional vacuum-forming machinery. It maintains its original reflectivity and electrical integrity after forming. FMP products require no touch-up metallizing or protective spraying after being formed.

metal powders (see powder metallurgy)

metal rolling (see cold rolling and hot rolling)

metals, allotropic (see allotropic metals)

metals, melting points of

mercury	−38°F	(−39°C)	silicon	2570°F	(1410°C)
gallium	85.5	(29.7)	nickel	2651	(1455)
sodium	208	(97.8)	iron	2804	(1540)
lithium	367	(186.1)	titanium	3074	(1690)
tin	449	(231.7)	zirconium	3326	(1830)
bismuth	520	(271.1)	thorium	3348	(1842.2)
cadmium	609	(320.6)	vanadium	3452	(1900)
lead	621	(327.2)	boron	3812	(2100)
zinc	787	(419.4)	hafnium	3866	(2130)
magnesium	1204	(651.1)	columbium	4379	(2415)
aluminum	1224	(662.2)	molybdenum	4752	(2622.2)
copper	1918	(1047.8)	tantalum	5425	(2996.1)
uranium	2071	(1132.8)	tungsten	6152	(3400)
beryllium	2343	(1283.9)	carbon	6600	(3648.9)

meters, flow (integrators). Total flow can be measured with flow meters that incorporate integrators. An integrator is a calculating device that combines multiplication and addition. The changing flow rate is multiplied and the product is added continuously to the previous total on the counter. If differential pressures are used as input, a means of converting differential pressure to flow rate must be included in the device. Such integrators can be mechanical or electrical, continuous or intermittent.

meters, orifice plate (flow rate meter). An orifice plate is a thin, circular metal plate with a sharp edged hole in it. It is held in the pipe between two flanges, which are called orifice flanges. The shape and location of the holes are the distinguishing features of the three kinds of orifice plates. The concentric plate has a circular hole located below its center; the segmental plates have a hole that is only partly circular, located below its center; the eccentric plates has a hole located below its center.

meters, positive displacement. Meters operating by positive displacement admit fluid into a chamber of known volume and then discharge it. The number of times the chamber is filled during a given interval is counted. Two chambers are used. They are arranged so that as one is filling, the other is being emptied. This provides continuous flow. The total flow for a given interval can be determined by multiplying the number of times the chambers are filled by the volume of the chambers. The average flow rate is computed by dividing this total flow by the time units in the periods.

meters, velocity. A velocity meter converts the velocity of a flowing liquid to a rotary shaft motion. The liquid enters the meter and drives a turbine wheel or propeller at a speed that varies with the flow rate. The turbine wheels drive a gear train, which is connected to a counter. The counter then registers the total quantity of the liquid that has passed through the meter.

meters, Venturi tube (flow rate meter). A specially shaped length of pipe resembling two funnels joined at their smaller openings. The Venturi tube is used for large pipes. It is more accurate than the orifice plate.

mica. Micas are a group of complex silicates of aluminum and the alkalies. All contain hydroxyl and most contain one or more of the elements iron, magnesium, lithium, and fluorine. The dark iron–magnesium variety, biotite, is the most common in rocks, but it is of little or no economic importance. Muscovite and phlogopite, like other micas, crystallize in the monoclinic system, in six-sided crystals that are pseudohexagonal in habit. Most commercial muscovite is distinctly colored, especially in sheets thicker than $^1/_{16}$ in. (1.6 mm) or so, in various shades of red and green, especially green. Thin films of mica possess great flexibility and mechanical strength. Another property of value is its transparency, with both flexibility and transparency varying inversely with thickness. Mica is infusible at ordinary temperatures. Mica is highly heat resistant, being able to undergo high temperatures, and sudden thermal shock, without appreciable physical or chemical change. It also has low electrical conductivity, very high dielectric strength, high dielectric constant, and low power loss. Sheet mica is especially suitable for a number of electrical and electronic applications. Large sheets are employed in electrostatic "memory tubes" in some computers. As an insulating material, mica is made into disks, tubes, washers, bushings, plates for use in such equipment as condensers, transformers, rheostats, radio and electronic tubes, and radar circuits. Mica of lower quality is utilized as the nonconducting element in toasters, irons, and similar appliances.

microcomputer. Microcomputers are systems based on the uses of microprocessors. They have limited flexibility but decided cost and operational advantages. Microcomputer memory is usually of two types. The first has a fixed content and is read only (ROM), which is used to store the microcomputer's operating program. The second type of memory has a variable content and is called read/write or random-access memory (RAM). It is generally used to store the variable data on which the microcomputer is to act. Random-access memory can also be used to store frequently changed programs

microelectronics. The size of electronic devices has been continuously decreasing, and it is now possible to connect more than 1000 transistors into a circuit only a tenth of an inch square. Microelectronic processes are grouped into (1) semiconductor microelectronics, which are the components of transistors, diode, *pnpn* switches, and resistors, and (2) thin-film microelectronics, which are used as resistors, capacitors, and interconnections of electronic circuits. Thin-film resistors may be made by depositing a metallic film of required thickness and area. Conductors are usually made of thick-film of high-conductivity metals such as aluminum or gold; film capacitors are made by sandwiching a dielectric film between two conductive electrodes. The materials used for substrates are ceramics and glasses or glazed ceramics. Thin films can be deposited by one or a combination of the following methods: vacuum deposition, sputtering, anodization, silk screening, epitaxy, vapor plating, plasma decomposition, and electron beam decomposition.

microprocessor (computer). A programmable large-scale integrated circuit chip containing all the elements required to process binary encoded data. A microprocessor can perform basic arithmetic and logical as well as control functions equivalent to the central processing unit of a conventional computer. The microprocessor differs from a conventional central processing unit by occupying only a single chip, or at most a few chips of silicon. Microprocessors have a wide range of actual and potential applications including control functions for automobiles, household appliances, and factory machinery. When a microprocessor is supplemented with power circuitry, input–output control interfaces, and memory, it becomes a fully operational microcomputer.

microhardness. The hardness of a material as determined by forcing an indenter such as a Vickers or Knoop indenter into the surface of a material under very light load; usually, the indentations are so small that they must be measured with a microscope. Capable of determining hardnesses of different microconstituents within a structure, or of measuring steep hardness gradients such as those encountered in case hardening.

microwave drying (see drying, microwave)

Miller indices (metallurgy). As it is essential to devise a method for describing the face of a crystal or the atomic planes within a crystal or space lattice, and in a quantitative way, sets of numbers that identify given planes are used, these sets of numbers being known as Miller indices of the plane. These special planes and directions within the crystal of metals serve an important function in the study of hardening reactions, plastic deformation, and other areas of the behavior of metals under various conditions. These numbers are also useful in x-rays and other investigations of metals.

milling, chemical (see machining, chemical and electrochemical)

milling, grains. The process of grinding grain into flour or meal. A pair of large, flat round millstones are used to grind so-called "stone ground" grain in a mill building. Today, most grain is ground with steel cutters.

milling machines. Normally these are classified into three groups: the bed type, the column and knee type, and special milling machines. The bed type is used for production manufacturing. The general characteristics of this type of machine are that the height of the table is fixed and the adjustments for height are made with the spindles. In construction, these machines are very rigid, permitting greater metal removal. They are either fully automatic or semiautomatic, and most of them are capable of running through a complete cycle from start to finish. A machine may have one spindle mounted on the single column or two spindles mounted on both sides of the table on two columns. The planer type of milling machine is a bed-type machine which is usually very large. The table moves under an arch very much as the table of a planer. The cutter heads are mounted above and at the sides of the table. The column and knee type of milling machine is used because of its flexibility. Because of the ease with which it can be set up and its versatility, it is more adaptable for quick single-piece setups. The knee carries the feed mechanism and mounts the saddle and the table. The column of the machine is the machined front of the main casting. The knee is mounted on this column and fastened by dovetail ways, which permit movement of the knee in a vertical plane. In general there are two types of column and knee milling machines, the horizontal and the vertical milling machines, which may be either plain or universal.

milling machines, special-purpose. The number and design of special-purpose millers is probably as unlimited as the design of machine parts themselves. The counterpart of the automatic screw machine—the bar-stock automatic miller—for mass production of surfaces, forms, shapes, etc., adapted primarily to milling, utilizes bar stock of almost any cross section desired, mills it to a specified shape or shapes, and cuts it off. The part at this point is either completely finished or ready for final finishing. **(See also milling.)**

milling, metals. This is a machining process in which metal is removed by a rotating multiple-tooth cutter, each tooth removing a small amount of metal with each revolution of the spindle. Because both workpiece and cutter can be moved in more than one direction at the same time, surfaces having almost any orientation can be machined. The principal differences between milling and other machining processes are: (a) the interruptions in cutting that occur as the teeth of the milling cutter alternately engage and leave the workpiece, (b) the relatively small size of chips in milling, and (c) the variation in thickness with each chip. Chip thickness varies during the cut of any individual tooth, because feed is measured in the direction of table motion (workpiece moving into the cutter), whereas chip thickness is measured along the radius of the cutter. Milling is most efficient when the work is no harder than Rockwell C 25. However, steel at Rockwell C 35 is commonly milled and steel as hard as Rockwell C 56 has been successfully milled.

mini-load. A storage and retrieval concept whereby materials are accessed by bringing the container to the operator. The term is typically used in small parts applications and/or where the weight of the container does not exceed 750 lb (341 kg).

Misch metal. A mixture of the metals cerium, lanthanum, and didymium which occur naturally. The waste matter from monazite sand after the extraction of the thorium oxide often contains large quantities of cerium oxide, and the rare-earth metals lanthanum and didymium, yttrium, and other substances. Conversion of the oxides to chlorides allows removal of the metal by electrolysis. The alloy obtained contains about 50% cerium and 45% lanthanum and didymium and is called Misch metal. **(See also ceramic magnets.)**

modulus of elasticity. The ratio, within the elastic limit of the material, of stress to corresponding strain. It is the slope of the stress–strain curve in the elastic range, and is expressed in pounds per square inch. It is nearly the same value in tension and compression for most metals. The shear modulus of elasticity is lower than tensile or compressive moduli. The modulus of elasticity is expressed by an equation, which is also an expression of Hooke's law:

$$E = \frac{s}{\sigma}$$

where E = modulus of elasticity, pounds per square inch,
s = stress, pounds per square inch,
σ = strain, inches per inch.

molding, bag (see bag molding)

molding, blow (see blow molding)

molding, ceramics. Various methods are used for forming ceramic bodies prior to firing and the various forming methods are of either the wet or the dry process type. Ceramics to be formed by one of the wet methods is given the desired plasticity through the use of a suitable quantity of plastic clay together with sufficient water to provide a relatively soft, plastic mass. When the physical requirements of the fired material make it impossible to secure the required plasticity by use of clay and water, the mix may be plasticized by means of organic binders and plasticizing agents. With low-clay and clay-free ceramics, however, the dry process is often used, semimoist granular powder having 2–12% water by weight being compressed in dies to the desired form. Ceramics to be molded by casting are made fluid by the addition of deflocculants such as sodium silicate, sodium carbonate, or solutions of salts which yield hydroxyl ion and do not ordinarily contain any more water than those prepared for ordinary wet methods. Oldest and probably most diversified general process, plastic wet forming consists of five methods or preliminary shaping—extrusion, throwing, jiggering, pressing, and casting. Method of preliminary forming to be used depends on the part design features and the economical manufacturing considerations. After preliminary forming, the molded pieces are subjected to a drying period under controlled heat and humidity. Degree of drying depends on the dimensional accuracy desired in the finished part. Any final shaping required is performed at this stage and, if glazing is necessary, the piece is dipped or sprayed with a mixture of glazing materials suspended in a relatively high percentage of water and again dried. To develop the final physical, electrical and mechanical

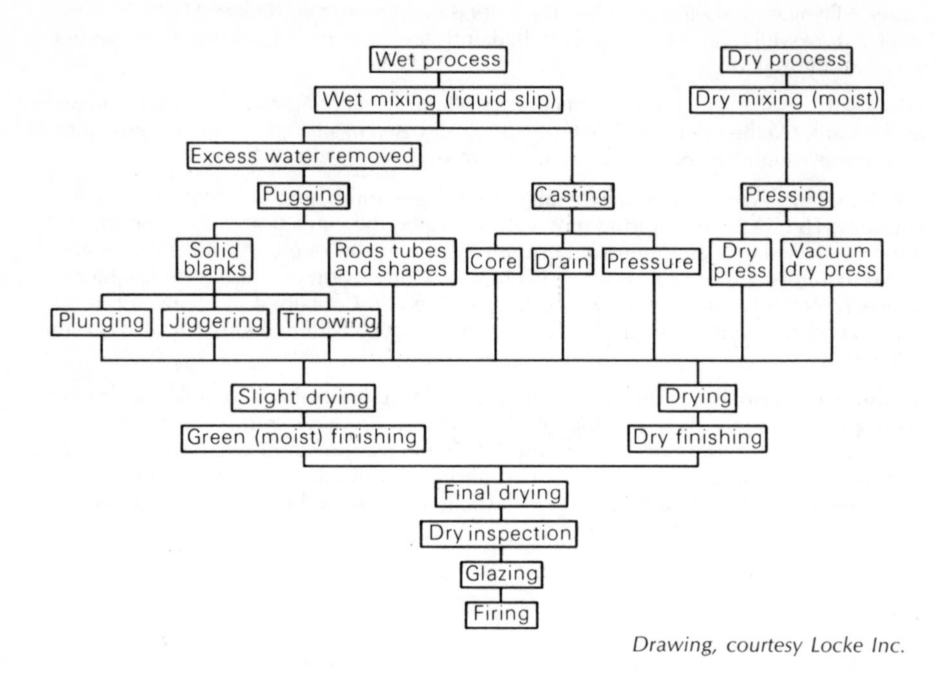

Drawing, courtesy Locke Inc.

Chart of ceramics manufacturing processes and procedures utilized today.

characteristics the formed pieces are fired in a kiln at approximately 2200°F (1204.4°C). Differing from the plastic process of forming in that only a semimoist granular powder is used, the dry process is subject to much less dimensional variation in manufacture. Pressing in metal dies is the only method of forming used. Typical parts possible are: Flat or ribbed plates of any form or curvature with or without holes, slots, recesses, and barriers; U and other shapes; intricate box shapes with ribbed partitions and side holes or slots. Dry pressing is most generally limited to relatively small parts in the neighborhood of 2–4 in. (5.08–10.16 cm). Maximum length or width possible is 14 in. (35.6 cm), height is limited to about 7 in. (17.8 cm) and projected areas to 196 in.^{2} (1265 cm^2). Dry pressing is well adapted to high production, and automatic presses may be used.

molding, compression (see compression molding)

molding, extrusion process (thermoplastics). As applied to plastics, is very much like the process that the butcher uses to grind meat. The plastic material feeds from a hopper into a screw chamber, from where it is conveyed by the action of the rotating screw. It is heated and compressed and forced through a heated die onto a conveyor belt. As the plastic passes onto the belt it is cooled by air or water sprays so as to harden it sufficiently to preserve the shape imparted to it by the die. It continues to cool as it passes along the belt. It is either cut to length, in the case of rigid plastics, or coiled, in the case of flexible plastics. In addition to providing a cheap and rapid method of molding, the extrusion process makes it possible to produce tubing and shapes having reentrant angles. **(See also extrusion machines, plastics.)**

molding, injection (thermoplastics). In this process raw material is fed by gravity from a hopper into a pressure chamber ahead of a plunger. As the plunger advances, the plastic is forced into a heating chamber where it is preheated. From the preheating chamber it is forced through the torpedo section where it is melted and the flow regulated. It leaves the torpedo through a nozzle, which seats against the mold and allows the molten plastic to enter the die cavities through suitable gates and runners. In this process the die remains cool, so that the plastic solidifies almost as soon as the mold is filled. To ensure proper filling of the cavity, the material must be forced into the mold rapidly under considerable pressure, since premature solidification would cause a defective product. In actual production, one complete cycle requires only a few seconds. **(See also injection molding machines.)**

molding machines, jet. In a jet molding process the plastic is preheated in the feed chamber to about 200°F (93°C) and then further heated to the polymerization temperature as it passes through the nozzle. The mold is held at an elevated temperature to complete the setting process. As soon as the charge for one cycle has nearly filled the die cavity, the nozzle is cooled so as to prevent the plastic in the nozzle from hardening and clogging the machine.

molding machines, transfer. Used to avoid the turbulence and uneven flow that often results from the nonuniform, high pressures in hot compression molding. In transfer molding the raw material is placed in a plunger cavity, where it is heated until it is melted. The plunger then descends, forcing the molten plastic into the die cavities. Since the material enters the cavities as a liquid, there is little pressure until the cavity is completely filled. As a result, excellent detail, good tolerances and finish, and fine sections can be obtained. An undercut on the plunger causes the sprue to be withdrawn with the plunger as the mold is opened.

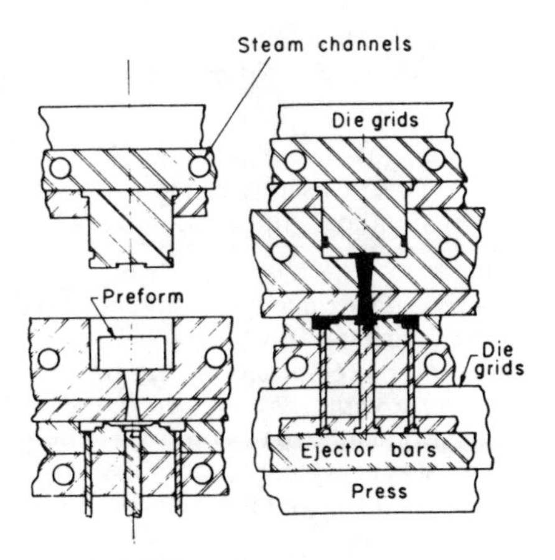

Die setup for transfer molding.

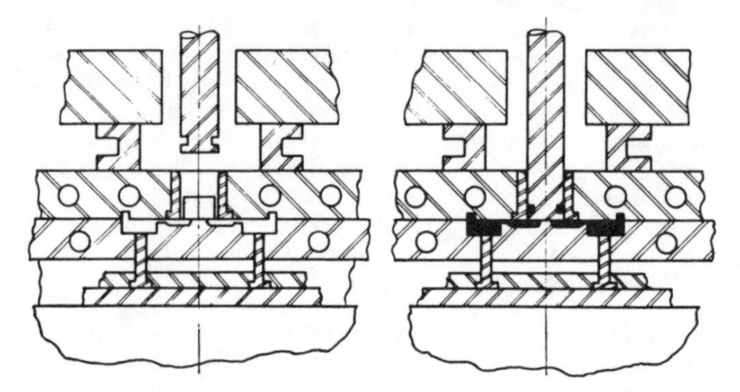

Duplex or plunger arrangement for transfer molding.

molding machines, vacuum forming. Used extensively to form shapes from thermoplastic materials. A sheet of plastic is placed over a die to form and heated until it becomes soft. A vacuum then is drawn between the sheet and the form so that the plastic takes the desired shape. It is then cooled, the vacuum dropped, and the part removed from the mold. The entire cycle usually requires only a few minutes. This process is used for producing a wide variety of products, ranging from panels for light fixtures to pages of Braille for the blind.

molding, rotational. A method used to make hollow articles from plastisols and latices. Plastisol is charged into hollow mold capable of being rotated in one or two planes. The hot mold fuses the plasticol into a gel after the rotation has caused it to cover all surfaces. The mold is then chilled and the product stripped out.

molding, rubber. Oldest and still most widely used method for producing finished rubber parts is simple compression molding using mechanical presses. With some of the newer methods of manufacture now in use, however, there are five practicable molding processes: compression molding, transfer-injection molding, full-injection molding, extrusion molding, and form-dip molding. **(See also rubber processing and also various molding methods.)**

molding, slush. Method for casting thermoplastics in which the resin in liquid form is poured into a hot hollow mold where a viscous skin forms. The excess slush is drained off, the mold is cooled, and the molding stripped out.

molybdenum disulfide (lubricants). Has a laminar structure similar to graphite, but, unlike graphite, adheres to a metal surface quite well. One method of using this substance is to form a surface film by spraying or brushing it on the part to be lubricated. Molydisulfide is an excellent extreme-pressure lubricant, being able to withstand very high pressures. It is effective over a wide temperature range and has high chemical stability. It is used either as a dry lubricant or as a component of oil, grease-consistency mixtures, and, in solvents. The most effective application of molysulfide as a dry lubricant are those where liquid films cannot be maintained and in industrial atmospheres.

molybdenum steels. Molybdenum dissolves in both gamma and alpha iron, but in the presence of carbon it combines to form a carbide. The behavior of molybdenum in steel is about the same as tungsten, but twice as effective. Molybdenum improves the hot strength and hardness of steels. The complex carbide formed by the addition of molybdenum acts to stabilize martensite and resist softening upon heating. Similar to the action from additions of tungsten. Molybdenum reduces the tendency toward grain growth at elevated temperatures and slows up the transformation rate from austenite to pearlite. This characteristic of molybdenum steels, increases hardenability, allows the use of air quenching of many of the alloy tool steels instead of oil quenching in the hardening operation. Molybdenum is often combined with other alloying elements in commercial alloy steels. Molybdenum–chromium, molybdenum–nickel, molybdenum–tungsten, and molybdenum–nickel–chromium are common combinations used. Molybdenum structural steels of the SAE 4000 series find many applications intended for heat-treated parts. The molybdenum in these steels strengthens the steel both statically and dynamically, eliminates temper brittleness almost completely, refines the structure, and widens the critical range. Molybdenum is alloyed with both structural and tool steel. The low-carbon molybdenum steels are often subjected to case hardening heat treatments. Medium carbon molybdenum steels are heat treated for parts including such applications as gears, roller bearings, and aircraft and automobile parts that are subject to high stresses. Tool steels for general use and for tools for hot working are often alloyed with molybdenum.

monel metal. A so-called natural alloy in that it is made by smelting a mixed nickel–copper ore. It contains 65% nickel, 28% copper, and the balance consisting chiefly of iron, manganese, and cobalt and impurities. Monel metal is harder and stronger than either copper or nickel in the pure form; it is also cheaper than pure nickel and will serve for certain high-grade purposes to better advantages than the pure metal. Monel is a substitute for steel where resistance to corrosion is a prime requisite. Monel metal can be cast, hot and cold worked, and welded successfully. In the forged and annealed condition it has a tensile strength of 80,000 psi (5.5×10^5 kPa) with an elongation of 45% in 2 in. (5.08 cm), and is classified as a tough alloy. In the cold worked condition it develops a tensile strength of 100,000 psi (6.9×10^5 kPa) with an elongation of 25% in 2 in. (5.08 cm). Monel metal has many uses including sheet, rod, wire, and cast form. Included in the applications for monel metal are sinks and other household equipment, containers, valves, pumps, and many parts of equipment used in the food, textile, and chemical industries.

monomers (organic polymers). A characteristic feature of a monomer is that its molecule must be bi- or polyfunctional. This means that the molecule that is to undergo polymerization must contain two or more reactive or functional groups. These reactive groups may be hydroxyl or amino acids, di- or polyalcohols, and diamino acids, or diacids. Molecules containing double bonds are also considered bifunctional because, under the influence of heat or light energy, the double bond may undergo intermolecular rearrangement, thus becoming available for a bifunctional reaction of the molecule. Depending on the functionality of the monomer, two main structural types are formed; linear polymers or chain polymers and three-dimensional network or cross-linked polymers. If a monomer is bifunctional, either having two groups at the end or a double bond, it will form linear polymers. Linear polymers can be straight-chain or branched-chain polymers, depending on the course of the polymerization reaction, which can be controlled by temperature, pressure, or catalyst. If a monomer is three or more functional or if there is a mixture of bifunctional and trifunctional monomers, three-dimensional network polymers are obtained in which all structural units are connected with each other by covalent bonds.

motorized hand/rider truck. A dual-purpose industrial truck that is designed to be controlled by a walking or by a riding operator.

movement/operation ratio (M/O ratio). This measures the relative efficiency of an overall materials handling plan. M/O Ratio = Total number of moves ÷ total number of productive operations.

moving-beam bar code reader. A device that dynamically searches for a bar code pattern by sweeping a moving optical beam through a field of view. **(See also bar code.)**

MRP (see materials requirements planning)

MTBF (see mean-time-between-failures)

MTTR (see mean-time-to-repair)

multiple cold saw (see cut-off machines)

multiple-insertion assembly (see automatic assembly)

Muntz metal (see brasses)

narrow aisle truck. A self-loading industrial truck primarily intended for right angle stacking in aisles narrower than those normally required by counterbalanced trucks for the same capacity.

natural aging (see aging)

natural gas. Consists mainly of methane, with smaller quantities of other hydrocarbons, particularly ethane, although carbon dioxide and nitrogen are usually present in small amounts. Sometimes there are also appreciable amounts of hydrogen sulfide present. This is generally removed before transmission in the field. Oxygen is present only when there has been an infiltration of atmospheric air. The characteristics of a natural gas are influenced by the underground conditions existing in the localities where it is found. Most frequently these areas contain oil deposits and as a result, the gas may be impregnated with heavy saturated hydrocarbons, which are liquid at ordinary pressures and temperatures. This "wet" gas is then dried by stripping from it those liquids which constitute the so-called "casing-head" gasoline. When sulfur is present in the oil deposit, the gas will frequently contain hydrogen sulfide. "Dry" natural gas produced by wells that are remote from oil-bearing areas, is gas containing less than 0.1 gal (0.38 liter) of gasoline vapor per 1000 ft^3 (28.32 m^3). The terms wet and dry, when associated with natural gas, reer to its gasoline content, not to its natural moisture. Natural gases are also classified as either "sweet" or "sour." The sour gas is one that contains some mercaptans and a high percentage of hydrogen sulfide, while the sweet gas is one in which these objectionable constituents have been removed.

natural rubber—natural polyisoprene (NR). Crude natural rubber is found in the juices of many plants, including the shrub guayule, Russian dandelion, goldenrod, and dozens of other shrubs, vines, and trees. The principal source is the tree *Hevea brasiliensis*, which is native to Brazil. Petroleum oils are the greatest enemy of natural rubber compounds. The synthetics have all but completely replaced natural rubber for seal use.

NC (see numerical control)

NC lathes and machining centers. Today, the most versatile automatic lathe is the numerically controlled machine. Changeover is rendered simple and tooling arrangements offer everything from the single-point cross slide tool to entirely automatic tool changers wherein as many as 16 or more tools in a magazine are available. Thus, without transfer of the piece part, hole patterns can be drilled, keyways milled, steps milled, or boring, reaming, and tapping perpendicular or parallel to the centerline can be done. These machines fill the production gap where small to moderate lots of parts of some complexity are required.

neoprene. Chloroprene rubbers were among the first synthetics produced in the United States. Following purification by fractionation, the chloroprene monomer is emulsified in water and polymerized at about 100°F (38°C) under atmospheric pressure in the presence of sulfur, with potassium persulfate as a catalyst. Neoprene is a particular versatile elastomer, because it has a combination of properties suited for many and varied applications, including those involving extreme service conditions where the product must withstand a combination of deteriorating conditions such as oil and heat or extreme exposure to weather.

nepheline syenite (minerals). The syenites are a comparatively rare group of phaneritic igneous rocks, in which the essential mineral is feldspar. Both albite and microcline are generally present. Nepheline is usually in smaller grains, intercrystallize with the feldspar. It has a glassy appearance and a greasy luster, somewhat like that of quartz. The valuable properties of nepheline are of the same type as those of feldspar and both minerals are put to the same uses. Nepheline syenite, being a mixture of the two, is therefore utilized as a bulk product. It is used as an ingredient of glass, especially container glass, in whiteware, and in ceramic glazes and enamels. The presence of abundant alkalies lowers the temperature of fusion, making nepheline syenite an excellent vitrifying agent or flux. Its use decreases fuel costs and wear on refractories in the burning of ceramic ware.

nickel. An important source of nickel is Ontario, Canada, where it occurs in the form of pentlandite, an iron–nickel sulfide. Copper, gold, silver, and the platinum group metals are also present in considerable quantity. The average content is 3% nickel, so the ores must be concentrated. Recovery processes begin with the roasting of the concentrate in an open-hearth furnace to eliminate most of the sulfur and other impurities in the slag. More sulfur and most of the iron are eliminated in a Bessemer converter. Copper and nickel are then separated by the Orford process. This consists of adding sodium sulfate to the converter matte and allowing the molten mixture to solidify. The sulfides of nickel and copper are formed, the nickel sulfide settling to the bottom and the copper sulfide forming a layer on top. When cooled, the two layers are easily separated. The nickel sulfide "bottoms" are further refined and converted to nickel oxide; this is reduced in a reverberatory furnace to metallic nickel, which is then electrolytically refined. Nickel is a silvery white metal which is capable of retaining a high polish. It is almost as hard as low-carbon steel and has an ultimate strength, after rolling and annealing, of from 65,000 to 75,000 psi (4.5×10^5 to 5.2×10^5 kPa). When nickel contains a small amount of carbon, it is quite malleable, but small amounts of arsenic and sulfur make it brittle. Nickel is resistant to corrosion by the atmosphere and salt water or freshwater, and resists most acids, except nitric. It is resistant to neutral and alkaline salts and to alkalies. Nickel is one of the few metals that has ferromagnetic properties. Most nickel is used in the production of heat- and corrosion-resistant ferrous materials. Large quantities are made into such alloys as "Nichrome," "Monel," and coinage metal. In pure form it is used as a plating material for other metals.

nickel alloys. Available in heat-treatable and nonheat-treatable types. Nickel and copper are solid solutions soluble in each other in all proportions, with an infinite number of nickel–copper and copper–nickel alloys possible. If the percent nickel exceeds the percent copper, the alloy may be termed a monel. If the nickel content is in the range of 12–30%, the alloy will be termed a cupronickel. The Monel may be considered to be two-thirds nickel and one-third copper. Monels are more resistant to reducing chemicals than nickel and more resistant to oxidizing chemicals such as chlorides and nitrates, than copper, hence they have better corrosion resistance than either constituent. Monels are used as piping and pumps for brine and seawater, steam condensers, ships' propellers, and shafting. R Ronel (Monel R-405) contains 66% Ni, 31.5% Cu, and 0.05% S. The sulfur addition is made, as in the case of sulfurized steels, for machinability. The Monel then is selected for parts that must be machined. Other Monels may contain from 44.5% to 84% nickel. The standard Monel is Monel 400 containing 66% Ni, 31.5% Cu, and about 1% each of iron and manganese. The presence of half a percent or more of aluminum or titanium or both in a high nickel alloy indicates a precipitation-hardening alloy. The nature of the compound precipitated is a little uncertain; it may be an intermetallic nickel–aluminum and nickel–titanium, $Ni_3(Al,Ti)$. Permanickel, 98.6% Ni, 0.50% Ti, and Duranickel, with 4.5% Al, 0.50% Ti, are thus heat-treatable nickels. The 500 Monels, K-500 and 501, both contain 2.80% Al and 0.50% Ti. Inconel superalloys are all heat treatable and contain even more aluminum and titanium.

Nickel Alloys

	Ni	Cr	Fe	Al	Ti	Mn	C	S	Cu	Other
A Nickel	99.5		0.15			0.25	0.06	0.005	0.05	
Permanickel	98.6		0.10		0.50	0.10	0.25	0.005	0.02	0.35 Mo
Duranickel	94.0		0.15	4.50	0.50	0.25	0.15	0.005	0.05	
Monel 400	66.0		1.35			0.90	0.12	0.005	31.5	
Monel R-405	66.0		1.35			0.90	0.18	0.050	31.5	
Inconel 600	76.0	15.8	7.20			0.20	0.04	0.007	0.10	
Inconel 700	46.0	15.0	0.70	3.00	2.20	0.10	0.12	0.007	0.05	28.5 Co, 3.75 Mo
Inconel X-750	73.0	15.0	6.75	0.80	2.50	0.70	0.04	0.007	0.05	0.85 Cb
Incoloy 800	32.0	20.5	46.0	0.30	0.30	0.04	0.04	0.007	0.30	
Rene 41	balance	19	2.5	1.65	3	0.30	0.10			11 Co, 9.75 Mo
Hastelloy A	53		22			2				22 Mo
Hastelloy C	balance	16	5							16 Mo, 4 W
Constantan	45								55	
Alumel	95.3		0.1	1.6		1.75				1.2 Si
Chromel P	90	9.5	0.2							0.4 Si
Inco-Weld A	71	16.4	6.6		3.2	2.3	0.03	0.007	0.04	

nickel–cadmium cells. Used in small devices which formerly used carbon-zinc dry cells and in other devices where carbon-zinc dry cells cannot meet the load requirements. An advantage of nickel–cadmium secondary cells is that it can stand in a discharged condition indefinitely at normal temperatures without deterioration. In nickel–cadmium dry cells the negative electrode consists of a metallic cadmium and the positive electrode is nickel hydroxide. When the cell is discharged, the negative electrode becomes cadmium hydroxide and the positive electrode becomes less oxidized, that is, it is reduced. The most common electrode designs for nickel–cadmium cells consist of perforated steel pockets to hold the active materials or sintered nickel plates into which the active materials are impregnated. During the discharge of a nickel–cadmium cell, electrons are released in the negative material as chemical changes take place. These electrons flow through the outer electrical circuit and return to the positive electrode. Positive ions in the electrolyte remove the electrons from the positive electrode. During charge, the reverse action takes place, and the negative electrode is restored to a metallic cadmium state.

nickel–iron alloys. An alloy of 36% nickel is known as "Invar," and has a very low coefficient of expansion so that it is used for standards of length. The alloy of iron and nickel that contains 46% nickel is known as "Platinite" since it has the same coefficient of expansion as glass and may be used in place of platinum to seal into glass. The alloy of 78½% nickel and 21½% iron is known as Permalloy and has very high magnetic permeabilities at low fields so that it has been used in the construction of submarine telegraph cables.

nickel–silver alloys. Also called German silver, is a brass alloy containing 10–30% nickel and 5–50% zinc, the balance copper. These alloys are very white, similar to silver, hence the name. They are resistant to atmospheric corrosion and to the acids of foodstuffs. Their use depends largely on their resistance to corrosion and pleasing appearance.

nickel steels. Nickel, when added to steel, dissolves to form a solid solution with iron, lowering the critical range to a marked degree, and thus forms a steel that may be made pearlitic, martensitic, or austenitic by simply varying the percentage of nickel. The lower carbon pearlitic–nickel steels contain from 0.5% to 6.0% nickel, and are particularly suited for structural applications because of greater toughness, strength, and resistance to corrosion. Martensitic–nickel steels, because of brittleness and hardness, are little used. They contain from 10% to 22% nickel.

nitrates. Compounds of nitrogen are used extensively in industry, chief of these is anhydrous ammonia. Others include ammonium sulfate, ammonium nitrate, nitric acid, and sodium nitrate. Nitrogen compounds are used chiefly in the fertilizer and chemical industries, and in the manufacture of nitroglycerin and other explosives. There are three main sources for ammonia and its relatives: the atmosphere, coal, and natural nitrate deposits. The most important of the nitrogen compounds used in fertilizers are ammonium nitrate, sodium nitrate, ammonium sulfate, and ammonia. These substances may be applied directly to the soil, but normally they are mixed with other plant foods, especially phosphate and potash. Ammonia is the dominant nitrogen compound in the chemical industry, and is also used in the manufacture of wood pulp, plastics, and synthetic fibers, in refrigeration and other industrial processes. Nitric acid is valuable as a solvent and in dye manufacture. Practically all explosives (other than atomic) are nitrobodies of one form or another. Nitric acid is used to produce nitroglycerin, nitrocellulose, trinitrotoluol (TNT), and other explosives.

nitric acid process, petrochemicals. A process for the continuous production of nitric acid from ammonia and air. Gaseous ammonia is combined with air and passed over platinum gauze for conversion into nitrous oxide. After heat interchange the nitrous oxide is compressed, again heat exchanged, and passed through an absorber for conversion into nitric acid. The nitric acid produced is bleached with air and sent to storage. Off-gas from the bleaching operation is recycled to the absorber.

nitriding, metal. Steel is gas nitrided in a furnace at 950–1050°F (510–566°C) with an atmosphere, commonly ammonia, that permeates the surface with nascent nitrogen. As an indication of the long period required, with SAE 7140 steel at 975°F (524°C) case depth reaches 0.02 in. (0.508 mm) at 50 h and 0.04 in. (1.02 mm) at 200 h. Liquid nitriding is done also at 950–1050°F (510–566°C) in a bath of molten cyanide salts. Quenching is not needed because the case consists of inherently hard metallic nitrides. For efficient results nitridable steels alloyed with aluminum, chromium, vanadium, and molybdenum to form stable nitrides are used. Various modifications of nitriding are used to speed up the process. Chapmanizing or liquid-pressure nitriding entails passing anhydrous ammonia through the nitriding salt bath under pressure while the work is being treated. Pressure nitriding is done with the work in a sealed retort holding an ammonia atmosphere under pressure. Glow discharge nitriding or ionitriding is done with the work as the cathode in an anodic retort. An electric current heats the work and produces a glow discharge that ionizes the nitrogen atmosphere.

nitroaromatics process, petrochemicals. A process for the continuous nitration of aromatic hydrocarbons using mixed acid. The formation of nitroaromatics takes place according to the following general equation: $Ar + x\ HNO_3 = Ar(NO_2)_x + x\ H_2O$. The aromatic hydrocarbon and the required nitric and sulfuric acid are conveyed separately through precisely working proportioning equipment into the nitration vessels, where reaction takes place, with optimum intermixing. This means that the production of a mixed acid from the components in a separate plant prior to entering nitration equipment is not required. After having passed the nitration step, the reaction mixture is separated in centrifugal separators; the acid nitroproduct is washed alkaline and neutral in several washing units, the washing water is separated and the final product is dried in a drying column. The spent acid separated after reaction is extracted with the aromatic in order to recover remainders of nitric acid and nitro compounds dissolved in the spent acid. The aromatic hydrocarbons, in this way already slightly nitrated, enters into the main nitration process. The resulting spent acid can be returned into a concentration unit without denitration or can be used in the production of fertilizers.

nitrogen. The properties and constants of nitrogen must necessarily be nearly those of air. Nitrogen is a rather inert gas. It boils at −320.3°F (−195.7°C) at atmospheric pressure, and cannot be liquefied above −232°F (−146.7°C) (critical temperature). It is available in standard gas cylinder and is chiefly used as liquid nitrogen for refrigeration and as an inert or purging gas.

nondestructive testing (see eddy-current testing, liquid penetrant inspection, magnetic-particle inspection, magnetic reluctance testing, sonic testing, ultrasonic testing)

nonfill. Failure of metal to fill a forging die impression.

non-heat-treatable alloy. An alloy that can be strengthened only by cold work.

nonwoven fabric. Fabric materials produced without weaving on a needle punch loom.

normalizing (see annealing)

notched-bar impact testing (metals). Provides a quick way of loading and measuring toughness of a notched bar, i.e., the ability to absorb energy. The test results are useful in comparing, for a given composition, the effects of prior history on toughness. Of particular interest is the effect of heat treatment on properties of steel. Overheating steel coarsens the austenite grain size and impairs toughness. This can readily be detected by a notched bar impact test and by examining the appearance of the fractured surface. Another important application is to study the effects of tempering cycles on hardened steels. The most common kinds of impact test uses notched specimens loaded as beams. The beams may be simply loaded or loaded as cantilevers. The notch is usually a V notch cut to specifications with a special milling cutter. The specimen is held in a rigid vise or support, and struck a blow by a pendulum traveling about 17 fps (5.2 m/s). The energy input is a function of the height of fall and the weight of the pendulum. The energy remaining after fracture is determined from the height of rise of the pendulum.

NR-16, nitrile barrier resins. A rigid, transparent copolymer of high nitrile content, with a low level of permeability to gases such as carbon dioxide and oxygen; while it was specifically developed for use in carbonated beverage bottles, its combination of barrier capabilities and physical properties make it suitable for typical engineering thermoplastic uses in which the need for extreme toughness is subordinate to other properties. It has excellent melt stability and can be processed in conventional equipment. It can be blow molded into tough bottles by both extrusion and injection blow molding.

numerical control, machine tools. A method of controlling the motions of machine components by means of numbers. In its simplest application it is used to position the work relative to the tool. By combining the principal of numerical control with punched control tapes or punched cards and sometimes an electronic computer it can also start and stop the machine operations, control the relative motion (path) between the work and the tool during cutting, and cause the work (or tool) to move from the position at the end of one cut to that for the beginning of the next cut. As an example, after a workpiece is set up on a table of the machine, the work tool must be brought into proper relationship for each of the holes drilled. If this is to be done automatically, a means must be available for precisely specifying, measuring, and controlling the relative motions of the machine table and the spindle carrier so that the location dimensions specified on the drawing will be reproduced on the workpiece. To accomplish this requires feedback signals that will tell exactly the location of a machine component, such as a table. Such feedback signals can be supplied by transducers, actuated by the feed screw which moves the machine component or by the actual movement of the component. These transducers provide either digital or analog information (signals). Digital information usually is in the form of electric pulses. There are two major types of digital transducers. One supplies incremental information and tells how much motion of the input shaft or table has occurred. The second type of digital information is absolute in character, with each pulse corresponding to a specific location of the machine component. In analog type of information, the signal may be in the form of an electric voltage which varies as the input shaft is rotated or the machine component is moved; a variable output is produced which is a function of movement. Movement is evaluated by measuring or matching the voltage or by measuring the ratios between the applied and the feedback voltages.

nylon. A generic term for a long-chain synthetic polymeric amide with recurring amide groups as an integral part of the main polymer chain. There are two types of nylon: (1) those prepared from a diamine and a diacid and (2) those prepared from an amino acid or amino acid derivative. Nylon 66 is an example of the first type. It is prepared from the six-carbon atom diamine, hexamethylene diamine, and the six-carbon atom diacid, adipic acid. It is a general purpose resin available in a range of formulations for extrusion and injection molding. Other nylons of this type are nylon 610 (from hexamethylenediamine and sebacic acid) and nylon 612 (from hexamethylenediamine and dodecanedioic acid). These are generally used where the engineering properties of nylon are required but where lower moisture absorption and the resultant better dimensional stability are needed. Nylon 6 is an example of a nylon of the second type. It is prepared from the six-carbon amino acid derivative, caprolactam. Other commercially nylons of this type are nylons 11 (from amino undecanoic acid) and 12 (from laurolactam). Nylon 11 has very low moisture absorption and can readily be plasticized for industrial and automotive hose applications. Nylon 12 has the lowest absorption of any commercially available nylon. Nylons prepared from a lactam can also be anionically polymerized with a variety of catalysts.

nylon 6. Process for the production of nylon 6 (polycaproamide 6-PA6) starts from caprolactam monomer. Caprolactam (LC) is polymerized in the presence of water at elevated temperature by the Iventa VK-tube method (VK = *vereinfacht kontinuierlich*). Prepolymerization steps are added to accelerate the hydrolysis of LC by elevated water content to reduce the total reaction time, especially desirable in large capacity lines. Additives, e.g., dulling agents, chain terminators, catalysts, and some polymer modifying agents can be added to the LC solution. The raw polymer at reaction equilibrium is extruded into strands and cut into granules. The 10% nonreacted monomers and oligomers are removed in a continuously operating extraction column by hot water in counterflow to the chips. The extracted polymer is dewatered and dried continuously in the drying column with pure, hot, dry nitrogen circulating in counterflow to the chips. The product is cooled to room temperature and blown to the storage silos. The relative viscosity of polymer dissolved in sulfuric acid from this process can range from 2.3 to 3.4 or up to 7 when post poly condensation conditions are applied to the drying stage.

nylon 11. A polymer of aminoundecanoic acid; it is marketed as Rilsan. The material is a specialty item, and, because of its high cost, has limited sales. The advantage of nylon 11 for some uses is its low rate of water absorption. It absorbs only 1% by weight of water in air at 65% relative humidity, and 68°F (20°C), and only about 2% when immersed in boiling water. The commercial method of preparation of the monomer begins with castor oil, which is converted by methanolysis to the methyl ester of ricinoleic acid. This is then split thermally into *n*-heptaldehyde and methylundecylenate. The ester is sapoinified to undecylenic acid and converted to 11-bromoundecanoic acid by adding HBr across the double bond between the two terminal carbon atoms. Nylon is continuously produced from a suspension of the 11-aminoundecanoic acid in water. Polymerization is carried out in a reaction column with three compartments similar to the continuous unit for nylon 66. The reaction proceeds at 392–428°F (200–220°C). In other respects the manufacture of nylon 11 is very similar to the manufacture of other nylons.

nylon 66. Nylon 66 polymerization proceeds by two steps. The first reaction is the neutralization of the adipic acid with the base, hexamethylenediamine, to form the salt, hexamethylenediammonium-adipate. In the second step, heat is necessary to dehydrate the salt and initiate the polycondensation reaction to form the polymeric molecule, polyhexamethyleneadipamide. The formation of nylon 66 proceeds only by polycondensation with a molecule of water removed for each molecule of nylon salt added to the chain. The polymerization rate and the degree of polymerization are affected by temperature, agitation, and pressure. The activity of the functional groups is independent of the length of the chain.

nylon 610. Produced from hexamethylenediamic and sebacic acid. The manufacturing process is identical to that of nylon 66, except that polymerization proceeds at a somewhat lower temperature owing to the lower melting point of the 610 polymer. It is not necessary to vacuum strip or extract either the nylon 11 or nylon 610 polymer or to remove unreacted monomer. Both reactions proceed nearly to completion. The critical point in nylon 610 manufacture is the cost of sebacic acid. The conventional method for production of the acid involves fusion of castor oil with sodium hydroxide where the ricinoleic acid content of the castor oil splits to yield sebacic acid and a mixture of octanol-2 and methyl hexyl ketone. The caustic fusion is carried out at 482–572°F (250–300°C), and the product is neutralized to the point where monobasic acids are precipitated and monosodium sebacate remains in solution.

octane number, gasoline. The ability of a gasoline to perform well in an internal combustion engine is given by its octane number. In order to standardize the performance of a gasoline, 2,2,4-trimethylpentane, $(CH_3)_3C{-}CH_2CH(CH_3)_2$, a very fine motor fuel with no tendency to premature explosion under high compression, is assigned an octane number of 100. Normal heptane, $CH_3CH_2CH_2CH_2CH_2CH_2CH_2CH_3$, an extremely bad "knocker," is assigned a value of zero. The octane number of any fuel is determined from a comparison of its performance to that of a blend of 2,2,4-trimethylpentane and *n*-heptane. Both the fuel in question and the prepared blend are tested in a specially instrumented engine. The percent of 2,2,4-trimethylpentane in the blend that gives the same performance as the fuel in question establishes the octane number of the latter. Motor fuels can be prepared that have octane numbers well in excess of 100, that is, they will perform even better as motor fuels than 2,2,4-trimethylpentane.

Octane Ratings (Number) Of Some Hydrocarbons

Hydrocarbon	Formula	ml TEL[a]/gallon	
		0.0	3.0
Methane	CH_4	> 120	—
Ethane	C_2H_6	118.5	—
Propane	C_3H_8	112.5	—
n-Butane	C_4H_{10}	93.6	101.6
n-Pentane	C_5H_{12}	61.7	88.7
2-Methylbutane (Isopentane)	C_5H_{12}	92.6	102.0
n-Hexane	C_6H_{14}	24.8	65.3
Methylcyclopentane	C_6H_{12}	91.3	105.3
n-Heptane	C_7H_{16}	0.0	43.5
Methylcyclohexane	C_7H_{14}	74.8	88.2
Methylbenzene (Toluene)	C_7H_8	103.2	111.8
n-Octane	C_8H_{18}	−19.0	25.0
2,2,4-Trimethylpentane (Isooctane)	C_8H_{18}	100.0	115.5
2,4,4-Trimethyl-2-pentane	C_8H_{16}	103.5	105.7
Isopropylbenzene (Cumene)	C_9H_{12}	113.0	116.7
1,3,5-Trimethylbenzene (Mesitylene)	C_9H_{12}	> 120	—

[a]Tetraethyl lead.

off gage. Deviation of thickness or diameter of a solid product, or wall thickness of a tubular product, from the standard or specified dimensional tolerances.

off-line. Refers to operations that function independently of a large central computer control unit.

oil gas fuel. Oil gas is manufactured by heating checker brickwork and then passing a mixture of oil and steam through passages in the brickwork. Its great use on the Pacific Coast resulted from the relatively high cost of coal and coke there. Oil gas is similar in composition, heating value, and flame temperature to carbureted water gas.

olefins (alkenes). The olefins, or alkenes, are characterized by the presence of double bonds between adjacent carbon atoms. The general formula C_nH_{2n} corresponds to an open-chain olefin if only one double bond is present in the molecule. The same general formula also represents a cycloalkane. In the first, or simplest, member of the series, $n = 2$, and the formula C_2H_4 is that for ethylene, or ethene. As a class these compounds are commonly referred to as the olefins, the name of which originated because ethene, in its reaction with chlorine, forms, 1,2-dichloroethane, an oily liquid. The name alkene is used in systematic nomenclature when referring to the olefins. Systematic names for members of the olefin family are formed by replacing the suffix, -ane, of the corresponding alkane with, -ene. Common names usually are employed to name the simplest members of the alkenes. Such common names are formed by replacing the suffix ane of the corresponding alkane with xylene.

Olsen ductility test (see ductility test)

on-line. Pertaining to devices under direct control of the central processing unit (CPU). Operation where input data are fed directly from the measuring devices into the CPU, or where data from the CPU are transmitted directly to where it is used. Such operation is in real time.

open-hearth furnace. The open-hearth furnace is rectangular and rather low, holding from 15 to 200 tons (13.6 to 181.6 tonnes) of metal in a shallow pool. It is heated either by gas, oil, or tar, and flames come from first one end and then the other. Waste gases pass through regenerators corresponding to the stoves of the blast furnace. When the flame is reversed, these regenerators, through which the hot gases have just been passing, heat the air and gas entering the furnace, whie those at the other end begin to be reheated by the passage through them of the waste gases. The regenerative chambers are filled with a checkerwork of brick, around which the gas and air pass. Most of the furnaces are stationary, but some very large furnaces are of the tilting type. The basic open-hearth furnace has a thick bottom and walls of sintered magnesite or dolomite, with an arched roof of silica brick. The bottom of the acid furnace is formed from sand. The front of the furnace, facing the charging platform, has from three to seven water-cooled and hydraulically operated doors through which the furnace is charged and melting and refining processes are observed. At the middle portion of the back of the furnace is the tap hole, tightly plugged in the stationary furnaces and loosely closed in the tilting furnaces. The basic furnaces have another hole at slag level, through which part of the first slag formed is removed. It takes from 6 to 12 h to produce steel in these furnaces, depending on the furnace, its age, the fuel used, and the character of the charge.

open-hearth process. In a typical installation the furnace converts pig iron into steel by oxidizing out the surplus carbon, but, instead of using air like the Bessemer converter, the oxygen is supplied by throwing raw iron ore into the furnace. The oxygen from the iron ore combines with the carbon, silicon, and manganese of the bath, thus lowering the carbon in the pig iron and, at the same time, reducing the iron ore to metallic iron. The resulting products mix to form an average composition. In the open-hearth process air and fuel are passed through a honeycomb of hot firebrick called checkers, which preheats the air and fuel so that they are ready for combustion when they enter the hearth. At the same time the products of combustion pass through the checkers at the other end of the furnace. The furnace is raised on a stiltlike structure (bricked in) with the charging platform, at the rear, also raised so that the charge may be put into the furnace. The melt is tapped off the front into large ladles. The chemical composition of the end product depends on the lining, and the charge, and the control impurities added during the melt or after the melt has been tapped off into the ladle. In the basic (magnesite) lined furnace, the charge consists of pig iron, limestone, and scrap iron. The limestone forms a slag, which combines with the oxygen in the air to remove impurities. The lining plays a major role in control of impurities. Oxygen is one of the most important elements used in the reduction of the molten metal. Rust, scale, slag, and limestone are some of the sources of oxygen. Oxygen is introduced into the furnace with oxygen lances through the roof of the furnace.

open-loop control. Control achieved by driving control actuators with a sequence of preprogrammed signals without measuring actual system response and closing the feedback loop.

open-seam tube (see tube)

operator platform. A platform or area from which a standing person controls the functions of a lift truck or other material-handling device.

optical code reader (OCR) (see bar code reader)

orange peel. Surface roughening on formed products resulting from the use of coarse-grained material.

order picking. The selection of less-than-unit-load quantities of material for individual orders.

organic halogen compounds. The earth's crust contains inorganic halides of every kind. There are relatively few naturally occurring organic compounds in which halogen atoms are covalently bonded to carbon. Many organic halogen compounds are synthesized simply to be used as chemical reagents. Others are prepared as useful commodities in their own right and are employed as propellants in aerosol sprays, as refrigerants, insecticides, herbicides, fire extinguishing agents, dry cleaners, and plastics. The double and triple bonds in the olefins and acetylenes are the structural features that bestow a characteristic reactivity upon these compounds. All organic halogen compounds may be classified either as aliphatic or aromatic. If aliphatic, they are alkyl halides that may be classed further according to structure; they are either primary, secondary, or tertiary alkyl halides. The classification is based on the nature of the carbon atom to which the halogen is bonded. Common names frequently are used for the simpler members of the family. In such names the alkyl group to which the halogen is attached is given first, and this is then followed by the name of the halogen. The halogen (as halide) is a separate word. Alkyl halides of five carbons or greater and the aromatic halogen compounds are named more conveniently as substituted hydrocarbons according to IUPAC rules.

organic polymers. Most of the industrial polymers are plastics and elastomers. Plastics are those synthetic resins characterized as a group by plastic deformation under stress; elastomers or rubbers are those which are capable of extensive elastic deformation. However, there are many exceptions to it. Some plastics are highly elastic and any rubber can be deformed plastically under certain conditions. Plastics themselves are classified into two groups; thermoplastics and thermosets. The thermoplastics, of which polyethylene, polystyrene, and asphalts are examples, can be softened repeatedly with heat and will reharden when cooled. Thermosetting plastics harden or "cure" with heat and then cannot be resoftened by heating. Typical thermosets are Bakelite, epoxy, silicones, and rubber. There are three general methods of polymerization: (1) addition polymerization, (2) condensation polymerization, and (3) copolymerization. In addition polymerization, the polymer is produced by adding a second monomer to the first, then a third monomer to the dimer, a fourth to the trimer, and so on until the long polymer chain is terminated. Polyethylene is produced by the addition polymerization of ethylene monomers. Addition polymerization always results in a thermoplastic. Condensation polymerization occurs in a combination of a compound with itself or other compounds, accompanied by the elimination of some simple compound such as water as a result of the polymerization. Condensation products may be either thermosets or thermoplastics. Copolymerization is the addition polymerization of two or more different monomers. Many monomers will not polymerize with themselves, but will copolymerize with other compounds. One of the more important copolymers is butadiene–styrene, a rubber used in tires.

oscillators. Generators of signals used extensively to originate the various low-frequency, intermediate-frequency, and high-frequency signals required in the operation of electronic equipment. In radio and television receivers oscillators are used in the tuning stages. In transmitters, oscillators are used to generate the fundamental signal that is to be sent for many miles to various receivers. Tape recorders also use oscillators for erasure of recorded tape, and electronic organs employ oscillators for generating the fundamental musical tones. Oscillators are also widely used in radar, electronic computers, and other electronic devices. Some oscillators are employed for the generation of audio-frequency sine-wave signals, while others are used to generate signals having square or rectangular shapes. Oscillators are also designed to produce RF sine-wave signals ranging from a few hundred kilohertz to well above several thousand megahertz. Some oscillators employ resonant circuits, while others use resistance and capacitance combinations for the generation signals having specific frequencies. Certain oscillators have provisions for varying the frequency of the output signal by manual adjustment and others have a fixed frequency that cannot be changed readily without circuit modifications. Generally, all oscillators have low signal output and must be followed by amplifiers to raise the signal level to that desired for practical applications.

oscilloscope. The cathode-ray tube (CRT) is the indicating device for an oscilloscope. The tube consists of an "electron gun" and the deflection plates combined in a vacuum tube with a fluorescent screen on the enlarged end. The purpose of the electron gun is to produce a beam of electrons that is focused to a sharp point at the fluorescent screen. The end of the tube is coated with various phosphors which emit visible phosphorescent radiation at the point of bombardment with electrons. The intensity of the emitted visible light depends on the number of electrons per unit time striking a given area of the screen; the large the number the more intense the light. By partially surrounding the cathode with a metal grid, and applying a voltage so that the grid is negative with respect to the cathode, the rate of electron flow through the grid can be regulated. An oscilloscope circuitry normally includes amplifiers for vertical and horizontal signals, a saw tooth signal generator which may be a relaxation oscillator or a multivibrator, a synchronizing circuit, a power supply, a cathode ray tube, and all the necessary switches and adjustments to provide for a wide variety of applications. An oscilloscope may be used to analyze any type of recurring electrical value and is most valuable in determining where trouble exists in a complex electronic circuit. In tracing the trouble in an electronic circuit, the oscilloscope may be used to study the wave shapes at various points throughout the circuit. When a wave shape is found that does not match what is normal at that point, then the trouble may be located.

output, analog. Not all processes can be controlled with contact closures. Some processes are of an analog nature and require a variable-type control. If the equipment is suitably adapted or manufactured to accept an external control voltage, then a computer-controlled robot with a programmable voltage source could be taught the appropriate settings, and they would become a part of the robot's routine. This greatly increases the versatility of the robot by eliminating the undesirable single or limited multiple set point compromise.

output contacts. Output contacts (switch contacts operated by robot) provide a robot with some control over the application, such as, turning on or off motors, heaters, grippers, welding equipment, etc. Controlling output contacts becomes part of the robot routine. The robot is taught to close (or open) a contact at a particular point in the routine. Whenever that routine is replayed, that contact is closed (opened) at the same point at which it was taught. In addition to turning on or off individual output contacts, computer-controlled robots allow simultaneous operation of several contacts, pulsing of contacts, and "handshaking". The latter term means that after an output contact is turned on or off, the robot waits for a specific input signal to acknowledge that action before performing the rest of the routine. This is especially useful when controlling a tool.

oxide discoloration. Discoloration of the metal surface due to oxidation during thermal treatment.

oxides. Six common oxides are the basic materials for a wide range of industrial products: silica, SiO_2; alumina, Al_2O_3; lime, CaO; magnesia, MgO; iron oxide, Fe_2O_3; and rutile, TiO_2. Iron has two oxides. The red oxide of iron rust is FeO_3, while the black oxide found on hot-rolled steel sheet is Fe_3O_4. Except as iron ore, iron oxide has few industrial uses, its chief one being as a flux in the manufacture of cement. It is a fluxing impurity in clays, minerals, and other deposits. White clays such as kaolin are white because of the absence of iron oxide, and this characteristic makes such clays valuable. The iron oxides are unusual in that they melt at approximately the same temperature as iron, 2800°F (1538°C). Lime and magnesia are bsic (alkaline) in their chemical reactions. Lime is produced from limestone, $CaCO_3$, and magnesia from magnesite, $MgCO_3$. In addition the mineral dolomite is a solid solution of about half limestone and half magnesite. Magnesia is chiefly employed as a furnace refractory in metal melting and cement burning furnaces. Silica is found naturally in its quartz phase, most often encountered as quartz rock, quartz sand, or sandstone. At temperatures above 1600°F (871°C) the phase changes to tridymite, and at 2678°F (1470°C) to crystobalite. The melting point of pure silica is 3130°F (1721°C). Silica may be used at temperatures very close to its melting point, for like carbon, it has the unusual property of maintaining a very high strength almost to its melting point. Alumina has a wide range of uses in ceramic technology such as nickel does in metals technology. The melting point of pure alumina is 3720°C (2049°C). Its many uses include spark plug bodies, ceramic cutting tools for machining metals, grinding wheels, and firebricks. Rutile is one of the more common constituents of the earth's crust. It is a white oxide, used to provide opacity and whiteness to paints and porcelain enamels.

oxychloride cements. Formed by mixing magnesia, MgO, with an aqueous solution of magnesium chloride, $MgCl_2$, to produce a hydrated magnesium oxychloride. Oxychloride cements are hard and strong, but under the action of water there is a tendency for the chlorides to be washed out. These cements are comparatively expensive, hence their uses are restricted. Mixed with sawdust and other ingredients they are used as flooring cements. They are used as binders in the manufacture of artificial stone, grinding wheels, and tiles. Building stones made with these cements are strong and very hard, and can be highly polished.

oxygen. A colorless, odorless, tasteless gas, slightly magnetic, weighing 11% more than air. Its boiling point at atmospheric pressure is −297.3°F (−182.9°C). It is probably the most dangerous of the industrial gases. It is very active at ordinary pressures and even more so at higher temperatures and pressures. It combines violently with oil or grease even at ordinary temperatures. Oxygen is produced mainly from liquid air. Since air is approximately one-fifth oxygen, it can be liquefid, then separated into oxygen and nitrogen, argon, helium, and its other components by a process of rectification. From the separation plant oxygen goes into storage holders and then into cylinders, where it is compressed as a gas and made ready for use. Oxygen is available both as therapy oxygen for hospital use and industrial oxygen in 2400 psi (1.7×10^4 kPa) cylinders and in larger quantities. The two types of oxygen are supplied in cylinders with a different color coding, and a cylinder of oxygen for the one purpose is never permitted to be used for the other purpose, the reason being that industrial oxygen may become contaminated with other gases through the piping system to which a cylinder of industrial oxygen is attached. Liquid oxygen or lox is used as a rocket fuel. Oxygen is also used as a welding gas for both gas welding and certain types of arc welding, and for torch cutting of steels.

P

painted sheet (see sheet)

painting, airless cold spray. Hydraulic force rather than compressed air is used to produce atomization of paint material in airless spray painting. The coating material is pumped at high pressure through a hose to a spray nozzle or tip in an airless gun. As the material leaves the nozzle, the shearing action of the liquid passing at high velocity over the sharp edge of the nozzle orifice causes the material to atomize into tiny particles instantaneously. The momentum of each particle carries it to the surface being painted. Pumping pressure of the material is usually between 500 and 3000 psi (3.45×10^4 and 2.1×10^4 kPa) with most systems operating at about 2000 psi (1.4×10^4 kPa). The degree of atomization and the spray pattern are determined by two variables: size and shape of the orifice in the spray nozzle and pressure of material at the nozzle. Spray nozzles are available in equivalent orifice diameters of from 0.007 to 0.109 in. (0.178 to 2.77 mm) and in spray angles ranging from 5° to 95°. Some nozzles are available with preorifices to obtain a better feather edge of the spray pattern and to help reduce the tendency of the material to spit when the gun is triggered. An airless gun is either full on or full off. Any adjustment to flow rate of material, fan pattern, or degree of atomization can only be made by changing the hydraulic pressure or by changing the spray nozzle. Spray nozzles are usually made of tungsten carbide and are very accurately machined to produce the desired spray pattern. If worn to any degree by abrasive pigments in the paint, spraying results will begin to deteriorate and cannot be corrected except by changing the spray nozzle. In contrast, compressed air spraying has infinite adjustment of fluid pressure and atomizing pressure to adjust for varying conditions in a manufacturing plant.

painting, compressed air cold spray. This technique is known as conventional spraying and has been used to apply almost every sprayable material. The basic elements of cold air spraying are an air compressor, spray guns, paint supply by pressure tank or pump, and hoses. Dozens of gun designs and hundreds of cap/tip combinations offer a great range of flexibility and adaptability to automatic production requirements. Compressed air is used to feed paint from a tank or a pump to the gun nozzle. A separate line feeds air to the nozzle. When the gun is actuated, the material and air streams join together either inside or outside the gun nozzle where the stream of air atomizes the material into tiny particles of paint. The velocity of the air stream carries the material to the surface to be painted.

painting, curtain coating. Curtain coating is a modern, high-speed production technique for applying smooth films of lacquers, paints, varnishes, adhesives, and other materials to flat, curved, concave, fluted, or molded surfaces of wood, metal, and plastic products. It uses the flow coating principle of conveying the product through a continuous falling stream of coating material. However, the stream is accurately controlled in thickness and rate of flow so that there is no excess material runoff from the product. Two methods are used to accomplish curtain coating. In one method, material is pumped from a reservoir to a coating head where it is uniformly distributed and allowed to flow over a weir or dam. The material then drops from the skirt of the weir by gravity an falls either onto the product or into a gutter for return to the reservoir. The product is conveyed through the curtain by a flat bed conveyor which extends horizontally away from two sides of the gutter. In a second method, material is pumped from a reservoir to an air-tight coating head, which has an adjustable orifice across its V-shaped bottom. The width of the orifice opening, the speed of the pump, and the viscosity of the coating material control the pressure of an air cushion formed in the upper part of the coating head. When the pressure is greater than atmospheric, the curtain of coating material is forced through the orifice at a velocity greater than that caused by gravity alone. When the pressure is less than atmospheric, the curtain of material is restrained.

painting, dip coating. Dip coating is used to apply prime or finish coats to products that require thorough coverage. It is capable of producing a smooth, reasonably uniform coating for a wide variety of products ranging in size from small parts of complex shape to automobile frames and bodies. In most applications of dip coating, an overhead conveyor is used to carry the products into and out of a tank containing the coating material. Dip coating can also be accomplished by raising and lowering the dip tank to and from the products, by raising and lowering the level of material in a stationary tank, and by lowering and raising groups of polelike products in and out of a vertical dip tank. After the products have been dipped, they are withdrawn at a predetermined rate and allowed to drain excess material under controlled conditions. The drained material can be reused in the painting system.

painting, electrocoating. Electrocoating is a relatively new dipping technique for applying organic coatings through the use of electrical energy. The process is known by many other names such as electrophoresis, electropainting, electrodeposition, electric dipping, and anodic hydrocoating. Basically, the product to be coated is immersed in a specially formulated, low solids content, water-soluble paint. The product is connected to a direct current power supply and is generally made the anode. The cathode is the reservoir or electrodes in the tank. When voltage is applied (usually 50–500 Vdc with a current density from 2–5 A/ft^2), resin and paint pigment migrate to the product and form a uniform film. Coating time varies depending on product configuration, film thickness desired, pH of solution, temperature, and solids content of bath. It may take 15 s to coat a flat panel or up to 3 min for complex enclosed products. When the product is withdrawn from the bath, it is covered with a uniform coating that has been irreversibly deposited. The film is virtually dry and well adhered to the product. It could be scratched off, but would not leave paint on a finger touching the film. The low solids content of the bath ensures that runs, sags, and heavy edges are almost nonexistent. The product is rinsed to remove any undeposited particles of paint. Drain-off of water is rapid, and the product may be baked immediately or may be overcoated wet-on-wet with a conventional paint and then baked.

painting, electrostatic powder spraying. This technique is similar to electrostatic spraying except a dry powder is used. The product to be coated must have adequate electrical conductivity to serve as a collecting electrode for charged powder. Products may be preheated or not, depending on type of powder used and film thickness desired. Powder does not fall off prior to fusing or curing since the electrostatic charge bonds the powder to the product, despite handling vibrations or air currents. Powder coating materials are available in a variety of colors. Organic powders most commonly used for electrostatic powder spraying are epoxy resins, polyethylenes, nylons, polyvinyl chlorides, cellulose acetate, Teflon, and polyesters.

painting, electrostatic spray. Electrostatic spray painting is based on the principle that unlike electrical charges attract each other. At the point of atomization, the paint material is given a high negative charge. By electrically grounding the product, the charged material is attracted to its bare or primed surfaces. There are many different techniques used in commercial equipment to produce an electrostatic spray. Basically, atomization of material in an electrostatic gun is achieved by compressed air (conventional spray), by hydraulic force (airless spray), or by rotating disks or balls. In compressed air and airless guns, the atomized material is usually charged by being exposed to a high voltage needle extending through the nozzle orifice. Application techniques are similar to those discussed under conventional and airless spraying. In guns tht use a rotating disk or ball to atomize the material and apply the electrical charge, paint is fed to the disk or ball and atomizes from the periphery of the device under the influence of an electrostatic field. If a flat disk is used, conveyorized products must circle around the disk to receive a full coating. If a rotating ball is used, the products can be conveyed past it in a straight line. Charging voltages for electrostatic systems vary from about 60,000 to 150,000 V with currents limited to milliampere levels. Some electrostatic systems are adaptable to spraying heated materials which permit heavier film builds per coat.

painting, flow coating. Flow coating is an automatic operation in which products to be painted are conveyed through a chamber equipped with low-pressure nozzles that completely flood the products with paint. Depending on the configuration of the product, the nozzles can flood downward from the top and sides of the chamber, can flood upward from an H-pattern in the bottom, or can flood upward from a bottom manifold tube that oscillates around and reciprocates along its lineal axis. After being flooded, the products are in the same general state as dip-coated products. Vapor drain tunnels are usually required to control drain and flow-out times. Excess paint is collected and reused by the system.

painting, fluidized bed coating. Parts to be coated are preheated in an oven to a temperature above the melting point of the coating to be applied. The parts are then immersed with suitable motion in a fluidized bed or tank. A rising current of air, passing through a porous plate at the bottom of the tank, fluidizes the dry powder material. As the powder particles contact the heated surface, they fuse and adhere. After removal from the fluidized bed, the parts are often briefly postheated to completely fuse the coating or to cure the resin if it is a thermosetting compound. This technique can apply most of the materials mentioned under electrostatic powder spraying.

painting, hot spray. One of the major variables in painting is the change in viscosity of painting material due to temperature variation. For example, if room temperature changes 15°F (8.4°C) between morning and midday, material viscosity may change as much as one-third. Therefore, cold material that is at the right viscosity for spraying in the morning may be too thin by noon. By heating the material, the paint can be delivered to spray guns at a uniform viscosity throughout an operating shift. Hot spray can be applied to both conventional and airless spraying systems. For viscosity control only, the material is heated to about 100°F (37.8°C). For viscosity and hot spray application, the paint is heated from 120 to 170°F (48.9 to 76.7°C), with most applications being in the 150–160°F (65.6–71.1°C) range. Several methods are used to heat the paint material. In air heating, an electrically heated hot air interchanger heats material circulating in coils surrounding the heater unit. The material is then fed to spray guns through a dual hose assembly with paint in the inner hose and hot air in the outer hose. Another arrangement immerses the coils in an electrically heated hot water tank, and heated material is pumped to spray guns through a single hose assembly. In more elaborate systems, a separate hot water heater circulates water to a heat exchanger where cold paint is heated to proper temperature. The material is then pumped through a dual hose assembly to the spray guns. Hot water is constantly recirculated through the outer hose to maintain material temperature.

painting, roller coating. Two techniques are employed in roller coating—direct and reverse. In direct roller coating, the product is coated as it moves between a coating roll and a drive roll. The coating roll may be either on the top or bottom of the product for single coat jobs, or the drive roll may be a coating roll for painting top and bottom surfaces simultaneously. In any case, the coating roll rotates in the same direction as the product moves. In reverse roller coating, the product is driven independently of the coating roll, which can rotate in the same or opposite direction to product travel. Direct coating is generally used for sheet stock, while reverse coating is used for continuous web or strip materials. Both direct and reverse coating use three or more rolls to transfer material from a manually or pump-fed reservoir to the product. In direct coating, the reservoir is formed by the space between two counterrotating rolls having dams at each end. In reverse coating, the reservoir is a coating pan from which material is picked up by a single roll. In either case, after the material is adhered to a roll, it is either transferred directly to the product by the roll or transferred to one or more intermediary rolls before reaching the product. In direct coating, material film thickness is controlled by adjusting the gap between the two counterrotating rolls. In reverse coating, the thickness is controlled by adjusting the gap between the paint pickup roll and its adjacent transfer roll.

painting, trichlorethylene finishing. This technique is based on using room-temperature trichlorethylene as a solvent for black lacquer. Color is obtained by a dye in the resin, so there is no tendency for the pigment to settle out. Finish is a high gloss black. The trichlor solvent evaporates within 60 s, leaving a relatively dry film. Products are baked after coating. The material can be sprayed but products usually are dipped. DuPont has developed another trichlorethylene process offered by several companies. Its process is designed for cleaning, phosphatizing, and painting parts continuously on a production line basis in a single medium of trichlorethylene. In the DuPont process, the vapor degreasing, phosphatizing, and painting is accomplished at the boiling point of the solvent [188°F (86.7°C)]. Either dipping or spraying can be used in the phosphatizing and painting sections of a system. Parts leave the system completely dry when lacquer is used, or ready for a short baking step when curing paints are used. The trichlorethylene is nonflammable, providing plant safety, lower insurance rates, and reduced cost for fire protection equipment.

paints. The basic ingredients of oil-base paints are drying oils, resins, pigments, and volatile solvents. Secondary materials which are sometimes present are driers, plasticizers, metallic soaps, and antioxidants. The oils and resins act as principal film formers; the pigments act as coloring agents and film toughners; the solvents are used to produce a workable consistency. Driers accelerate the rate of film formation. The drier acts somewhat like a catalyst, increasing the rate at which the binder is oxidized. It also increases polymerization and thereby influences the tensile strength and elasticity of the paint film. Chemically, driers are classed as metallic soaps. The organic radical is usually a resinate, linoleate, or naphtheate, and the metal is lead, cobalt, iron, zinc, manganese, or calcium, alone or in combination. Water-base paints are known by a number of names, including rubber-base paints, emulsion and resin-emulsion paints, and synthetic latex paints. The function of the drying oil in oil-base paints is performed by emulsions of styrene-butadiene latex, acrylic resins, and polyvinyl acetate. Vehicle is the name given the mixture of binder and solvent or thinner. Binders are oils which, on drying, leave a thin protective film. The most common binder is linseed oil. Thinners or solvents control viscosity, increase penetration, and permit the paint to flow and form a flat outer surface. The most common thinners are turpentine, solvent naphtha, and volatile petroleum hydrocarbons. Pigments are used to add color and opacity to paints. They also help produce a much tougher and more durable film than would be produced by the binder alone. Pigments increase the imperviousness, strength, and viscosity of the film. Pigments are divided into three groups: prime, or body pigments, color pigments, and extenders. White body pigments include white lead, zinc oxide, lithopone, and titanium dioxide. Other body pigments include red lead, zinc chromate, zinc metal dust, and aluminum.

pallet loader. An automatic or semiautomatic machine, consisting of synchronized conveyors and mechanisms to receive objects from a conveyor and place them on pallets according to a prearranged pattern.

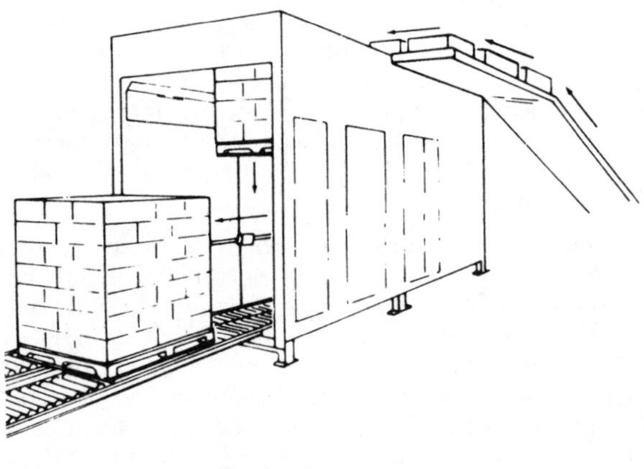

Pallet loader.

pallet truck. A self-loading, low lift truck equipped with wheeled forks of dimensions to go between the top and bottom boards of a double-faced pallet and having wheels capable of lowering into spaces between the bottom boards so as to raise the pallet off the floor for transporting.

panel flat sheet (see sheet)

panel saw. These automatic machines are designed for long and extra-long saw cuts on panels of wood, metal, plastics, rubber, stone products, fiberglass, and fabricated stock. Speed of the saw head is controlled in both directions and cuts can be made in both directions up to 12.5 ft (3.8 m). Cuts can be straight, at angles, at compound angles, and beveled.

paper. Most paper is made from wood pulp. The cellulose of wood is separated from a noncellulosic substance called lignin. The latter is converted into an alkali-soluble substance by treatment with calcium hydrogen sulfide, and then removed from the insoluble cellulose fibers. The cellulose fibers, after removal of lignin, are washed and removed from the mixture as a matting on a large, flat filter. Compression of the matting, followed by drying, produces paper. Additional treatment produces paper products for every purpose. In the manufacturing process, paper is made on a woven wire belt in a long paper machine. The pulp stock is pumped onto the woven wire of the paper machine in the form of a thin flat ribbon from a suitable nozzle. Moisture in the stock is drained through the supporting mesh, after which a series of pressure rolls and a dryer removes further moisture. The paper sheet is then calendered, that is, passed through a series of rolls stacked vertically. The calendering operation controls the thickness and smoothness. Finally the sheet of paper is reeled.

paperboard. Paperboards may be classified according to the type of machine on which they are made; viz. Fourdrinier, cylinder, wet machine, insulating board. Fourdrinier board is a single-ply product formed on a level, moving wire screen. This does not exclude boards made with a secondary head box where a second layer of stock is deposited on the first, these are still considered Fourdrinier board. Shipping case board is an example. Cylinder boards are made up of several plies, each ply being contributed by a cylinder. Usually the top ply (liner) differs in composition from the middle plies (filler) and the back; thus the liner may be made of high-grade pulp and the filler and back of less expensive materials. Wet machine board is formed in the same way as cylinder board, but pressing and drying are carried out in a different manner. On the cylinder machine the sheet moves continuously through presses to the driers. The wet machine winds the newly formed wet sheets on a roll until a certain thickness is reached. It is then cut off, pressed in a stationary press, and hung in an oven to dry. Insulating board may be formed on a cylinder or Fourdrinier. It is pressed continuously and cut into large sheets before drying. The machine differs from the conventional Fourdrinier or cylinder in that the sheet is kept flat until it is dried. The wet-cut sheet is conveyed into a large oven called a Coe dryer, containing a number of decks of rollers (eight decks in most cases) which carry the sheet slowly through. Fans circulate hot air over the sheets, which evaporate the moisture.

paperboard containers. The major types of paperboard containers are folding cartons, rigid set-up boxes, fiber cans, trays, platforms, and cardpacks. Folding paperboard cartons are precision-made, low-cost packages supplied in knock-down form or ''blanks''. When erected, they become three-dimensional rigid packages that meet many requirements of packagers, distributors, and end users. They can be handled by high-speed automatic machinery for companies operating on a mass-distribution scale, by semiautomatic or hand-operated equipment, or by hand for smaller production. Rigid paperboard boxes, once commonly referred to as a set-up box, is a three-dimensional container that is noncollapsible and delivered to a customer for product loading in a fully erect state. Paper cans are made from plies of paper, and often a ply of foil or plastic is combined in the winding of the body to meet specific requirements. The ends can be paperboard, plastic, or metal. The typical shape is cylindrical, but square and oblong convolute wound shapes are available. Wood fiber combined with ingredients to impart wet strength and oil and water resistance is vacuum formed in open molds to produce trays, plates, cups, and formed cushioning. Cardpacks are a combination of printed card with a transparent plastic dome or skin. Such packages are then known as blister and skin packs.

paper, book. Covers a wide group of grades rather than a single one. Defined as any kind of printing paper except newsprint. The distinction between low-grade book paper and newsprint is not a sharp one, but the book paper will normally carry more chemical pulp and some loading materials, such as clay. At the upper end of the scale, book papers contain rag, alpha, and bleached chemical pulps. In between a large volume of papers used for printing magazines and books contain bleached sulfite or bleached sulfate with various percentages of soda pulp and/or deinked waste papers. The latter is likely to contain small amounts of groundwood. While some book papers must be strong to withstand considerable use such as for text reference books, they are primarily made with printing quality in mind, strength being a secondary factor. Book papers are made with great variation in degree of surface smoothness. Antique papers are favored for book printing. Machine finish covers a range of smoothness imparted by the calender rolls of the paper machine. In includes most offset and lithographic printing papers, uncoated magazine papers, and papers for stiff bound books. English finish is the smoothest paper that can be made with machine calenders as the sole finishing equipment. The highest finish secured in uncoated paper is accomplished by supercalendering. Coated book papers consist of a base sheet of the quality of medium-grade book paper but generally with a lower amount of filler and with longer fibers than would be found in uncoated book.

paper, cigarette. These are very thin products and requirements are most exacting. They must not impart off-flavor to the cigarette and must burn at a controlled rate. Strength an opacity are also important. They are generally made from flax fibers, which are pulped and bleached to a high white color. Calcium carbonate is added to control burning rate and sometimes titanium dioxide is included in the furnish to improve opacity. Great care is exercised to see that the paper has no pinholes or other flaws.

paper converting. Converting operations on paper include laminating, coating, or forming into bags. Coating is required for many types of paper to enable the paper to receive the printing ink and to reproduce the printed design with sharpness, fidelity, and attractive appearance. Quality white clay (kaolin) in a starch binder is the usual coating for this purpose. Titanium dioxide is also used. A waxed paper is first coated, printed, and then waxed. Building papers and roofing felts are made of paperboard with asphalt. Lamination provides a paper with a combination of properties or surfaces, or is used to produce a heavier or stronger paper or paperboard. The lamination of paper and aluminum foil is familiar in the packaging of chocolate bars. Many papers are creped, especially decorative papers, sanitary papers such as towelling, and packaging papers. Crepe begins by wetting the paper in a thin sizing solution and passing it through a heated roll. The paper is peeled from the roll with a blade called a creeping knife. The setting of the knife, its contour, the speed of production, and the amount of size, all control the amount of creeping or stretch produced in the paper. Corrugated paperboard is made of three webs of kraft paperboard, the middle one being corrugated on rolls and the outside laminations being coated to it with a suitable adhesive such as starch. The chief types of paper bags are the multiwall shipping bag of the type used to bag Portland cement, and the brown kraft paper grocery bag. These are made on high-speed machines, where they are first formed into tubes, after which bottoming machine closes the bottom.

paper, glassine and greaseproof. These grades of paper are semitransparent and differ from each other in that glassine has greater transparency. They are very dense papers and are resistant to the penetration of greases and oils. They are produced from a high hemicellulose-content pulp with beating until the stock is very slow. Formerly, stone roll beaters were used, but steel bars are now commonly employed. Paper machine operation is normal, but very good formation must be achieved which limits machine speed. Only light calendering is given on the paper machine. The paper is converted into glassine by dampening the paper while rewinding it and allowing it to stand in the roll until the moisture becomes evenly distributed. It is then pressed through steam-heated supercalenders, which impart a glossy surface and increase the transparency. Glassine blisters when held near a flame, because of the sudden escape of water in the form of steam, and this effect is used to determine the quality of the sheet. The high water content is doubtless a factor in resistance to oil and grease. However, glassine and greaseproof are not good barriers to moisture vapor.

paper, newsprint. In newsprint the furnish is largely mechanical wood pulp with some chemical wood pulp. The paper is machine finished and slack sized and has little or no mineral loading. Usually some blue dye is added to change the color of the pulp from yellowish white to bluish white. Newsprint is made in weights in the range of 30–35 pounds (24 × 36–500). By this is meant that 500 sheets measuring 24 by 36 inches weighs 30–35 pounds. Most newsprint has a basis weight of 32 lb (14.5 kg). Newsprint is made at very high speeds on wide machines. A speed of 2000 fpm (610 m/min) has been exceeded and may become common.

paper processes. The kraft or sulfate process uses alkaline caustic soda and sodium sulfide to remove lignin. The pulp produced by this method is a stronger pulp, hence the name "kraft," the German name for "strong." The cooking is done in digesters of much the same construction as those used for the sulfate process. The sulfite and sulfate processes produce chemical pulp. The groundwood process produces mechanical pulp by grinding the wood in the presence of water. Groundwood pulp is produced for newsprint, tissue, towelling, and board paper, and molded packaging materials such as egg containers and paper plates. The raw pulp produced by these methods may require bleaching. In addition the pulp must be put through a beating process, as it is called; actually it is a brushing process. This improves the strength and other properties of the finished paper. The pulp also requires various additives, such as sizing agents, dyes, and clays to improve brightness and printing characteristics.

paper, roofing. This is a class of paper used after saturating with tar or asphalt for cheap roofing and sometimes as an underliner in high-grade roofing. An essential characteristic is saturating properties; that is, some strength and a high degree of porosity. Formerly cotton and wool rags were used for this, but it has now been augmented with specially prepared wood pulp. In the preparation of this pulp, the wood is chipped, soaked in hot water, and ground in an attrition mill under steam pressure. This yields long fiber bundles, which produce paper of good saturating properties. However, this wood pulp is generally used in admixture with the rags.

paper, towelling. In paper towelling the grades vary considerably. Some are made from unbleached kraft pulp only and some from bleached kraft or sulfite. A large volume is made from cheaper pulp, such as unbleached sulfite and groundwood or waste papers. All are dry-creped to increase softness and improve absorbency. A certain amount of strength is necessary, and the better towels are resin treated for wet strength. Absorbency is all-important and where necessary, rewetting agents are used to achieve it.

paper, twisting. Refers to papers that are cut into narrow strips, which are then twisted into twines that are in turn woven into fabrics. These fabrics serve as mesh bags, automobile seat covers, floor rugs, and the like. A certain softness is required to give a smooth twist, although softness is partially achieved by moistening the tape before twisting. One of the outstanding characteristics is unidirectional tensile strength, as the twines require strength only in the linear direction. The paper is therefore made of long strong fibers, sometimes bleached, sometimes unbleached. Some papers are dyed so that color patterns may be achieved in weaving. Some engine sizing is provided and many grades are wet-strength treated.

paper, writing. Normally includes not only paper used for stationary but papers used for documents and records known as bond and ledger. They are made of bleached wood pulps, alpha pulp, and rag stocks. The properties of these papers vary widely, but they mostly have in common brightness, either white or in color, and a good sizing to writing ink. The lower grades of bonds are made on rather large machines, at fairly high speeds. Furnishes are mostly bleached sulfite, rosin size, and alum. Considerable refining is given to the stock to impart stiffness and "rattle" to the finished paper. The higher-grade bonds contain up to 100% rag stock, with bleached sulfite or alpha as the rest of the fiber furnish. Practically all are tub sized with either glue, starch, or a mixture of the two. Some are machine dried, some are air dried, and some loft dried. Rosin size is added at the beater, which helps control the uptake of the tube size. These grades are made on slow-running machines, under the greatest of care to ensure high-quality products. They are generally watermarked. Tablet paper is made in various qualities, but generally resemble the lower-grade bond papers. They are usually made of sulfite pulp, but the cheaper grades may contain some groundwood. Strength and firmness are important and good engine sizing with rosin is necessary to prevent feather (spreading) of writing ink.

paramagnetic materials (electronics). Materials which become slightly magnetized, even though under the influence of a strong magnetic field, as with the diamagnetic materials. The difference between the two, however, is that the paramagnetic materials become magnetized in the same direction as the external magnetizing field. Permeability of the materials is greater than unity, though low compared to the ferromagnetic materials. Paramagnetic materials include aluminum, chromium, manganese, platinum, and air.

parent coil. A coil of sheet that has been processed to the final temper as a single unit and subsequently cut into two or more smaller coils or into individual sheets to provide the required width and length.

parent plate. A plate that has been processed to final temper as a single unit and subsequently cut into two or more smaller plates to provide the required width and/or length.

Pareto's Law. A common phenomenon also known as the ''80/20 Rule'' in which a 20 percent cause produces an 80 percent effect and/or an 80 percent cause produces a 20 percent effect.

partial annealing. Thermal treatment given cold worked metal to reduce the strength to a controlled level. **(See also heat treating.)**

patterned or embossed sheet. Sheet on which a pattern has been impressed on one or both sides.

P&D stations. A location at which a load entering or leaving AS/RS storage is supported in a manner suitable for handling by the S/R machine. (Prior usage has also called this the transfer/station, I/O, pickup and delivery station, feed/discharge station, etc.)

pearlite (metals). Steel containing 0.90% carbon; an important iron alloy. It starts to freeze when the molten solution is cooled to 2690°F (1477°C) and is completely frozen at about 2500°F (1371°C). No change occurs in the austenite until the low temperature of 1333°F (723°C) in the region is reached. This is a minimum in a solid solution comparable to a eutectice in a liquid solution and is called a eutectoid. At this point the gamma turns to alpha iron, and the iron carbide is forced out of solution if cooling is slow. The eutectoid alloy increases in volume on transformation and the resulting formation consists of a series of plates of iron carbide interspersed with plates of ferrite in each grain. This lamellar structure is known as pearlite. Quite slow cooling produces a coarse pearlite, while faster cooling rates cause closer spacing of the plates in what are known as medium and fine pearlite. Hardness increases from coarse to fine pearlite. The interlayers of the two phases in pearlite reinforce each other. Ferrite has a tensile strength of about 42,500 psi (2.9×10^5 kPa) and elongation of 40% and cementite a strength of 5000 psi (3.45×10^4 kPa) and negligible elongation. The pearlite combination has a tensile strength of 120,000–125,000 psi (8.3×10^5–8.6×10^5 kPa) and elongation of 10–15%. **(See also austenitic.)**

pearlite (minerals). A volcanic glass of rhyolitic composition, containing 2–5% of combined water and characterized by a system of concentric, spheroidal cracks called perlitic structure. Most pearlite possesses a property of commercial value, namely, the ability to expand suddenly or "pop" when rapidly heated. The tendency is to apply the term perlite to any glass rock with the capacity to expand greatly on heating. A feature common to most volcanic glasses is an appreciable content of combined water. When grains of crushed perlite are abruptly heated to the temperatures of incipient fusion, their contained water is converted into steam and they form light, fluffy, cellular particles, which are in fact artificial pumice. A volume increase of at least 10 times is not uncommon. Variations in composition of the glass affect softening point, type and degree of expansion, size of the bubbles and wall thickness between them, and porosity of the product. The temperature at which expansion takes place lies between 1400 and 2192°F (760°C and 1200°C) and the time required ranges from a fraction of a second to many seconds. About 70% of the expanded perlite is utilized as aggregate in plaster. Premixed perlite–gypsum plaster and wallboard are manufactured and sold. Fifteen percent of the output is used as aggregate in lightweight concrete, especially for roof decks and prefabricated structural panels. The remainder is put to a variety of uses, among which are loose-fill insulation, filtration medium, soil conditioner, paint filler, constituent of oil-well drilling muds and concretes, and inert packing materials.

performance evaluation reporting technique (PERT). This project management technique was developed by the Special Projects Office of the United States Navy's Bureau of Ordnance. PERT contains a feature which the critical-path technique lacks, i.e., probabilities for the estimates of job durations. However, the PERT System lacks the critical path's cost–time function, which is the most important basis in maintaining complete project control. **(See also critical-path technique.)**

permanent mold casting (see casting)

permeability (electronics). A measure of the conductivity of magnetic flux through a material. Permeability is symbolized by the lower-case Greek letter mu (μ). It is the ratio of the flux that exists when using a certain material to the flux that would be present if air were used instead. The permeability of air is thus considered as unity (one), and all other materials have varying degrees of permeability above one. Soft iron, for instance, has better conductivity for lines of force than steel because it has greater permeability. Other materials with high permeability include cobalt, ferrite, and certain alloys. High permeability materials find extensive use as cores of transformers.

PERT (see performance evaluation reporting technique)

petrochemical intermediates. Chemicals such as ethylene, manufactured from petroleum and natural gas, which are further processed into finished products. They are the raw material for most rubbers and plastics, and numberless other chemicals for industry. Petrochemical intermediates may also be manufactured from coal, which is rich in aromatic hydrocarbons. The most important of the intermediates is ethylene. In addition to such products as engine antifreeze (ethylene glycol), many plastics and rubbers are produced form ethylene. The chief feedstocks used for cracking into ethylene are ethane and propane, but butane, naphthas, and gas oils may also be used. In principle, any organic chemical can be made from any fraction of petroleum or natural gas. Other olefin intermediates are acetylene, propylene, butylene, isobutylene, and butadiene. Most of these are used in the production of rubber, among other things. Nylon may be produced from butadiene. Acetylene can be converted into such plastics as polyvinyl chloride and Orlon. Next to the olefins the most important group of intermediates is the aromatics, chiefly the BTX trio—benzene, toluene, xylene. Of the cyclic naphthenes the most significant intermediate is cyclohexane, used in the production of nylon. Insecticides, detergents, solid rocket fuels, photographic films, pharmaceuticals, solvents, explosives, and alcohols are some of the end uses of the olefin, aromatic, and naphthene intermediates mentioned.

petroleum, catalytic cracking. Cracking is the process by which fractions boiling in the gasoline or light naphtha range are produced from heavier fractions. The larger molecules are divided into fragments of lower molecular weight basically by means of heat, but with the assistance of pressure and catalysts. A catalyst is a material that promotes a chemical reaction without being consumed in the reaction. Without the catalyst, excessive amounts of methane and ethane are produced by cracking, that is, the cracking occurs too close to the ends of the chain of carbon atoms. Petroleum hydrocarbons are cracked at temperatures above 650°F (343°C). Higher temperatures are required to crack larger molecules. The resulting cracked product is a mixture with much the same boiling range as the crude from which it is made, containing fractions from gases to tar and coke. The cracking material will contain perhaps twice as much material boiling in the gasoline range, that is, pentane, C_4, and heavier. The cracked material has no fractions suitable for kerosene, lubricating oils, waxes, or asphalts. These products therefore are not cracked; gasoline, diesel fuel, and furnace oil are cracked. The cracked hydrocarbons are more olefinic than the original crude hydrocarbons. If a C_{12} (dodecane) paraffin is cracked into two parts, one part will probably be a hexane paraffin and the other a hexane olefin. Besides olefins, the cracked material will contain aromatics. All these nonparaffinic components improve the octane rating of the cracked product. Straight-run distillate gasolines made of the pentane, hexane, heptane, octane, etc., of crude oil cannot be burned in the modern automobile. A special type of cracking process used to upgrade the quality of gasoline is called reforming. Reforming produces greater amounts of naphthenics and aromatics, converting hexane, for example, into benzene. The products of the reforming process may also be converted into petrochemicals.

petroleum coke. The residue from the destructive distillation of petroleum: it may be found in car engines that "carbon up". Petroleum coke is not pure carbon, but contains up to 16% of volatile petroleum compounds, and is oily in appearance. It is sometimes used as a fuel, although it requires special furnace designs and has a very high ignition temperature of over 1000°F (538°C). Its chief use is in the manufacture of carbon electrodes for aluminum refining and other such metallurgical processes. For blast-furnace operations, coke is made by distilling coal in coke ovens.

petroleum refining. Crude petroleum represents a mixture of a wide range of oil and gas components that must be converted into products of fixed characteristics and standard quality, regardless of variations in characteristics of the crude oil used. The first refinery operation is the separation of the crude into various "cuts" using distillation methods. For flexibility of operation, two other basic refinery processes are needed. First, there must be a process for "cracking" or breaking up the larger molecules into shorter gasoline molecules. Second, there must be a method of polymerizing small gas molecules into longer molecules. The first process is called cracking, and the second polymerization. These processes provide the basic operational flexibility to meet changes in marketing demands for petroleum products and variations of type of crude oil received. Other refinery processes are incorporated with these either to improve product quality or to produce special products such as greases, coke, or chemical intermediates.

phase equilibrium (metals). A phase can be defined as a homogeneous bond of matter existing in some prescribed physical form. Thus the chemical substance water, H_2O, can exist in three different phase forms, solid, liquid, and vapor; while pure iron can exist as four different phases:

Many solid substances are multiphase in character, such as ordinary concrete, which contains cement, sand, and aggregate as identifiable phases in the structure while, on the microscopic scale, the structure of steel consists of two phases, a solid solution—ferrite and iron carbide, Fe_3C. The various types of phases likely to be encountered in metallic alloys can be either pure metals, intermetallic compounds, or solid solutions,

Phases in Pure Iron

Temperature range [°F (°C)] over which the phase is stable at 760 mm pressure	Phase
Above 5558 (3070)	Gas
2797–5558 (1536–3070)	Liquid
2552–2797 (1400–1536)	bcc solid (δ)
1670–2797 (910–1400)	fcc solid (γ)
Below 1670 (910)	bcc solid (α)

for despite the effects of cooling, a solid solution is a single continuous phase that will become homogeneous if it is allowed to reach thermodynamic equilibrium by a process of diffusion. In any multiphase structure two phases have a common boundary separating them. Thus a drop of oil placed in water shares a common phase boundary with the water which surrounds it.

phase (metallurgy). A phase is defined as any substance that has a homogeneous structure bounded by definite surfaces; a heterogeneous system consists of two or more homogeneous systems, and any homogeneous part of a heterogeneous system is called a phase. An example is melting ice, which is a heterogeneous system consisting of two or three phases, i.e., either ice and water, or ice, water, and water vapor, respectively. A system is a combination of components, and the number of components of any system is the smallest number of independent variable constituents that are sufficient to define the composition of each phase under equilibrium conditions in a heterogeneous system. A crystal in the solid state with a definite lattice structure is considered as a phase. A knowledge of the structure of an alloy is helpful in the prediction of its properties, for the properties of a material are dependent on the nature, number, and distribution of the phases that compose it. Phase diagrams, also known as equilibrium diagrams and constitution diagrams, graphically indicate the form the constituents of the alloys take, with respect to the temperatures involved. Such an equilibrium diagram is actually a plot of the composition of the phases as a function of temperature in any system of alloys under equilibrium conditions.

phenolic foam. Reaction-type phenolic foam is an open-celled, brittle thermoset material made by the simultaneous condensation and foaming of a phenolic resin that can be made at densities ranging from 0.2 to 65 lb/ft^3 (3.2 to 1040 kg/m^3). The foam is characterized by high dimensional stability, heat resistance up to 250°F (121°C), and resistance to burning. A mineral acid catalyst is used with a phenolic resole resin in the preparation of phenolic foam. The exothermic condensation reaction produces steam, which expands the polymerizing phenolic resin. Volatile liquids can be incorporated into the formulation as blowing agents to regulate cell size and density. A protective hard phenolic skin can be formed around a core of more flexible, rigid foam. This skin can be formed by lowering the temperature of the mold in which foaming takes place; the cooler the mold, the thicker the skin. Phenolic foam properties are strongly influenced by such structural variables as density, cell size, geometry, and overall polymer composition. A limitation of phenolic foam for thermal applications is partly open-cell structure. Phenolic foams are much more friable than urethane foams, and exhibit relatively high water-vapor transmission and absorption. The thermal conductivity of phenolic foam is low compared to other materials, except for closed-cell materials containing fluorocarbon gas. The outstanding characteristic of phenolic foams is resistance to burning and low smoke emission. Its excellent fire resistance makes phenolic foam more suitable in applications requiring a high degree of fire retardation and shock absorption. One application is glass-reinforced phenolic foam as insulation material in steel containers for uranium hexafluoride. Phenolic foam is useful for form-fitting packaging. The reaction-type phenolic foams can be cut, carved, and shaped with conventional hand tools. **(See also foams.)**

phenolic plastics. The phenolic family of thermosetting materials represents the original engineering plastic. Phenolics are characterized by their ability to deliver predictable properties over a wide range of applications at low cost. Phenol and aqueous formaldehyde are reacted in the presence of alkaline or acid catalysts to produce both liquid and rigid resins. These resins are used to bond friction materials for automotive brake linings, clutch parts, and transmission bands. They serve as binders for wood particle board used in building panels and core material for furniture, as the water-resistant glue for exterior-grade plywood, and as the bonding agent for converting both organic and inorganic fibers into acoustical and thermal insulation pads, batts, or cushioning for home, industrial, and automotive applications. One of the most important uses of phenolic resins is for the production of reinforced molding materials. They have excellent dimensional stability heat resistance superior to most other thermoset or thermoplastic materials, high heat deflection temperatures, outstanding creep resistance, good flame resistance, plus water and chemical resistance.

phenols. Considered as hydroxyl-substituted hydrocarbons of the general formulas *R*—OH, and *Ar*—OH. The hydroxyl group (—OH) is the functional group that characterizes phenols. Compounds that have hydroxyl groups joined to carbon atoms like those found in the aromatic ring compounds are phenols. Phenols are commonly named as derivatives of the parent substance, and simplest member of the family, phenol. Other substituents on the phenol ring are located by number or by ortho, meta, or para designations. Sometimes phenols are named as hydroxy compounds or, as in the case of polyhdroxyl benzenes, as di- or triols. Phenols and other polyhydroxyl benzenes are colorless, low melting solids only slightly soluble in water. Perhaps the most distinctive property of phenols and the one that sets them apart from the alcohols is the acidic character they possess. Phenols thus dissolve readily in hydroxide bases to form phenoxides. Phenol and its related compounds are of great industrial importance. They are starting materials for important pharmaceuticals such as aspirin and adrenalin. Certain derivatives of phenol are important photographic developers. In the plastics industry phenol is used as starting material for bakelite. The herbicides, 2,4-dichlorophenoxyacetic acid and 2,4,5-trichlorophenoxyacetic acid, popularly known as 2,4,-D and 2,4,-T have their origin in phenol.

phenoxies. Phenoxy resins are high-molecular-weight (about 30,000) polyhydroxyethers made by reacting bisphenol A and epichlorohydrin. They differ from epoxy resins since they contain no epoxy groups and are higher in molecular weight. They are thermoplastic resins with a linear structure. Phenoxy has a high rigidity and impact strength. It has very good creep resistance, high elongation, low moisture absorption, and very low gas transmission, and it is also self-extinguishing. It can be injection or blow-moled, extruded, and applied as a coating from solution or in a fluidized bed. Although normally used as a thermoplastic it can be cross-linked with isocynates, anhydrides, triazines, and melamines. Phenoxy resins are excellent adhesives for wood and several metals. They are used in various protective coatings for both wood and metals. In the molded form they are used in electronic parts, sporting goods, and appliance housings. They are extruded into pipe, which is particularly suitable for handling gas and crude oil.

phenylene-oxide-based resins (plastics). A patented process for oxidative coupling of phenolic monomers used in formulating Noryl (General Electric Co.) thermoplastic resins. This family of engineering materials is characterized by outstanding dimensional stability at elevated temperatures, broad temperature use range, outstanding hydrolytic stability, and excellent dielectric properties over a wide range of frequencies and temperatures. Available in eight standard grades, providing a choice of performance characteristics to meet a wide range of engineering applications. Among their principle design advantages are: (1) excellent mechanical properties over temperatures from below −40°F (−40°C) to above 300°F (149°C); (2) self-extinguishing, nondripping characteristics; (3) excellent dimensional stability with low creep, high modulus, and low water absorption; (4) good electrical properties; (5) excellent resistance to aqueous chemical environments; (6) ease of processing the materials with injection molding and extrusion equipment; and (7) excellent impact strength.

phosphor bronze (see bronzes)

photoelectric controls. These systems are used for detecting presence, counting, or actuating operations. Systems consist of four parts; a light source, a photocell, an amplifier, and an output/control device such as a relay. The light source and photocell

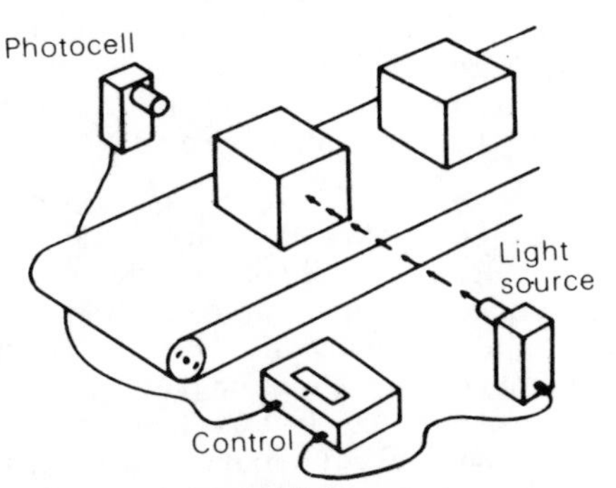

(a) Interrupted-light system using dark-operated counter or control.

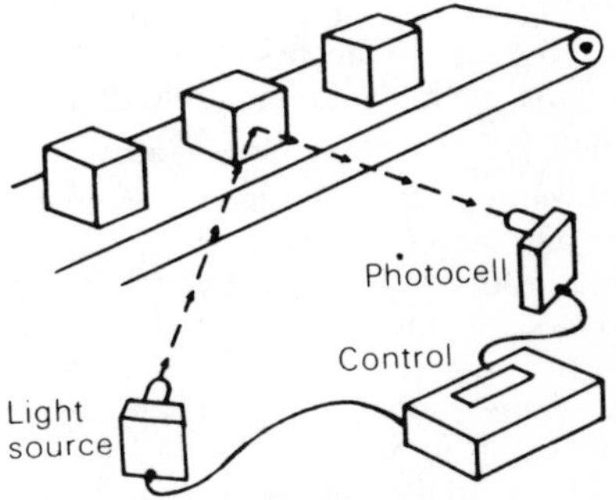

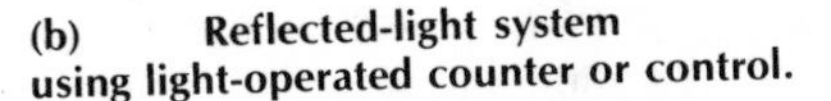

(b) Reflected-light system using light-operated counter or control.

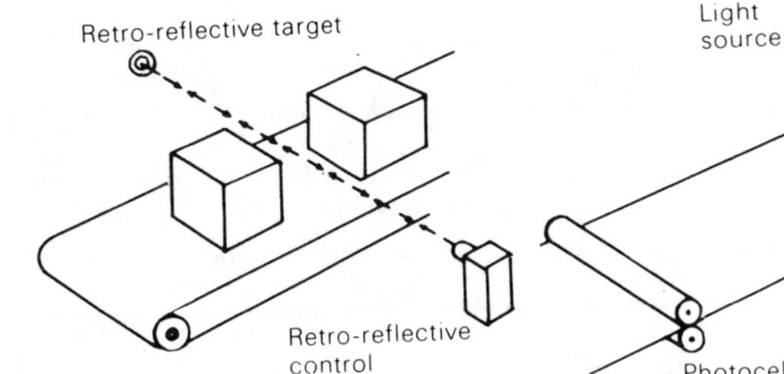

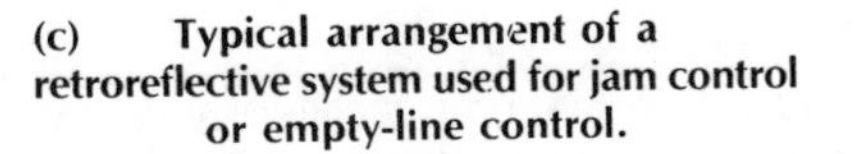

(c) Typical arrangement of a retroreflective system used for jam control or empty-line control.

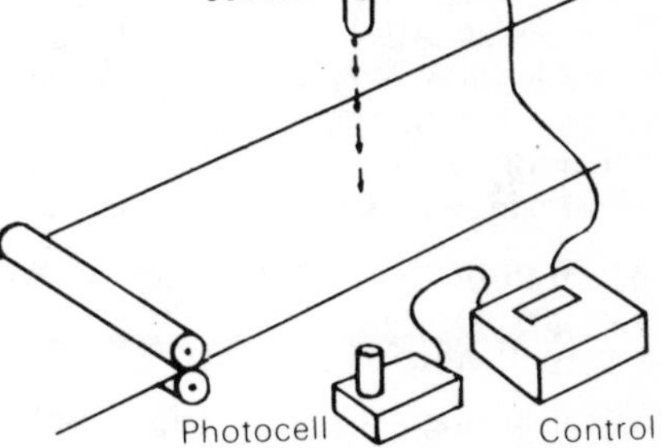

(d) Interrupted-light system using light-operated control to detect break in material.

Courtesy of Scanning Devices Co.

Examples of photoelectric controls.

may be arranged in a number of different ways. The most common arrangement is the interrupted light system in which the light source and photocell are mounted on opposite sides of an object to be detected and the object breaks the light beam as it passes through and triggers subsequent actions. The reflective light system consists of a light source and photocell mounted on the same side as the object to be detected. The retroreflective light system is distinguished by the light source and photocell mounted coaxially; by means of a special retroreflective target, the light beam returns on the same path from which it arrived.

photoengraving. The process of reproducing graphic material by preparing a reverse image on a coated surface photographically and acid etching away the areas not protected by the acid-resistant coating. Removal of the coating results in a relief pattern, i.e., as used for plates in printing and for etched circuit board wiring in electronics manufacturing. **(See also etching.)**

photoetching. Utilizes photoengraving techniques practiced in printing. Parts may be made from flat 0.0001 in. (0.00254 mm) foil up to sheets $^{1}/_{16}$ in. (1.6 mm) thick of some plastics and most metals, including alloys of aluminum, copper, and steel. The sheet is covered with a photosensitive resist on one or both sides and exposed in a camera to an image of the part or parts desired. The coating is developed to expose the lines or areas to be eaten away subsequently in an acid bath. The resist may also be applied by printing methods. Pieces up to 20 in. (50.8 cm) square have been made, but sizes less than 4 in. by 6 in. (10.2 by 15.2 cm) are most successfully fabricated.

photoforming. Basically an extension of the technology of photoengraving, this process utilizes a photoresist mask made by photographic processes to permit selective removal of metal. Metal is removed by etching away the areas of metal not protected by the photoresist coating. The process lends itself to production of parts from very thin, very brittle, or very soft materials. Generally, photoforming is competitive with stamping or blanking operations but is best suited to small or medium-size runs on parts too small or too intricate for successful stamping. Dimensional tolerances are exact and fine detail is easily reproduced in production. Practical quantities may run fom a few pieces to thousands. Parts may be made from flat 0.0001 in. (0.00254 mm) foil up to sheets $^{1}/_{16}$ in. (1.6 mm) thick of some plastics and most metals, including alloys of aluminum, copper, and steel. Pieces up to 20 in. (50.8 cm) square have been made, but sizes less than 4 in. (10.16 cm) by 6 in. (15.2 cm) are most successfully fabricated. **(See also chemical milling.)**

photogrammetry. A relatively new inspection process, based on principles long in use for aerial reconnaissance and mapping. Photogrammetry is the extraction of 3D dimensions from a set of photographs of that object. At least two photographs are taken from two vantage points, and the data observed in each of these two-dimensional photographs are merged and reduced to data of the actual three-dimensional object. That merging and data reduction are carried out by an appropriate computer program.

physical properties. The properties, other than mechanical properties, that pertain to the physics of a material, for example, density, electrical conductivity, heat conductivity, thermal expansion.

physical testing. Determination of physical properties.

picket fence code (see horizontal bar code)

pickling. A method of removing rust, stain, and oxide from various metals by immersion in certain acids. Surface compounds are released when the acid attacks the metal, or when the oxides are soluble in the acid. In the descaling process, an inhibitor is added to the acid to minimize the attack on steel. Copper dissolves readily in cold nitric acid, but is hardly affected by nonoxidizing acids. Steel is pickled in warm 5% sulfuric or hydrochloric acid. Nickel and stainless steel react slowly to pickling solutions, but aluminum, magnesium, and zinc react readily.

	Copper	Steel	Stainless Steel	Aluminum	Magnesium	Zinc	Nickel
Sulfuric acid	●	●					
Hydrochloric acid		●					
Nitric acid	●						
Hydrofluoric acid			●		●		
Nitric and sulfuric acid mix	●	●			●		●
Nitric and hydrofluoric acid mix			●	●			●
Nitric and chromic acid mix					●		
Sulfuric and chromic acid mix				●	●	●	

Acids used for pickling of metals.

piezoelectricity. Analogous to magnetostriction. Certain ceramic materials such as quartz and titanates produce the piezoelectric effect. When a crystal of such material is compressed, a small voltage is generated across the opposite faces of the crystal and conversely, a potential difference across the opposite faces will cause the crystal to expand or contract. Piezoelectricity is the preferred method of instrumentation for recording pressure in engine cylinders and gun barrels. For this purpose a small mounted crystal is screwed into the wall of the cylinder as a spark plug is, and the pressure read out as a voltage on an amplifier.

pig iron. The basic ingredient in the manufacture of ferrous metals and alloys. Pig iron is produced in a blast furnace. Iron ores including hematite (Fe_2O_3), magnetite (Fe_3O_4), and limonite (Fe_2O_3) are used in the manufacture of pig iron. The principal iron ores used are oxides of iron, or ores that may be transformed into oxides by heating. The process of making pig iron consists of the removal of oxygen and impurities from iron ore. The iron ore is reduced to pig iron by heating it with carbon, usually in the form of coke. The carbon combines with oxygen from the iron ore to produce carbon dioxide (CO_2). A flux, such as limestone, combines with the impurities in the ore to form a molten slag. In operating a blast furnace to produce pig iron, the charge is first fed into the top of the furnace. The charge consists of about 60% iron, including scrap iron, 25% coke, or some other high-carbon fuel, and about 15% flux in the form of limestone. The molten limestone gradually works toward the bottom of the furnace, combining with the impurities. Air, preheated to about 930°F (500°C) is forced under pressure through the charge from the bottom of the furnace. The oxygen in the air combines with the coke to form carbon monoxide, increasing the temperature at the bottom of the furnace to about 2910°F (1600°C). As the ore works down to the bottom of the stack, it comes in contact with the carbon monoxide gas, which reduces the iron oxides. That is, the carbon monoxide (CO) removes oxygen (O) from the iron ores to produce carbon dioxide (CO_2). The molten reduced iron ore called pig iron is removed through an outlet at the base of the blast furnace and is then cast into ''pigs'' weighing about 100 lb (45.4 kg). Cast iron is produced by remelting pig iron with coke and heated air in the cupola or air furnace. This process further reduces the pig iron by removing both oxygen and impurities and yields a more uniform product. **(See also acid-Bessemer process.)**

Piobert lines (see Lüder lines)

pipe. Tube in standardized combinations of outside diameter and wall thickness, commonly designated by ''Nominal Pipe Sizes and ANSI Schedule Numbers''.

pipe, seamless. Pipe produced from hollow extrusion ingot.

pipe, structural. Extruded pipe, which may contain an extrusion seam, suitable for applications not involving internal pressure.

pit. A sharp depression in the surface.

pitting corrosion (see corrosion)

pixel (computer). A picture element. A small region of a scene within which variations of brightness are ignored. The pixel is assigned a brightness level that is the average of the actual image brightness within it. Pixels are usually arranged in a rectangular pattern across the scene, although some research has been done with hexagonal grids.

planers, metal. The principle of planing is very similar to that of shaping in that both are reciprocating. On the shaper the tool oscillates over the work and the table hitch-feeds at 90° to the oscillating motion. On the planer the platen (table) oscillates under the tool and the tool-hitch feeds at 90° to the oscillating motion. Planers may have mechanical as well as hydraulic drives. The mechanical planer uses gear arrangements for both the platen drive and the feed. The gears are powered with electric motors. The hydraulic planer platen is actuated with a cylinder (or cylinders). The piston rod is attached directly to the platen. Planers also may be fitted with a tool-block lifting mechanism for raising the tool at the end of the cutting stroke in preparation for the return stroke. Multiple heads for the surface cutting and heads mounted to cut the sides of large workpieces are also used. The most common types of planers are the double-housing planer and the open-side planer, which has one rigid column to support the crossrail and heads. Another adaptation of a planer is the milling planer, which makes it possible to plane and mill at the same time. Pit planers are special machines wherein double columns are mounted on separate beds. The planer table is stationary, but the column stradles the platen. The columns move the length of the bed while the work remains stationary.

plaster, cement. Cement or hard-wall plasters are combinations of partially hydrated gypsum and either certain natural impurities or added agents, such as hydrated lime or clay, in amounts of up to about 15%. These plasters have greater plasticity and longer setting times than plaster of Paris, thus improving working qualities. They have better sand-carrying capacity, and sand and other ingredients that are used for the construction of interior walls. Cement plasters are sometimes mixed with cinders or crushed stone to form gypsum concrete, which is used for floors and roofs. Mixed with sawdust or ground cinders, cement plaster is molded into building blocks or tile. Wall board is composed of laminations of cement plaster and cardboard or wood. This material differs from plaster board, which is made from ''second settle plaster,'' gypsum calcined at 400°F (204°C) or above. Plaster board is given subsequent plaster coatings in construction, whereas wall board is decorated directly, once the joints have been sealed by tape. Gypsum cements also appear in building construction as precast partition tile.

Plaster is applied to walls in more than one coat; hence the plasters used may be classified as basecoat or finish plasters. When three coats are used, they are referred as scratch, brown, and finish coats.

plaster gypsum. The mineral gypsum, $CaSO_4 \cdot 2H_2O$, which is the basic raw material for gypsum plasters, occurs chiefly as rock gypsum. It is a soft material, white or pearly gray in color, and has a specific gravity of about 2.4. Upon calcination either $CaSO_4 \cdot 2H_2O$ or $CaSO_2$ is produced, each with the property of combining with water to reform the original compound as a hardened mass. The setting time of gypsum plasters varies from as low as a few minutes for plaster of Paris to several hours for hard finish plaster. With the use of additives the time of set may be altered. Glue, sawdust, certain acids, and stockyard organic compounds are some of the materials used to retard setting; certain salts accelerate it. Fine grinding and continued mixing with water also decreases setting time. The important properties of strength and workability are both dependent on the water–plaster ratio. Strength increases but workability decreases as this ratio decreases. Additives like hair and fiber tend to increase strength, whereas those used to alter setting time may cause the plaster to become weaker. Workability is increased with finer grinding of the plaster. Other important properties of gypsum plasters are firmness, lightweight, and resistance to fire.

plaster, hard-finish. Hard-finish plasters, including flooring plaster, are obtained by calcination of gypsum at red heat temperatures, well above 400°F (204°C). The gypsum is almost completely anhydrous and must be calcined in lumps if it is to retain its ability to recombine with water. The hardening time of such plasters is from 1 to 4 h and accelerators may be used to reduce the setting period. A hard-finish plaster forms a very hard surface, which may be ground and polished, and is used for both walls and floors. Keene's cement, of which there are several commercial varieties, is a type of hard-finish cement that is made by adding alum to the calcined gypsum and recalcining. Mack's cement is similarly made, sodium or potassium sulfate being the additive. Heating finely ground gypsum at temperatures above 900°F (480°C) drives off all water of crystallization and leaves a product that does not have the ability to recombine with water. This finely divided material is known as "dead-burned" gypsum and is used as a filler for plastics, rubber, paper, paint, and other products.

plaster mold casting (see casting, plaster mold)

plaster-mold casting molds. Mold material is, basically, gypsum plaster with a small amount of talc added, to prevent cracking, and other materials, such as terra alba or magnesium oxide, to reduce the setting time. Lime or cement may be added to control expansion during baking. Often about 25% of fibers are added to increase the strength. When the wet mix is poured around the pattern and allowed to harden, great accuracy of detail and dimensions is obtained. The mold is baked in an oven for a number of hours to remove all of the moisture.

plaster of Paris. Pure half-hydrated gypsum is known as plaster. Because it sets in from 5 to 15 min after the application of water, it is not suitable for construction purposes. It is used for architectural ornaments and as a casting material, since it expands while hardening. It is finding increasing use in the precision casting of metallic parts. Plaster of Paris is also used for molds in the ceramic industry. Stucco is practically the same material as plaster of Paris, but is not as pure or as finely ground.

plastic deformation. Many materials when stressed beyond a certain minimum stress, show a permanent, nonrecoverable deformation. This is called plastic deformation, and is the result of permanent displacement of atoms or molecules or groups of atoms and molecules from their original positions in the lattice. The displaced atoms and molecules do not return to their original positions after the removal of stress. If the material subjected to a constant load of a sufficient magnitude shows a continuously increasing deformation, the phenomenon is called flow. The flow is a characteristic phenomenon of liquids and gases that deform immediately when subjected to a shearing stress. The mechanism of plastic deformation is essentially different in crystalline materials and amorphous materials. Crystalline materials undergo plastic deformation as the result of slip along definite crystallographic planes, whereas in amorphous materials plastic deformation occurs when individual molecules or groups of molecules slide past one another.

plasticizers. High-boiling-point liquids, usually organic esters, or low-melting-point solids, such as camphor, which are added to plastic polymers to modify their physical properties. Originally as the name implies, their primary purpose was to render the resin mixture more plastic. They are frequently necessary in thermoplastic resins to produce improved flexibility in such end products as vinyl raincoats or hose. Sometimes they are necessary in both thermoplastic and thermosetting plastics to promote better molding properties, such as improved flow, during fabrication. Solvents may be used in the initial compounding of thermoplastic molding compounds or in the formulation of surface-coating products such as lacquers. Primary plasticizers can be used alone without incompatibility. Secondary or extender plasticizers must be used with a primary plasticizer. Copolymerization is sometimes referred to as an internal plasticization since the introduction of the second monomer, even in small quantities, may make considerable change in the properties of the finished compound. In addition to improving flexibility, plasticizers may modify one or more of the following characteristics: elongation, tensile strength, toughness, softening point, melting point, heat sealing characteristics, resistance to flow under pressure, flame resistance, electrical properties, water absorption, oil resistance, abrasion resistance, and toxicity. Several phthalates, including the dimethyl, dibutyl, dioctyl, butyl benzyl, and diisodecyl, are used in thermoplastic materials to improve flexibility. Triphenyl and tricresol phosphates are used for their flame-proofing properties. Among other plasticizers are camphor, vegetable oils, adipates, sulfonamides, phthalyl glycollates, ortho nitrobiphenyl, butyl oleate, and methoxyl ethyl stearate.

plastics. Synthetic plastics are chemical compounds produced from such basic chemical elements as carbon, hydrogen, nitrogen, and oxygen. These atoms, when combined in specific ways to build long chains of molecules, result in a solid substance called a polymer. (Other chemicals, notably chlorine, fluorine, silicon, and sulfur are also important for producing some polymers.) By using these few chemicals as basic building blocks, chemists are able to fashion giant molecules in an unlimited number of combinations. At least 500,000 substances based on carbon compounds are known, but only a few are capable of forming polymers of commercial significance. Chemical names used to describe most synthetic plastics are often preceded by the prefix poly to indicate particular polymers. Some polymers are familiar to the public (e.g., polyester, polyethylene, polystyrene, and polyurethane); most are not. Seldom does a polymer reach the marketplace in its natural state. It may be modified to make it more flexible or self-extinguishing, or to have improved weathering. Plastics are divided into three major categories: (1) Thermosetting types. These are plastics that undergo chemical change during processing to become permanently insoluble and infusible. Phenolic, amino, epoxy, and polyester resins are typical. They are also called thermohardening plastics. (2) Thermoplastic types. These are resins that soften repeatedly when heated and harden when cooled. Most thermoplastics are soluble in specific solvents and support flame to some degree. Softening temperatures vary with polymer type and grades. Nylon, acrylic, vinyl, and cellulose acetates are typical thermosoftening plastics. (3) Elastomers. Materials that can be stretched repeatedly to at least twice their original length (at room temperature) and which will return with force to their approximate original length when an immediate stress is released. Polymers ranked as elastomers include modified thermoplastics as well as natural and synthetic rubbers. Most elastomers attain their properties by a process of vulcanization that renders them incapable of reprocessing, as is possible with thermoplastics. Examples of elastomeric polymers are silicone, urethane, nitrile rubbers, and chlorinated polyethylene. **(See also specific types.)**

plastics, resin alloying. Combining plastics resins to achieve new property balances is referred to as interpenetrating polymer networks (IPNs). The first product of IPN technology, Shell Chemical Company's Kraton styrenic thermoplastic rubber, is used to "compatibilize" blends of two or more otherwise incompatible polymers.

plastics, reinforced. The term reinforced plastics, or composites, indicates that a resin is structurally or otherwise upgraded by incorporating fibers into a resin matrix. Laminates in which fibrous felts or woven goods are superimposed in either flat or contoured design belong in this category. Also included are processes in which chopped strands are laid down in jackstraw arrangement or in which continuous filaments are found in a predetermined pattern or joined in parallel configurations. Compression and injection molding play an important role in mass production of many reinforced plastics.

plate. A rolled product rectangular in cross section and form of thickness 0.250 in. (6.35 mm) or more, with either sheared or sawed edges. **(See also alclad plate.)**

PLC (see programmable logic controller)

plywood (see veneer and plywood)

polishing. A mechanical finishing operation for the purpose of producing a gloss or luster on the surface of a product.

polyaldehydes. One type of plastic material, polyaldehyde, is an unbranched polyoxymethylene chain formed by polymerizing formaldehyde in an inert solvent such as hexane with a catalyst such as an amine or cyclic nitrogen-containing compound. Since such a chain with a terminal OH radical would be unstable, terminal methyl or acetyl radicals are inserted instead. Another type is produced by the reacting of vaporized trioxane and a gaseous catalyst such as boron trifluoride and cooling. About 2% of a second monomer such as ethylene oxide is copolymerized into the chain for better stability. The polyacetals are thermoplastics and can be formed in injection molding, blow molding, extrusion, and machining. They have good tensile strength and creep resistance over a considerable range of temperature and humidity and withstand continuous flexural loading. They have good electrical properties and excellent chemical resistance. The acetals are characterized as "engineering plastics," which can be used to replace die-cast and stamped metals. Applications include high-speed gears, pump impellers, truck bearings, automobile instrument cluster housings, showerheads, aerosol containers, automotive window cranks, and refrigerator door handles.

polyallomer copolymer. Two monomers are polymerized in the presence of catalysts to produce polyallomer copolymers. A typical propylene–ethylene polyallomer is a crystalline copolymer containing polymerized segments of both propylene and ethylene. Infrared spectra indicate that the polymer chains exhibit crystallinity normally associated only with the stereo-regular homopolymers of propylene and ethylene. Variations in the basic propylene–ethylene polyallomer can be made resulting in polymers that exhibit different physical properties such as high stiffness, medium impact; moderately high stiffness, high impact; and extra impact. Special formulations that have added heat endurance or added resistance to ultraviolet radiation, as well as formulations that can legally be used in contact with food under regulations of the FDA, are available. Good thermal properties permit the use of polyallomers in the production of articles that are useful over a wide temperature range. Packages for frozen food as well as heat-sterilizable containers can be made with polyallomer plastics. Although the surface hardness of polyallomer plastics is slightly less than those of polypropylene, resistance to abrasion is greater. They have outstanding flow characteristics, excellent moldability, high softening point, good surface hardness, excellent stress crack resistance, and low mold shrinkage.

polyamides. Polymide resins are known by the generic name "nylon." For the most part, nylons are produced by the interaction of adipic acid and hexamethylene diamine. The resulting polymerization product is polyhexamethylene. This is better known as adipamide-type nylon. A second type called polymide is produced by th condensation of amino carboxylic acids or their derivatives. The nylons are definite compounds and melt without decomposition. They are crystalline in structure, and mechanical working leads to high tensile strength through crystal orientation. It is this property that makes extruded nylon filaments so strong. The filaments are manufactured into textiles, hosiery, rope, and bristles.

poly (amide–imide). Amide–imide polymers are based on the combination of trimellitic anhydride with aromatic diamines. In the amic acid form, poly (amide–imide) can be dissolved in a number of polar–solvent systems, applied to a substrate, and thermally cured to insoluble heat-resistant coatings. Molding resins can be produced by thermally curing and modifying amide–imide polymers. Amide–imide polymers have found wide commercial acceptance as high-temperature magnet wire enamels, high-performance decorative coatings, laminates, and adhesives. Amide–imide wire enamels have received an IEEE 57 thermal stability rating of 437°F (225°C) for 20,000 hr and exhibit excellent resistance to thermoplastic flow, heat shock, and burnout. Decorative coatings exhibit excellent resistance to high temperatures and can be compounded in a variety of colors. Amide–imide laminating varnishes can be used to produce thermally stable void-free glass laminates, and adhesives based on amide–imide polymers exhibit high-temperature lap shear strengths. Solid parts can be fabricated from Torlon (a registered Amoco trademark) poly (amide–imide) molding resins by compression molding, transfer molding, extrusion, and injection molding. Parts molded from amide–imide resins exhibit high tensile, compression, and flexural strengths and maintain these properties at high temperatures. They are resistant to aliphatic, aromatic, and chlorinated or fluorinated hydrocarbons as well as many acidic and basic systems. However, physical properties are generally degraded by high temperature caustic systems.

polyaramid fiber. A "miracle fiber" developed by Du Pont that is five times stronger than steel of the same size. Under the trade name Kevlar, this fiber is used as a metal substitute in aircraft, tires, bulletproof vests, etc.

polyaryl ether. A unique combination of three important properties makes polyaryl ether an outstanding plastics material. The material has (1) a heat deflection temperature that is exceeded only by two other engineering plastics, 300°F (149°C), measured at 264 psi (1820 kPa) fiber stress on a 1/4 in. (0.64 cm) by 1/2 in. (1.3 cm) by 5 in. (12.7 cm) bar; (2) the highest impact strength of all engineering plastics except polycarbonate in thin sections, Izod impact strength of 8 ft-lb/in. notch at 72°F (22°C) and 2.5 ft-lb/in. notch at −20°F (−29°C); (3) excellent chemical resistance, resisting organic solvents except chlorinated aromatics, esters, and ketones. Fabricating advantages include a low shrinkage factor that permits precision molding of void-free parts. Modified polyaryl ether resists the tendency toward sinks and even thick sections can be molded without voids or sinks. It is not moisture sensitive and requires only nominal drying before molding. It can be solvent welded, chrome plated, and painted with commercial industrial finishes without pretreatment or priming.

polyaryl sulfone. Available in both pellet and powder form. Its thermoplastic nature allows it to be fabricated by injection molding or extrusion techniques on standard commercially available equipment. Consists mainly of phenyl and biphenyl groups linked by thermally stable ether and sulfone groups. It is distinguished from polysulfone polymers by the absence of aliphatic groups, which are liable to oxidative attack. This aromatic structure gives it resistance to oxidative degradation and accounts for its retention of mechanical properties at high temperatures. The presence of ether oxygen linkages give the polymer chain flexibility to permit fabrication by conventional melt processing techniques. Polyaryl sulfone is amorphous rather than crystalline. Characterized by a very high heat deflection temperature, 525°F (274°C) at 264 psi (1820 kPa). The resistance to oxidative degradation is indicated by the ability of polyaryl sulfone to retain its tensile strength after 2000 hr exposure to 500°F (260°C) air-oven aging. Polyaryl sulfone has good resistance to a wide variety of chemicals including acids, bases, and common solvents. It is unaffected by practically all fuels, lubricants, hydraulic fluids, and cleaning agents used on or around electrical components. High polar solvents such as *N,N*-dimethylformamide, *N,N*-dimethylacetamide, and *N*-methylpyrrolidone are solvents for the material. Polyaryl sulfone has been successfully used for electrical connector bodies that are required to meet severe structural and environmental conditions. The material's unique properties enables construction of one material rather than bonding together several materials, and production in a mono-block that provides the basis for connector designs. Other applications include retention members for electrical connectors, integrated circuit test sockets, high-temperature bobbins used in relays and solenoids, and sleevings for high-temperature solderless terminals.

polycarbonates. Considered to be special types of polyesters in which groups of dihydric phenols are linked through carbonate linkages. They can be produced by a variety of methods. Additives may be used to enhance the thermal, color, and ultraviolet stability of the polymer, while fatigue and some mechanical properties may be improved by glass-fiber reinforcement. Polycarbonates have high impact strength, water-clear transparency, excellent creep resistance, wide temperature limits, high-dimensional stability, good electrical properties, and self-extinguishing behavior. Other advantages are nontoxicity, chemical resistance, colorability, abrasion resistance, hardness, and rigidity despite their ductility. They have high impact strength. Izod impact values on $^1/_8$ in. (3.2 mm). ASTM standard bars average 16 ft-lb/in. of notch. On $^1/_4$ in. (0.64 cm) by $^1/_2$ in. (1.3 cm) bars, the Izod value is between 2 and 3 ft-lb/in. of notch. They have exceptionally wide temperature limits and low, uniform shrinkage from mold dimensions allows design and production of precision parts at tolerances only polycarbonate can supply.

polychlorotrifluoroethylene (CTFE). Replacing a fluorine atom with chlorine results in the formation of the fluorplastic CTFE. This thermoplastic is extruded and injection molded on conventional equipment and is also compression molded. It is characterized by chemical inertness, high-temperature resistance, and low-temperature flexibility. CTFE is transparent in thickness to 0.125 in. (3.2 cm), becoming more translucent as sections increase. In transparent sheets, it is used for windows in infrared tracking missiles, flexible printed circuits, and instrument faces. Extruded over wire it is used in computer circuits, chemical processing equipment, and other applications where thin-wall insulation, corrosion resistance, or low-temperature flexibility is required.

polyester. Commercial unsaturated polyester resins are comprised of the following essential ingredients: a linear polyester resin, a cross-linking monomer, and inhibitors to retard cross-linking until the resin is to be used by the fabricator. The linear polyester is typically the condensation product of an unsaturated dibasic acid as the source of α, β-ethylenic unsaturation and a glycol. Usually a saturated diabasic acid is employed with the unsaturated acid to modify the degree of unsaturation and thereby the reactivity of the resulting resin. The unsaturated intermediates are commonly maleic anhydride and fumaric acid; the saturated acids are phthalic anhydride, isophthalic acid, and adipic acids; and the glycols are propylene glycol, ethylene glycol, diethylene glycol, and dipropylene glycol. The usual cross-linking monomers are styrene, vinyl toluene, methyl methacrylate, methyl styrene, and diallylphthalate. Conventional inhibitors are hydroquinone, auinone, and *t*-butyl catechol. These basic resins can be polymerized into infusible materials by the action of free radicals. The usual free radical source is a peroxide catalyst. Metallic soaps and/or tertiary amines are added to polyesters to act as catalyst activators at room temperatures, the combination of methyl ethyl ketone, peroxide and cobalt octoate, and benzoyl peroxide and diethyl aniline, being commonly employed. Polyester resin is very versatile. Simulated wood door fronts, drawer fronts, trim elements, lamp shades all finish like wood are made with inexpensive tooling. Other end uses are lamp bases, wall plaques, and simulated slate table tops. Polyester is also used for manufacturing bathroom components such as shower stalls and unitized tube segments. By layup and sprayup techniques large and/or short run items are fabricated such as boats of all kinds, dune buggies, custom auto bodies, truck cabs, motor homes, housing modules, concrete forms and playground equipment.

polyethersulfone. A high-temperature thermoplastic with long-term resistance to creep at temperatures up to 302°F (150°C). Capable of being used continuously under load at temperatures of up to about 356°F (180°C) and in some low-stress applications up to 392°F (200°C). Polyethersulfone consists essentially of a polymer of repeat unit. The structure gives an amorphous polymer that possesses only bonds of particularly high thermal oxidative stability. While the sulfone group confers high-temperature performance, the ether linkage contributes toward practical processing by allowing mobility of the polymer chain when in the melt phase. Polyethersulfone has a tensile strength of 12,200 psi (8.4×10^4 kPa) and an Izod impact strength of 1.6 ft-lb/in. notch. At 356°F (180°C) it has a tensile strength of 9400 psi (6.5×10^4 kPa). Exhibits low creep, and is especially resistant to acids, alkalies, oils, greases, and aliphatic hydrocarbons and alcohols. It is attacked by ketones, esters, and some halogenated and aromatic hydrocarbons. Retains good electrical properties at temperatures in excess of 392°F (200°C).

polyethylene (PE). Ethylene polymers are available as hundreds of compounds of widely different structures: copolymers, polymer mixtures, filled compositions, cross-linkable products, ect. As a general class of materials, PE is noted for (1) outstanding dielectric properties; (2) excellent chemical resistance to solvents, acids, and alkalies; (3) toughness; (4) good barrier properties; and (5) their relative adaptability to various processing techniques. The main polymer structural features that influence the properties of PE are: (1) the degree of crystallinity, (2) average molecular weight, and (3) the molecular-weight distribution. Polyethylene is partially crystalline and partially amorphous, and the percentage crystallinity has a marked effect on physical properties. Side-chain branching is the key factor that controls the degree of crystallinity. The branching in low-density PE (LDPE) homopolymer occurs primarily as ethyl and butyl groups (short-chain branching) at a concentration of about 10–20 branches/1000 carbon atoms. Some long-chain branches also are present. When a low-pressure process is used for LDPE, essentially all branches are short chain, either ethyl or butyl, with the result that processing characteristics and some physical properties are different for the high-pressure and low-pressure types. Polyethylene consists of a mixture of molecules of varying sizes; consequently, their structure must be described in terms of size (molecular weight) and distribution of sizes (molecular-weight distribution). The properties of PE can be adjusted by the use of additives such as antioxidants and ultraviolet light stabilizers. The various classes and types of antioxidants used in P.E include phenols, amines, and phosphites. The degree of crystallinity influences the fabricating properties of PE. Polyethylenes and their copolymers are essential raw materials in such major industries as packaging, appliances, transportation, communication, electric power, construction, and housewares. Some major uses of LDPE film are food packaging, garment bags, construction coverings, agricultural film, refuse bags, and shipping bags.

polyimide. Typical attributes of polyimides include high heat resistance [500°F (260°C) and up], good mechanical properties, wear resistance, low friction, chemical inertness, low outgassing, radiation and cryogenic temperature stability, and inherent nonflammability. Two major classes of polyimides are in commercial production: condensation resins and addition resins. Condensation polyimides are predicated on the reaction of an aromatic diamine with an aromatic dianhydride. The resultant tractable polyamic acid is converted during cure to the infusible and intractable polyimide with loss of water. Commercial products derived from condensation polyimides include DuPont Vespel molded parts and diamond abrasion wheels, DuPont Pyralin laminating varnishes, DuPont Pyre M1 wire enamel, and DuPont Kapton insulating film. Polyimide parts are dimensionally stable. Creep is almost nonexistent, even at high temperatures, and coefficients of thermal expansion are closely matched to those of metals (5 to 10×10^{-6} in./in./°F for glass- and graphite-filled material). Polyimides intrinsically possess good wear resistance and low coefficients of friction. Polyimide objects are resistant to dilute acids and organic solvents. They are unaffected by exposure to aromatic and aliphatic hydrocarbons, esters, ethers, alcohols, Freons, hydraulic fluids, JP-4 fuel, and kerosene. Polyimides are attacked by dilute alkali and concentrated inorganic acids. Exposure to steam or water at 212°F (100°C) decreases tensile and flexural strength. Prolonged boiling in water results in a maximum water absorption of 2.5% with accompanying dimensional changes of up to 0.006 in./in. Polyimide moldings are nonflammable and generate low smoke on exposure to fire.

polyisoprene (cis 1,4). A synthetic "natural" rubber made by polymerizing isoprene, using stereospecific catalysts such as titanium or a Ziegler type. The monomer is commonly separated by extractive distillation from the other thermal-cracking products of gas, oils, and naphtha. The isoprene, dissolved in *n*-pentane, is polymerized in a stainless-steel jacketed reactor at approximately 120°F (49°C). To ensure purity both the isoprene and solvent are redistilled and dried before use. The reaction is exothermic and requires the removal of considerable heat. The mixture of polymer and solvent is very viscous, causing mixing and handling problems. At the end of the reaction the batch is pumped into a holding tank where catalyst deactivator and antioxidants are added. The solvent is then vaporized off and condensed for reuse.

polymerization, addition. A reaction which results in the bonding of two or more molecules without the elimination of any by-product molecules. Two general types of reactions are recognized: (1) those involving a single molecular species and (2) those involving more than one molecular species. Of the first group the reactions with compounds having carbon to carbon unsaturation are common. Olefinic compounds as $CR_2 = CR_2$, where *R* represents hydrogen or a substituted radical, are the simplest of this type. Examples include ethylene, styrene, vinylidene chloride, and methyl methacrylate. In the second group are the copolymers. Copolymerization results when two different monomers react to form a polymer at relatively the same energy level rates. The ratio of the two monomers making the polymer may vary considerably among different copolymers. The arrangement of the monomer units may be random as *XXYXXXXYYXY*, or alternating as *XYXYXYXY*. Terpolymers may be formed from three different monomers. A special type of copolymer known as a block polymer is formed by joining a prepolymerized short chain of one monomer to prepolymerized chains of another monomer to form a long chain.

polymerization, condensation. A reaction in which the bonding together of molecules results in the elimination of an element such as nitrogen or hydrogen, or of a simple compound, such as hydrogen chloride, water, and sodium chloride. If both reactants are monofunctional, as in the reaction between ethyl alcohol and acetic acid, a relatively small molecule with no plastic properties is formed. If either the alcohol or acid is bifunctional and the other monofunctional, a larger molecule is produced. An example is diethylphthalate, which is a plasticizer but not a plastic polymer. If both acid and alcohol are polyfunctional, it is possible to produce a plastic polymer. For example, phthalic anhydride and glycerol react to form the alkyd resin, which is made up of a long chain of repeating units. The remaining OH group could then react with one hydrogen of another acid molecule to put on a side chain. This would then cross-link to another chain. Phenol, which has three reactive hydrogens, reacts with formaldehyde to form chains which cross-link and setup thermally. If one of the three reactive groups has been previously replaced by another radical as in paracresol, only a chain can be formed.

polymerization (plastics). Polymerization is achieved by uniting molecules in such a way that one of the double bonds is eliminated. One double bond remains available for a second union. In this manner it is possible to construct large chainlike molecules of the binder; those of some plastics attain a molecular weight of 20,000. This process is known as addition polymerization. Polymerization may occur very slowly, but is accelerated by heat and pressure. Catalytic agents are frequency added to increase the reaction rate. The product of polymerization are called polymers. It is possible to polymerize two different polymers to form copolymers.

polymers. High polymers are formed by polymerization reactions that occur by two main mechanisms: addition polymerization and condensation polymerization. The name of polymer is formed by adding the prefix "poly" to the monomer generic name, for example, polyethylene. When the monomer has a substituted parent name, or a multiword name, the parenthesis are used after the prefix "poly". Thus we can write poly(vinyl chloride), poly(propylene oxide), and poly(chlorotrifluoroethylene). Addition polymerization is defined as the reaction that yields a production in an exact multiple of the original monomeric molecule. Such a monomeric molecule usually contains one or more double bonds that, by intermolecular rearrangement, may make the molecule bifunctional. Addition polymerization reactions usually proceed by a chain mechanism involving either free radicals or ionic catalysis. Polymers cover a wide range of structural arrangements and may vary in their consistency from viscous liquids to rigid solids. Being organic compounds, mostly compounds of carbon with hydrogen and oxygen, they are not resistant to high temperatures and are easily decomposed or burned under such conditions. In this respect they differ greatly from inorganic compounds. Organic polymers are high-molecular-weight substances consisting of large molecules of colloidal dimensions. These are called macromolecules, and their molecular weight may vary from 10,000 up to several million. Macromolecules are built up from small molecules, called monomers, by polymerization reaction.

polymers, amorphous (organic polymers). An amorphous state of polymers is characterized as in inorganic glasses by a completely random arrangement of molecules. Highly cross-linked polymers are always amorphous, since their rigid three-dimensional network does not permit any changes in the molecular arrangement, which is always at random. However, linear (thermoplastic) polymers can easily assume different molecular configurations under relatively weak external forces or small temperature changes. Thus thermoplastics can exist in an amorphous, crystalline, or semicrystalline state, depending on the structure, the chemical nature of the polymer chain, and the processing conditions of the polymer. Polymer chains with a long repeating unit or with a low degree of symmetry having irregular chains do not crystallize easily and generally form amorphous structures. The presence of bulky, irregularly spaced groups prevents the chain segments from fitting into a crystal lattice and prevents the laterally bonded groups from approaching each other to the distance of best interaction. Polystyrene, poly(methyl methacrylate), and poly(vinyl acetate), all having bulky side groups attached at random to the main carbon chain, are typical amorphous polymers, also called polymeric glasses.

polymers, linear. Linear polymers form the largest group of plastics covering a large variety of diversified products used in different forms and applications. Linear polymers can be obtained either by addition polymerization or condensation polymerization of bifunctional monomers. Being thermoplastics, they can be easily worked into required forms and shapes at elevated temperatures. The structure and properties of linear polymers depend on the chemical nature of the monomer, the geometry of the polymer chain, and the magnitude of the intermolecular forces between the chains. These intermolecular forces depend on the molecular weight or the chain length, the presence of the polar groups and their spacing and regular distribution along the backbone chain, the possibility of the formation of the hydrogen bond, and the distance between the chains. The structural regularity of the chain determines the degree of packing of the chains and its state, either amorphous, crystalline, or semicrystalline.

polymers, organic. Most of the industrial polymers are plastics and elastomers. Plastics are those synthetic resins characterized as a group by plastic deformation under stress; elastomers or rubbers are those that are capable of extensive elastic deformation. However, there are many exceptions to it. Some plastics are highly elastic and any rubber can be deformed plastically under certain conditions. Plastics themselves are classified into two groups: thermoplastics and thermosets. The thermoplastics, of which polyethylene, polystyrene, and asphalts are examples, can be softened repeatedly with heat and will reharden when cooled. Thermosetting plastics harden or "cure" with heat and then cannot be resoftened by heating. Typical thermosets are Bakelite, epoxy, silicones, and rubber. There are three general methods of polymerization: (1) addition polymerization, (2) condensation polymerization, and (3) copolymerization. In addition polymerization, the polymer is produced by adding a second monomer to the first, then a third monomer to this dimer, a fourth to the trimer, and so on until the long polymer chain is terminated. Polyethylene is produced by the addition polymerization of ethylene monomers. Addition polymerization always results in a thermoplastic. Condensation polymerization occurs in a combination of a compound with itself or other compounds, accompanied by the elimination of some simple compound such as water as a result of the polymerization. Condensation products may be either thermosets or thermoplastics. Copolymerization is the addition polymerization of two or more different monomers. Many monomers will not polymerize with themselves, but will copolymerize with other compounds. One of the more important copolymers is butadiene–styrene, a rubber used in tires.

polymers, straight chain. The long linear chains formed by addition polymerization may be bundled together in a haphazard manner like straws in a stack. The atoms in the chains are held strongly together by primary valences. The bonds between the chains are secondary bonds or van der Waal's forces resulting from the state of imbalance or unsaturation at the surface. These bonds, while they may vary considerably in strength, are normally much weaker than most primary bonds. They are effective only where the surfaces of adjacent chains are in contact. This unoriented haphazard structure produces an amorphous material. Amorphous solids do not have sharp melting points, break with a conchoidal fracture, and tend to flow under stress. The magnitudes of the physical properties are not dependent on direction. If the chains are parallel, the polymer tends to be crystalline. The symmetry of the chains also affects the crystallinity. Sometimes the polymer is made up of both crystalline and amorphous areas. The crystallinity may vary with the conditions of solidification. It is frequently possible to orient the chains into parallel positions by stretching, thus changing from an amorphous to a crystalline product. These straight chain polymers, whether amorphous or crystalline, are thermoplastics; that is, they may be repeatedly melted by heat and solidified by cooling. They are, in general, soluble in certain organic solvents. The higher the degree of polymerization, the longer are the chains, the lower the solubility, and the higher the melting point. As crystallinity increases, penetration of solvent between the chains is less, resulting in lower solubility. The melting point is also higher since the molecules are held more tightly together.

polymers, thermoplastic. Compounds based on ethylene in which one or more of the hydrogen atoms have been replaced by a different atom or group of atoms are known collectively as vinyl compounds. Some of these vinyl compounds will polymerize spontaneously at ordinary temperatures while others polymerize when heated in the presence of a catalyst. A typical vinyl compound may exist as a very mobile liquid. The liquid flows as easily in water because only weak van der Waal's forces operate between the relatively small molecules. As polymerization proceeds the molecules reach a size where the van der Waal's forces become effective so that the liquid becomes relatively viscous. As polymerization nears completion the molecules become very long so that the sum of van der Waal's force acting between adjacent molecules becomes considerable and the material becomes so viscous as to be regarded as a solid. It no longer flows unless it is heated. This increases the energy of the molecules so that they are able to overcome the van der Waal's forces acting between them. Such a substance is said to be thermoplastic since it can be softened repeatedly by the application of heat. Sometimes two different monomers are used in the synthesis of the large molecule. The latter is then known as a copolymer. As an example, a mixture of vinyl chloride and vinyl acetate can be polymerized to produce the copolymer, polyvinyl chloride–acetate.

polymethylpentene. Methylpentene polymer is manufactured by Ziegler-type catalytic polymerization of 4-methylpentent-1 at atmospheric pressure. After rigorous extraction of catalyst residues, the polymer is obtained as a free-flowing dry powder that is normally blended with stabilizers and other additives, and compounded into granular form. In TPX polymer, comonomers are included to enhance optical and mechanical properties. TPX polymer has four outstanding and distinguishing properties: (1) a high crystalline melting point of 464°F (240°C), coupled with useful mechanical properties at 400°F (204.4°C) and retention of form stability near melting. (2) Transparency with a light-transmission value of 90% in comparison with 88–92% for polystyrene and 92% for PMMA. (3) A density of 0.83, which is close to the theoretical minimum for thermoplastic materials. (4) Excellent electrical properties with power factor, dielectric constant (2.12), and volume resistivity of the same order as PTFE.

polyolefins. Produced by the polymerization of unsaturated hydrocarbons, and includes the plastics polyethylene and polypropylene, and the rubbers polyisoprene and polyisobutylene. The polyolefins belong to the large group of vinyl and other polymers produced by the addition polymerization of relatively simple monomers in a straight carbon chain. The regular and simple monomer of polyethylene, ethylene, a gas boiling at −155°F (−104°C), permits a considerable degree of crystallization, about 60% in low-density polyethylenes. The crystal structure is orthorhombic. The angle between the carbon bonds in the polymer chain is 109°. The degree of crystallinity decreases with temperature, until at about 250°F (121°C) no crystallinity can be detected in low-density polyethylene. The presence of branching inhibits crystallinity. Polyethylene contains more hydrogen atoms per unit volume than water. This makes it an excellent material for shielding against neutrons. Neutron-shielding capacity is improved blending about 1% boron compounds into the polyethylene. Polyethylene has poor strength, high creep, low elastic modulus, and high expansion and mold shrinkage. Polyethylene is produced in three grades: low density, intermediate density, and high density. The largest outlet for polyethylene is film. The construction and agricultural industries use the film as a building vapor barrier, for temporary heated enclosures to protect construction workers in cold weather, as protection for wet concrete, for protection of plants against frost, and other uses. Propylene is a gas boiling at −54.4°F (−48°C) at atmospheric pressure. Its polymer polypropylene is the lightest plastic so far produced, with a specific gravity of 0.90. General physical and electrical properties of polypropylene are similar to those of high-density polyethylene. Polypropylene, however, is harder and has a softening point and lower shrinkage. Polypropylene is molded into domestic hollow ware, toys, bottles, and automobile distributor caps.

polyphenylene oxide (PPO). Made by oxidative coupling of 2,6-dimethylphenol in a solvent, such as *o*-dichlorobenzene, and a nonsolvent, such as propyl alcohol, with a catalyst consisting of an amine complex of copper salts, and oxygen. PPO has a brittle point below −275°F (−171°C) and a heat deflection point of about 375°F (191°C). It is highly resistant to acids and bases but soluble in certain chlorinated hydrocarbons and aromatic solvents. It has excellent mechanical and electrical properties. PPO may be fabricated by injection molding and extrusion. Applications may include electrical and electronic products, household appliances, food-handling equipment, pumps, hospital equipment, and plumbing.

polyphenylene sulfide (PPS). A crystalline polymer, polyphenylene sulfide (PPS), has a symmetrical, rigid backbone chain consisting of recurring parasubstituted benzene rings and sulfur atoms. This chemical structure is responsible for the high melting point [550°F (288°C)], outstanding chemical resistance, thermal stability, and nonflammability of the polymer. The polymer is characterized by high stiffness and good retention of mechanical properties at elevated temperatures, which provide utility in coating as well as in molding compounds. Polyphenylene sulfide is available in a variety of grades suitable for slurry coating, fluidized bed coating, flocking, electrostatic spraying as well as injection and compression molding.

polypropylene (PP). Chemically classified as a polyolefin, PP is manufactured from propylene obtained by cracking of petroleum products. The polymerization reaction involved results in the formation of an ordered arrangement of the individual monomeric propylene units into chainlike molecular configurations. Close packing of these chains encourages the development of a crystalline molecular structure. PP is used in the appliance, automotive, fiber, and packaging markets. Properties particularly useful in appliance applications include resistance to deterioration at elevated temperatures and to exposure to detergent solutions. Reinforced grades are used in load-bearing parts.

polystyrene (PS). Related to the polyvinyl plastics, being made of vinyl benzene, commonly called styrene. Styrene is found in coal tar, but most of it is produced synthetically. Polymerization occurs from the application of heat and catalysts. The resin is crystal clear, light in weight, odorless, and tasteless. It is unaffected by acid and alkalies. It has the lowest water absorptivity of all plastics and has outstanding dimensional stability. Polystyrene is somewhat brittle and has only moderate tensile strength but shows no loss in strength down to −40°F (−40°C). It is available in an unlimited color range, but has a tendency to discolor when exposed to sunlight. Polystyrene is commercially prepared by reacting benzene and thylene in a gaseous phase to obtain ethyl benzene, which is dehydrogenated at 1112°F (600°C). The resultant styrene monomer (vinyl benzene) is then polymerized to the transparent solid thermoplastic polystyrene.

Many grades of resin are available to suit a variety of end uses. Styrene is also copolymerized or blended with other thermoplastics to improve specific properties or modified with additives to aid processing. Housewares, packaging, toys, wall tile, knobs, and novelties are molded in a wide choice of solid and mottled colors. General-purpose grades are brittle, making a sharp, metallic sound when dropped. Styrene copolymerized with methyl methacrylates is used for extruded sheet or molded panels for reflective or diffusion lighting applications. A copolymer of styrene with methyl styrene permits use in continuous service at 150–210°F (66–99°C). A copolymer of styrene and acrylonitrile improves chemical resistance. Called SAN types, these resins are molded, extruded and, thermoformed into products requiring resistance to acids, alkalies, mineral oils, and detergents. Injection molded products have thousands of uses, from toys, containers, and housing to appliance components.

polystyrene (PS) foam. Practical PS foam densities may range from as low as 0.5 to more than 50 lb/ft^3 (8 to 800 kg/m^3). The many methods of foaming resins can be classified into three basic types: (1) mechanical, which involves beating air into the polymer matrix; (2) physical, which involves dissolving a gas or liquid into the resin; (3) chemical, which involves compounding a blowing agent with the resin that will release an inert gas at elevated temperatures. The physical and chemical processes are predominantly used. Polystyrene used in closed molds is normally referred to as expandable PS beads. Expandable beads are spheres ranging in size from 10 to 45 mesh. The beads contain from 5 to 7% by weight of a saturated parafinic hydrocarbon. The normal processing sequence is preexpansion, aging or conditioning, filling the closed mold with loose, dry preexpanded beads, introducing heat, cooling, opening the mold, ejecting the molded part, and drying. The SPI defines structural foam as "a plastic product having an integral skin, a cellular core and having a high enough strength-to-weight ratio to be classified as structural". Products made from PS structural foam are up to three times as rigid as those made from equivalent weights of solid polymer. Three basic processes in use include low-pressure injection molding, high-pressure expansion molding, and foundry-type casting. Items molded by low-pressure injection molding are stress free and have little tendency to warp. Sink marks can be eliminated, even opposite heavy ribs and bosses. **(See also foams.)**

polysulfone. A heat-resistant, transparent thermoplastic. In its completely natural and unmodified form polysulfone is a rigid, strong thermoplastic. It can be molded, extruded, or thermoformed (in sheets) into a wide variety of shapes. Polysulfone's heat-deflection temperature is 345°F (174°C) at 264 psi (1820 kPa) and long-term use temperature of 300–340°F (149–171°C). Polysulfone is produced by the reaction between the sodium salt of 2,2,-bis(4-hydroxyphenol) propane and 4,4,'-dichlorodiphenyl sulfone. The sodium phenoxide end groups react with methyl chloride to terminate the polymerization. The molecular weight of the polymer is thereby controlled and thermal stability is assisted. The diaryl sulfone grouping is a feature of the molecular structure of polysulfones. Thermal gravimetric analyses show polysulfone to be stable in air up to 932°F (500°C). This thermal resistance, together with outstanding oxidation resistance, provides a high degree of melt stability for molding and extrusion. Molded and extruded products are self-extinguishing. Resistance to oil and gasoline enables usage in automotive, under the hood application, as well as gasoline pump and valve parts. Heat resistance to 300°F (149°C) and ability to withstand deformation after thousands of hours under stress conditions make it applicable for home appliances (dishwashers, dryers), kitchen-range hardware, and wire insulation. Extruded sheets are used in thermoformed aircraft parts where heat resistance, self-extinguishing properties, and low-smoke generation are of paramount importance. Polysulfone is resistant to dilute or concentrated acids and alkalies, but is attacked by ketones and chlorinated and aromatic hydrocarbons.

polytetrafluoroethylene (PTFE). May be considered as ethylene in which fluorine has been substituted for all the hydrogen. It is polymerized in the presence of water and a peroxide catalyst in a stainless-steel autoclave. The polymer (PTFE) may be produced in either a granular or dispersion form. PTFE is completely resistant to all chemical attack except by molten sodium, fluorine, and some fluorine compounds. It may be used in a temperature range of −450–550°F (−268–288°C). It has a very low coefficient of friction and excellent antistick properties and electrical properties. Although classified as a thermoplastic, PTFE cannot be molded by conventional methods. Techniques similar to those used in powder metallurgy, involving performing at high pressures followed by sintering, are used. It is also possible to extrude it in the form of rods, tubes, and fibers by using an organic extrusion aid, which is later vaporized off, and the product is then sintered. PTFE dispersion may be applied to surfaces by dipping or spraying, followed by drying and sintering. Films may be cast from the dispersions. Tape is made by shaving from molded pieces or by extrusion. It is used in electrical insulators, wire covering, valve seats, seals, and gaskets. It is used in thin coatings as a nonstick surface for objects such as cookware. It is fixed with fillers, such as glass, carbon, and metals, for use where greater wear might otherwise occur, as in bearings.

polyurethane. Urethane polymers are the product of reacting isocyanates with compounds containing hydroxyl groups. Among the various forms they can take are rigid and flexible foams, adhesives, spandex fibers, coatings, and elastomers. Urethanes are known for their physical properties and versatility as protective coatings, while the elastomers have been replacing natural rubber and other elastomers, plastics, and even metals with their high degree of toughness and abrasion resistance. Urethane coatings exhibit rapid drying ability, high gloss, mar and abrasion resistance, toughness, flexibility, and overall versatility. Polyurethanes are used in housing, transportation, furniture, and apparel industries. Polyurethanes produced as flexible foams are used for mattresses and cushions, as automotive dashboards, armrest and visor safety features, and as clothing interliner. Rigid and semirigid versions are used to seal mine shafts and to insulate houses, storage tanks, rail cars, refrigerated trucks, warehouses, and pipe lines. Complete lines of furniture, modeled to any design replace traditional wooden items. Methods of producing foams vary depending on the scope of operation and item to be made. Foamed urethanes are produced by reacting an isocyanate with a polyol. The most common isocyanate is actually a mixture of isomers of toluene diisocyanate (TDI). A polyol is an alcohol containing more than two hydroxyl groups; fewer hydroxyl groups provide fewer reactive sites for a given quantity of isocyanate resulting in a flexible foam. Typical polyols are polyether glycols, castor oil, and polyester resins. Rigid urethanes produced in densities to 60 lb/ft^3 (961 kg/m^3) can be made by reacting the mixed isocyanate isomer with a polyol such as polyoxypropylene derivative of sorbital or sucrose and from other polyethers. A foaming agent, either in the form of small amount of water or a fluorocarbon is used to develop cell structure. Fluorocarbons are used for rigid foam of lower density.

polyvinyl acetate (PVAc). Although similar to PVC, polyvinyl acetate is quite different in physical properties. It is transparent and resists ultraviolet light, but is brittle at low temperatures. It is seldom used alone for molding since it becomes tacky at elevated temperatures. PVAc is formed by the reaction between acetylene and acetic acid, which is carried out in either a solvent solution or water emulsion. Water dispersions are solid as the familiar "white glues" and are also used for sizing agents in textile processes. Coated on paper, these resins provide grease and waterproof characteristics. Polyvinyl acetate is soluble in many organic solvents and is compatible with nitrocellulose for lacquers. Filled with wood flour, the solution is sold as patching compounds for furniture and woodwork. Toughness and adhesion are evidenced by the fact that the patch can be sandwiched, nailed, or sawed. In addition to its use as adhesive for paper tape, bread wrap, bookbinding, and wood, PVAc is the major ingredient in emulsion-type paints. These are low cost, odorless, and fast drying, and tools can be cleaned by water rinsing.

polyvinyl alcohol (PVA). Vinyl alcohol cannot exist in monomeric form since it undergoes a molecular rearrangement to form acetaldehydes. The polymer must therefore be made by acid or alkaline hydrolysis of polyvinyl acetate. PVA is water soluble to the degree of polymerization, which is, in turn, dependent on that of the original polyvinyl acetate. PVA resins are mainly used as adhesives that can be remoistened, as grease-resistance coatings, and as sizing for loom operations. PVA solutions are available for brush or spray release agents that form a tough wrinkle-free continuous coating on tooling surfaces. PVA films are used to package soaps, swimming pool chemicals, and dry bleaches. The package is simply tossed, unopened into water. Limited quantities of polyvinyl alcohol are used as molding or extrusion compounds. Its excellent resistance to gasoline, hot oils, and greases allows its use as tubing and hoses to convey these materials.

polyvinyl. One of the important and diversified groups of thermoplastics made up of vinyl polymers and copolymers. Available in a variety of chemical types such as polyvinyl chloride (PVC) and copolymers, chlorinated PVC, polyvinyl acetate used in paint and adhesives, polyvinylidene chloride used in packaging films and coatings, polyvinylidene fluoride, polyvinyl alcohol, and polyvinyl butyrl used as a laminate in a safety glass. Cellular PVC can be rigid or flexible and density can be varied widely. The impact resistance of PVC can be modified by compounding and various fillers, processing aids and stabilizers can be added to reduce cost, to increase production speed, and to reduce the affects of heat and sunlight. Vinyl chloride polymerizes easily. Rapid polymerization is obtained with many free radical initiators at 140°F (60°C). The molecular weight can be varied over a wide range by changing the polymerization temperature.

polyvinyl chloride (PVC). In the preparation of polyvinyl chloride the monomeric vinyl chloride is polymerized in emulsion form in autoclaves at a pressure of 50 atm and a temperature of 149°F (65°C). The polymer is spray dried at 266°F (130°C). In the chlorination of PVC the polyvinyl (PVC) is made into a 25% solution in tetrachloroethane and chlorinated for 30 h at 176°F (80°C) in a vessel cooled by a water jacket. After chlorination, the polymer is sprayed-dried, the solvent being removed by vacuum and recovered. The PVC polymer has softening point (it does not melt sharply) of 176–185°F (80–85°C). In spining, a 28% solution of PVC polymer is made in dry acetone and, after filtration, is extruded through spinnerets into cold water. The yarn has a tenacity of 1.8–2.2 grams per denier and an elongation at break of 40%. Its softening temperature of 212°F (100°C) renders impracticable its use for textile materials that have to be laundered. The fiber has good nonflame properties and good chemical stability. It has been used for filter pads for the chemical industry, fishing nets, and mosquito nets. Its uses are similar to those of Vinyon and Saran. It has also technical applications such as filter cloths, wadding and braid, protective clothing, flying suits (nonflame), canvas, blinds, curtains, and fairings. It is molded or extruded to form materials for gaskets, shoe soles, floor tiles, etc. Pipe, tubes, brush, bristles and textile fibers are also extruded.

porcelain enamels. These enamels are ceramic surface coatings applied to steel, iron, and aluminum for protection against corrosion and for decorative appearance. The enamel frit is a low-melting mixture of oxides resembling small particles of glass. Quartz and feldspar are blended with fluxes such as borax or soda ash (sodium oxide) and with materials such as titanium oxide or inorganic colors for color or opacity. Enamel is applied after first cleaning the metal surface with an alkaline cleaner. The glassy frit is ground fine, mixed with water, and applied by dipping or spraying. The enamel coat is allowed to dry and is then "burned" or fired at temperatures in the range of

1000–1500°F (538–816°C). Porcelain enamel is applied to a wide range of household appliances and equipment. It is also used for architectural purposes such as curtain wall sheets or high-rise buildings. A variant of the enameling process is the application of ceramic coatings to parts requiring abrasion resistance or protection against high temperature or oxidation. Materials such as silica, rocket motors, and gas turbines may require ceramic coatings.

powdered metals (metallurgy). Powdered metals are used to produce metal coatings, as in sherardizing, where a coating of fine zinc particles is bonded to another metal. They find wide application in paint pigments, such as aluminum and bronze powders. The manufacture of tungsten wire for use in electric light bulbs depends upon the manufacture of tungsten bars from compressed and sintered tungsten powder. These bars are then wrought into smaller diameters and finally drawn into wire, cemented carbide dies being used for this purpose. The manufacture of powdered metal parts involves two fundamental steps: (1) The metal particles are compressed into the desired shape. The particles thus form a compacted mass strong enough to withstand ordinary handling. (2) The compacted mass is heated, during which time its strength is increased. The step is called sintering; it usually results in dimensional changes such that there is an increase in density.

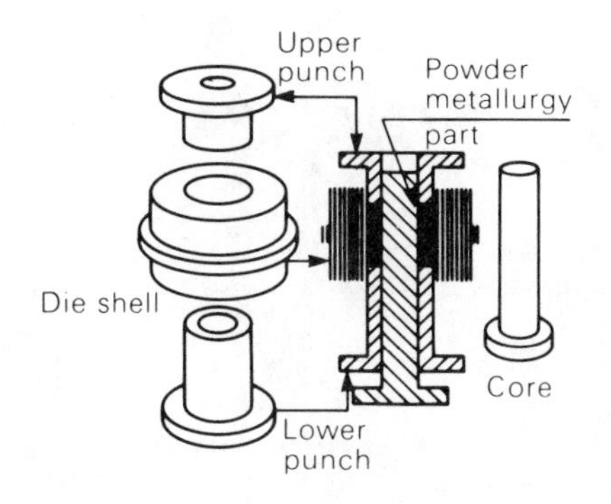

Courtesy, Metal Powder Industries Federation.

Simple punch and die arrangement for a cylindrical part. Telescoping punches are used for more complex parts.

powder metallurgy process. The shaping of articles from metallic powders is accomplished by compacting and sintering. The finer the particles, the greater the strength of the compact will be. Usually polydispersed powders of the proper size range give stronger compact and higher sintering rates than monodispersed powders of the same particle size. Powder metals are often compacted in mechanical or hydraulic presses, where a vertical stroke exerts a force to consolidate the loose powder into a compact shape. The pressure between two rotating rolls can also be utilized for powder compaction. Sintering causes the metal powders to bond or fuse together. During sintering, control of the atmosphere prevents undesired chemical reaction of the powder, such as formation of various oxides or carburization. Very often sintering operations in metallic systems leads to the formation of a homogeneous alloy. By blending different powders an average composition of the material can be maintained. Depending on the cooling rate, different microstructures—pearlitic, martensitic, or bainitic—can be produced in the product. Mechanical strength of the sintered products and their porosity and permeability depend on the sintering temperature and on the sintering time. This is also affected by the degree of compacting of the green product before sintering. Rapid sintering rates and high densities can be obtained in the presence of a liquid phase. Structures and properties of the sintered materials are determined by residual porosity and permeability. Grain size is second in importance to porosity with respect to uniqueness of structure. Extra strength can be obtained by infiltration with other metals during or after sintering. Sintered metal powder parts are sometimes coined for better finish or greater density, or sized to hold closer tolerances. Other subsequent operations may include heat treating, plating, and chip-removal processes. Advantages of the powder metallurgy process include close control over physical properties, the fabrication of parts from materials not practical by conventional means, the high production rates attainable, the low labor costs, and the elimination or reduction of scrap in the form of chips as the result of the need for little or no subsequent machining.

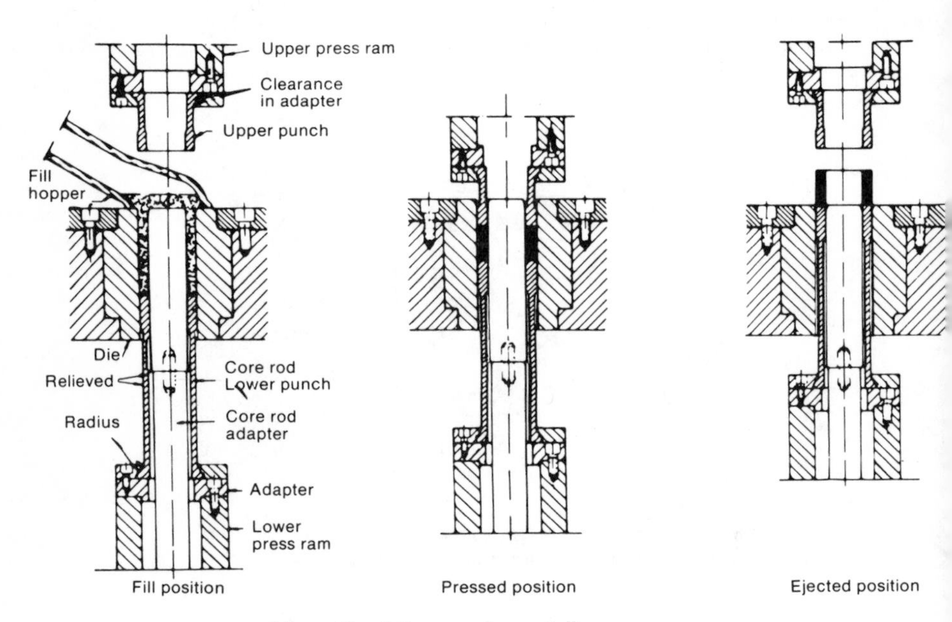

Schematic of the powder metallurgy process.

power brushing (see brushing)

pozzolans. Siliceous material, not cements in themselves, which combine with lime in the presence of water to form compounds with cementing properties. A wide range of materials offer possibilities as pozzolans: calcined clays and shales, pumicites, blast-furnace slags, brick, and flyash from boilers. Pozzolans, like cement, must be finely ground. As much as 40–50% of pozzolans may be added to cement, and generally produce improved workability, lower heat of hydration, improved watertightness, improved resistance to sulfates, and reduced alkali-aggregate reaction. Because of the lower heat of hydration, pozzolan additions are made to concrete used in large pours such as dams.

precipitation hardening (see aging)

precipitation number (lubricants). Indicates the amount of oil-soluble material present. This test is applied to used oils and to unused cylinder oils and black oils. One volume of oil to be tested is diluted with nine volumes of petroleum naphtha and the solution is centrifuged. Oil-insoluble materials are precipitated, and the amount is measured. In used oils, the quantity of sediment is indicative of the degree of contamination. In most new oils the precipitation number is small, except in the case of black oils. A high number is not detrimental for some applications.

press brakes. For producing such components as shells, frames, tubular sections, either uniform or tapered in length, cabinets or housings, or structural sections of almost any cross-section design, press brakes have a long, narrow press bed with a relatively short

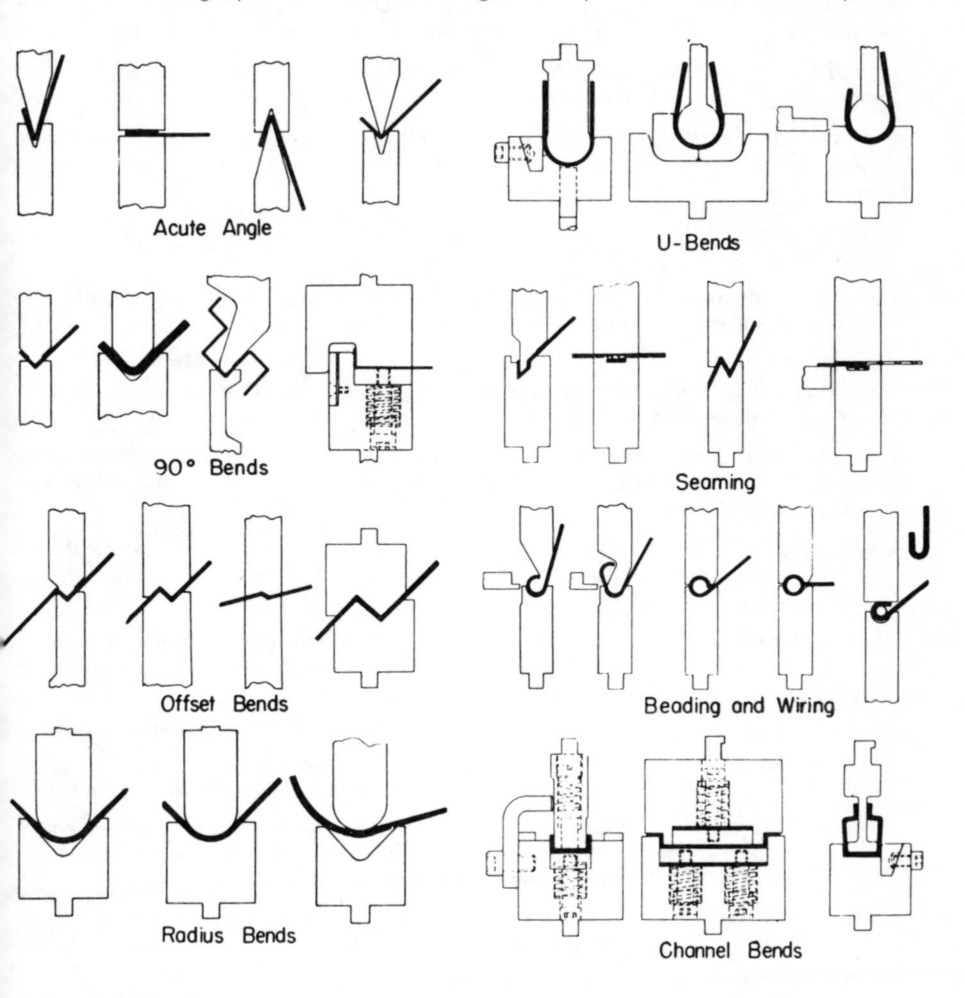

Drawings, courtesy Cincinnati Inc.

A group of representative press brake forming die sets, which illustrate the multitude of forms which can be produced.

ram stroke. Many types of press brakes are now available, some being more accurately termed open-front presses. Press sizes range from small machines with a capacity around 20–25 tons for light bending—up to 4 ft (1.2 m) in length with 14-gage sheet—to around 2000 tons for heavy work. Bed lengths, respectively, run from 6 to about 30 ft (1.8 to 9.1 m). A standard 900-ton press with V-die bending will handle 1 in. (2.54 cm) plate 12 ft (3.7 m) long, 3/4-in. (1.9 cm) plate 20 ft (6.1 m) long, or 1/2-in. (1.27 cm) plate 30 ft (9.1 m) long. Where extra length is necessary, presses are often used in tandem to achieve the desired results. Ram stroke required in the forming of parts must be relatively small and generally should not be over 6 in. (15.2 cm); the longest stroke available in present machines being approximately 8 in. (20.3 cm). Standard brakes, however, are usually made with a 3- or 4-in. (7.6- or 10.2-cm) stroke. Simple adjustment provided on the ram and means for exactly recording such adjustment for duplication makes possible the forming of many parts not practical by any other method. This is especially true with work that must rise in front of or behind the dies during forming.

printed-circuit assembly. A printed-circuit board on which separately manufactured component parts have been added.

printed-circuit board. A board for mounting of components on which most connections are made by printed circuitry. Also, a board having printed circuits on one or both sides.

printed wiring. A portion of a printed circuit comprising a conductor pattern for the purpose of providing point-to-point electric connections only.

printing, decalcomania. A preprinted design is released from backing material with water, other solvents, or heat and applied to the desired surface.

printing, electrostatic. In this process the ink particles, attracted from a positively charged stencil to a negative plate, are intercepted by the paper to be printed and are "fixed" by heat or solvent vapor. There is no plate contact.

printing, flexographic. A relief method especially adapted for package printing. Image carrier is flexible rubber or composition plate. Inks are of low viscosity with a high proportion of solvent.

printing, gravure. This is called Intaglio printing. Fast drying, solvent inks are transferred from etched plates or cylinders on which the minute cells vary in size and depth, depending on gravure method used.

printing, heat-transfer. Preprinted designs on a carrier web are transferred by heat, pressure, or both. It may or may not require a film coating, surface pretreatment, or postapplication cure.

printing, hot-stamping. Thermoplastic dyes or metallics suspended with temperature release waxes on a cellophane or Mylar backing tape are transferred to film sheet or flat objects through pressure and dwell time of heated metal or silicone rubber dies.

printing, letterpress. Relief printing. The face of the printing image is raised above open areas. Ink is applied to the face and is transferred to paper or other print surfaces directly.

printing, letterset. Indirect relief printing. Uses low-relief metal, plastic, or rubber plates. The ink image is transferred from plate to an offset blanket, then from the blanket to paper stock.

printing, lithographic offset. Lithographic printing does not use relief. The print and nonprint areas are in same plane of the plate. The plate is dampened each cycle and the water-repellent areas accept ink, the nonprint areas accept dampness, thus reject ink. The image transfer to paper is by means of a third rubber-covered "blanket" cylinder.

printing, screen. Porous stencil printing. A rubber blade (squeegee) forces ink through unblocked print areas of screen (silk, wire, synthetic). Nonprint areas are closed to ink passage.

process control. The application of automatic-control theories and hardware to operations found in the process industries. These industries fundamentally handle three forms of materials: (1) fluids, including a vast variety of liquids and gases, which are transported in pipes and ducts using pumps, blowers, and gravity as the transportation energy sources and various forms of valves, dampers, and other final controlling elements to control flow; (2) bulk solids, which are transported pneumatically or by means of various types of conveyors; and (3) sheeted and webbed materials, such as steel and aluminum sheets, paper, and textile fabrics. The process industries include the chemical, petroleum, metal producing, and pipeline industries, and portions of other industries, such as various food products. Sometimes a distinction is made between dry and wet processing. The most important variables measured and controlled in the process industries include temperature, pressure/vacuum, flow, liquid level, bulk-solids level, density and specific gravity, and chemical composition, including pH, oxidation-reduction potential, and many forms of spectrometry and spectrophotometry. From the standpoint of instrumentation and control, the process industries differ rather dramatically from the discrete-piece-handling industries, such as metalworking, electrical machinery, electronic components, and apparel manufacturing where the most important variables are dimension, shape, count, speed, position, and other physical properties, such as hardness, ductility, and color—with much emphasis on nondestructive inspection.

processor (computer). A machine language program that accepts a source program written in a symbolic form and translates it into an object program acceptable to the machine for which the source program was written. A processor frequently called a compiler or translator is a program usually supplied by the equipment manufacturer for creating machine language program. A processor previously stored in the computer receives the source program written in symbolic language by the programmer and translates the instructions into machine language instructions that are acceptable to the computer. This machine language program is called an object program.

producer gas (fuels). Manufactured in a generator by blowing streams of air and steam through a bed of fire. Incomplete combustion of the coal or coke used as the fuel results in the formation of carbon monoxide. Decomposition of the steam releases hydrogen. Because of the use of air, the nitrogen content of the gas is high, generally 50–55% by volume, and the heating value is rather low, ranging from about 140 to 180 Btu/ft^3 (5140 to 6606 kJ/m^3). The ranges of component gases in this fuel are given below.

Component	Volume (%)
CO	20–30
H_2	8–20
CH_4	0.5–3
CO_2	3–9
N_2	50–56
O_2	0.1–0.3

production equipment utilization (PEU). This provides an indication of the extent to which production equipment is being utilized to its fullest potential: PEU = Actual output ÷ theoretical output.

productivity. The ratio of output to input; popularly defined as output per man-hour.

product layout or assembly line layout. A plan showing the arrangement of equipment according to the processing sequence required to manufacture a product.

programmable logic controller (PLC). A stored program device intended to replace relay logic used in sequencing, timing, and counting of discrete events. Instead of physical wiring relay, pushbuttons, limit switches, etc., a PLC is programmed to test the state of input lines, to set output lines in accordance with input state, or to branch to another set of tests. The instruction sets of these machines generally exclude all arithmetic and Boolean operations, but do include vital decision instructions such as skip, transfer, unconditional, transfer conditional, and even transfer and link.

programmable manipulator. A device that is capable of manipulating objects by executing a stored program resident in its memory.

proximity switch. A method of object detection in which the source and detector are located on the same side; the detector senses energy from the source, which is bounced back by the object being detected.

puddling process (metals). In this process about 600 lb (272.4 kg) of pig iron is put into a furnace with a shallow hearth. Iron oxide, in the form of mill scale, is added. Virtually all the carbon, manganese, phosphorus, silicon, and sulfur is oxidized leaving almost pure iron. Because the melting point of the iron increases as it becomes purer, the furnace no longer provides a high enough temperature to keep it molten, and it becomes a pasty mass of sponge iron impregnated with slag, at about 2600°F (1427°C). The pasty mass is gathered into balls and removed from the furnace and squeezed into blooms. The squeezing ejects the surplus slag but leaves thin strings of slag embedded in almost pure iron. These blooms are then rolled or forged into various products. This process was discontinued after 1930 when the Aston process was used in its place.

pulp, alkaline papermaking. Alkaline pulps are chemical pulps cooked by an alkaline. The preparation of the wood in chip form for cooking and the basic principles of digestion for sulfite pulp also apply to alkaline pulping. Soda pulp refers to pulp obtained by cooking chips of the hardwood trees in caustic soda or sodium hydroxide. The alkali, that is sodium hydroxide, reacts with the lignin, carbohydrates, resins, and organic acids in the wood rendering them soluble and thus freeing the cellulose fibers. Because hardwood trees are used, soda pulp has short fiber length and low strength. It is almost always bleached. Soda pulp produces papers having high opacity, very uniform formation, high bulk, outstanding softness, and printing cushion. Because of the shortness of the fiber, very smooth flat papers can be made by using soda pulp. Soda pulp is not used alone, since such papers would be too weak. It is blended with longer and stronger pulps, such as sulfite, in papermaking. **(See also paper.)**

pulping, chemical papermaking. Pulping processes are of three essential types, namely, mechanical, chemical, and semichemical. All wood is composed of cellulose fibers, lignin, and hemicelluloses, with primary concern with cellulose, because this fibrous substance composed of the elements carbon, hydrogen, and oxygen is the most important. Cellulose fibers are held together in wood by the complex chemical substance called lignin. The objective of chemical pulping is to render soluble by a combination of heat, pressure, and chemical treatment the lignin and other substances that encrust the cellulose fibers without, at the same time, harming the latter. Consequently, the process of cooking wood represents a complicated chemical process requiring careful attention and control. Wood for chemical pulping is prepared differently than it is for mechanical pulping. First, the bark must be removed from the pulpwood. The barked wood to be cooked is reduced to chip form by sending the pulp sticks through mechanical chippers. The chips can vary from about 0.5 to 1.75 in. (1.27 to 4.5 cm) in length, again depending on the particular pulping requirements. The chips are charged to the digester, which is a large upright cylindrical-shaped container so lined inside and constructed as to withstand the chemicals used, as well as the pressure of cooking. After the digester has been charged with chips, it is closed, whereupon cooking liquor and steam are admitted. The first stage in cooking is one of complete preparation of the cooking liquor into the wooden chips. This must first take place before actual cooking takes place. Cooking is allowed to take place under controlled pressure and temperature and for a definite period of time.

pulping, mechanical (groundwood). In the straight mechanical process, no effort is made to separate the lignin from the cellulose. The wood after it is stripped of bark, is merely subjected to an abrasive action, which reduces it to pulp. The groundwood pulp is then blended with the necessary quantity of chemical pulp before conversion to paper. Newsprint is composed of approximately 75–80% groundwood, blended with 20–25% chemical pulp. Book papers also use a large percentage of groundwood.

pulping, semichemical papermaking. Semichemical pulp wood is first reduced to chip form. The first basic treatment is one of softening the chips or a partial dissolving of the lignin that holds the fibers together. The chemical treatment will vary considerably depending on the characteristics desired in the final pulp. Any of the regular chemicals employed in pulping such as acid sulfite and sodium hydroxide may be used; the chemical treatment found most satisfactory is neutral sulfite. Sodium sulfite is buffered with the alkalies sodium carbonate or sodium bicarbonate to a pH range near seven, hence the terminology "neutral sulfite semichemical pulp." By maintaining a neutral condition during cooking, the wooden chips are not darkened so much as a lighter colored pulp results. After chemical treatment, the chips are discharged. They may or may not be washed at this point. The next step is fiberizing. After fiberizing, the pulp will be washed if washing has not occurred before. Next usually follows screening and cleaning operations to remove shives, dirts, and other unwanted materials in the pulp. If the pulp is intended for white papers, bleaching will be the final step. Semichemical pulp is more suitable for pulping of hardwood trees than softwood trees. Hardwood species pulped by the semichemical process are stronger than when cooked by the conventional sulfite process and generally equal in strength to what they would be when cooked by the sulfate process.

pulp mills. Wood pulp is prepared from a wide variety of hardwoods and softwoods obtained in many geographical areas. Although pulp is the raw material of the paper-maker, not every pulp mill produces paper. Conversely, there are many paper mills that do not produce pulp from which they manufacture paper and related products. The principal function of the pulp mill is to separate the cellulose fiber from other substances in the wood. After this separation is completed, the pulp mill then cleans, refines, and otherwise treats the fibers so that they can be used to make paper or paperboard. The rejected portion of the wood, mostly ligneous in character, is either discarded or may be treated and used as a fuel. The first step in a pulping operation is to prepare the logs after their delivery to the mill. If they are transported by truck or railroad, removal of noncombustible surface dirt and sand is desirable. Some mills accomplish this as a part of log-handling system in which the logs are dropped into a water sluiceway and transported for a short distance, thereby being cleaned while being moved. Where logs are floated to the mill site or stored in water at the site, special provisions for surface cleaning are not always required. Debarking is the second step in the preparation of the log. One is hydraulic, in which the high-pressure water jets are directed at the logs and the bark is separated by impact of the water. A second dry drum debarking, where logs are fed into rotating horizontal drums and bark is removed by physical impact as the logs tumble over one another. A third is wet drum debarking, which is similar to the foregoing except that water in the form of low-pressure sprays is used in the drums as a soaking agent.

pulp, sulfate papermaking. The sulfate process is an alkaline method of preparing chemical pulp and is the outgrowth of the original kraft process. The sulfate process uses sodium hydroxide and sodium sulfate. The combined action of these two chemicals produces a different pulp than does either the sulfite or the soda process. The sulfate process gets its name from the fact that sodium sulfate is added as a make-up chemical in the pulping procedure and which, in the process, is reduced to sodium sulfide. The chemical action of the sulfur compounds from the sodium sulfide provides a very important contribution to the pulping action that is better than using alkali alone. Lignin is more rapidly dissolved, cooking conditions are less drastic, and the cellulose fibers are more protected against any harmful action of the sodium hydroxide. Consequently, very strong and long fibers are obtained and the percentage yield is higher.

pulp, sulfite papermaking. To make sulfite liquor, sulfur is burned in a controlled amount of air to form sulfur dioxide gas. This gas is allowed to react with limestone and water. This is commonly done by the tower system. There are two towers each several stories high. The limestone rock rests on grates at various levels up the towers. Water is admitted at the top of the towers and allowed to trickle down over the limestone. At the same time, the sulfur dioxide gas enters at the bottom of the tower and rises upward. Water and sulfur dioxide react to form sulfurous acid. Some of the sulfurous acid then reacts with the limestone to produce calcium bisulfite. The liquid that reaches the bottom of the tower will contain sulfur dioxide, sulfurous acid and calcium bisulfite. Sulfite pulp has medium strength as compared to the low strength of mechanical pulp and the high strength of sulfate pulp. Sulfite pulp refines easily during fiber preparation for papermaking and produces excellent formation in paper. Papers made from sulfite pulp have good bursting strength, but not as high tearing strength as papers made fom sulfate pulps. Sulfite pulps are of various papermaking properties depending on the species of wood used and the many variables in cooking, which are temperature, time, pressure, and concentration of cooking liquor. Strong unbleached sulfite is used in newsprint, wrapping tissue, and other grades where strength is most important. Soft bleached sulfite is used for papers such as book, blotting, and tissues where softness is necessary. Sulfite pulp for glassine paper is especially made so that it will hydrate easily. Strong bleached sulfite pulp combining high brightness and strength is used for manufacture of papers such as bond and bristols. There are specially purified and chemically modified sulfite pulps known as alpha pulps that are manufactured for dissolving purposes, that is, pulps that must have the requisite chemical purity for being dissolved chemically and made into products, such as rayon and explosives. Alpha pulps are also used for making specialty papers like photographic paper and saturating papers.

pultrusion. A plastics-processing method so named because of the combination of pulling a reinforced extrusion through suitable dies in lieu of merely pushing the material through under pressure. Shapes must be of constant cross section, and continuous lengths of fiber-reinforcing material in predetermined amounts are pulled through a bath of liquid-catalyzed thermosetting resin, then through a performer where they are shaped to the desired cross section. Mat or woven roving can be combined with the continuous strands to provide cross strength. The wetted, performed, resin-reinforcement next is pulled through a long set of heated dies, which initiates resin cure. After emerging from the dies, the cured pultrusion is cooled by air or water, then cut to specified lengths by a cut-off saw. Section lengths are limited only by handling capacities and shipping regulations. Beams, channels, bars, tubes and various other structural shapes can be made.

pulverizers (see coal)

pumice and pumicite (minerals). Light colored, silicic volcanic glasses. They are produced from viscous lava by explosive volcanism, and they accumulate as pyroclastic rocks. The distinction between them is arbitrary, being based on particular size. Fragments larger than 0.08 or 0.12 in. (2 or 3 mm) (course-sand or fine-gravel size) are generally called pumice and those from this size down to powder are termed pumicite. Pumice is highly vesicular. Its countless cells formed by expansion of gas, chiefly water vapor, during the rapid chilling that accompanies fragmentation of semimolten lava. Although the true specific gravity of glass is about 2.5, the cellular structure of pumice gives it an apparent specific gravity of less than 1.0. Since each cell is sealed from its neighbors by a membrane of glass, pumice has a low permeability and fragments will float on water for long periods. Pumicite, on the other hand, consists largely of angular shards and cell walls of highly fragmented pumice. Pumice and pumicite are commonly intermixed. Pumice may be nothing but frothy glass, or it may contain hard streaks of noncellular glass or minute crystals of quartz, feldspar, or mafic minerals. Pumice and pumicite are used as abrasives. They are utilized in hand soaps and household cleaners, and the finest grades of air-floated material are used commercially in finishing silverware, polishing metal parts to be electroplated, and woodworking. Crushed or ground pumice are also used in place of sand in acoustic plaster; and as loose-fill insulation filter aids, poultry litter, soil conditioner, sweeping compound, insecticide carrier, and blacktop highway dressing.

punched tape control (see numerical control)

pyrometallurgical process. Involves the use of fire, and in most instances accomplished by smelting in large smelting furnaces that resemble the open-hearth furnace of the steel industry. The pyrometallurgical process is used, for example, in the reduction of the compound of the copper sulfide ores, and also for low-grade manganese from the oxidized ores; however, in the case of the oxidized copper ores, and manganese ores of higher purity, the hydrometallurgical method of reduction is found to be more suitable. In the smelting operation used to reduce compounds to the metallic state, two things are accomplished; first the reduction of the metal, and second the separation of the metal from the impurities. The usual smelting operations, the ore, ore-concentrates, reducing agents, and flux are placed in the smelting furnace. A high degree of heat during smelting must be maintained either from an outside source or from the combustion of the reducing agent added.

pyrometers. The pyrometer is an instrument that measures temperatures, especially beyond the range of a mercury-in-glass thermometer, directly. For example, pyrometers can be made which use a comparison of electric currents produced by heating dissimilar metals (thermocouples). Since there is a direct relationship between voltage and temperature, the pyrometer is scaled in degrees Celsius (or centigrade) or fahrenheit. Radiation thermometers can be used for measuring temperatures in the pyrometric range.

Q

QC (see quality control)

quality. (1) The totality of features and characteristics of a product or service that bear on its ability to satisfy a given need (fitness-for-use concept of quality). (2) Degree of excellence of a product or service (comparative concept). Often determined subjectively by comparison against an ideal standard or against similar products or services available from other sources. (3) A quantitative evaluation of the features and characteristics of a product or service (quantitative concept).

quality circles. The main purpose of a quality control circles program is improved productivity and quality. Quality Circles is a program developed by Dr. W. Edwards Deming that focuses on prevention of quality problems and quality as a prerequisite for true productivity. Implementation of the Quality Circles concept is essentially the same in Japan and the United States, but there are some subtle differences. According to Donald L. Dewar of the Quality Circles Institute, the Japanese use the brainstorming approach that involves circle leaders asking the participants to volunteer ideas. In the United States everyone is involved; there are no spectators. The Quality Circles concept is not a ''cure-all,'' but it is a unique tool that can create an atmosphere in which people are able to solve their own quality and other problems.

quality control (QC). Although they are often used interchangeably, quality control, strictly speaking, is a technique of management for achieving quality, whereas inspection is simply one of the tools employed as a part of that technique. Any quality-control program relies on inspection together with the reporting, collecting, sorting, and analyzing of inspection results statistically to indicate wherein a lack of quality control exists. Such a program generally embraces four separate phases: (1) Inspection to locate flaws in the raw materials and in the processing of that material, which will cause trouble at subsequent operations; (2) inspection to segregate defective products; (3) investigation of inspection results so as to locate those points in the manufacturing process where control is required; and (4) the correction or salvaging of rejected material. The first three phases are concerned with *prevention*: preventing defective goods from reaching the customer, preventing manufacturing trouble caused by defects, and preventing the defects themselves through better control. The fourth phase, however, deals with the rehabilitation of defective goods after they have been located and segregated.

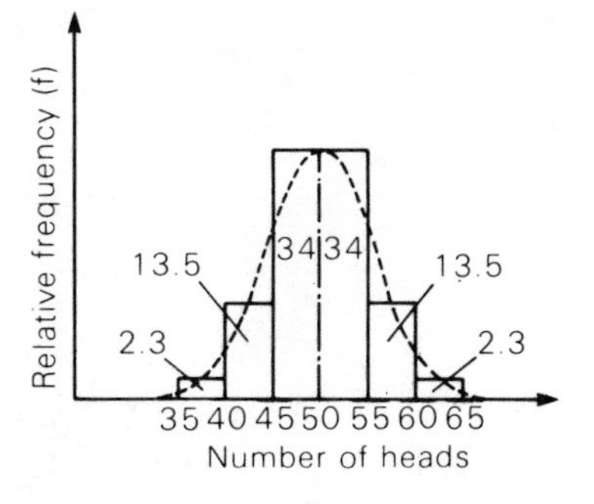

Frequency histogram of expected distribution of number of heads appearing in 100 tosses of a coin, $N = 100$, $\bar{p} = 0.50$

quenching, metal. The hardening of steel requires the formation of martensite. This is accomplished by heating to a temperature high enough for steel to become austenitic, then cooling fast enough, usually by quenching in water or oil, to secure complete transformation to martensite. The composition of the steel to be hardened, the quenching technique used, and the design for heat treatment are all very important factors affecting the properties of the final product. In the fully quenched state steel containing more than 0.2% carbon has such low ductility as to render the material useless for engineering applications, and it must be softened to some desired degree by tempering before use.

queue. In manufacturing, a line formed by loads or items while waiting for processing.

R

rack supported building structure. A complete and independent load storage system in which the storage rack is the basic structural system.

radiation, alpha rays. Alpha rays were found to consist of a stream of particles of matter travelling at an average of about 1/20th the velocity of light. The particles themselves are helium nuclei, that is, helium atoms lacking their two electrons. Hence an α-particle consists of a close association of two protons and two neutrons, and since it is traveling at a high velocity, possess considerably kinetic energy. Consequently, an α-particle is capable of producing considerable local effects on collision. Nevertheless, it rarely travels more than 4 in. (100 mm) in air due to collisions with molecules of nitrogen or oxygen. Similarly α-particles are effectively trapped by a sheet of thin paper. Since an α-particle is in fact a helium nucleus, it carries a resultant positive charge.

radiation, beta rays. Beta rays consists of particles similar to ordinary electrons, as far as mass and magnitude of charge are concerned, and projected at enormous velocities approaching closely that of light. However, the electrical charge carried may be either negative or positive. When the charge is negative, the beta particle is identical with the ordinary electron, but when a beta particle carries an equal positive charge, it is called a positron. It may be regarded as a positive electron. Since a positron and electron tend to annihilate each other on contact, the positron has little effect on ordinary chemical properties.

radiation, gamma rays. Gamma rays, like both light and x-rays, are electromagnetic vibrations. They are of extremely short wavelength and have great powers of penetrating matter. This makes them of value in the radiography of metals. At the same time biological tissue is generally damaged by penetration of gamma-rays so that for the sake of safety all gamma-ray sources must be adequately screened.

random-access memory (RAM). An auxiliary memory device on which the programmer can directly access each separate data area without having the search through the whole data file. RAM has a larger memory space than ROM, commonly 4096 bits, often expressed as 4K bits. RAM is different from ROM because it can be changed: new data can be "written" in or stored. RAM can be static where the contents of memory words will not disappear in computer operations, or dynamic, where the contents of memory words have to be constantly refreshed. RAM is used to store the macroprogram, the program designed by the computer programmer.

reaction injection molding (see RIM)

read-only memory (ROM). The ROM often has a space of 1024 eight-bit data units. Once bits are stored in ROM they cannot be erased and will not disappear when the power running the computer is turned off. They are prefabricated circuitry and are acquired in a unit. Their configuration cannot be changed in computer operation. Generally ROM carries the basic instructions to control a robot.

random sample. A sample of pieces drawn from a lot in such a manner that all pieces in the lot have equal likelihood of selection as part of the sample.

rayon, cuprammonium (plastics). A solution is produced by dissolving cellulose in Sweitzer's reagent, an ammoniacal solution of copper oxide. A 9% solution is extruded into warm water to produce rayon fibers. Dilute sulfuric acid is used to remove any remaining copper salts, which are recovered for reuse. "Cupra" rayon sells for a higher price than viscose rayon. It is usually produced in very fine filaments for use in high-grade fabrics.

rayon, viscose. The most common type of rayon is produced by extruding viscose through fine openings to form fibers that are coagulated in a bath in much the same manner as cellophane. Delustrants such as titanium dioxide or certain oils may be added to the viscose to produce a dull fiber. "High-strength" rayon for tire use is made from a higher alpha cellulose, such as cotton linters, and then given a cold stretching to orient the molecules.

reaming. Reamers are made with one end of the tool having shallow spiral or straight flutes sharpened so as to remove very small amounts of metal from slightly undersized holes. Common practice is to drill the hole about 1/64-in. (0.4 mm) undersize. The allowance generally depends on hole size and the amounts usually allowed for removal range from 0.010-in (0.254 mm) for 1/4-in. (6.4 mm) holes, 0.015-in. (0.38 mm) for 1/2-in. (0.127 mm) holes, graduating on up to 0.025-in. (0.64 mm) for 1 1/2-in. (38.1 mm) holes. For hand-reamed holes the allowance seldom exceeds 0.001–0.003 in. (0.025–0.76 mm). Where through holes cannot be provided, allowance at the bottom of blind bores for the chamfered tip of the reamer will obviate additional operations with shouldering or bottoming reamers to completely finish the entire length of a hole. Reaming is widely used to size holes more accurately than is possible with drills.

recrystallization, metal. If a metal is heated to a high enough temperature after being cold worked, new equiaxed, unstrained crystals will be formed. This is known as recrystallization. The temperature at which recrystallization takes place is different for each metal and varies with the amount of cold deformation. In general, the greater the amount of deformation, the lower is the recrystallization temperature. However, there is a practical lower limit below which recrystallization will not take place in a reasonable length of time. Metals fracture if the crystals are deformed too much. It is possible to carry deforming processes to greater lengths without danger of fracture by heating a metal to allow recrystallization to take place after some initial cold working has been done. The material is then ready for further deformation. This is of importance in many manufacturing operations. In order to obtain new, stress-free small grains in metals which do not undergo allotropic change, it is first necessary to work them while cold. Thus, if a metal has a coarse-grain structure and one wishes to change it to a fine-grain material, it must be plastically deformed while cold and then heated to bring about recrystallization.

redraw rod (see rod)

redraw tube (see tube)

refinery gas. Gaseous fuels encountered in refinery operations when crude oil is processed into gasoline and other similar products. These gases are "rich" but variable in composition as a result of differences in the characteristics of the oil refined and the extent of cracking to which the oil has been subjected. Their heating value is higher than natural gas, owing to the larger percentage of heavier hydrocarbons present. There are also present some illuminants, or unsaturated hydrocarbons, from the cracking operations.

refractories. Special materials of construction capable of withstanding high temperatures in various industrial processes and operations. The main bulk of the commercial refractories are complex solid bodies consisting of high-melting oxides or a combination of oxides of elements such as silicon, aluminum, magnesium, calcium, and zirconium, with small amounts of other elements present as impurities. Common refractory materials represent the main bulk of commercial refractories used in high-temperature processes and operations because of their relatively low price and ready availability. They consist of crystalline or partly amorphous constituents held together by a more or less glassy matrix of variable composition. One of the most widely used refractories is based on alumina–silica compositions varying from nearly pure silica, through a wide range of alumina silicates, to nearly pure alumina.

refractory alloy. (1) A heat-resistant alloy. (2) An alloy having an extremely high melting point. **(See also refractory metal.)** (3) An alloy difficult to work at elevated temperatures.

refractory metal. A metal having an extremely high melting point, for example, tungsten, molybdenum, tantalum, niobium (columbium), chromium, vanadium, and rhenium. In the broad sense, it refers to metals having melting points above the range of iron, cobalt, and nickel.

regenerator waste, gas. The lighter hydrocarbons or gas oils produced in the petroleum coking processes are further refined in catalytic cracking units. These units fall into two general types: fluid units in which fine powdered catalyst flows through the equipment with flow characteristics resembling a liquid, and moving bed units, which use either spherical or pelleted catalyst circulated by elevators or gas lifts. The cracking of the feed occurs in what is termed a chemical reactor vessel. The preheated catalyst is maintained in a fluid state, and during the cracking reaction the catalyst becomes coated with a coke deposit. This material must be removed in order to maintain catalyst activity. The spent catalyst is continuously removed from the reactor and transported to a regenerator vessel. In this unit compressed air is used to fluidize the catalyst and burn off the carbon. To keep compression costs to a minimum, as well as to keep the temperature inside the regenerator restricted to a level that will not destroy catalyst activity, the smallest amount of air is used that will effectively clean the catalyst. This combustion process, therefore, normally produces an appreciable percentage of carbon monoxide. A typical analysis of regenerator waste gas and the range of the various constituents is given below. The gas, although at temperatures as high as 1125°F (607°C), has a heating value of not over 40 Btu/ft^3 (1470 kJ/m^3):

	Volume (%)	Range (%)
Carbon monoxide	6.9	3–9
Carbon dioxide	8.1	7–11
Oxygen	0.8	0.3
Nitrogen	65.8	56.75
Water vapor	18.4	7.3

remelt ingot. A cast form suitable for remelting.

reroll stock, foil stock, sheet stock. A semifinished rolled product of rectangular cross section in coiled form suitable for further rolling.

resilience, of materials. A measure of the ability of a material to absorb and store energy under stress in the elastic range. The modulus of resilience (in. lb. per cu. in.) is the energy stored up in a body at the elastic limit of the material and is equivalent to the area under the stress–strain diagram up to the elastic limit.

resins, copal. Copals are fossil resins. Of these the Congo resins from Africa are one of the hardest natural resins. Kauri is a similar resin from New Zealand. Fossil grades of Manila resins are also found. These resins are heat cracked to make them soluble and are used in various coatings. Amber, a yellow to almost black fossil resin, is also used in jewelry.

resins, damara. Secured by tapping a pine tree that grows in Indonesia and on the Malay Peninsula. It varies from colorless to yellow, bleaching upon exposure to ultraviolet light. While it can be used in molding powders, its main uses are in paint, enamels, and varnishes. A soft grade of Manila resin results from tapping Agathis trees. It is used as a sizing material and a shellac substitute.

resins, plant. Plant resins originate as solutions of complex saturated organic compounds called resenes in essential oils, such as terpenes present in plants. The solutions originally exuded from trees as a result of accidental breakage of twigs or limbs or as the result of insect attack. On exposure to the air they gradually harden to form the solid resin, partly by evaporation of the essential oil and partly by oxidation and polymerization. The physical and chemical properties of resins frequently change slowly so that the freshly formed product might vary considerably from that which had exuded months or years before. Some of the resins become covered with soil and develop into the so-called fossil resins. Some resins are now recovered by tapping the trees so as to obtain a greater yield. Four major groups of natural resins are commonly recognized: (1) soft natural resins such as rosin and dammars; (2) semifossil resins, such as East India and batu; (3) fossil resins, such as copals, kauri, and Congo; and (4) miscellaneous group, including such resins as elemi, mastic, and sanarac.

resistance. That property of a conductor which tends to hold back, or restrict the flow of an electric current; it is encountered in every circuit. Resistance may be termed electrical friction because it affects the movement of electricity in a manner similar to the effect of friction on mechanical objects. The unit used in electricity to measure resistance is the ohm.

resistors. A resistor is a circuit element designed to insert resistance in a circuit. A resistor may be of low value or of extremely high value. Resistors in electronic circuits are made in a variety of sizes and shapes. They are generally classed as fixed, adjustable, or variable, depending on their construction and use. The value of the resistance for each resistor is determined by the makeup and size of the carbon compound and may vary from only a few ohms up to several million ohms. The two important values associated with resistors are the value in ohms of resistance and the value in watts which represents the capacity of the resistor to dissipate power. The two types of fixed resistors are called axial-lead resistors and radial-lead resistors because of the way the electrical leads are attached to the resistor. Resistors required to carry a comparatively high current and dissipate high power are usually of the wire-wound ceramic type. A wire-wound resistor consists of a ceramic tube wound with fine resistance wire which is then covered with a ceramic coating or glaze.

resonance. In many electronic circuits, the inductive reactance and the capacitive reactance are often equal in value. When this happens, the total reactance is zero because the inductance would cause the voltage to lead by 90° and the capacitor would cause the voltage to lag by 90° and these two conditions oppose each other. Hence, the voltage would neither lead nor lag the current and the voltage and current would be in phase. This in-phase condition, when an inductance and capacitor are in the circuit, is known as resonance. This phenomenon permits the selection of desired and specific signal frequencies while rejecting those not wanted. Thus, resonant circuits are extensively used in virtually all branches of electronics.

resonant circuits. In the design and operation of electronic systems, resonant circuits provide the key to frequency control. When a certain frequency is to be produced, it is necessary to establish a circuit which is resonant at that frequency. Also, when a certain frequency is to be passed through a circuit and others eliminated, it is necessary to have a circuit which is resonant at the frequency to be passed. When a certain frequency is to be blocked, it is then necessary to place a circuit a resonant tank circuit, which will block the frequency for which it is resonant. As can be seen, resonant circuits are most essential in radio and television receivers and transmitters.

reverberatory furnace. A furnace with a shallow hearth, usually nonregenerative, having a roof that deflects the flame and radiates heat toward the hearth or the surface of the charge.

rheostat. In practical electric circuits it is often necessary to insert resistances which may be adjusted to reduce the voltage applied to a load. This is usually accomplished by means of varible resistors called rheostats. A rheostat is usually constructed by winding high-resistance wire radially on a circular form made of a nonconducting material. A sliding contact arm is mounted on a shaft located in the center of the circular resistance, with one end of the contact arm resting on the barb wire. The contact arm is often called the wiper. When one terminal of a circuit is connected to one end of the resistance and the other terminal is connected to the sliding contact arm, it is possible to vary the resistance of the circuit by rotating the shaft and moving the contact arm along the resistance. When the arm is moved in one direction, it places additional coils of resistance wire ih the circuit, thus increasing the resistance. If the sliding arm is moved in the opposite direction, it removes a part of the resistance from the circuit.

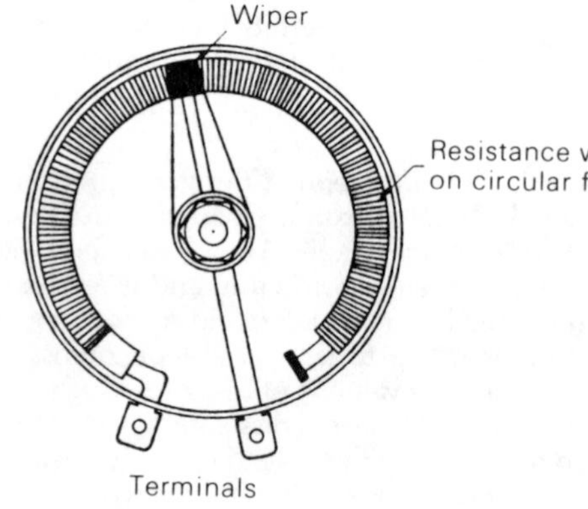

A rheostat.

rider truck. An industrial truck that is designed to be controlled by a riding operator.

RIM (reaction injection molding). A low-pressure molding process for polyurethane cellular foams that are foamed in the mold. It produces large structural components, with or without inserts, weighing up to 100 lb (45.4 kg) and having an integral skin which can be textured. Reinforcement, usually in the form of milled glass fiber, increases the stiffness and dimensional stability of RIM parts. Maximum glass content of reinforced reaction injection molding (RRIM) is about 25%.

riveting machines. The tools and machines designed for upsetting and forming rivet heads include several types. These special tools or machines for riveting may be divided into several general types. First, they may be classified according to the method of forming the rivet head, which may be either by compression, a succession of rapid blows, combined compressive and rolling or spinning action, rapid blows accompanied by rotary motion of the rivet set, or the application of pressure to heated rivet. Riveting machines differ from riveters in that the riveting operation with a machine is effected by a succession of blows or by a compressive rotating action, whereas a riveter merely subjects the rivet to compression. These compression riveters may be classified according to the power used for operating the riveting plunger. Thus there are hydraulic, pneumatic, hydropneumatic, and other types.

rivet rod (see rod)

rivet wire (see wire)

robot, bang-bang. A robot in which motions are controlled by driving each axis or degree of freedom against a mechanical stop. **(See also fixed-stop robot, pick-and-place robot.)**

robot, continuous-path. This robot operates, in theory, through an infinite number of points in space that, when joined, describe a smooth compound curve. This curve is usually developed during the programming or "teaching" phase, which is carried out by an operator. Load capacities of point-to-point continuous-path robots are essentially a function of width, motion, inertia, and other factors. Although the load capacities differ by individual manufacturers, as do sizes, a present-day capacity range of between 300–500 lb (136–227 kg) on a fully extended medium technology robot arm with ±0.008 in. (0.2 mm) accuracy on repeated positioning has been reported. Continuous-path control is used where the path of the end effector is of primary importance to the application, such as when used for spray painting. The unit is generally not required to come to rest at unique positions and perform functions as is common in applications employing a point-to-point control. Typically, a robot using this type of control is taught by the operator physically grasping the unit and leading it through the desired path in the exact manner and at the exact speed that the operator wishes the robot to repeat. While the device is moved through the desired path, the position of each axis is recorded on a constant time base, thus generating a continuous time history of each axis position. Every motion that the operator makes, whether intentional or not, will be recorded and played back in the same manner. Since the operator must physically grasp the robot, it must be designed to be essentially counterbalanced and free under no power so that the operator can perform the task; therefore, this control is generally limited to light duty robots. Since the operator is manually leading the robot through the desired sequence, the teaching is very instinctive, and there is no concern for the position of each axis. The programming is direct. Another characteristic of this type of control is that considerable memory capability is required to store all the axis positions needed to record the desired path smoothly. For this reason, magnetic-tape-storage means are generally used.

robot, controlled-path. The controlled-path type of robot is less common and utilizes a computer control system with the computational ability to describe a desired path between any preprogrammed points. Each axis or degree of freedom can be controlled and actuated simultaneously to move to those points. The computer calculates both the desired path and the acceleration, deceleration, and velocity of the robot arm along the path.

robot, cylindrical-coordinate. This robot type is represented by such models as the Pacer, Versatran, and Auto-Place. Its configuration consists of a horizontal arm mounted on a vertical column that, in turn, is mounted on a rotating base. The horizontal arm moves in and out, its carriage moves up and down on a vertical column, and these two members rotate as a unit on the base Thus, the working area or envelope is a portion of a cylinder.

robot, fixed-stop. A robot with stop-point control but no trajectory control; that is, each of its axes has a fixed limit at each end of its stroke and cannot stop except at one or the other of these limits. Such a robot can therefore stop at no more than two locations (where location includes position and orientation). Often very good repeatability can be obtained.

robot, generation 1. This is the robot in use today. It is characterized as being a programmable, memory-controlled machine with several degrees of freedom. It can be equipped with grippers of special-handling attachments, which can hold and operate hand tools, welding guns, and power tools, as well as perform workpiece and material-handling, -manipulation, and -transfer functions.

robot, generation 1.5 This is the robot that will be sensory controlled and have capabilities to perform "make" and "test" functions. It will work on principles of electrooptics, pressure, torque, force-sensitive touch, and proximity. It will be capable of recognizing and manipulating workpieces, parts, and tools. The motion paths of the robot will be memory controlled with overrides of preprogrammed control depending on sensor inputs.

robot, generation 2. This is the future robot that will have hand-and-eye coordination control through machine-vision concepts. The robot will see objects and will be able with hand interactions to perform manipulative functions.

robot, generation 3. A "factory-intelligence-controlled" robot that will provide artificial intelligence to help solve "factory" problems.

robot, hydraulic. This robot type normally includes a hydraulic power supply as either an integral part of the manipulator or as a separate unit. The hydraulic system generally follows straightforward industrial practices and consists of an electric-motor-driven pump, filter, reservoir, and, usually, a heat exchanger (either air or water). These robots normally operate on petroleum-based hydraulic fluid; however, most are available with special seals for operation on fire-retardant fluid.

robot, industrial, types of. Design variations existing in various industrial robots can be classified broadly into three different types. These are the rectilinear type, polar configuration type, and the anthropomorphic type. The rectilinear-type design has basic movements that are linear in nature, excluding the wrist and rotating movements. A robot designed with the polar configuration has prime movements with polar co-ordinates, except the arm-extend and -retract movement. The main difference between the rectilinear- and polar-type configurations is limited to the method of achieving the vertical up-and-down movement. A robot designed to be anthropomorphic in nature more closely simulates human movements; it has additional movements in the arm that simulate the human elbow.

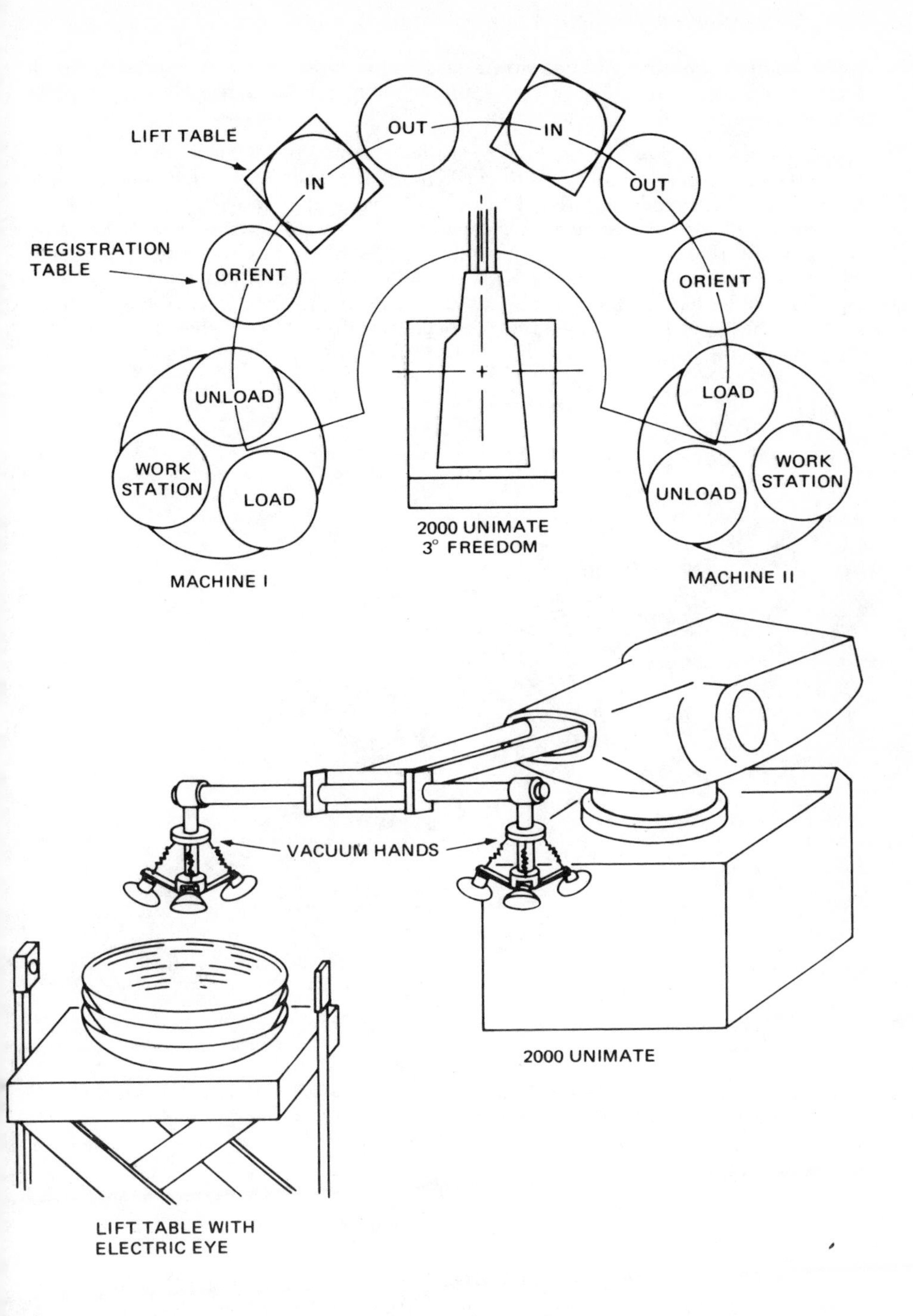

A generation 1 robot.

robot, jointed-spherical (jointed-arm). This robot type has rotary joints in several places along the arm that roughly correspond to the human shoulder, elbow, and wrist. It is usually mounted on a rotary base.

robot, manipulator type. A method of classifying a robot by the manipulative function that it is capable of performing: (1) the pick-and-place type, where only three degrees of freedom are generally required, with a very simple gripper design for part clamping–unclamping; (2) the special-purpose type, which is designed to perform a specific function such as paint spraying, press load–unloading, or die-casting-machine load–unload and which usually employs unique features that make it suitable for its intended function; (3) the universal type, in that it can be used to perform any manipulative type of function and which is usually designed with five degrees of freedom, excluding special gripper designs or tools that can be attached to the robot's arm.

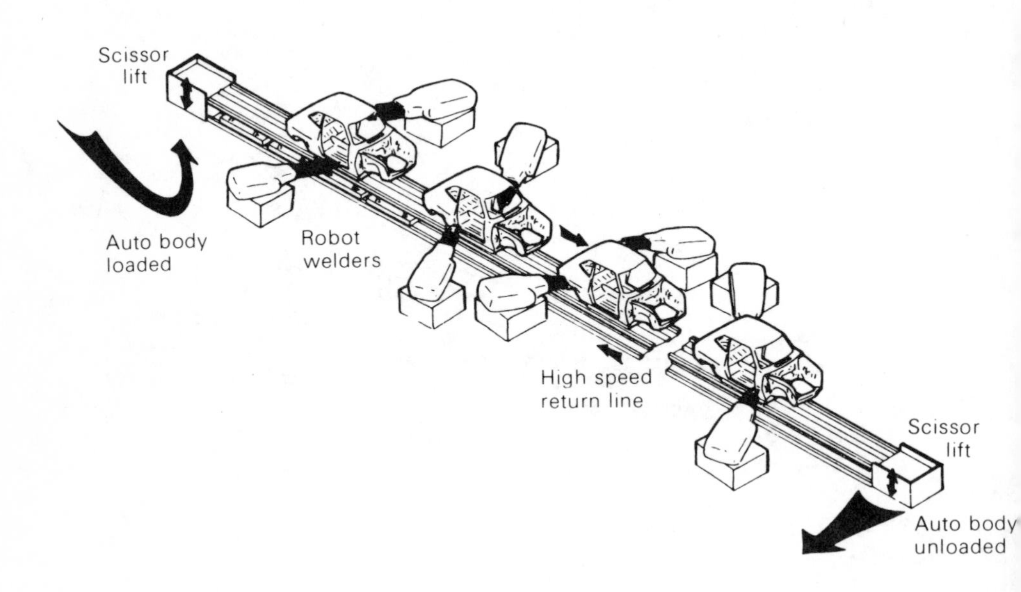

A sketch of an automatic welding line using robots. The scissor lifts at each end are to lift and lower the body cars to transfer them to and from the high-speed return line.

robot, material-handling. A robot designed, programmed, or dedicated to grasping, transporting, and positioning materials in the process of manufacture.

robot, material-processing. A robot designed, programmed, or dedicated to cutting, forming, heat treating, finishing, or otherwise processing materials as part of manufacture.

robot, medium-technology. A medium-technology robot consists of a basic mechanical and electrical control package. Most units are self-contained and can be moved easily from one jobsite to another. Programming is fast and uncomplicated. This robot type is adaptable to most jobs since it can carry payloads of up to 100 lb (46 kg) and operate within a reach of up to 10 ft (3.1 m) across its sphere of influence.

robot, mobile. A robot mounted on a movable platform.

robot, pick-and-place. A simple robot, often with only two or three degrees of freedom, that transfers items from place to place by means of point-to-point moves. Little or no trajectory control is available. Often referred to as a bang-bang robot.

robot, point-to-point. This control method is perhaps the simplest and most frequently used. Teaching is done by moving each axis of the robot individually until the combination of axis positions yields the desired position of the robot end effector. When this desired position or point is reached, it is programmed into memory, thereby storing the individual position of each robot axis. In replaying these stored points, each axis runs at its maximum or limited rate until it reaches its final position. Consequently, some axes will reach their final value before others. Furthermore, because there is no coordination of motion between axes, the path and velocity of the end effector between points is not easily predictable. For this reason point-to-point control is used for applications where only the final position is of interest and the path and velocity between the points are not prime considerations.

robot, record–playback. A robot for which the critical points along desired trajectories are stored in sequence by recording the actual values of the joint-position encoders of the robot as it is moved under operator control. To perform the task, these points are played back to the robot servo system.

robot, rectilinear–Cartesian. A continuous-path extended-reach robot that offers the versatility of multiple robots through the use of a bridge and trolley construction that enables it to have a large rectangular work envelope. Being ceiling mounted, such devices can service many stations with many functions, leaving the floor clear. *X* and *Y* motions are performed by bridge and trolley, the vertical motions are performed by telescoping tubes, and additional axes can be used.

robot, rectilinear-coordinate. This robot is mounted on a fixed base. All motions of the manipulator are in a straight line, either in-out, up-down, or side-to-side. Even though a pivoting wrist joint may be provided, the work envelope of this robot conforms to either a cube or a rectangular solid.

robot, sensory-controlled. A robot whose control is a function of information sensed from its environment.

robot, sequence. A robot whose motion trajectory follows a preset sequence of positional changes.

robot, servo-controlled, continuous path. Typically the positioning and feedback principles are the same as in a servo-controlled robot. There are, however, some major differences in control systems and some unique physical features. During programming and playback data are sampled on a time base, rather than as discretely determined points in space. The sampling frequency is typically in the range of 60–80 Hz. Owing to the high rate of sampling of position data, many spatial positions must be stored in memory. A mass-storage system, such as magnetic tape or magnetic disk, is generally employed. During playback, owing to the hysteresis of the servovalves and inertia of the manipulator, there is no detectable change in speed from point to point. The result is a smooth continuous motion over a controlled path. Depending on the controller and data-storage system used, more than one program may be stored in memory and randomly accessed. The usual programming method involves physically moving the end of the manipulator's arm through the desired path, with position data automatically sampled and recorded. The speed of the manipulator during program execution can be varied from the speed at which it was moved during programming by playing back the data at a different rate than that used when recording. Continuous-path servo-controlled robots share the following characteristics: These robots generally are of smaller size and lighter weight than point-to-point robots. Higher end-of-arm speeds are possible than with point-to-point robots; however, load capacities are usually less than 22 lb (10 kg). Their common applications are to spray painting and similar spraying operations, polishing, grinding, and arc welding.

robot, servo-controlled, point-to-point. A typical servo-controlled robot is used in a wide variety of industrial applications for both part-handling and tool-handling tasks. Significant features are for those robots using the "record–playback" method of teaching and operation, initial programming is relatively fast and easy; however, modification of programmed positions cannot be accomplished readily during program execution. Those robots using sequence/potentiometer controls tend to be more tedious to program; however, programmed positions can be modified easily during program execution by adjusting the potentiometers. The path through which the various members of the manipulator move when traveling from point to point is not programmed or directly controlled in some cases and may be different from the path followed during teaching. Common characteristics include high-capability control systems with random access to multiple programs, subroutines, branches, etc., and great flexibility provided to the user. These robots tend to lie at the upper end of the scale in terms of load capacity and working range. Hydraulic drives are most common, although some robots are available with electric drives.

robot, servo-controlled, significant features. The manipulator's various members can be commanded to move and stop anywhere within their limits of travel, rather than only at the extremes. Since the servovalves modulate flow, it is feasible to control the velocity, acceleration, and deceleration of the various axes as they move between programmed points. Generally, the memory capacity is large enough to store many more positions than a nonservo-controlled robot. For some sophisticated units this means access capability at as many as 4000 points in space. Given programs select and sequence activity points for a particular operating scheme. Programs can be varied to maintain the scheme while changing the activity points. Both continuous-path and point-to-point capabilities are possible. Accuracy can be varied, if desired, by changing the magnitude of the error signal, which is considered zero. This can be useful in "rounding the corners" of high-speed contiguous motions. Drives are usually hydraulic or electric and use state-of-the-art servo-control technology. Programming is accomplished by manually initiating signals to the servovalves to move the various axes into a desired position and then recording the output of the feedback devices into the memory of the controller. This process is repeated for the entire sequence of desired positions in space.

robot, special-purpose. Offering more versatility in accommodating frequent changeovers and ease of installation and maintenance, a special-purpose robot, designed specifically for parts extraction and orientation, fills the gap between integrated custom machine equipment and a general-purpose industrial robot.

robot, spherical-coordinate. Mounted on a rotary base and resembling the turret of a tank, the arm of a spherical-coordinate robot cannot only extend and retract, but is pivoted so that it can swing vertically, allowing rotary motion about a horizontal plane. The end effector moves in a volume of space that is a portion of a sphere.

robot, supervisory-controlled. A robot incorporating a hierarchical control scheme, whereby a device having sensors, actuators, and a computer, and capable of autonomous decision-making and control over short periods and restricted conditions, is remotely monitored and intermittently operated directly or reprogrammed by a person.

Rockwell hardness test (see hardness test, Rockwell)

rod. A solid product that is long in relation to cross section which is 3/8 in. (9.5 mm) or greater in diameter. Alclad Rod is rod having on its surface a metallurgically bonded aluminum or aluminum alloy coating that is anodic to the core alloy to which it is bonded, thus electrolytically protecting the core alloy against corrosion. Cold-finished rod is rod brought to final dimensions by cold working to obtain improved surface finish and dimensional tolerances. Cold-finished extruded rod is rod produced by cold working extruded rod. Cold-finished rolled rod is rod produced by cold working rolled rod. Cold-heading rod is rod of a quality suitable for use in the manufacture of cold-headed products such as rivets and bolts. Extruded rod is rod produced by hot extruding. Redraw rod is coiled rod of a quality suitable for drawing into wire. Rivet rod, see cold-heading rod. Rolled rod is rod produced by hot rolling.

rolled bar (see bar, rolled)

rolled-in metal. An extraneous chip or sliver of metal rolled into the surface.

rolled-in scratch. A scratch that occurs during the fabricating process and is subsequently rolled over.

rolled ring. A cylindrical product of relatively short height circumferentially rolled from a hollow section.

rolled rod (see rod)

rolled shape (see shape)

rolled special shape (see shape)

rolled structural shape (see shape)

roller burnishing. A method of improving finish and dimensional accuracy and work hardening a surface by pressure rolling without removing metal. By this process, hole diameters can be increased by 0.0005–0.002 in. (0.0127–0.051 mm) without damage to the surface. Size increase, however, usually is a secondary objective, primary aims being improved accuracy and finish, and work-hardened surfaces. In addition to the finishing of inside cylindrical surfaces, roller burnishing is applicable to the finishing of tapered holes and to outside cylindrical surfaces and circular flat surfaces. Machines suited to roller burnishing include drill presses, lathes, boring machines and automatic bar or chucking machines. Roller burnishing is sometimes used instead of reaming, but it is most often used to supplement reaming or boring. The process is applicable to metals softer than about Rockwell C 40. However, metals that work harden rapidly must be at lower hardness before roller burnishing than plain carbon steel or other steels that work harden slowly. Roller burnishing increases surface hardness to a depth of 0.005–0.030 in. (0.127–0.76 mm).

roll forming machines. These use a series of matched rollers that progressively form sheet material into the desired shapes. Rollers at the side shape the sides of the forms. If needed, the forms may be seam welded with rollers acting as electrodes. Straightening rollers complete the operation.

rolling ingot. A cast form suitable for rolling. **(See also fabricating ingot.)**

rolling metals (see cold and hot rolling)

rolling mills, metals. Hot or cold rolling is done in stages on a series of rolling stands. Hot ingots first are rolled into blooms. Blooms are large bars having a minimum thickness greater than 6 in. (15.2 cm), and are usually square in cross section. Blooms then are rolled into billets or slabs. Billets are rolled bars having dimensions between 0.5 and 6 in. (1.27 and 15.2 cm), but not a finished shape. Slabs are much wider than they are thick; usually more than 1½ in. (3.8 cm) thick and at least 10 in. (25.4 cm) wide. Billets and slabs are the raw material from which finished hot-worked shapes are made. Rolls on rolling mill stands may be arranged in six configurations. Blooms are rolled from ingots on two-high reversing or three-high mills. The three-high type is equipped with an elevator on each side of the stand for raising or lowering the bloom and mechanical manipulators for turning the bloom and shifting it to the various roll passes as it is rolled back and forth to final size. Four-high and cluster roll arrangements are used in hot rolling very wide plates and sheets, and in cold rolling, where the deflection of the center of the roll would result in a variation in thickness. The surfaces of hot-rolled products are slightly rough and covered with a tenacious mill scale, owing to the heating involved. Dimensional tolerances of hot-rolled products vary with the kind of metal and the size of the product.

rolling slab. A rectangular semifinished product, produced by hot-rolling fabricating ingot and suitable for further rolling.

roll mark. A raised area formed during rolling by the imprint of a depression in a roll.

roll planishing. Primarily a method of smoothing and flattening weld beads to the thickness of the parent metal by roll compression. The weld is passed between the driving and idler rollers of the planishing machine, with pressure being applied pneumatically, hydraulically, or mechanically. In addition to smoothing and flattening the cold working improves the surface grain structure and relieves shrinkage stresses. The process is faster and less expensive than machining, grinding, or hammer planishing.

roofing sheet. Coiled or flat sheet in specific tempers, widths, and thicknesses suitable for the manufacture of corrugated or V-crimp roofing.

RP molding, pultrusion (see pultrusion)

rosins. A product of several varieties of pine trees, composed of about 90% complex organic acids, largely abietic and related acids, and about 10% nonacids such as resenes and terpene. When the pine tree is tapped, a mixture of about 75% rosin, 20% turpentine, and 12% water is obtained. The turpentine and water are distilled off, leaving what is known as gum rosin. Wood rosin is extracted from pine stumps remaining after lumbering operations. The stumps are removed by bulldozers and chipped and the rosin extracted by a light aromatic aliphatic mixed solvent. The coloring matter is partially removed from the solution by adsorption by a bleaching of earth or clay, or by liquid–liquid extraction with furfural. Tall oil rosin is separated from crude tall oil obtained from kraft cook liquor used in the production of paper from pine wood. The rosin acids can be esterfied by heating with various alcohols. The methyl ester is a heavy, viscous liquid used in mastic compositions in asphaltic impregnants, as a tackafier in adhesives, as a drying oil extender in linoleum, and as a varnish constituent. The glycerol ester, known as ester gum, melts at 197.6°F (92°C) and is used as a lacquer and varnish resin. When pentaerythritol is substituted for glycerol, a resin with a melting point of 239°F (115°C) is produced. Varnish made with this resin dries more quickly than that made with ester gum and has greater resistance to water and alkali. Hydrogenated methyl esters are used as plasticizers.

rotary swaging and hammering. Basically a fast mechanical hammering or impacting method, rotary swaging affords many of the desirable characteristics inherent in the various hot and cold forming methods. Perhaps more accurately termed rotary reducing, the swaging process should not be confused with swedging which consists of forming or shaping with but a single squeeze or blow of a solid die. Rotary swaging is, broadly speaking, confined to the reducing of a symmetrically cross-sectioned metal blank to a desired shape or form. In swaging, the blank is fed by hand or by mechanical means into the die openings where thousands of hammer blows per minute conform the blank to the shape of the dies used. Blanks ordinarily may be of any symmetrical cross section, although the actual finished swaged form, being produced by a series of fast hammer blows accumulatively applied about a circle, must be substantially of circular cross section. Along the length of the part, however, an almost unlimited variety of diameters or forms can be produced. Swaged parts may be either hollow or solid. In either case the working of the metal through swaging refines the grain structure and improves metal quality. Parts produced by cold swaging exhibit the highly desirable characteristics which normally accompany cold-worked metal. Extent of the effects of cold working depends largely on the machine used. High-quality finish and increased strength are largely sacrificed when swaging hot, but hot working is necessary with highly work-hardening materials or extremely severe reductions. Normally, therefore, hot swaging should be avoided whenever possible for maximum economy in production and where fine surface finish is desirable without machining. An interesting adjunct to the swaging process is that known as hammering. The principle of operation is very similar to that of swaging and like swaging offers the advantage of cold working and economy of material. Hammering, however, is done by two dies operating in a vertical plane, the upper die reciprocating and striking the work held in the lower die, which is either mechanically or hydraulically fed to suit the work being produced.

rotational molding (plastics). A molding process intended for the manufacture of hollow objects from thermoplastic and also to a limited extent to process thermosetting materials. The solid or liquid polymer is placed in a mold; the mold is first heated and then cooled while being rotated about two perpendicular axes simultaneously. During the first portion of the heating stage when molding powdered material, a porous skin is formed on the mold surface. This gradually melts as the cycle progresses to form a homogenous layer of uniform thickness. However, when molding a liquid material, it tends to flow and coat the mold surface until the gel temperature of the resin is reached, at which time all flow ceases. The mold is then indexed into a cooling station, where forced air, water spray, or a combination of both cool the mold. It is then positioned in a work zone where the mold is opened, the finished part is removed, and the mold is recharged for the following cycle.

Roto-Flo process (see cold rolling)

rottenstone (see abrasives)

RP molding equipment. The filament winding process consists of feeding continuous stand roving reinforcement from a creel impregnated in a resin bath and wound on a mandrel having a surface of revolution. Machines similar to a lathe lay down the glass-fiber reinforcement in a precisely determined pattern until the required number of layers has been applied. After the part is cured at room temperature or in an oven, it is removed from the mandrel. Filament-wound objects have the highest specific tensile strength of any available material. Unlike metals, glass filaments have no yield points. They exhibit perfect elastic behavior from no load to rupture. They do not creep under stress. The electrical characteristics, resistance to chemical attack and nonmagnetic properties are valuable attributes of fiberglass-reinforced plastics. **(See also filament winding.)**

RP molding, high-pressure. High-pressure molding processes for fibrous-glass-reinforced plastics have been reserved for high production runs where detailed configuration and finished surfaces on both sides are required. The principal RP pressure-molding systems in use are compression molding, injection molding, and high-pressure laminating. The most popular mass-production method for fibrous-glass-reinforced plastic parts is compression molding. In this process reinforcement is combined with resin mix either at the press or in a separate combining process to make a molding compound. Compression molding using mat reinforcement is most economical for large flat or simple shapes such as machinery cabinets, trays, electrical flat sheet, and printed circuit boards. Fabric reinforcement is used to gain higher physical strength through high glass content. It is commonly used for aircraft and missile components and for electrical laminates for insulators, breakers, and bus bars.

RP molding, low-pressure. Especially suited for the production of large complex shapes. Resin and glass fibers in the form of fabric, woven or mat, are placed in molds manually. Entrapped air is removed with squeegees or serrated metal rollers. Successive layers of glass and resin can be added to build the part to the desired thickness. If a smooth colorful surface is required, a pigmented gel coat can be sprayed on the mold before layup. The mold side of the part becomes the finished surface. The resin is in contact with the air and the wet layup normally hardens at room temperature. Heat may be used to reduce hardening time. A smoother exposed side may be achieved by wiping on a film such as cellophane, which can be removed after hardening.

rubber, acrylic. Copolymers of ethyl, methyl, or butyl acrylate. A typical rubber is about 95% ethyl acrylate and 5% chlorethylvinyl ether. The acrylic rubbers are produced by emulsion polymerization. Another possibility is a copolymer with ethylene. Acrylic rubbers are vulcanized or cured with amines such as triethylenetetramine, and tetraethylenepentamine. Carbon black is used as a reinforcing filler in amounts from 35 to 60 parts per 100 parts of the polymer. White silica types of pigments can also be used in similar amounts. Softeners or plasticizers are used only to a limited extent. "Tempering" of vulcanized articles at around 300°F (149°C) for 24 h improves the properties of finished articles. Acrylic rubbers have excellent resistance to high temperatures and to oils. They can be used at temperatures from −10 to 400°F (−23 to 204°C) They have very good resistance to oxygen, ozone, and sunlight. They are decomposed by alkaline solutions and have poor resistance to acids. The tensile strength of the acrylic rubbers is from 500 to 2400 psi (3448 to 16,550 kPa) and elongation 100–400%. The acrylic rubbers are used mainly in applications where their excellent resistance to heat and to oils, especially sulfur containing oils is needed. These applications include oil hose, automobile gaskets, O-rings, belting, tank linings, and cements. Acrylic rubbers are also available in latex form. Uses include sizes and binders for textiles, paper, and leather, and adhesives where the bond is exposed to oil.

rubber, butadiene. The compounding and processing of butadiene–styrene rubbers are similar to those operations using natural rubber. The butadiene–styrene rubbers are less unsaturated than the natural rubber and are slower in curing. Milling requires somewhat more power. The unvulcanized butadiene–styrene rubber is not tacky like natural rubber. In tire manufacturing this makes it necessary to use natural rubber cement on the surfaces to be joined. Butadiene–styrene rubber is used for practically all uses of natural rubber. Butadiene–styrene hard rubber is very similar to natural hard rubber.

rubber butyl. Butyl rubber is an all petroleum product made by copolymerizing isobutylene and just enough isoprene to obtain the desired degree of unsaturation necessary for vulcanization. Brominated and chlorinated butyl rubber are also available and are prepared by select replacement of hydrogen with bromine or chlorine. Until the introduction of ethylene propylene rubber, butyl was the only elastomer that was satisfactory for phosphate-ester-base hydraulic fluids service over a temperature range of −65–+225°F (−54–+107°C). In addition, butyl has excellent resistance to gas permeation, which makes it particularly useful for vacuum applications and accounts for its wide use in the manufacture of inner tubes and the inside layer of tubeless tires.

rubber, chloride. Chlorine is bubbled into a solution of natural rubber in a solvent such as carbon tetrachloride at 175–230°F (79–110°C). The granular product is separated by running into hot water, distilling off the solvent, and centrifuging. It is used in special-purpose paints and as a bonding agent for metals. A similar product can be made from synthetic polyisoprene and butadiene–styrene rubbers.

rubbers, chlorosulfonated polyethylene. This rubber is made by substituting chlorine and sulfonyl groups into polyethylene with a molecular weight of about 20,000. Approximately one chlorine atom to each seven carbon atoms and one sulfonochloride group for each 90 carbon atoms are added. The saturated rubber is cured with litharge (PbO) or magnesia (MgO) and an accelerator such as mercaptobenzethiozole. This rubber can be extruded, molded, or formed to sheets by calendering. It is also available as a coating (paint) material. It has excellent temperature resistance [−65–+250°F (−54–121°C)] and resistance to many chemicals, including oxidizing acids. It is used for lining chemical tanks, for hoses handling acids, for conveyor belts, and in a variety of molded products where resistance to heat and chemicals is important. It is used as a coating over other rubbers to protect against ozone. Coated fabrics are used in a variety of products such as industrial pump diaphragms, camera bellows, and automobile topping.

rubber, epichlorohydrin. Epichlorohydrin is a recent addition to the oil-resistant polymer class. Compounds of this type are aliphatic polyethers with chlorofunctional side chains. Two basic classes, homopolymers (CO) and copolymers (ECO), are available. Both have excellent resistance to hydrocarbon oils, fuels, and ozone. High-temperature resistance is good, but compression set at 300°F (149°C) is only fair. This property, plus the corrosive nature of epichlorohydrin, are limiting factors in some applications. Copolymers give very good low-temperature properties, providing a temperature range of −65–+275°F (−54–+135°C), where corrosion is not likely to be encountered and where compression set is not a problem. The homopolymers are useful through a temperature range of approximately −40–+275°F (−40–+135°C) under the same conditions.

rubber, ethylene. Mixture of propylene with 20–80% ethylene can be copolymerized in a solvent at between 50 and 150 psi (345 and 1035 kPa) with a catalyst such as dialkylaluminum combined with vanadium oxytrichloride to form an elastomer (EPR). This can be vulcanized with a peroxide such as dicumyl peroxide. The terpolymer (EPT), which can be vulcanized with sulfur, is made by adding about 3% dicyclopentadiene to the thylene and propylene and polymerizing in hexane as a solvent. The copolymer and terpolymer are readily processed in conventional rubber machinery and are compatible with oils, carbon black, and mineral fibers. These rubbers are highly resistant to ozone degradation, having good electrical properties, excellent abrasion resistance, and low density. The terpolymer is being used in tires, wire insulation, and hose.

rubber, ethylene propylene. Ethylene propylene rubber is an elastomer prepared from ethylene and propylene monomers (ethylene propylene copolymer) and at times with a small amount of a third monomer (ethylene propylene terpolymers). Although EP is a relatively new introduction to the rubber industry (1961), it has won acceptance because of its excellent resistance to phosphate-ester-type hydraulic fluids. Ethylene propylene has a temperature range of −65–+300°F (−54–+149°C) for most applications.

rubber, fluorocarbon. Fluorocarbon elastomers were first introduced in the mid-1950s. Since then they have grown to major importance in the seal industry. Owing to wide spectrum chemical compatibility and temperature range, fluorocarbon rubber is the most significant single elastomer development in recent history. Its working temperature range is considered to be from −20 to +400°F (−29 to +204°C) but it will take temperatures up to 600°F (316°C) for short periods of time. It has been known to seal at −65°F (−54°C) in some static applications. Newer developments have tremendously improved the compression set characteristics of this very useful material. Compounds made from fluorocarbon elastomers should be considered for use in aircraft, automotive, and other mechanical devices requiring maximum resistance to deterioration by test and functional fluids.

rubber, GRS or SBR. Styrene and butadiene rubber, formerly known as Buna-S or GRS (government rubber styrene), is the synthetic most like natural rubber in processing and performance characteristics. It is a copolymer made by combining butadiene and styrene to form a compound which is then polymerized. GRS can be vulcanized with sulfur and rubber accelerators and can be cured to a hard condition. It resists atmospheric deterioration slightly better than natural rubber. The pure gum stocks of GRS are of inferior quality and must be reinforced with carbon black in order to secure the best physical properties. Because of carbon compounding it is available only in black. GRS can be used to replace natural rubber or to be compounded with natural or reclaimed rubber. Its chief use is in the manufacture of automobile tires.

rubber, hydrochloride. Made by bubbling hydrogen chloride into a solution of rubber or by treating rubber with a liquid hydrogen chloride at −45–−26°F (−43–−32°C). This is used largely as a packaging film.

rubber, isomerized. When rubber is heated with acids, such as sulfuric, benzenesulfonic, or *p*-toluene sulfonic, a rearrangement occurs in the molecule, producing "thermoprene," used in a cement for bonding rubber to metals and in chemically resistant paints. A similar compound is produced by reacting rubber with certain metallic chlorides such as tannic or titanium chlorides. It is used as a chemically resistant tank lining, in paints, and as a thermoplastic molding compound.

rubber, natural. Natural rubber is made from latex, a milky juice found in the bark of a number of tropical trees and other plants. Latex is collected from trees in the same manner as maple sap and turpentine. The rubber is separated from the latex by evaporation or by coagulation. Formic acid is the ingredient usually added to produce coagulation. After preparation, the crude rubber is washed and rolled into thin sheets. It is then drained, dried, and smoked leaving it in a condition ready for shipment. Crude rubber has undesirable qualities including low strength. Vulcanization of the rubber by addition of sulfur at about 160°F (71°C) improves its properties. The degree of vulcanization modifies the properties of rubber. If no sulfur has been added, the rubber has the tendency to become tacky with time and of being affected by temperature changes and by certain reagents. When sulfur is added, rubber becomes much stronger and can be stretched to eight to ten times its original length before breaking. Rubber containing a high percentage of sulfur is hard. Soft rubber, with a lower percentage of sulfur, has unusual stress–strain diagrams with reverse curves. At the higher stresses, soft rubber tends to stiffen with increasing load. Hard rubber may have ultimate strengths as high as 10,000 psi (6.9×10^4 kPa). Fillers such as carbon black, zinc oxide, and clay vary the properties of rubber. In amounts up to 30% these fillers increase the strength of rubber. Soft rubber has excellent energy-absorbing capacity, since large deformations can be produced before fracture. Under cycles of stress, the energy loss per cycle, or hysteresis, is large. The loss of energy appears as heat, which tends to deteriorate the rubber if the hysteresis losses are high. Rubber has low electrical conductivity and is used as an insulating material.

rubber, neoprene (chloroprene, CR). Neoprenes are homopolymers of chloroprene (chlorobutadiene) and were among the earliest of the synthetic rubbers available to industry. Neoprene can be compounded for service at temperatures to −65–300°F (−54–149°C). Most elastomers are *either* resistant to deterioration from exposure to petroleum lubricants or oxygen. Neoprene is unusual in having limited resistance to both. This, combined with broad temperature range and moderate cost, accounts for its desirability in many applications. Neoprene is processed by natural rubber procedures. Metallic oxides, such as a combination of magnesium and zinc oxides, are superior to sulfur as vulcanizing agents. Fillers used are mainly those used in natural rubber. Petroleum oils and polyester plasticizers may be added. Neoprene has good strength and resistance to weathering. Its high chlorine content imparts a high resistance to heat and flame. This, in addition to excellent resistance to oil and chemical, made neoprene suitable for many industrial applications such as gasoline hose and special conveyor belts. The different varieties combined with various fillers and other modifying agents make possible special properties for individual applications. Neoprene latices are used in production of gloves, balloons, industrial parts, and foam items. They are also used in adhesives.

rubber, nitrile or Buna N (NBR). Nitrile is a copolymer of butadiene and acrylonitrile. Acrylonitrile content is varied in commercial products from 18% to 48%. As the nitrile content increases, resistance to petroleum-base oils and hydrocarbon fuels increases, but low-temperature flexibility decreases. Owing to its excellent resistance to petroleum products, and its ability to be compounded for service over a temperature range of −65–250°F (−54–121°C), nitrile is the most widely used elastomer in the manufacture of seals and sealing devices. The copolymer can be modified for the addition of methacrylonitrile, styrene, vinylidene chloride, methyl methacrylate, and the acrylic acids. Process is similar to that for natural or butadiene–styrene rubbers. The nitrile rubber is less plastic than natural rubber and develops more heat in milling. Desirable characteristics of nitrile rubbers include excellent oil and heat resistance. Resistance to paraffinic oils is greater than to aromatic oils. Since nitrile rubbers are higher in close than natural rubber, they are used mainly in special applications such as bullet-sealing tanks and fuel hose where oil resistance is needed. They are used in blends with polyvinyl chloride which have greater sunlight and ozone resistance than the rubbers alone. These blends are used in upholstery fabric, insulation, and leather substitute. Phenolic resins are blended with nitrile rubber to produce a stiffer stronger product. High percentages of the resin produce thermosetting adhesives used for bonding metals, plastics, and glass. Nitrile rubber also blends with natural rubber, high-styrene copolymers, and polysulfide rubbers. In general, it tends to add toughness.

rubber, polyacrylate. This material has outstanding resistance to petroleum fuel and oil. In addition, it possesses complete resistance to oxidation, ozone, and sunlight, combined with an ability to resist flex cracking. Compounds from this base polymer have been developed, which are adaptable for continuous service in hot oil over the temperature range 0–350°F (−18–177°C). Resistance to hot air is slightly superior to nitrile polymers, but strength, compression set, and water resistance are inferior to many of the other polymers. There are several polyacrylate types available commercially, but all are polymerization products of acrylic acid esters. Greatest usage of polyacrylate is by the automotive industry in automatic transmissions and power steering gears.

rubber, polybutadiene. In cis-polybutadiene, butadiene is polymerized into a chain polymer using a stereospecific cobalt catalyst of the Ziegler type. Polybutadiene can be processed by the conventional rubber methods and can be blended effectively with natural and styrene–butadiene rubbers. Oils, fatty acids, and resin are used as plasticizers. Carbon black is used as a filler and sulfur as the vulcanizing agent. The addition of polybutadiene to natural rubber or styrene–butadiene may give a longer-wearing tire tread. It is also being used alone as tread stock and for conveyor belt covers and other uses where high abrasion and flex-cracking resistance are desirable.

rubber, polysulfide. Condensation products of an organic dihalide with a polysulfide. Typical combinations are ethylene dichloride and sodium polysulfide, and dichlorethyl formaldehyde and sodium sulfide. Properties of the rubber may be modified by using two or more chlorides together and other sulfides may be substituted. Bromides may also be substituted for the chlorides. The constituents are reacted in water solutions in stainless-steel agitated reactors. A latex results from which rubber may be coagulated by the addition of acetic or sulfuric acid. Polysulfide rubbers vary in properties but in general have lower tensile strength and abrasion resistance than natural rubber, but excellent weathering, oxygen, and oil resistance. Most solvents do not dissolve them, although some cause swelling. The polysulfides are used for O-rings, gaskets, printing rolls, hose and as lining in bulletproof airplane gasoline tanks. They are used as protective coatings, applying either in the latex form or by flame spraying. Polysulfide rubber was one of the earliest commercial synthetic polymers and is prepared from dichlorides and sodium polysulfide. It has a remarkable combination of solvent resistance, low-temperature flexibility, flex-crack resistance, and oxygen and ozone resistance. However, heat resistance, mechanical strength, and compression set are not outstanding. Other seal compounds are more versatile from the performance standpoint, hence polysulfide rubber is recommended only for specific applications that cannot be satisfied by any other elastomer. Temperature range is −65–225°F (−54–107°C). Seals of polysulfide are recommended for service involving contact with solutions of petroleum solvents, ketones, and ethers.

rubber, polyurethane. Polyurethanes exhibit outstanding mechanical and physical properties in comparison with other elastomers. Over a temperature range of −65–200°F (−54–93°C), resistance to petroleum oils, hydrocarbon fuels, oxygen, ozone, and weathering is good. However, polyurethanes quickly deteriorate when exposed to acids, ketones, and chlorinated hydrocarbons. Certain types of polyurethane are also sensitive to water and humidity. The inherent toughness and abrasion resistance of polyurethane seals is particularly desirable in hydraulic systems where high pressures, shock loads, wide metal tolerances, or abrasive contamination is anticipated.

rubber processing. Rubber products are made by several processes. The simplest is where they are formed from a liquid preparation or compound. These are commonly known as latex products. Dipped products can be made by immersing a form repeatedly into the latex compound, causing a certain amount of the solution to adhere to the surface of the form each time. The film is allowed to dry, usually in air, after each dipping. Dipping is continued until the desired thickness is obtained. After vulcanization, usually in steam, the products are stripped from the forms. Most latex products are made by the anode process. This process utilizes the electrical charges on the latex particles. The charges are neutralized by being associated with a coagulant that has previously been deposited and which releases positively charged ions when dipped into the latex. The positive ions neutralize the charges on the adjacent latex particles and thus cause them to be deposited on the form. This process goes on continuously so that any desired thickness can be deposited. When products are to be made from solid rubber, the first step is the compounding of the rubber, or elastomers, and the vulcanizers, fillers, antioxidants, accelerators, and other pigments. This is usually done in a Banbury mixer, which breaks down the rubber and mixes in the pigments to form a homogeneous mass. The mix next is put on a rubber mill, the rolls on this mill being made of chilled iron and rotate toward each other at different speeds. Mill rolls are cooled by the circulation of water through their interiors. The sulfur and accelerators usually are added at this stage. Rubber compounds and also plastics in sheet form are produced on a calender. The sheet coming from a calender is rolled into a canvas liner to prevent the adjacent layers from sticking together.

rubber, silicone. Compounds containing silicon, oxygen, and one or more organic groups. The silicones with elastomeric properties are of four types. The general-purpose rubber is the polydimethysiloxane made up of alternating silicon and oxygen molecules with two methyl groups attached to each silicon. The second type has a small percentage of vinyl groups substituted for part of the methyl groups. The third type has phenol groups substituted for a small part of the methyl groups. This lowers the brittle point improving the low-temperature properties. The fourth type is a fluorosilicone, polytrifluoropropylmethyl siloxane. It has physical properties similar to the other silicone rubbers but has superior solvent resistance. Processing of the silicone rubbers is similar to processing natural rubber except that no accelerator, softener or plasticizer, or antioxidant is required. The vulcanizing or curing agent is benzoyl peroxide which breaks down above 185°F (85°C) to produce free radicals. These remove a hydrogen from a methyl group, producing reactive —CH_2 units which will cross-line with other like units. The rubber sets up in a mold after 5–10 min at 260°F (127°C). It is then necessary to heat it for several hours at 480°F (249°C) to complete the cure and drive off any volatile impurities. Various manufactured and natural silica products are used as fillers. Carbon black may be used with the unsaturated methylvinyl rubber. The rubber can be compression molded, extruded, or formed into sheets. Silicone rubbers are used mainly in applications at extremely high or low temperatures [−175–700°F (−115–371°C)]. Typical uses are wire and cable insulations, gaskets and seals for electric and electronic equipment, and seals in outdoor floodlights. A large amount of silicone rubber is used as seals, gaskets, and o-rings in aircraft.

rubber, sponge. Produced from dry natural or synthetic rubber, using a blowing agent such as sodium bicarbonate and a fatty acid. If an open-cell structure is desired, the compounding and curing are regulated to produce the gas before vulcanizing, thus rupturing the cell walls and producing an interconnecting structure. For a closed-cell structure the rubber is partially cured before gas expansion. Vulcanizing is in suitable frames or molds in either batch or continuous operation. Open-cell types of sponges are used in gasketing, sealing, heat insulating, cushioning, and shock absorption. The closed-cell types have similar uses in addition to those requiring buoyancy in water. The hard rubber product is used in heat insulating, sandwich construction, and floats. Foam rubber is produced by adding gas to rubber latex and vulcanizing. It is used largely in pillows, furniture cushions, mattresses, automotive pads, and gaskets.

rubber, urethane. These elastomers result from the reaction of isocyanites with active hydrogens in polycols, such as polyesters, polyethers, glycols, or castor oil. They contain the urethane linkage ($RNHCO_2R'$). Commonly used isocyanates are toluene diisocyanate and hexamethylene diisocyanate. Most of the urethane rubbers can be reinforced with carbon black similarly to other rubbers. They have high strength and load-bearing capacity, very good resistance to abrasion, tear, and oils. Uses include solid tires for industrial trucks, vibration mounting, belts, shoe soles and heels, seals, gaskets, and gears. Large amounts are used in foams.

rub mark. A minor form of scratching consisting of a large number of very fine scratches or abrasions.

S

sample. One or more units of product (or a relatively small quantity of a bulk material) that is withdrawn from a lot or process stream, and that is tested or inspected to provide information about the properties, dimensions, or other quality characteristics of the lot or process stream. Not to be confused with specimen.

sand, abrasive. A general term for quartz sand used in stone sawing, glass grinding, and metal polishing, and also for sandblasting to clean castings; to remove paint, rust and stain; and to carve designs on stone. Requirements for most of these uses are not strict, but all demand clean, hard, tough grains, and most require certain grade sizes. There is little uniformity in grain-shape specifications. Sand for sandpaper is mostly crushed quartz or quartzsite, and so does not fall under the heading of true sand.

sand casting (see casting, sand casting metals)

sand, filter. Municipal water departments chiefly use filter sand, which is utilized to remove sediment and bacteria from water supplies. The sand must be free from lime, clay, and organic matter, and must be at least 98% insoluble in hydrochloric acid. Grain shape seems to make little difference, but uniformity and size distribution of the grains are important. Most filter sand is produced from the same deposits that furnish glass and molding sands.

sand, foundry. Includes molding sand, used for making molds into which molten metal is poured, and core sand, which makes forms that occupy space within the mold and produces hollows in the casting. For these uses sand must have the following properties: (1) Sufficient cohesiveness to hold together when moist. Clay is the commonly used bond. (2) Sufficient refractoriness to withstand the heat of the molten metal. The required heat resistance varies with the type of metal being cast; for steel, high-silica sand with a fire-clay bond is needed to withstand pouring temperatures of 2444–2732°F (1340–1500°C). (3) Sufficient strength to resist the pressure of the metal. (4) Enough permeability to permit water vapor and gases, generated during cooling of the metal, to escape outward from the mold instead of inward into the metal. (5) Proper texture and composition, so that the mold will produce a smooth surface on the casting and will not react with the hot metal.

sand, glass. Quartz sand makes up 52–65% by weight of the mix from which glass is made (the other constituents being soda ash and lime). The first requirement for glass sand is a high silica content, at least 93% for optical glass. There is a small but very important percentage of impurities in glass sand. Iron oxide, for example, must not exceed 0.06% since iron is a discoloring agent in glass. Rigid restrictions also exist on other elements, including chromium and cobalt. Alumina needed in small amounts in the mix may be present in the sand up to 4–5% (in feldspar grains). There is no industry-wide specifications as to grain size but glass sand should pass the No. 30 sieve and be retained on the No. 140 sieve.

sanding machine. A hand-operated or production machine having a powered abrasive-covered disk or belt, used for smoothing, polishing, or finishing wood and other materials for furniture, plywood, etc. Also called "sander."

sand molding. Molding for casting may be done by four different methods. These are bench molding, floor molding, machine molding, and pit molding. With the exception of machine molding, these methods differ essentially only in respect to the size of the molds which are made. Machine molding differs from the others in that the sand is packed around the patterns by some semiautomatic mechanical device, thus greatly reducing the amount of labor required. The molder uses some simple tools and equipment. Some type of flask must be provided into which the sand may be packed. These flasks are usually rectangular in shape and constructed either of wood or metal. Most flasks are made in upper and lower halves to permit the mold to be opened for removal of the pattern. The halves are provided with lugs or pins on one half and holes on the other half to ensure proper alignment when the two halves are assembled. Flasks are made so that they may be removed from the mold after it is completed. One type has a hinge on one corner, and the opposite corner is provided with a latch which permits the flask to be fastened shut or opened. After removable flasks are removed, it usually is necessary on all but very small molds to put a slip jacket around the mold before pouring the metal in order to resist the pressure of the metal. Heavy weights often are placed on top of the molds to keep the sections from being separated by the hydrostatic pressure of the molten metal.

sand molding, bench mold. In making a bench mold, the molder places the lower half of the flask, known as the drag half, upside down on a mold board, which is on the molding bench. The mold board is a smooth board, slightly larger than the area of the flask, usually provided with two cleats about 1 in. (2.54 cm) thick which raises the board off the bench sufficiently so that the molder can easily take hold of it for handling and turning. If a split pattern is used, the lower half of the pattern also is placed upside down on the mold board in the center of the drag half of the flask. A light coating of parting dust is sprinkled over the pattern to prevent sand from sticking to it and thus make it easier to withdraw the pattern. Molding sand or facing sand is then put into the mold by sifting it through a riddle [a sieve having $^1/_8$–$^1/_4$ in. (0.3–0.64 cm) mesh openings, usually circular in shape]. After sufficient facing sand has been put in, regular tempered molding sand is shoveled into the flask. The sand is then packed to the proper degree by means of a hand rammer. After the drag half of the mold is completely filled with sand and rammed, the sand is cut off even with the flask by means of a strike-off (a flat steel bar that is swept across the mold, resting on the sides of the flask, to remove the excess sand).

sandpaper. Heavy paper coated on one side with sand (originally) or aluminum oxide or silicon carbide abrasives. Used on sanding machines and for hand smoothing, polishing and finishing.

sandstone (building stone). Composed of quartz grains cemented together by iron oxide, calcium carbonate or clay. Graywacke is a sandstone with a variable quartz content, generally less than 75%; 25% or more consists of sand-size grains or rock and feldspar. Cement is a detrital clayey material, not a mineral precipitate. In arkosic sandstone the grains are dominantly quartz and feldspar, and cement, if present, is a mineral precipitate. The variety with more than 25% feldspar is arkose, that with less is feldspathic sandstone. The name lithic sandstone is applied when rock particles exceed feldspar. At various times all these varieties of sandstone have been used commercially. The present day, however, the most important are protoquartzite and orthoquartzite, on account of their high content of silica. Much of the sandstone used commercially contains little cementing material, is loose and crumbly, and commonly referred to as sand. Dimension sandstone is used for exterior facing and trim on large buildings, for flagstones and curbstones, and bridge abutments and retaining walls. Well-cemented sandstone may be crushed and used for concrete aggregate, railroad ballast and riprap. Refractory stone is orthoquartzite suitable for manufacture into superduty silica brick, a high-temperature refractory used in lining coke ovens and metallurgical furnaces. Pulverized quartz, some of which is made by grinding high-silica sand, is used in the manufacture of whiteware, glazes and porcelain enamel, as an inert filler or extender in paint, and as an abrasive in mechanics' soap and cleanser. For these purposes it is ground to minus 200 mesh (silt size and finer).

saponification (soap preparation). Glycerides, like other esters may be hydrolyzed by heating with a solution of sodium or potassium hydroxide. The hydrolysis products are glycerol and the alkali metal salts of long-chain fatty acids. The latter are called soaps and alkaline hydrolysis is called saponification, whether the term is applied to fats, oils, or simple esters. The best soaps are those in which the hydrocarbon segment is saturated and is from 12 to 18 carbon atoms in length. Sodium soaps are hard soaps and commonly are used as cake soaps. Potassium soaps are soft soaps, produce a finer lather, and usually are employed in shaving cream. Natural or "hardened" fats are saponified by heating them in open kettles with a solution of sodium hydroxide. When the reaction is complete, the thick curds of soap are precipitated by the addition of sodium chloride. The water layer is drawn off and the glycerol it contains is recovered by concentration and distillation. The crude soap, which contains salt, alkali, and some residual glycerol, is heated with water to dissolve these impurities. The soap is again reprecipitated by the addition of salt. The washing and precipitation procedure is repeated several times. Finally, the soap is heated with sufficient water to give a smooth mixture, which on standing separates into a homogeneous upper layer of kettle soap. Pumice is incorporated for scouring soaps and antiseptic (phenol, hexachlorophene) for medical or deodorizing soaps. When air is blown into molten soap, the specific gravity of the solidified product is lowered to 0.9. This treatment produces a floating soap. Sodium and potassium soaps are soluble in water and are the common household soaps. The long-chain fatty acid salts of the heavier metals are not water soluble and may be blended with mineral oils to form lubricating greases.

Saran. Vinylidene chloride polymerizes readily in the presence of peroxides. While it can be mass polymerized, emulsion polymerization using redox technique allows better control and is a more satisfactory method. Because of difficulties in plasticizing and molding the pure polymer, the commercial product known as Saran is a copolymer with either vinyl chloride or acrylonitrile, ranging upward from 73% vinylidene chloride. The softening range of Saran varies from about 160 to 350°F (71 to 177°C). It is commonly formed by extrusion and injection molding. When cooled rapidly, it is amorphous, soft, weak, and pliable. Cooling in the mold at about 200°F (93°C) or cold drawing develops a crystalline structure. Saran has good resistance to most organic solvents and to common acids and alkalies. It is nonflammable and has good mechanical properties, toughness, and durability. It has good stability to aging and is tasteless, odorless, and nontoxic. Saran films show very low water vapor transmission rates and remains flexible at low temperatures. Saran tubing is used in laboratories and plants to handle solvents and chemical solutions. Rigid Saran and Saran-lined steel pipe is used industrially. Saran is also used in fibers largely for outdoor upholstery and automobile seat covers. Saran film is used in the food industry.

scalped extrusion ingot (see extrusion ingot)

scalping. Mechanical removal of the surface layer from a fabricating ingot or semifinished wrought product so that surface imperfections will not be worked into the finished product.

scanner. An electronic device that optically converts coded shop floor information into electrical control signals.

scanning wand (see wand scanner)

scleroscope test. A hardness test where the loss in kinetic energy of a falling metal "tip," absorbed by indentation upon impact of the tip on the metal being tested, is indicated by the height of rebound.

scratch. A sharp linear indentation in the surface of the metal. (See also gouge and rub mark.)

scratch brushed foil (see foil)

screw machines (see automatic bar and chucking machines)

screw machine stock. Bar, rod, and wire in certain standard alloys, tempers, sizes, and shapes suitable for automatic screw machine applications.

seam. A line juncture resulting from the deliberate bonding of two or more edges by pressure, fusion, or mechanical interlocking. (See also extrusion seam.)

seamless tube (see tube)

semiconductors. Like carbon, both silicon and germanium are Group IV elements and resemble the carbon allotrop, diamond, in that they crystalize in structure in which each atom is covalently bonded to four similar atoms. However, both silicon and germanium differ from diamond in that the energy gap between valency and conduction bands is relatively small compared with that in the diamond atom so that it is easier to free an electron into the conduction band. The transfer of this electron leaves a vacancy such that one atom in the lattice possesses only three valency bonds, instead of four. This deficiency is referred to as an electron hole. Another valency electron is able to "jump" into this "hole" and so cause the movement of the hole to proceed in the opposite direction to that in which the electrons are moving. Consequently, since the electron hole is migrating in the opposite direction to the flow of electrons its movement can be regarded as equivalent to the motion of a positive charge. Materials in which conduction occurs in this manner are termed intrinsic semiconductors.

semiconductor diodes. A semiconductor diode is usually a single crystal of semiconductor material which is artificially created. One-half of the crystal is made *n*-type and the other made *p*-type by the addition of the appropriate impurities during processing. If an external voltage source is applied to a diode so that the negative lead is attached to the *n*-type half and the positive lead attached to the *p*-type half, current flows through the diode. The negative charged electrons are attracted to the positive *p*-type side, and the positive holes are attracted to the negative *n*-type side. However, if the externally applied voltage is reversed, making the lead to the *n*-type side positive and the lead to the *p*-type side negative, no current flows through the diode. The electrons are repelled from the negative *p*-type side and the positive holes are repelled by the positive *n*-type side. The practical value of the semiconductor diode lies in its ability to conduct electricity when the *n* side is wired negatively and the *p* side positively, and to stop conducting when the *n* side is wired positive and the *p* side negatively.

semiconductors, intrinsic. Conduction that arises from thermally or optically excited electrons is called intrinsic and such materials are called intrinsic semiconductors. The conductivity of intrinsic semiconductors usually takes place at elevated temperatures, since sufficient thermal agitation is necessary to transfer a reasonable number of electrons from the valence band to the conduction band. The most important elemental semiconductors are silicon and germanium. In intrinsic semiconductors the number of mobile holes equals the number of mobile electrons. The resulting electrical conductivity will be the sum of the conductivities of the valence band (holes) and the conducting band (electrons).

semiconductor (*n*-type). Germanium acquires the diode property of rectification and the transistor property of amplification through the presence of certain impurities in the crystal structure. The two types of impurities are the donor and the acceptor. Arsenic and antimony are typical donor elements. The atoms of each have five valence electrons. The minute traces of antimony are added to germanium, each antimony atom joins the crystal lattice by donating one electron to the crystal structure. Four of the five valence electrons are paired, but the fifth electron is relatively free to wander through the lattice like the free electrons of a metallic conductor. The detached electron leaves behind an antimony atom with a unit positive charge bound into the crystal lattice. Germanium with this type of crystal structure is called *n*-type, or electron-rich germanium. *n*-type germanium consists of germanium to which is added equal numbers of free electrons and bound positive charges so that the net charge is zero.

semiconductors (*p*-type). Atoms with three valence electrons such as those of aluminum and gallium, will act as acceptors. When minute traces of aluminum are added to germanium, each aluminum atom joins the crystal lattice by accepting an electron from a neighboring germanium atom. This leaves a hole in the electron pair bond from which the electron is acquired. As an electron fills the hole it leaves another hole behind into which another electron can fall. In effect the hole (positive charge) detaches itself and becomes free to move, leaving behind the aluminum atom with a unit negative charge bound into the crystal lattice. Germanium with this crystal structure is called *p*-type or hole-rich germanium. *p*-type germanium consists of germanium to which is added an equal number of free positive holes and bound negative charge is zero.

semiconductors, *pn* junctions. Semiconductor crystals can be made in such a way that one region will be *n*-type and an adjacent region will be *p*-type. This can be accomplished by doping during the growth of a single crystal, by fusing the opposite type of impurity in the form of a metallic contact, or by changing the growth speed impurity in the form of a metallic contact, or by changing the growth speed of the single crystal. The boundary between such regions within a single crystal is called the *pn* junction. The *pn* junction has important applications, e.g., in rectifiers, photocells, solar energy converters, and transistors.

semihollow drawn shape (see shape)

semihollow extruded shape (see shape)

semihollow shape (see shape)

sensors, electrooptical imaging. Until recently, electrooptical imaging sensors have provided the most commonly used "eyes" for industrial robots and visual inspection. Standard television cameras, using vidicons, plumbicons, and silicon target vidicons, have interfaced with a computer and have provided the least expensive and most easily available imaging sensors. These cameras scan a scene, measure the reflected light intensities at a raster of approximately 320 × 240 pixels (picture elements), convert these intensity values to analog electrical signals, and feed this stream of information serially into a computer, all within 1/60th of a second. These signals may either be stored in the computer core memory for subsequent processing or be processed in real time "on the fly" with consequent reduction of memory requirements. Recently, solid-state area array cameras have become commercially available. These small, rugged, and potentially reliable cameras are fabricated using modern large-scale-integration silicon technology and may become the dominant electrooptical sensors for industrial applications. The photoactive chip of an area-array camera consists of photodiodes, usually charged-coupled devices, whose number at present varies from 32 × 32 to 320 × 512 for different requirements of resolution. These cameras operate in a raster-scan mode, similar to that of the vidicon television cameras, and produce two-dimensional images of scenes. A one-dimensional solid-state camera, using a linear array that varies from 16 to 1872 elements, is also available commercially. This device can perform a single linear scan and is very useful for sensing objects that are in relative motion to the camera, such as workpiece moving on a conveyor belt. Another large class of electrooptical sensors, which differ in several important characteristics from the above-mentioned cameras, has been used primarily in advanced "hand–eye" artificial intelligence research projects. These sensors include the image dissector camera, the cathode-ray flying spot scanner, and the laser scanner. These electrooptical sensors can be programmed to image selected areas of the field of view in a random-access manner, as contrasted with the prescribed "raster-scan" acquisition of the ordinary television camera. In many instances this method of operation permits the acquisition, storage, and processing of only the relevant data in a field of view. The image dissector has low sensitivity requiring high levels of illumination and is relatively expensive.

sensors, photoelectric proximity noncontact. A version of the photoelectric tube and light source, these sensors appear to be well adapted for controlling the motion of a manipulator. They consist of a solid-state light-emitting diode (LED), which acts as a transmitter of infrared light, and a solid-state photodiode, which acts as a receiver. Both are mounted in a small package. The sensitive volume is approximately the intersection of two cones in front of the sensor. This sensor is not a rangefinder because the received light is not only inversely proportional to the distance squared but is also proportional to the target reflectance and the cosine of the incidence angle, both of which may vary spatially. However, if the reflectance and incidence angle are fixed, then the distance may be inferred with suitable calibration. Usually, a binary signal is generated when the received light exceeds a threshold value that corresponds to a predetermined distance. Furthermore, the sensor will detect the appearance of a moving object in a scene by sensing the change in the received light. Such devices are sensitive to objects located from a fraction of an inch to several feet in front of the sensor.

sensors, range-imaging. Measures the distance from itself to a raster of points in the scene. Although range sensors are used for navigation by some animals (e.g., the bat), hardly any work has been done so far to apply range image to control the path of a manipulator. Difference range-imaging sensors have been applied to scene analysis in various research laboratories. These sensors may be classified into two types, one based on the trigonometry of triangulation and the other based on the time of flight of light (or sound). Triangulation range sensors are further classified into two schemes, one based on a stereo pair of television cameras (or one camera in two locations) and the other based on projecting a sheet of light by a scanning transmitter and recording the image of the reflected light by a television camera. Alternatively, the second scheme may transmit a light beam and record the direction of the reflected light by a rocking receiver. The first scheme suffers from the difficult problem of finding corresponding points in the two images of the scene. Both schemes have two main drawbacks: missing data for points seen by the transmitter but not by the receiver and vice versa, and poor accuracy for points that are far away. These drawbacks are eliminated by the second type of range-imaging sensor using a laser scanner, which is also classified into two schemes, one based on transmitting a laser pulse and measuring the arrival time of the reflected signal and the other based on transmitting amplitude-modulated laser beams and measuring the phase shift of the reflected signal. The transmitted beam and the received light are essentially coaxial. Range-imaging sensors have been applied so far primarily to object recognition. They are also very suitable for other tasks, such as finding a factory floor or a road, detecting obstacles and pits, and inspecting the completeness of subassemblies.

shape. A wrought product that is long in relation to its cross-sectional dimensions and has a cross section other than that of sheet, plate, rod, bar, tube, or wire.

shape, cold-finished. A shape brought to final dimensions by cold working to obtain improved surface finish and dimensional tolerances.

shape, cold-finished extruded shape. A shape produced by cold-finishing an extruded shape. Also called drawn shape.

shape, cold-finished rolled shape. A shape produced by cold-finishing a rolled shape.

shape, drawn. A shape brought to final dimensions by drawing through a die.

shape, extruded. A shape produced by hot extruding.

shape, extruded structural. A structural shape formed by hot extruding.

shape, fluted hollow. A hollow extruded or drawn shape having plain inside surfaces and whose outside surfaces comprise regular, longitudinal, concave corrugations with sharp cusps between corrugations.

shape, helical extruded. An extruded shape twisted along its length.

shape, hollow. A shape any part of whose cross section completely encloses a void.

shape, hollow drawn. A hollow shape brought to final dimensions by drawing through a die.

shape, hollow extruded. A hollow shape formed by hot extruding: Class 1 is a hollow extruded shape the void of which is round and 1 in. (2.54 cm) or more in diameter and whose weight is equally distributed on opposite sides of two or more equally spaced axes; Class 2 applies to any hollow extruded shape other than Class 1, which does not exceed a 5-in. (12.7-cm) diameter circumscribing circle and has a single void of not less than 0.375-in. (0.95-cm) diameter or 0.110-in.^{2} (0.71-cm^2) area; and Class 3 is any hollow extruded shape other than Class 1 or Class 2.

shape, lip hollow. A hollow extruded or drawn shape of generally circular cross section and nominally uniform wall thickness with one hollow or solid protuberance or lip parallel to the longitudinal axis; used principally for heat-exchange purposes.

shape, pinion hollow. A hollow extruded or drawn shape with regularly spaced, longitudinal serrations outside and round inside, used primarily for making small gears.

shaper, metal. Shapers are made either with a bull wheel which operates a reciprocating arm or with a hydraulic drive. In either case the forward stroke is usually the cutting stroke. A ram pushes a tool bit forward while it is in contact with the work and thereby removes metal. The stroke may be as short as 6 in. (15.2 cm) or as long as 40 in. (101.6 cm). Shapers that cut on the return stroke are usually large and not very common. The reverse stroke of the shaper is used to return the tool bit to its initial position in preparation for the next forward stroke. During the return stroke the work also moves in a perpendicular direction to the direction of the tool. This is called the feed.

shape, rolled. A shape formed by hot rolling.

shape, rolled special. A rolled shape other than a structural shape.

shape, rolled structural. A structural shape produced by hot rolling. **(See also shape, structural.)**

shaper-planers, metal. Designed to fill the gap between the shaper and the planer, these machines provide shaperlike speed with planer accuracy. Made in three sizes [24 × 24 in. (61 × 61 cm), 32 × 24 in. (81 × 61 cm), 32 × 36 in. (81 × 91 cm)] and five lengths ranging from 4 in. (10.2 cm) to 144 in. (366 cm), these versatile machines are used for a tremendous variety of unusual work. Master forms mounted on the table can be reproduced by means of a duplicating attachment. Complex shape production for unusual machine elements is virtually unlimited within the dimensional capacity of the machines.

shape, semihollow. A shape any part of whose cross section is a partially enclosed void the area of which is substantially greater than the square of the width of the gap.

shape, semihollow drawn. A semihollow shape brought to final dimensions by drawing through a die.

shape, semihollow extruded. A semihollow shape formed by hot extruding.

shape, solid. A shape other than hollow or semihollow.

shape, solid drawn. A solid shape brought to final dimensions by drawing through a die.

shape, solid extruded. An extruded shape whose cross section changes abruptly in area at intervals along its length.

shape, stepped extruded. An extruded shape whose cross section changes abruptly in area at intervals along its length.

shape, streamline hollow. A hollow extruded or drawn shape with a cross section of tear-drop shape.

shape, structural. A shape, rolled or extruded, in certain standard alloys, tempers, sizes, and sections, such as angles, channels, tees, zees, I-beams, and H-sections, commonly used for structural purposes.

shape, tapered extruded. An extruded shape whose cross section changes continuously in area along its length or a specified portion thereof.

shaping, metal. A machining process for removing metal from surfaces in horizontal, vertical, or angular planes by the use of a single-point tool supported by a ram that reciprocates the tool in a linear motion against the workpiece.

shearing machines. For cutting or trimming steel plates and for cutting off bars and structural material a shearing machine is generally equipped with one fixed blade and one movable blade which receives motion from a mechanism designed to give a powerful cutting movement. Blades for different material being sheared also vary on different types of machines. Another design of shearing machine has the upper or movable shear bolted to a slide which is given a vertical reciprocating motion.

sheet. A rolled product rectangular in cross section and form of thickness 0.006 through 0.249 in. (0.15 to 6.3 mm) with sheared, slit, or sawed edges.

sheet, Alclad. Composite sheet comprised of an aluminum alloy core having on both surfaces (if one side only, Alclad one side sheet) a metallurgically bonded aluminum or aluminum alloy coating that is anodic to the core, thus electrolytically protecting the core against corrosion.

sheet, clad. Composite sheet having on both surfaces (if on one side only, clad one side sheet) a metallurgically bonded metal coating, the composition of which may or may not be the same as that of the core.

sheet, coiled. Sheet in coils with slit edges.

sheet, coiled-sheet circles. Circles cut from coiled sheet.

sheet, coiled sheet cut to length. Sheet cut to specified length from coils and which has a lesser degree of flatness than flat sheet.

sheet, flat. Sheet with sheared, slit, or sawed edges, which has been flattened or leveled.

sheet, flat-sheet circles. Circles cut from flat sheet.

sheet, mill finish (MF). Sheet having a nonuniform finish which may vary from sheet to sheet and within a sheet, and may not be entirely free from stains or oil.

sheet, odd-shaped sheet blanks. Sheet cut into shapes other than circles or rectangles.

sheet, one side bright mill finish (1SBMF). Sheet having a moderate degree of brightness on one side and a mill finish on the other.

sheet, painted. Sheet one or both sides of which has a factory-applied paint coating of controlled thickness.

sheet, panel flat. Sheet which has a higher degree of flatness than flat sheet.

sheet, standard one side bright finish (S1SBF). Sheet having a uniform bright finish on one side and a mill finish on the other.

sheet, standard two sides bright finish (S2SBF). Sheet having a uniform bright finish on both sides.

sheet stock coiled (see also reroll stock)

shellac. One of the most important natural molding resins or plastics, secreted by an insect scarcely 1/40 in. (0.064 mm) long, which is found by the thousands upon certain trees, such as the plum in India. Bleached or white shellac is made by bleaching it in a sodium carbonate solution with hypochlorite. The bleached shellac is precipitated from the solution with dilute sulfuric acid. Shellac is a thermoplastic resin which has properties that made it for many years an outstanding material for two major uses: phonograph records and high-voltage insulators. It has been replaced as a record material by other plastics such as vinyl polymers. Nonracking types of synthetic resins are replacing it in the insulation field. Shellac has been an ingredient in many products such as dental blanks, grinding wheels, novelties, poker chips, leather dressing, polishes, paper glaze, gasket cement, match heads, and scaling wax. A solution of shellac in alcohol is used as a clear sealer coat under varnish or as a finishing coat on wood.

shell molding (see casting, shell molding)

Shore hardness test. Same as scleroscope test.

shot peening (see cold working)

shrinkage. Contraction that occurs when metal or plastic cools from the hot-working temperature to room temperature.

shrink-wrap machines. For packaging products to be stored or shipped, ordinarily on pallets, these machines utilize plastic sheet to encase the pallet load of product. A variety of machines available wrap the pallet load in plastic sheet, seal the wrap, and cut it off after which the load is passed through a heating tunnel for a 2–3 s exposure which shrinks the plastic wrap tightly over the load. **(See also stretch-wrap machines.)**

side loader. An industrial self-loading truck, generally high lift, having load-engaging means mounted in such a manner that it can be extended laterally under control to permit a load to be picked up and deposited in the extended position and transported in the retracted position.

side set. A difference in thickness between the two edges of sheet or plate.

silica and silicates. Constitute the bulk of all rocks, soils, clays, and sands, which almost entirely form the crust of the earth. They are also main ingredients of all inorganic building materials, ranging from natural rocks such as basalt, and granite to manufactured products such as bricks, various ceramics, glasses, cements, and mortars. The fundamental structural unit for silica and all silicates is a silicon–oxygen tetrahedron in which the silicon atom is surrounded by four oxygen atoms placed at the corner of a tetrahedron. In order to complete its octet each oxygen atom requires one extra electron, thus becoming negatively charged. This results in the formation of a tetravalent negative ion. The different silicate types arise from the various ways in which the silicon–oxygen tetrahedra can be combined with each other in a given structure. Because of the polyfunctionality of the SiO_4^{4-} anion resulting from its four-valent nature, different polymerization processes are possible.

silicon carbide (see abrasives)

silicone rubbers. Mixed inorganic–organic polymers produced by the polymerization of various silanes and siloxanes. They have an outstanding resistance to heat, making them uniquely useful for high temperature applications. The chain is based on alternate atoms of silicon and oxygen, with no carbon. Silicones and their derivatives have a variety of uses, such as solubility in organic solvents, insolubility in water and alcohols, heat stability [−80–400°F (−62–204°C)], chemical inertness, high dielectric properties, relatively low flammability, solutions low in viscosity at high resin content and with low viscosity change with temperature, and nontoxicity. Because of these properties, silicons are useful in (1) fluids for hydraulics and heat transfer, (2) lubricants and greases, (3) sealing compounds for electrical applications, (4) resins for lamination and for high-temperature resistant varnishes and enamels, (5) silicone rubber, (6) water repellents, and (7) waxes and polishes, etc.

silicones. A family of unique polymers, which are partly organic and partly inorganic. They have a quartzlike polymer structure, being made up of alternating silicon and oxygen atoms rather than the carbon-to-carbon backbone that is characteristic of the organic polymers. Silicones can be classified as fluids, elastomers, and resins. Their ultimate physical form is determined by molecular weight, extent of cross-linking between polymeric chains, and type and number of organic groups attached to the silicon atoms. The synthesis of silicone begins with sand; the sand is first reduced to silicon metal in an electric arc furnace. The metal is then converted into organochlorosilanes, the basic intermediates used to produce silicones. Organochlorosilanes are materials that have organic and chlorine groups attached to silicon, R_xSiCl_{4-x}. The R groups are most commonly methyl, phenyl, vinyl, hydrogen, or trifluoropropyl. An almost unlimited variety of organic substituents can be attached to silicon. The hydrolysis of organochlorosilanes, either alone or as mixtures, yields polysiloxanes, or silicones which may be liquid or solids. Some properties that distinguish silicone polymers from their organic counterparts are: (1) relative uniform properties over a wide temperature range, (2) low surface tension, (3) high degree of slip or lubricity, (4) excellent release properties, (5) extreme water repellency, (6) excellent electrical properties over a wide range of temperature and frequency, (7) inertness and compatibility both in physiological and in electronic applications, (8) chemical inertness, and (9) weather resistance.

silicon steels. Silicon as an alloying element dissolves in both gamma iron and alpha iron and does not form a carbide. The solubility of silicon in the gamma iron is approximately 2% whereas it is soluble in the alpha iron up to 18% and remains highly soluble even in the presence of carbon. Silicon is present in all steels as a result of the manufacturing process. In low-carbon structural steel and machine steel, the silicon content runs around 0.25% to 0.35%. Silicon in this range does not influence the properties of the steel to any extent. With the additions of silicon above 0.50% to more than 2% the tensile strength is improved as is the yield point. Silicon has been alloyed to steels of the stainless and heat-resisting types, with improvements noted in their ability to resist certain types of corrosion and oxidation at elevated temperatures. Steels used for gas engine valves in the automotive and airplane industry often contain appreciable amounts of silicon. These steels contain from 1% to 4% silicon, 6% to 9% chromium, 0.15% vanadium, and approximately 0.5% carbon.

single deep storage. AS/RS loads that are stored one deep on each side of the aisle.

sintered carbide. A product of powder metallurgy, made of finely divided, hard particles of carbide of refractory metal sintered with one or more metals of the iron group (iron, nickel, or cobalt), forming a body of high hardness and compressive strength. The hard particles are tungsten carbide, usually in combination with lesser amounts of carbides. The additional carbides are those of titanium and tantalum, with some occasional specialized use being made of the carbides of columbium, molybdenum, vanadium, chromium, zirconium, and hafnium. The auxiliary or binder metal is usually cobalt; nickel or iron is used infrequently. The carbides are present as individual grains, and also as finely dispersed network resulting from the precipitation, during cooling, of carbide dissolved in the cobalt during sintering. After solidification, the cobalt is present in the interstices as almost pure metal with its original ductility. Solid cobalt dissolves only about 1% of tungsten carbide at low temperature. The limit of solid solubility of tungsten carbide in nickel is 25% and in iron, 5%. It is this low solubility of tungsten carbide in cobalt, compared with the solubility in nickel or iron, which accounts for the use of cobalt as the binder metal. The higher solid solubility of carbide in iron or nickel would result in a more brittle auxiliary or binder phase. The cobalt also has superior ability to wet the carbide at elevated temperatures, which is important during sintering.

sintering, metals. A process of heating finely divided metal particles in a furnace to approximately melting temperature. The metal particles weld together where they are in contact with other particles and this results in a porous material.

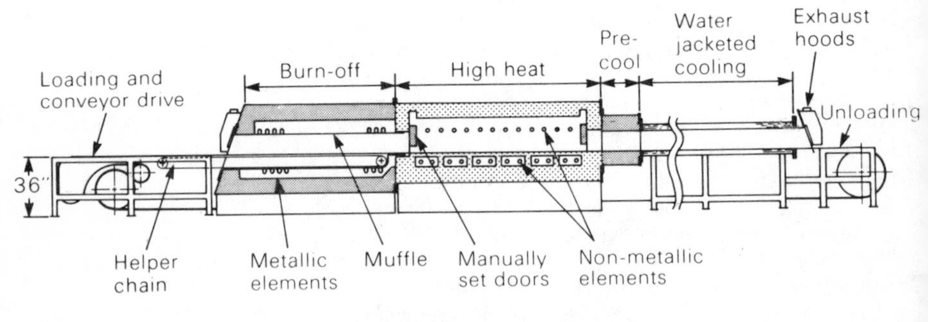

Sintering furnace.

sintering, plastics. In forming articles from fusible powders, e.g., nylon, the process of holding the press-powder article at a temperature just below its melting point for a period of time. Particles are fused (sintered) together, but the mass, as a whole, does not melt.

sized tube (see tube, sized)

skimming plant, petroleum. An oil refinery designed to remove and finish only lighter constituents from crude oil, as gasoline and kerosene. In such a plant, the portion of the crude remaining after the above products are removed is usually sold as fuel oil.

slag cements. Made by rapidly cooling the slag with cold water as it comes from the blast furnace, causing it to break up into small particles, which are dried, partially ground, and mixed with hydrated lime. The mixture is then ground finer. Slag cements set slowly, so accelerators may be required. Slag cement concrete is best suited to uses where the requirement is for bulk rather than for strength since it does not weather well when exposed to the elements. Slag cements are sometimes added to Portland cement to increase plasticity.

slate. Slate is a microcrystalline metamorphic rock, characterized by highly developed rock cleavage. Since many of the particles are less than 1 micron (0.001 mm) in diameter, the mineralogy and petrology of the slate are difficult to study. Slate is cut and fabricated into dimension form, and is also crushed to granules and pulverized to flour. In addition to roofing slates and flagstones, considerable dimension slate is produced as mill stock. This includes slate for switchboards, electrical panels, mantels, steps, baseboards and sills, and billiard table tops. Granules are used chiefly in the manufacture of composition roofing. Slate flour is utilized as a filler in paint, linoleum, and other products. Gray slate of various shades is utilized for roofing and for most of the mill stock and flour. The common colors of slate are black, gray, purple, green, and red. Mottled and variegated slate is produced in minor amounts. Being made of quartz and stable silicate, slate has very high durability.

slave pallets. A handling base or container on which a unit load is supported and which normally is captive to a system.

sliver. Slender fragment of splinter which is a part of the material but not completely attached thereto.

Slotters, metal. Differing radically from shapers and planers, the slotter carries a vertical ram which, with its tool or tools, reciprocates up and down to remove metal. Offering as standard a rotating round table with longitudinal and transverse feed in lieu of a reciprocating one, a wide variety of work can be done, especially with the new NC type machines. Standard strokes for slotters are 12 (30.5), 20 (50.8), 36 (91.4), and 48 in. (122 cm). Throat clearance on the 12-in. (30.5 cm) is 18 in. (46 cm) and maximum distance from the table to bottom of tool head is 30 in. (76.2 cm). On the 20 in. (50.8 cm) this is 24 in. (61 cm) and 40 in. (102 cm), on the 36 in. (91.4 cm), 40 in. (102 cm) and 54 in. (137.2 cm), and on the 48 in. (122 cm), 46, 55, or 59 in. (117, 140, or 150 cm) and 60, 69, or 73 in. (132, 175, or 185 cm), respectively. Work for slotters is often of large size and can be classed as follows: Work of irregular section where a clear pass over the face to be machined is difficult to set up on a planer; work requiring planing on internal sections such as splines, teeth, or keyways; and work such as ratchet or gear rings that require primarily rotary feed. Design possibilities are considerably increased by the fact that the ram may be tilted for travel at any angle from vertical to 10° forward.

slub. A metal blank for forging or impacting.

smelting furnace (blast). In a blast furnace the material to be melted is mixed with the fuel, usually coke, the heat for melting being obtained by combustion of the fuel when a blast of air is blown through the charge. Blast furnaces are taller than their cross section, and the charge is fed in near the top. As this charge travels through the furnace, the coke is burned and the metal is reduced and melted. The metal is drawn off at the bottom; the slag formed from the ash of the coke and the impurities in the ore together with added flux is drawn off from a layer immediately above the metal.

smelting furnace (reverberatory). A type of melting pot or hearth, long and narrow, with the charge heated by a flame directed over the top of the material to be melted, so that much of the heating is indirect. The hearth has a slight tip toward the end of the furnace away from the firing end; the molten metal or alloy, and the slag, if any, are drawn off here. Slags are molten glassy materials which are purposely formed in certain metallurgical furnace operations for several reasons; first, impurities in the ore and the ash of the fuel must be removed and prevented from contaminating the metal. The slag layer over the metal also prevents excessive oxidation. Thus for lead ores, with silica as an impurity, iron combines to form a glass molten slag at the temperature of working. If the iron oxides were not added, the silica would combine with lead oxide, and lead would be left in the slag. Reverberatory furnaces vary greatly in size and are used for a variety of purposes, such as copper matte, smelting, lead fire refining, etc.

sodium bisulfate (niter cake). Commonly called niter cake because it was obtained by the obsolete process of reacting sodium nitrate, or niter, with sulfuric acid. Major uses are in the manufacture of acid-type toilet bowl cleaners and for industrial cleaning and metal pickling. Minor uses are in dye baths, carbonizing wood and various chemical processes.

sodium bisulfite. Finds industrial use either in solution or as a solid. The solid is of the anhydrous form, and the pure reagent has the formula $NaHSO_3$. The commercial product usually consists almost entirely of $Na_2S_2O_5$ (sodium pyrosulfite), or sodium metabisulfite which is the dehydrated derivative of two molecules of sodium bisulfite. Major uses are as chemical intermediate, in pharmaceuticals, and in food preservatives. Minor uses are as an antichlor for pulp, paper, and textiles, and in water treatment and pollution control. It is also used in the tanning industry as a reducing agent for chrome solutions and in the photographic and organic chemical industries.

sodium cyanide. Used in the inorganic and organic chemical fields and has many metallurgical applications. It is used in treating gold ore, in the case hardening of steel, in electroplating, in organic reactions, in the preparation of hydrocyanic acid, and in making adiponitrile.

sodium hydrosulfite. An important chemical in the dyeing and printing industries. It is a very powerful reducing agent and has a specific action on many dyes, particularly vat dyes, reducing them to soluble form. This reducing agent is employed for stripping certain dyes from fabrics and for bleaching straws and soaps. The pulp and paper industry use it to meet EPA regulations for zinc discharge from their plants. It is also used in the clay industry, where "hydro" is used to remove the red iron color from clay used in manufacturing coated papers and china.

sodium hypochlorite. Employed as a disinfectant and deodorant in dairies, creameries, water supplies, sewage disposal, and for household purposes. It is also used as a bleach in laundries. As a bleaching agent it is used for cotton, linen, jute, artificial silk, paper pulp, and oranges. Much of the chlorine bought for bleaching cellulose products is converted to sodium hypochlorite before use.

sodium minerals. Two chemicals of fundamental industrial importance are sodium carbonate, known commercially as soda ash, and sodium sulfate, known as salt cake. The name soda ash is derived from the ancient practice of obtaining sodium carbonate from the ash of seaweed; salt cake is so-called because it forms an anhydrous cake in the standard process of manufacture. Most soda ash is manufactured from common salt, limestone, and coal by the Solvay process. Of the many natural sodium-bearing salts, by far the most abundant is halite. Others include nitrates and borax. Soda ash is used in making caustic soda, sodium bicarbonate (baking soda), and a host of other compounds. This group of minerals is widely used in the chemical industry. A major application of caustic soda and soda ash is in soaps, detergents, cleansers, and water softeners. Soda ash is one of the three main substances from which ordinary container glass and flat glass are made, the other being silica and lime. Other uses of soda ash are in nonferrous metallurgy, pulp and paper manufacture, and petroleum refining. The largest use of salt cake is in the kraft paper industry as a raw material of the "cooking" process by which wood fibers are made available for conversion into wrapping paper and carton board. Salt cake is the parent material of numerous chemicals, for example, sodium hyposulfite, the "hypo" used in photographic darkrooms. It is also used in glassmaking, the processing of textile fibers, dye manufacture, leather tanning, and other processes.

sodium nitrite. A very important chemical employed for the diazotization of amines in making azo dyes. It is used in meat processing as a preservative to prevent botulism, and when mixed with sodium nitrate it is employed in metal treatment. It is also used as a bleach for fibers, in photography, and in medicine.

sodium peroxide. A pale-yellow, hygroscopic powder which absorbs moisture from the air and forms a snow-white hydrate. When added to water, it forms sodium hydroxide and oxygen. Its principal use is as a powerful oxidizing agent, and in the bleaching of wool, silk, and fine cotton articles, and in chemical synthesis. Sodium peroxide reacts with carbon monoxide to form sodium carbonate, and with carbon dioxide to form carbonate with the liberation of oxygen.

sodium sulfate. A majority of sodium sulfate is consumed in the manufacture of kraft pulp. Salt cake, after reduction to sodium sulfide or hydrolysis to caustic, functions as an aid in digesting pulpwood and dissolves the lignin. It also goes into the compounding of household detergents, and the remainder into a variety of uses such as glass, stock feeds, dyes, textiles and medicine. Natural brines account for more than 52% of the sodium sulfate production.

sodium sulfide. An inorganic chemical that has attained an important position in the organic chemical industry. It is used as a reducing agent in the manufacture of amino compounds, and enters into the preparation of many dyes. It is also employed in the leather industry as a depilator. It is one of the necessary reactants for making Thiokol synthetic rubber. It is also used in the rayon, metallurgical, photographic and engraving fields.

sodium sulfite. A compound that is very easily oxidized. For this reason, it is employed in many cases where a gentle reducing agent is desired, such as bleaching wool and silk, as an antichlor after the bleaching of yarns, textiles and paper, as a preservative for foodstuffs, and to prevent raw sugar solutions from coloring upon evaporation. It is used in the preparation of photographic developers to prevent the oxidation of hydroquinone, and other agents. It also has application in the field of medicine as an antiseptic and as an antizymotic for internal use. The majority of its use is in the paper industry.

sodium thiosulfate. Crystallizes in large, transparent, extremely soluble prisms with five molecules of water. It is a mild reducing agent, like sodium sulfite. It is used as an antichlor following the bleaching of cellulose products and as a source of sulfur dioxide in the bleaching of wool, oil, and ivory. In photography it is used to dissolve unaltered silver halogen compounds from negative and prints, where it is commonly called "hypo." It is a preservative against fermentation in dyeing industries and serves in the preparation of mordants. It also has uses in the reduction of indigo, in the preparation of cinnabar, and in the preparation of silvering solutions, and in medicine.

solder, cadmium-zinc. This is used for joining aluminum to itself or to other metals. It is particularly suitable for applications where service temperature may reach 400°F (204°C) or higher.

soldered joints. The connection of similar or dissimilar metals by applying molten solder, with no fusion of the base metals.

solder, fusible alloys. Where soldering must be done at temperatures below 361°F (183°C) bismuth-containing solders are used. These solders creep during the long-time loading above room temperature. The higher-bismuth-content solders are not easily adaptable to high-speed soldering operations, nor do they easily wet base metals.

solder, indium. These are used for special applications, such as cryogenic products. The 50% indium, 50% tin alloy is particularly suitable for glass-to-glass and glass-to-metal soldering.

soldering. Low-melting point [under 840°F (449°C)] alloys are used to join metal components in the soldering process. The molten solder fills the space between surfaces to be joined, adheres to the surfaces, and solidifies. The usual steps in the soldering process are: clean the metal, apply flux, apply solder, heat, and, if necessary, clean the joint. Soldering is most frequently used for electrical and electronic assemblies, which account for almost half of all soldering applications. Sealing, such as tin can and radiator seams, is the second most common use. Flame heating is probably the most common way to reach the soldering temperature, although soldering with irons and bits follows closely in popularity. Conditions for choice of a heating method involve: cost of providing the heat for the time an assembly must be heated to achieve soldering temperature; efficiency in bringing joint to soldering temperature; sensitivity of assembly to heat; ability to automate.

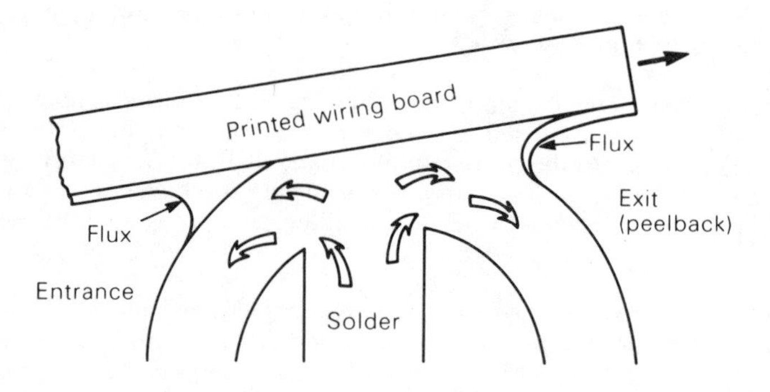

Wave soldering of a printed wiring board, seen in cross section. The board is treated with flux on its bottom side and preheated before passing through the wave at about one inch per second.

solder, lead–silver. This has high-tensile, creep, and shear strengths to 350°F (177°C). Fatigue properties are better than those of the nonsilver solder alloys. The lead–silver solders have poor wetting characteristics, however, and corrode in humid atmospheres. The addition of 1% tin, replacing silver, increases wetting and flow, and reduces susceptibility to corrosion. When tin is added, silver content should be limited to 1.5%. Above this percentage segregation occurs.

solder, tin-antimony. This 95% tin solder has the best electrical properties of the common solder alloys. It has high strength at temperatures to 300°F (149°C) and excellent flow characteristics.

solder, tin-lead. This is the largest single group and the most widely used soldering alloys. Tin-lead solders are compatible with all types of base-metal cleaners, fluxes, and heating methods. Most metals can be joined with these alloys. Joint clearances of 0.002 ± 0.0005 in. (0.05 ± 0.013 mm) are recommended. The highest-melting-temperature solder in this group is that classified as 5A. This alloy is particularly suitable where operating temperature of the assembly may reach 300°F (149°C). The general-purpose solders of the tin-lead system are the classifications 35A through 50A. They provide optimum soldering properties and good strength at operating temperatures below 250°F (121°C). The 60A solder is particularly suitable for delicate heat-sensitive electronic components. The solders with the higher tin content, e.g., 70A, can be used for soldering zinc. Wherever moisture might reach the assembly, however, a 95 zinc/5 aluminum solder should be used.

solder, tin-lead-antimony. These are used for the same general types of applications as the tin-lead solders. They are not recommended for use on aluminum, zinc, or galvanized steel. The addition of antimony—up to a maximum of 6% in place of tin—improves strength and does not seriously affect wettability or flow characteristics. Joint-clearance recommendations are the same as for the tin-lead solders.

solder, tin-silver. This has the same characteristics as in tin-antimony solders and is used for delicate instrument work and for high-strength applications.

solder, tin-zinc. This is used mainly for soldering aluminum, primarily where a lower soldering temperature than that of a zinc-aluminum solder is required.

solder, zinc-aluminum. This is designed specifically for soldering aluminum. It provides high joint strength and good corrosion resistance. A 95% zinc, 5% aluminum solder is used without flux in the ultrasonic soldering of heat exchangers.

solenoids. A coil of wire, when carrying a current will have the properties of a magnet. Such coils are frequently used to actuate various types of mechanisms. If a soft-iron bar is placed in the field of a current carrying cell the bar will be magnetized and will be drawn toward the center of the coil, thus becoming the core of an electromagnet.

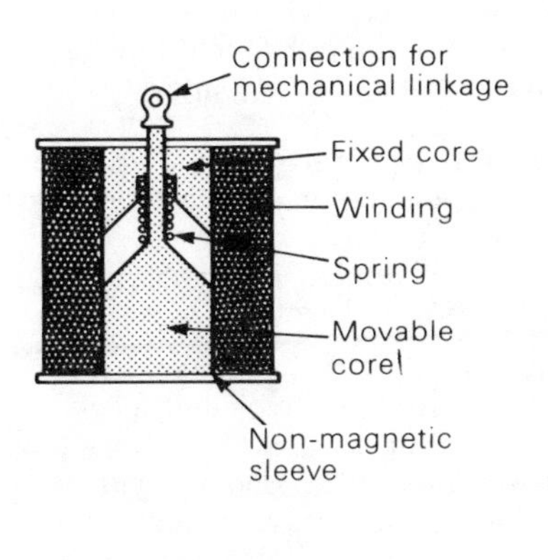

Solenoid

By means of suitable attaching linkage, the movable core may be used to perform many mechanical functions. An electromagnet with a movable core, or without any core, is called a solenoid. A solenoid is usually made with a split core. One part of the core is fixed permanently inside the coil and the other is left free to move. The two sections of the core are normally held apart by a spring; but when the coil is energized the fixed core has a polarity opposite to that of the adjacent face of the movable core, and hence the movable core is attracted to the fixed core. This imparts motion through a connection rod to the mechanical linkage. Solenoids are commonly used to operate switches, valves, circuit breakers, and several types of mechanical devices.

solid drawn shape (see shape)

solid extruded shape (see shape)

solid shape (see shape)

solid solutions. A fundamental characteristic of all solid solutions is that the complete intermingling of the atoms of both metals that prevail in the liquid solution is retained in the solid state. The range of compositions can vary in both liquid and solid solutions. Some materials are soluble in each other in all proportions, others only partially soluble one in the other. Alcohol and water are an example of two liquids that show complete mutual solubility in all proportions. Some liquids are partially soluble in each other. Thus water will dissolve a small amount of acetone giving a single liquid solution rich in water, while acetone will similarly dissolve a small amount of water giving a single liquid solution rich in acetone; but if approximately equal volumes of water and acetone are shaken together, two separate layers will remain. The upper layer will be acetone saturated with water and the lower layer will be water saturated with acetone. A great many pairs of metals show limited solid solubility in parallel manner. They are said to be partially miscible. A very few pairs of metals, mutually soluble as liquids, become completely insoluble as solids. Thus cadmium and bismuth, which form a single homogeneous liquid solution in all proportions, separate out, to all intents and purposes, completely on solidification so that the microstructure contains two phases, crystals of pure cadmium and crystals of pure bismuth. Solid solutions are either substitutional or interstitial. Interstitial solid solutions are generally produced only when the solute atom is small compared with the solvent atom. The solute ion is then able to fit into the interstices, or gaps, between the sites occupied by the solvent ions. In substitutional solid solutions atoms of the solute metal take up sites in the lattice structure of the solvent metal. This substitution can be either ordered or disordered. In the former, atoms of the solute metal take up certain fixed positions. Such solid solutions tend to be hard and brittle while those of the disordered type are tough and ductile and more useful as constituent phases of engineering alloys.

solid-state. An adjective used to describe a device, circuit, or system whose operation is dependent on any combination of optical, electrical, or magnetic phenomena within a solid. Specifically excluded are devices, circuits, or systems dependent on the macroscopic physical movement, rotation, contact or noncontact of any combination of solids, liquids, gases, or plasmas.

solid-state camera (robots). A TV camera that uses some type of solid-state integrated circuit instead of a vacuum tube to change a light image into a video signal. Solid-state cameras have the following advantages over vacuum-tube cameras: ruggedness; small size; no high voltages; intensitivity to image burn and lag (antibloom capability is possible with the proper readout technique); potentially very low cost, characteristic of solid-state technology; and a spatially stable, precise geometry that effectively superimposes a fixed, repeatable measurement grid over the object under observation without the pin-cushion or barrel distortion introduced by the deflection systems of tube cameras.

solution heat treating. Heating an alloy at a suitable temperature for sufficient time to allow soluble constituents to enter into solid solution where they are retained in a supersaturated state after quenching.

solvents. Three types of solvents are in use: paraffins, aromatics, and chlorinated solvents. The paraffin solvents are simply members of the paraffin family, such as diesel fuel or gasoline. These have rather limited solvency. The aromatic solvents include such chemicals as benzene and xylene. The chlorinated solvents such as trichlorethylene (C_2HI_3), methyl chloride (CH_3Cl), or carbon tetrachloride (CCl_4) are paraffins with the hydrogen atoms partially or wholly substituted by chlorine atoms.

solvents, aliphatic. Usually has a "gasoline smell," but through processing can be made odorless. The deordorized solvents are usually straight run naphthas in the 300–400°F (149–204°C) boiling range specifically treated by hydrofining, while the odorless type are synthetic hydrocarbons obtained as a by-product in the manufacture of aviation alkylate.

solvents, aromatic. Aromatic solvents such as benzene, toluene, and xylene have a much greater ability to dissolve resinous materials than aliphatic solvent. Catalytic reforming permits production of fractions very rich in aromatics, converting naphthene hydrocarbons into aromatic hydrocarbons. The mixture of aromatics produced by the reforming process is separated and purified to produce aromatic solvents of any desired purity.

solvents, dewaxing (ketone dewaxing). Process carried out by mixing waxy oil with one to four times its volume of ketone and heating the mixture until the oil is in solution. The solution is then chilled at a slow, controlled rate.

solvents, hydrocarbon. Composed of hydrocarbons of the paraffin, naphthene, and aromatic families. These hydrocarbons do not normally react within the materials they contact, nor do they decompose under moderate heat or in the presence of water, therefore their chemical stability; they are noncorrosive to metals.

sonic testing. Relatively new among NDT techniques, this is good for examining the internal structure of very thin materials and composites. It works on the principle of resonance in the range of 10–35 kHz. An expected response to excitation at the material's resonant frequency indicates a good product. Any other response, or no response, indicates a flaw or questionable product. In the sonic or resonance technique, a variable-frequency oscillator is swept through its tuning range, and its output is fed into a transducer in proximity to the product. When the frequency of the oscillator is at the resonant frequency of the product, a standing wave is generated and causes a momentary signal change in the transducer. Any change in signal that cannot be directly related to the product characteristic being measured would indicate some product change or defect. A dry probe eliminates the need to "couple" a NDT probe and the test materials with a liquid where wetting the product is not desirable. Because every solid structure has a resonant frequency regardless of what it is made of, sonic testing can determine the soundness of laminated materials, coated materials, painted materials, and bonding between materials. It also can detect built-in voids in materials such as honeycomb, cork, and foam rubber. Sonic testing can inspect a material from one side only. **(See also nondestructive testing.)**

space lattices (crystal structure). A space lattice can be considered as an infinite array of points in space, so arranged that it divides space into equal volumes with no space excluded. Every point, which is called a lattice point, has identical surroundings with every other point. The smallest volume that contains the full pattern of repetition is called a unit cell. Identical unit cells must completely fill the space when they are placed face to face, thus generating a space lattice. If a unit cell is so chosen that it contains lattice points only at its corners, it is called a primitive unit cell or simple unit cell. A primitive unit cell contains only one lattice point because each point at eight corners is shared equally with eight adjacent unit cells. The edge length of the unit cell, called a lattice constant or a lattice parameter, is a lattice translation in a given direction. There are only 14 possible different networks of lattice points, therefore, there are only 14 standard space lattices that are needed to describe all possible arrangements of points in space consistent with translational periodicity. Every crystal structure is based on one of the possible space lattices.

space lattices (metallurgy). All elements crystallize according to a particular symmetrical pattern or arrangement of atoms which is repeated at specific and regular intervals to form a crystal. An idealized representation of such a regular arrangement of atoms in the solid state is termed the space lattice. Those elements that have the characteristic of forming a three-dimensional, orderly arrangement of the atoms, in crystal formation may be considered as having a lattice of imaginary lines between the atoms. Metallic space lattices are, in general, highly symmetrical in character, although it is true that those of manganese, uranium, tin, plutonium, and antimony are rather complex and less symmetrical. In the case of the simple cubic space lattice, a single atom occupies each corner of each unit cube with each of the atoms in the eight corners being shared by seven other unit cubes. The smallest repetitive geometric prism that possesses the complete symmetry of the entire crystal is called a unit cell. By definition each unit cell in a space lattice is identical in shape and orientation with every other unit cell. Space lattices are characteristic of all crystalline materials. Most useful pure metals are found to crystallize in one of three distinctly different types of space lattices, as follows: (1) cubic system, three equal axes, mutually perpendicular; (2) tetragonal system, two equal axes perpendicular to each other and to a third (unequal) axis; (3) hexagonal system, two equal axes inclined at an angle of 120° and a third (unequal) axis perpendicular to their plane.

specialty sheet. Sheet product offered for specific end uses and usually designated by a name rather than by alloy and temper.

specimen. That portion of a sample taken for evaluation of some specific characteristic or property.

specular gloss. A measure of the amount of light reflected from a surface at an angle equal to that of the incident beam.

spheroidization (see annealing)

splice. The end joint uniting two webs.

spline rolling, metal. Straight, fine-pitch spur and helical splines and other forms such as serrations can be rolled on cylindrical thread rolling machines or on rack rolling machines, which use flat dies that are commonly called racks. For rolling splines and similar forms, the workpiece is held between centers or in a fixture. Tools for rolling helical and straight splines are similar. Because of the close fit required in assembly of the splined workpiece with its mating component, the blank diameter of the splined area must be made to a relatively close dimensional tolerance. When major diameter fit of a spline is critical, the major diameter can be rolled oversize and ground to final size after heat treatment. Grinding also can be used to correct the somewhat oval configuration sometimes observed in the splinal section as rolled. Sides of the spline teeth are controlled within dimensional tolerances, and the product is used as rolled in the side-fit assemblies that comprise the majority of spline applications. **(See also thread rolling.)**

spinning machines, metal. Variously known as Floturning, Hydrospinning, Rotoforming, spin-forging, shear spinning, etc. A chipless production method, in which metal is plastically deformed and progressively displaced under the compressive forces of a pressure roller as the roller traverses a path parallel to and at a preset distance from the surface of the mandrel. With the workpiece revolving with the mandrel, the required shape is actually spirally generated by the roller, and the metal is forced through the orifice between the mandrel surface and the forming roller. The metal undergoes a partial shear deformation as it is squeezed ahead of the roller and displaced parallel to the center line of the part being formed. Since metal flowing around the roller nose is initially uncontained, and cannot be extruded, it must be subjected to pure bending. Extrusion begins when the outer metal surface is trapped by the mandrel surface. Hollow parts with circular cross section and straight sides such as cones and cylinders, can be produced quickly, accurately, and cheaply in this way from blanks, drawn cups, welded cylinders, forgings, or castings.

sponge rubber. Produced from dry natural or synthetic rubber, using a blowing agent such as sodium bicarbonate and a fatty acid. If an open-cell structure is desired, the compounding and curing are regulated to produce the gas before vulcanizing, thus rupturing the cell walls and producing an interconnecting structure. For a closed-cell structure the rubber is partially cured before gas expansion. Vulcanizing is in suitable frames or molds in either batch or continuous operation. Open-cell types of sponges are used in gasketing, sealing, heat insulating, cushioning, and shock absorption. The closed-cell types have similar uses in addition to those requiring buoyancy in water. The hard rubber product is used in heat-insulating, sandwich construction, and floats. Foam rubber is produced by adding gas to rubber latex and vulcanizing. It is used largely in pillows.

spotfacing, metal. A machining operation for producing a flat seat for a bolt head, washer, nut, or similar element at the opening of a drilled hole, concentric with the hole at right angles to its longitudinal axis. Spotfacing usually follows drilling, either in combination with it in a single operation or as a separate operation. Sometimes, however, spotfacing precedes drilling, to provide a contoured workpiece with a flat surface so as to facilitate centering and starting of the drilled hole, which will then be 90° to the spotface. Spotfacing is similar to counterboring, except that the spotfaced surface is always at right angles to the axis of the hole, and depth of cut in spotfacing is shallower than in counterboring. Machines used for spotfacing are the same as those used for drilling.

squareness. Characteristic of having adjacent sides or planes meeting at 90°.

S/R machine. A machine operating on floor or other mounted rail(s) used for transferring a load from a storage compartment to a P & D station and from a P & D station to a storage compartment. The S/R machine is capable of moving a load both vertically and parallel with the aisle and laterally placing the load in a storage location. Common types of S/R machines are: mini-load S/R machines; unit load S/R machines; man-on-board S/R machines.

stabilizing. A thermal treatment to reduce internal stresses in order to promote dimensional and mechanical property stability. **(See also age softening and stress relieving.)**

stainless steels. The main alloying element in stainless steel is chromium, which must be present in sufficient quantity to make the steel corrosion resistant. Stainless steels are usually divided into ferritic, martensitic, and austenitic types, according to their characteristic microstructure. The ferritic and martensitic stainless steels are straight-chromium steels designated as the type-400 series. The austenitic stainless steels are chromium-nickel alloys designated as the type-300 series. Martensitic stainless steels contain from 11.5% to 18% chromium and from 0.15% to 1.20% carbon. Ferritic stainless steels are characterized by the chromium-to-carbon ratio that does not permit hardening by heat treatment. Chromium varies from 11.5% to 27% and carbon from 0.08% to 0.2% maximum. In ferrite stainless steel the main microconstituent is ferrite, present both at high and low temperatures. In the martensitic stainless steels the same phase changes take place as in carbon steels and low-alloy engineering steels during their hardening. Austenite is a stable phase at higher temperature and transforms on rapid cooling to martensite, making the steel hardenable by suitable heat treatment. Both ferritic and martensitic stainless steels are magnetic under all conditions. Generally, the corrosion resistance and oxidation resistance of ferritic stainless steels are higher than those of martensitic steels because of the higher-chromium and lower-carbon content. The martensitic stainless steels show their maximum corrosion resistance in the fully hardened condition because then all chromium carbides are dispersed uniformly in the matrix, but ferritic steels exhibit their maximum corrosion resistance in the annealed state.

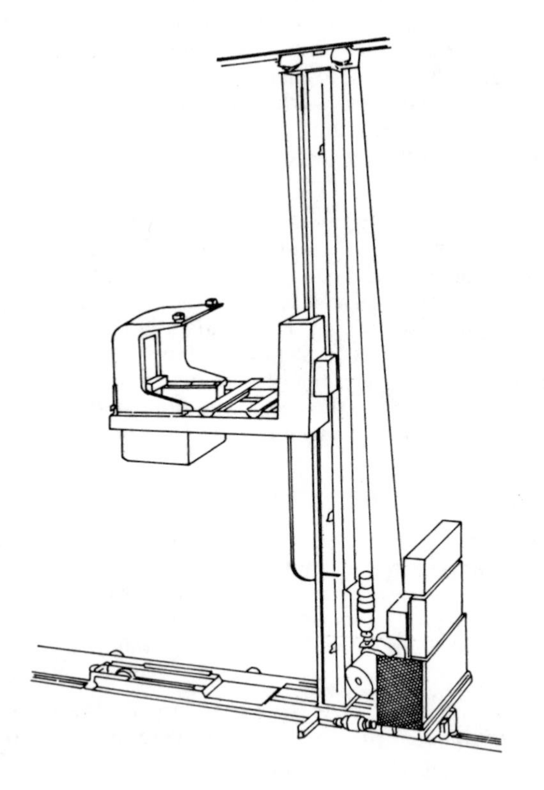

Man-on-board S/R machine.

stainless steels, numbering code. Stainless steels are not included in the AISI-SAE steel coding system. Instead, the stainless steels have a numbering code of their own, which uses three digits instead of four. Basically, all stainless steel compositions are assigned numbers in either the 300 or 400 range. The 400 series are straight chromium steels. The 300 series are chromium-nickel low-carbon steels. Many of the 300 series contain approximately 18% chromium and 8% nickel. The 300 series have the following characteristics: (1) they are low carbon, none containing more than 0.25% carbon; (2) almost all contain 16–20% chromium, 8–13% nickel; (3) they are austenitic face-centered cubic, paramagnetic; (4) they are not heat treatable; (5) they work-harden when formed, shaped, or cut; (6) they have an expansion coefficient a third higher than carbon steels. If the 300 series steels are used in the range of 800–1600°F (427–871°C) there is a rapid combination of chromium and carbon to form chromium carbides. These carbides tend to precipitate in the region of the grain boundaries of the stainless steels, and the effect is called carbide precipitation. Such loss of chromium impairs the corrosion resistance of the steel. The 400 series stainless steels contains no nickel. The absence of nickel, the austenite promoter, and the presence of chromium, a ferrite promoter, indicates that these steels cannot be austenitic. They are ferromagnetic and do not work harden to an unusual degree. These steels do not possess quite the high levels of corrosion and heat resistance that characterizes the richer 300 series. The 400 series actually includes two groups of steels, martensitic and ferritic. The most commonly used martensitic stainless steels are 410 and 416, containing 0.15% maximum carbon; others may contain as much as 1% carbon. Type 420, 0.20% carbon or higher, is used for stainless cutlery. The ferritic stainless steels cannot be heat treated. They are low carbon, high chromium irons containing 16% or more chromium, commonly used in sheet form.

stainless steels, series 302. This was the first of the stainless steels to come into common use, about 40 years ago, and is familiar in such articles as stainless-steel sinks and restaurant and hospital equipment. Because of the high (for stainless steels) carbon content, 0.15%, there is a tendency to replace 302 with 304 now. This steel also contains 17–19% Cr and 8-10% Ni.

stainless steels, series 304. Identical to 302 except that its carbon content is 0.08% maximum. There is a general-purpose 300 stainless steel.

stainless steels, series 301 and 304. The 300 series austenitic stainless steels all have elongation of about 55% in a 2 in. (5.08 cm) gage length. In sheet form, therefore, these steels are suitable for drawing and forming operations. They work harden rapidly and often require annealing (softening) between and after forming operations. Type 301, used for wheel covers on automobiles, has an unusually high elongation of 60% in 2 in. (5.08 cm). Where work hardening is a problem, type 305 is employed. This alloy has an increased nickel content, 10–13%, to produce a lower work-hardening rate.

stainless steels, series 303–303Se. Free-machining grades. Type 303 contains 0.15% sulfur or more; type 303Se contains about 0.25% selenium.

stainless steels, series 308. Type 308 is used in stainless-steel welding rods. There is normally a loss of chromium due to oxidation when steel is melted; the same loss will occur during fusion welding. Type 308 therefore contains a little extra chromium to compensate for chromium loss in welding operations.

stainless steels, series 316. This alloy contains 2–3% molybdenum for greater resistance to the attack of acids. This steel is used in the meat packing, food, and pulp and paper industries for such equipment as pumps, piping, and valves. In addition, the internal fittings of automatic control valves and other control instruments are often made of type 316.

stainless steels, types 321, 347, 348. Type 321 contains a small amount of titanium; type 347, columbium or tantalum. Type 348, a variant of 347, is suited to the nuclear power industry. Its content of tantalum and hafnium (hafnium always accompanies columbium as an impurity), is held to low levels because of the high neutron absorption cross section of these two metals.

stamping presses. A great variety of presses for performing stamping operations is available, usually with a stationary lower platen and a moving upper platen to which the dies are fastened for pressing sheet metal into desired shapes. These range from extremely small tonnage bench presses for tiny parts to very large single and transfer presses. For many simple, single-operation piece parts the inclinable-type C-frame presses offer many possibilities for gravity feed and ejection. These various machines may range from hand-fed types to those with completely automatic dial feeds, transfer arrangements, or multiple-station die sets devised for completion of the product. From relatively slow hand-fed and/or the extremely large mechanical or hydraulic presses, production speeds may range up to as high as 800–1000 strokes per minute (pieces per minute) with inverted presses. Inverted presses are ideally suited for multiple-station dies and today have up to 500-ton capacity. By use of multiple dies on high-speed presses, parts requiring blanking, piercing and shallow drawing operations can be produced at rates as high as 10,000 pieces per minute. Equally as important as production speed, multiple-station dies permit a sufficient number of operations to virtually complete parts ready for assembly.

stamping process. The process of stamping sheet metal between dies generally covers that portion of press working and forming that includes the following operations: punching (blanking, piercing, perforating and lancing); bending; forming (beading or curling and embossing); shallow drawing; extruding; swedging; coining; shaving; trimming.

standard one side bright finish sheet (SISBF) (see sheet, standard one side bright finish)

staple. A measurable, relatively short fiber. All natural fibers are staple fibers. Man-made filament fibers are often cut into short, staple lengths.

statistical quality control (see quality control)

steel. Steel is an alloy of iron and carbon. It results from the refinement and treatment of pig iron and/or scrap iron. The carbon content of steel containing no alloying elements may vary from 0.08 to 2%. If the carbon content is greater than this, the material is a cast iron. Other elements may be added in order to impart certain desirable characteristics or to remove undesirable characteristics. These elements also have the effect of changing the limiting carbon content of 2%. Steel is produced in wrought and cast forms. It has sufficient plasticity to be formed either hot or cold. The strength and hardness of steels may be closely controlled by one or more of the following methods: (1) varying the carbon content, (2) introducing other alloying elements, (3) suitable heat treatment. **(See also specific types.)**

steel, aluminized. Sheet steel with an aluminum coating on both sides, produced by a hot-dip process. In appearance this product somewhat resembles aluminum. Two grades are produced. Heat-resistant aluminized steel has an aluminum coat 0.001 in. (0.0254 mm) thick on each side, while corrosion-resistant aluminized steel has a coating of 0.0002 in. (0.0051 mm) per side. Aluminized steel may be found in automobile mufflers and exhaust piping, ducting, and heating appliances.

steel, annealing. The term annealing refers to any heating and cooling operation that is usually utilized for the softening of steel or iron. The main purposes of annealing are to relieve cooling stresses induced by cold or hot working, and to soften the steel to improve its ability to be more easily machined or formed, subsequently. At times it involves only a subcritical heating to relieve stresses, to recrystallyze cold worked material, or to spheroidize the carbides; in other cases it may involve heating above the critical temperature, with subsequent transformation to pearlite or directly to a spheroidized structure when cooled.

steel, austempering. The properties of lower bainite are usually similar in strength and are somewhat superior in ductility to those of tempered martensite. In order to overcome this, an isothermal heat treatment to lower bainite, in order to obtain optimum strength and ductility has been developed; this heat treatment is known as austempering. It involves quenching to the desired temperature in the lower bainite range, usually in molten salt, and then holding at this temperature until transformation is complete. The piece may be quenched or air cooled to room temperature after transformation is complete and may be tempered to a lower hardness level if desired. Because of its hardenability limitation, austempering has found its widest application in the heat treatment of plain carbon steels in small section sizes, such as sheet, strip, and wire products, in heat treatment of alloy steels, and cast iron applications in which it is important to hold distortion to a minimum.

steel, carburizing. A case-hardening method, applied to low-carbon, plain steels, and low-alloy, low-carbon steels. Involves increasing the carbon content of the surface, so that upon suitable heat treatment the surface will respond to the hardening. In pack carburizing the pieces surrounded with solid carbonaceous materials (often charcoal to which barium carbonate has been added as an energizer). Gas carburizing involves placing the piece in a furnace atmosphere having a suitable concentration of a hydrogen gas. Salt baths containing potassium and sodium cyanides as the active ingredients are also used (liquid carburizing). Temperatures of about 1700°F (927°C) are used for periods of time up to about 75 h depending on the depth of case required.

steel, flame hardening. In flame hardening, an oxyacetylene or similar gas flame is passed over the surface to be hardened. Quenching immediately after heating may be accomplished by a jet of water or air, or by immersion in conventional quenching liquids. The temperature developed at various locations in the place can be controlled by the speed of travel and intensity of the flame. The severity of quench can be adjusted by varying the time interval between heating and quenching, and by the selection of quenching medium. Flame hardening is particularly suited to surface hardening of parts whose size renders furnace treatment impractical. It also applies to any part where hardening throughout is not desired. It has been used for hardening gear teeth.

steel, flame and induction hardening. Consists of heating just the surface of the steel and then quenching it. It is a much simpler method for case hardening, since it can be applied directly to only the areas to be hardened. Steel hardened by this method must contain sufficient carbon to be hardenable. Although the heating may be done with a flame, the more practical method is to use induction heating, such as in the "Tocco" process. This method is best for mass production for such parts as automobile crankshafts, containing up to 0.40–0.60%. Cast irons with carbons of 2.50% can also be induction hardened successfully. The advantage of induction hardening is that heat can be localized by means of induction coils, and also that it can be done very quickly. Special prepared induction blocks are shaped around the surface to be hardened and a high-frequency current of from 1000 to 15,000,000 cycles per second is applied. The voltage is comparatively low and the amperage quite high. Most of the heating is done by eddy currents.

steel, hypoeutectoid and hypereutectoid. Steel containing less than 0.80% carbon is called hypoeutectoid and with more than 0.80% carbon it is called hypereutectoid.

Hypoeutectoid steel starts to freeze at around 2740°F (1504°C) and delta iron is precipitated until 2718°F (1492°C) is reached. The delta iron changes over to gamma iron, and freezing is complete in austenite at 2630°F (1443°C). No further change occur with falling temperature until 1430°F (777°C) is reached. Then ferrity precipitates as the temperature is lowered to 1333°F (723°C), at which point the remaining austenite transforms to pearlite. The resulting structure consists of about 50% grains of ferrite interspersed with 50% grains of pearlite. A hypoeutectoid microstructure is such that magnification is not sufficient to reveal the laminations in the pearlite grains. The proportion of ferrite to pearlite depends on the carbon content. A hypereutectoid alloy may be exemplified by a steel containing 1.4% carbon. Solidification commences at about 2600°F (1427°C) with austenite separating out until freezing is complete at about 2300°F (1260°C), and the structure becomes all austenite. No change occurs in the austenite until the temperature is lowered to about 1700°F (927°C). Iron carbide is rejected below that point; at 1600°F (871°C) for instance, the austenite contains only 1.25% carbon, and at 1500°F (816°C) about 1.1% carbon. The excess iron carbide is rejected by the gamma iron. At 1333°F (723°C) the remainder of the austenite is transformed to pearlite on slow cooling with the iron carbide interspersed in the structure and in the grain boundaries. Iron carbide exists in hypereutectoid steel in proportion to the amount of carbon above the eutectoid point. Iron containing more than 2.0% carbon is usually not considered to be steel.

steel, induction hardening. This process accomplishes much the same results as flame hardening. However, instead of using a flame for heating, an induction coil carrying high-frequency (600 to 2,000,000 cycles per second) current is used. The coil surrounds the piece to be hardened but does not touch it. Induction heating can produce a very rapid raise in temperature. For most surface hardening, the time involved is in the order of 2–3 s. When the proper temperature has been achieved to the desired depth, the power is shut off and the part is quickly cooled. Quenching is accomplished by dropping the part into a conventional quench tank or by using a quenching spray. It may be used for heating nonferrous as well as ferrous metals.

steel martempering. The transformation to martensite during rapid cooling through the martensite temperature range, with the accompanying sharp temperature gradient, results in high stresses in the workpiece. As a result of this a modified quenching procedure, known as martempering is helpful in lowering these stresses after quenching. It is normally carried out by quenching the piece in a molten-salt bath at a temperature just above the M_s temperature, holding it in this bath long enough to permit the piece to reach the temperature throughout and then air cooling to room temperature. Transformation to martensite then takes place during the relatively slow air cooling, and, inasmuch as the temperature gradient characteristic of the conventional quench is not present, the stresses set up by the transformation are much lower than in the conventional quenching and tempering. In addition to the lower stresses much greater freedom from distortion is also achieved and the likelihood of cracking is reduced. After martempering, the workpiece is usually tempered to the desired strength level. Martempering is used in the treatment of tools, bearings, dies, etc., where difficulty was formerly encountered with quench-cracking or distortion, when the usual heat treating by conventional quenching and tempering were used.

steel, normalizing. A term referring to the heating or rather reheating of steel to about 100°F (40°C) above its critical temperature (Ac_3), and then cooling in air. The two main purposes of normalizing are, first to refine the grain size, and second, to obtain a carbide size and distribution more favorable for carbide solution, on subsequent heat treatment, than the "as-rolled" structure was prior to this treatment. The as-rolled grain size depends upon the finishing temperature during the rolling operation. Normalizing process or operation serves to refine a coarse grain size resulting from a high finishing temperature, and to ensure a uniform, relatively fine-grained microstructure.

steel tempering. When particular mechanical properties are desired, the method used most often is to quench the steel in a suitable medium from just above Ac_3 temperature and to follow this operation with a tempering quench at the correct temperature. It is customary to quench in water steels containing less than 0.3% carbon, to quench in oil or water those containing from 0.35% to 0.55% carbon, and to quench in oil those plain carbon steels containing in excess of 0.55% carbon. One of the important mechanical properties of steel may be obtained by means of tempering which is the process of heating a hardened steel to any temperature below the lower critical (A_1) temperature in the case of a steel with less than 0.83% carbon, and the $A_{1,3}$ where the carbon is in excess of 0.83%, then cooling the steel at a desired rate. The purpose of tempering is to reduce hardness and to relieve the internal stresses of a quenched steel, thereby affording greater ductility than is possible with a high-hardness steel that has been quenched.

stepped drawn tube (see tube)

stepped extruded shape (see shape, stepped extruded)

still, petroleum. A device for evaporating liquids by the application of heat, the vapor afterwards being condensed to a liquid state by means of a condenser. In petroleum refining, the still is a sort of boiler for heating crude oils or raw distillate and vaporizing them at different temperatures so that various hydrocarbons can be separated and collected in the order of their boiling points. The still may consist of a cylinder boilerlike tank or of oil-filled tubes heated on their outer surface by fire or steam.

stone. Stone may be classified on a chemical or physical basis. On the chemical basis, calcium carbonate ($CaCO_3$), silicates of alumina, and silica (SiO_2), are of engineering value. Granite and sandstone contain silica. Shale and slate are primarily silicates of alumina. The principal building stones of the calcium carbonate group are limestone and marble. On a physical basis, classification of stone is igneous, sedimentary, and metamorphic, based on the geological process involved in its formation. Igneous rocks (granite, basalt, and lava) are formed by cooling of molten material. Sedimentary rocks (sand and limestone) are formed by cementation of particles deposited in layers by wind or water. Metamorphic rocks (marble and slate) are formed when igneous or sedimentary rocks are subjected to pressure or heat.

storage module. Those items such as pallets, containers, boxes, etc., containing, holding, or constituting the unit load.

storage structure. A system of storage locations or compartments plus any other members that may be required to support and/or guide an S/R machine.

straddle truck. A general class of cantilever industrial trucks with horizontal, structural wheel-supported members extending forward from the main body of the truck, generally high lift, for picking up and hauling loads between its outrigger arms.

strain hardening (metals). The slippage that occurs between any two adjacent atomic planes is very small in magnitude. As slippage occurs, a phenomenon results that increases the resistance to further slip. This phenomenon is known as strain hardening, which means that as resistance to slip increases, the metal becomes harder. Stain hardening appears to be connected with roughening of the atomic planes, probably due to localized distortions of the crystal lattice at the atomic plane where slip occurs. This results in increased resistance to slip. Whether or not this is the correct explanation, the fact is that such strain hardening does occur. This has an important bearing on the behavior of metals subjected to mechanical working. This phenomenon of strain hardening often is referred to as work-hardening. As a result of strain hardening, after a small amount of slip has occurred on one crystal plane, its resistance to further slip is increased to a point where it is greater than that of other planes where slip has not yet occurred. Under these conditions slip on the first plane will stop and further deformation will take place through slippage on other planes. Thus the process of plastic deformation takes place through a series of small slippages between atomic planes of the metal crystal.

strength, material. The ability of a material to resist applied forces without yielding or fracturing. The strength of a material may vary markedly with respect to the way it is deformed. A material that is strong and ductile under static load may appear weak and brittle under cyclic or impact stresses. Materials also show differing strength when the method or the rate of stress application is varied. The elastic and plastic deformation of the material subjected to an external stress may terminate in an inhomogenous form of deformation, termed fracture.

stress, material. The intensity of the internal distributed forces or components of force that resist a change in the form of the body. Stress may be tensile, compressive, or shear and is usually expressed in pounds per square inch (psi.) A tensile stress occurs when forces tend to elongate the member. Tensile and compressive stresses act normal to the plane or section being considered while shear is a tangential stress thought of as acting on the plane.

stress corrosion cracking. Failure by cracking under combined action of corrosion and stress, either external (applied) stress or internal (residual) stress. Cracking may be either intergranular or transgranular, depending on metal and corrosive medium. **(See also corrosion.)**

stress relieving. Heating to a suitable temperature, holding long enough to reduce residual stresses, and then cooling slowly enough to minimize the development of new residual stresses. **(See also heat treating, annealing.)**

stress, static. Static stresses are of primary importance to the structural engineer. Any static load on a body can be described completely in terms of three mutually perpendicular principal stresses. The three most important cases are: (1) uniaxial tension or compression when stress is acting in one direction only, (2) bi- or triaxial tension or compression when stress is acting in two or three perpendicular directions, and, (3) shear, when the forces are parallel to an imaginary plane. The stress is called comprehensive if it tends to bring the material into closer contact; it is called tensile if it tends to separate the material. When the forces are parallel to an imaginary plane at a point, the stress is called a shear stress.

stress, working. Defined as the maximum stress permitted in a material under conditions of actual use. Synonomous terms are allowable stress, safe stress, and design stress. To ensure that the stress incurred by a material in actual use does not cause failure, the working stress is always much less than that which is considered the failure stress. The safe stress can be given by the following equation:

$$\text{Working stress} = \frac{\text{Strength of the material}}{\text{Factor of safety}}$$

stretch-wrap machines. These machines do a wrapping job similar to that of the shrink-wrapper but require no heat. Plastic sheet is stretched sufficiently as the pallet load is wrapped to hold the load securely. **(See also shrink-wrap machines.)**

suck-in. A defect caused when one face of a forging is sucked in to fill a projection on the opposite side.

sulfur (minerals). A nonmetallic element readily distinguishable on account of its bright yellow color and resinous luster. Native sulfur (brimstone) occurs in orthorhombic crystals, but is normally granular or massive. Hardness is 1.5–2.5, and specific gravity is 1.9–2.1. The mineral is brittle with an uneven to conchoidal fracture. The melting point of sulfur, about 230°F (110°C) is not far from the boiling point of water. Sulfur burns at 478°F (248°C) with a blue flame and the evolution of sulfur dioxide gas. The mineral is insoluble in water and in nearly all acids, but will dissolve in carbon disulfide (CS_2) and carbon tetrachloride (CCl_4). Sulfur is a very poor conductor of heat and a nonconductor of electricity. Sulfur is a mainstay of the chemical industry. Much of the output is converted into sulfuric acid. A great proportion is used in the manufacture of superphosphate fertilizers. It is also used in petroleum refining and in making chemicals, paints and pigments, iron and steel, rayon and film, industrial explosives, paper, and numerous other products. The major nonacid uses of sulfur are in the manufacture of insecticides, pulp, and paper dyes, rubber and explosives, and in the preparation of food products.

sulfur dioxide. May be produced by the burning of sulfur or by the roasting of metal sulfides in special equipment. It may be obtained also by recovery from the waste gases of other reactions. Sulfur dioxide is liquefied by compressing to about 7 atm and cooling. It is stored or put into cylinders. Sulfur dioxide is shipped as a liquid under 2 or 3 atm pressure. Its uses are numerous. A quite pure commercial grade, containing not more than 0.05% moisture, is suitable for most applications. A very pure grade, however, containing less than 50 ppm of moisture, is supplied for refrigeration. Sulfur dioxide also serves as raw material for the production of sulfuric acid. It finds application as a bleaching agent in the textile and food industries. Following the use of chlorine in waterworks and in textile mills, sulfur dioxide is an effective antichlor for removing excess chlorine. It is an effective disinfectant and is employed as such for wooden kegs and barrels and brewery apparatus and for the prevention of mold in the drying of fruits. Sulfur dioxide efficiently controls fermentation in the making of wine. It is used in the sulfite process for paper pulp, as a liquid solvent in petroleum refining, and as a raw material in many plants, e.g., in place of purchased sulfites, bisulfites or hydrosulfites.

superalloy (see heat-resisting alloy)

superplastic metals. A characteristic that defies the normal engineering perception of metals is superplasticity. Certain alloys, when deformed in tension at elevated temperatures, exhibit extensive plastic deformation exceeding 1000% without any necking. This phenomenon, called superplasticity, presents a new way of fabricating certain metals by methods so far used only for glass and plastics. Superplasticity depends on the microstructure and the deformation temperature, which may be as high as 1652–1832°F (900–1000°C) for zinc aluminum alloys. Super-plasticity occurs when the strain rate during deformation does not exceed the rate of recovery, about 0.1 s^{-1}. This requires an elevated temperature and a very fine micro-duplex grain structure [about 4×10^{-5} in. (0.001 mm)]. The presence of a finely dispersed second phase (eutectic or eutectoid) is necessary to inhibit grain growth at elevated temperatures. The requirement of both extremely fine grains and elevated temperature indicates that the mechanism of superplasticity may involve vacancy creep, dislocation creep, and grain boundary sliding. Proper heat treatment will change the microstructure from the very fine grains to coarser ones, and metals cease to be superplastic. Today, superplastic zinc and aluminum are being formed commercially. Even notoriously intractable titanium is now being formed superplastically for production aircraft. And superplastic steel may make its appearance in automobiles by the mid-1980s.

swaging (see rotary swaging and hammering)

swedging. A term used in stamping operations on metals. Where the walls of flanged holes are too thin for the purpose, swedging is employed to upset the metal and increase the thickness. Heads, keys, and other features can be swedged to shape during stamping.

Swiss automatic machine. The Swiss principle of automatic machining was originally developed before the turn of the century for the clock and watch manufacturing trades in Switzerland. The American counterpart of the Swiss high-precision automatic screw machine has now become a vital and necessary part of the production picture. It is particularly well adapted to the turning of pinion blanks, studs, worm gears, indicator staffs, shafts, and other minute or slender parts used in clocks, meters, radio or electronic equipment, calculating machines, and a wide variety of industrial and laboratory instruments. The method of turning employed in these automatics, which makes them ideally suited to the production of small and slender, finely finished parts, is that of feeding a revolving piece of stock through a guide bushing past radially fed tools. Unlike the turret-type single-spindle automatic screw machine, only single-point turning tools and narrow form tools are utilized. The tool slides are mounted in one working plane and disposed radially from the collet centerline in the upper half of a semicircle. Ability to produce long, slender parts to a high degree of finish, accuracy, and concentricity, therefore, may be attributed to the fact that the cutting faces of the tools are but a few thousandths of an inch from the stock support and at no time need be farther than $^{1}/_{32}$ in. (0.8 mm) away. Stock is fed through the guide bushing to the tools by means of a sliding headstock, which grips and carries it by means of a collet feed mechanism. Sliding headstock as well as the five radial tools are traversed by individual cam action. Independent cam control of each of these units eliminates the need for wide form tools since proper combination of headstock and tool movements can be had to generate pivot points, back shoulders, multiple diameters, tapers, etc. There is almost no limitation to the scope of the form generating, back shoulder, and back recessing work that can be accomplished within the capacity of the machines.

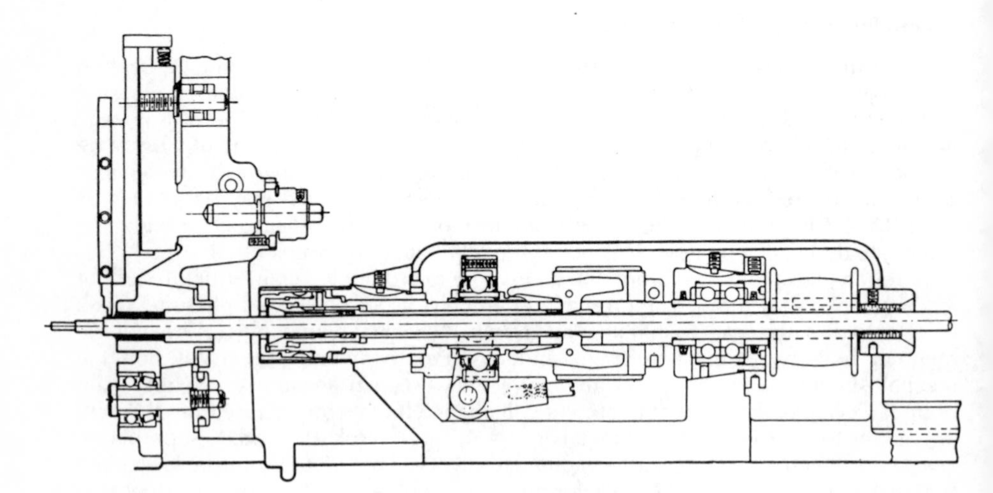

Drawing, courtesy George Gorton Machine Co.

Stock in the Swiss automatic is rotated within and fed to the tools through a guide bushing collet.

synthetic hydrocarbon fluids (SHF). Synthetic lubricants based on SHF have lubricating qualities that surpass those of mineral oils. They are now being used in many everyday industrial applications and are providing benefits in reduced power consumption, increased oil life, and longer equipment life. SHF lubricants by design have a uniform molecular structure consisting of only a few selected compounds. They do not volatilize easily or oxidize as components found in most mineral oils do. SHF lubricants are synthesized from pure hydrocarbons and have advantages over high-quality mineral oil lubricants. They have a high natural viscosity index suitable for a wide range of temperatures, excellent shear stability, good high-temperature oxidation stability, good low-temperature properties, and also are compatible with minerals oils and with common seal materials.

synthetic fibers. The manufacture of synthetic fibers basically involves the physical conversion of a linear polymer with the polymer chains in a relatively disordered state into a low denier continuous filament with molecular chains parallel to the fiber's longitudinal axis. The most important properties of polymers to be used for fiber production are: (1) High melting point, preferably above 212°F (100°C). (2) Linear and symmetrical structure and high molecular weight. The two major processes used for fiber production are melt spinning and solution spinning. Melt spinning is used for polymers that can be melted into a highly viscous liquid and that do not decompose during processing. The melt is extruded through spinnaret holes, either individually to form monofilaments or in bundles to form multifilaments. The filaments are then mechanically stretched in order to provide them with the required tensile strength. A solvent is used in solution spinning. After filtering and degassing the polymer solution, it is fed by use of a screw or gear pump through spinnarets while exposed to hot air to evaporate the solvent. The fibers are washed and subjected to mechanical stretching by passing them through rotating rollers. Technical developments in the processing of fiber polymers has created a new generation of fibers. Bulk and texture are improved by thermal or mechanical texturing using processing such as fast twist or jet texturing. Combining polymers with different shrinkage characteristics give fabrics with attractive luster and feel. New developments also include the processing and dyeing of polypropylene fibers.

systems engineering. In manufacturing, this is a discipline that demands viewing a problem in its widest relevant context so as to identify all the factors and interrelationships involved before launching into any automated manufacturing installation. In the integration of the various operations of a production process, systems engineering is imperative. The success of an automated production system depends on identifying all participating elements and factors, knowing how those elements and factors are related to one another, and understanding how they can be coordinated for optimum operating results. Systems engineering provides the coordination necessary to bring a complicated production process under effective, profitable control. **(See also computer process control.)**

T

taffeta weave. A plain, fairly tight weave, in which the warp and filling yarns cross each other at each row in both directions.

talc, mineral. A hydrous magnesium silicate known for its softness, greasy feel, pearly luster, and white to greenish color. Talc has perfect basal cleavage and occurs in compact foliated or massive aggregates. Talc in commercial usage refers to various rocks that are composed of magnesium silicates, in which talc, the mineral, may be dominant or minor. Tremolite is the most common of mineral talc deposits, occurring in fibrous or finely bladed form, often in tremolite-talc schist. Anthophyllite and serpentine are other usual constituents. Common nonsilicate minerals in talc deposits are calcite and quartz. Even though talc in lump form may be gray or greenish, it grinds to a powder that is brilliant white. Ground talc has great covering or hiding power, and being soft and smooth, has high lubricating ability. Other properties include low electrical conductivity and high fusion point, dielectric strength, and chemical inertness. White firing talc is used in the manufacture of whiteware, electrical porcelain, wall tile, and other ceramic products. Other valuable ceramic properties of talc include high specific heat, resistance to acids, content of magnesia that acts as a flux, and low firing shrinkage. Ground talc is used as an inert extender in paint. Finely ground talc of steatite or near-steatite grade is utilized as a lubricant in paper manufacture, as a filler in the highest grades of paper, and in face powders and pharmaceuticals. Talc of intermediate or lower grades is used as an insecticide carrier, asphalt filler, and dusting agent and filler in the rubber and roofing industries. There are many other applications ranging from foundry facings to crayons and rice-polishing powder.

tantalum. A refractory metal, melting at 5400°F (2982°C), used in electronic tubes, as alloying elements in steel, and as carbides in cemented carbides. Tantalum is a getter, i.e., it reacts with all but inert gases at temperatures above 600°F (316°C) in electronic tubes. It is also used for surgical implants. **(See also refractory metal.)**

tapping. A machinery process for producing internal threads. A tap is a cylindrical or conical thread-cutting tool having threads of a desired form on the periphery. Combining rotary with axial motion, the tap cuts or forms the internal thread. Most metals can be machined with single-point tools and can be tapped, but the cost of tapping usually rises sharply as the hardness of the work metal increases beyond Rockwell C 25. Although steel as hard as Rockwell C 52 can be tapped, efficiency is low and cost is high. Threads as fine as 360 threads per inch in 0.132 in. (3.4 mm) diameter holes, or as coarse as 3 per inch (1.2 per cm) in 24 in. (61 cm) diameter pipe fittings are tapped routinely.

TCP (see tool center point)

Teflon. The basic raw materials are fluorspar, CaF_2, from which anhydrous hydrofluoric acid is prepared, and chloroform. These two are treated together to give difluoromonochloromethane, a gas which boils at −42°F (−41°C). The gas is pyrolyzed at 1112–1472°F (600–800°C) to give tetrafluoroethylene. The tetrafluoroethylene is scrubbed and fractionated to remove hydrochloric acid and various fluorocarbons, which are formed during the pyrolysis. The purified tetrafluoroethylene is then polymerized in stainless-steel autoclaves in the presence of an aqueous solution of ammonium persulfate or other peroxy catalyst; the reaction is rapid and exothermic, and conditions have to be carefully controlled. The polytetrafluoroethylene produced is a granular white solid, which is washed and dried hot. Tenacity is moderate at about 1.5 grams per denier; elongation is medium at about 13%. Water absorption is zero, so that "wet" characteristics are the same as dry. Pliability is good and loop and knot tensile strengths are about 75% of the straight strength. Frictional properties are low, the dynamic coefficient being 0.28 and the static 0.20. Heat transfer is low and electrical properties are good. Teflon is nonflammable. The outstanding properties of PTFE (polytetrafluoroethylene) fiber are excellent temperature, chemical and solvent resistant, good mechanical and excellent electrical characteristics, very low frictional coefficients with slippery handle, no moisture regain, and almost no dyeability. Teflon is nontoxic, but when heated above 392°F (200°C) toxic gases are given off.

teleoperator. Although not classified as a robot in strict terms, the teleoperator unit covers industrial manipulators that are human operated, with the operator in the control loop on a real-time basis. Teleoperator manipulators use master-slave operator control with force feedback. Balanced use of human and machine, which permits the operator to transmit his or her inherent intelligence and dexterity through the machine to the task, is based on cybernetic anthropomorphous machine systems (CAMS) technology. Teleoperator manipulators are cybernetic in that a human is retained in the system, they are controlled by a human operator with his or her responsive, decision-making capability. They are anthropomorphous in that they resemble a human in form and duplicate the human's manipulative powers. Essentially, the machine system serves as a physical extension of human's strength, endurance, and task-performing functions. For this reason, such machines are said to have "instinctive" control, which has proved of interest in satisfying the relatively unstructured applications of the foundry. The first application of a teleoperator manipulator in the foundry industry was for a shakeout application in an automotive foundry. Other applications have included furnace slagging, shortblasting, ladle skimming, palletizing, and metal-pouring installations. A load capacity range of 125–7000 lb (56–3150 kg) is presently available for manipulators of this type. Six degrees of movement include horizontal extension, hoist, azimuth rotation, yaw, pitch, and roll. A servo-control system transmits a small proportion of the load force to the operator's hand, thus giving "instinctive control."

temper color. A thin, tightly adhering oxide skin (only a few molecules thick) that forms when steel is tempered at a low temperature, or for a short time, in air or a mildly oxidizing atmosphere. The color, which ranges from straw to blue depending on the thickness of the oxide skin, varies with both tempering time and temperature.

tempering (see heat treating, tempering)

tenacity. A measure of textile fiber strength. Expressed in g/denier, tenacity is also called "breaking strength."

tensile strength. The maximum tensile stress that a material is capable of developing before fracture occurs, calculated on the basis of original area. Two methods of mechanical testing by which it is possible to maintain the strains uniformly distributed throughout the specimen over a considerable range of strains are the tension test and the compression test. Ductile materials do fracture in compression, except through the action of parasitic tensile stresses developed in the surface of the skin after "barreling" has commenced. Tests for ductility are best accomplished by the tensile testing machine, which consists essentially of two components: (1) a device for straining the specimen, and (2) a means of measuring the resistance of the specimen to such straining. If it is desirable to construct the stress-strain curve for the specimen undergoing tests, a third device is necessary, a straining gage commonly known as an extensometer, designed to measure the strain over only that portion of the specimen in which it is substantially uniform. The final over-all elongation is obtained by means of the extensometer after the necking-down begins and the strain applied becomes more localized. A number of important properties may be determined by means of the tensile test including: (1) Yield point. The yield point of metals is of major importance since the design of structural members in machines, towers, beams, etc., is largely determined by this factor, which is relatively easy to measure. (2) Yield strength. In the case of many nonferrous materials and high-carbon and alloy steels which do not have a definite yield point, the most useful strength that can be determined is the yield strength, which is the stress at which a material is permanently elongated a small amount. At this point of stress the material shows a specified limiting deviation from the usual proportion of stress to strain.

terra cotta. Terra cotta products are in the form of hollow tile and "lumber." The hollow tile is extended in columns and cut to the desired length. The terra cotta lumber is made by adding sawdust or cut straw in the plugger before extrusion. These materials burn out during firing, leaving the finished product porous and capable of being cut by hand saws. Both forms are used in building construction for fire walls and roof sheathing, although terra cotta cannot support much weight.

testing (see eddy current testing, infrared testing, liquid penetrant inspection, magnetic analysis inspection, magnetic particle inspection, magnetic reluctance testing, sonic testing, ultrasonic testing)

textured metal. Unlike coining which weakens the metal or corrugating which increases the weight per foot and adds strength in only one direction, texturing redistributes the metal in a riblike pattern that increases strength in all directions without increasing the weight per square foot. Generally, no significant reduction of metal thickness occurs in texturing.

thermal cracking. Distillation under pressure whereby hydrocarbons of larger molecular size are broken down or "cracked" into mixtures of smaller molecules. It is a process for converting heavy petroleum such as gas oil, wax distillate, and fuel oil into lighter materials of greater value. It breaks down the molecular structure of hydrocarbons by the application of heat without employing catalysts.

thermal expansion (materials). Most materials expand as their temperature is increased and contract with a decrease in temperature. The rate of change in length is termed the coefficient of linear expansion and indicates the changes in length per unit length for each degree of temperature change. The coefficient of cubical expansion indicates the change in volume per unit volume for each degree of temperature change.

thermal reforming. Cracking converts heavier oils into gasoline; reforming converts or reforms gasoline into higher octane gasoline. The equipment for thermal reforming is essentially the same as for thermal cracking but higher temperatures are used. The products of thermal reforming are gases, gasoline, and residual oil or tar.

thermistors. A thermistor is a thermal resistor. It is a solid–state device of semiconductor ceramic materials made by sintering mixtures of metallic oxides, such as manganese, nickel, cobalt, iron, and copper. These thermistors have a "negative temperature coefficient" which means that as temperature increases, their resistance will decrease. They are made in a variety of forms such as beads, discs, washers and rods. Thermistors can be used in many electronic applications including temperature measurement, compensation and control, liquid level measurement, altimeters, switching, power measurement and voltage control. The most elementary circuit consists of a thermistor and an ammeter in series. Any change in temperature would produce a change in resistance and current flow. The ammeter could be calibrated in degrees of temperature rather than milliamperes.

thermoelectric pyrometer. A method of measuring temperature based on the principle that if two dissimilar wires are joined together at both ends, and one junction is heated, an electromotive force (emf), or voltage, is generated in the circuit. The apparatus consists of a millivoltimeter connected to two wires of different metals with the other ends joined together and placed in a furnace. The wires in question are called a thermocouple, one of the sides being the positive element, and the other the negative element. Although most thermocouples are made of two elements in the form of wires, at times they are made in the form of a wire element inside a tubular element.

thermometer, bimetallic. The operation of bimetallic thermometers is based on the principle that different metals have different coefficients of thermal expansion. Two metallic alloys with different physical characteristics are fused together and formed into a spiral or helix (coil). This is the bimetallic element. When the bimetallic element is heated, it unwinds because there is a different thermal expansion for each alloy. A pointer attached to the helix by a shaft moves as the helix unwinds and indicates the temperature on a calibrated circular scale. Bimetallic thermometers are available for industrial and laboratory use. Bimetallic thermometers are used to measure temperatures from −300 to +800°F (−184 to 427°C). They can indicate temperature as high as +1000°F (538°C) but not on a continuous basis because the bimetallic element (helix) tends to overstretch at this temperature, causing permanent inaccuracy.

thermometer, liquid-filled. Liquids such as alcohol are used to measure temperatures below the freezing point of mercury (−38.87°C or −37.96°F). The alcohol contains dye to enable the thermometer to be more easily read. Liquid filled glass, thermometers can be used for temperatures from −300°F to ± 600°F (−184 to 316°C). The scale can be etched directly on the glass tube or it can be engraved on a metal plate or tube to which the glass tube is attached.

thermometer, mercury-in-glass. The volumetric expansion of mercury is over six times greater than that of glass. The mercury-in-glass thermometer is based on this difference in volumetric expansion. Basically, this thermometer consists of a fine bore glass tube, joined to a reservoir and sealed at the top. A measured quantity of mercury is enclosed in the reservoir. When the thermometer is exposed to heat, the mercury expands more than the glass and is forced to rise up the tube. The mercury rises a certain amount for each degree of temperature. Reference marks are made on the thermometer, using a regulated bath which can establish and maintain temperatures very closely. The spaces between the reference marks are evenly subdivided. A large number of reference marks enables the thermometer to be read more accurately. Some mercury-in-glass thermometers are calibrated to be completely immersed; others are calibrated for partial immersion.

thermometer, radiation. The radiation thermometer picks up the radiant heat by means of a lens and focuses the heat wave on a thermocouple. The potentiometer is most often used with the radiation thermometer, and it operates just as it does when connected to a thermocouple that makes direct contact with a hot substance.

thermometer, resistance. The operation of the resistance thermometer depends on an electrical circuit. In this instrument the heat-sensitive element consists of a carefully made resistor. Platinum, nickel or copper wire, wrapped around an insulator is most often used for the resistance wire of the element. A properly protected resistance thermometer increases its resistance when it is exposed to heat. A resistance thermometer is used as part of a circuit in which its electrical resistance is compared with a known resistance. This type of circuit is known as a resistance bridge. The resistance thermometer does not generate its own voltage. An external source of voltage, like a battery must be incorporated into the circuit.

thermometer, vapor pressure. Unlike the liquid-filled and gas-filled systems which depend upon volumetric expansion for their operation, vapor pressure thermometers depend upon the vapor pressure of a liquid which only partially fills the system. The liquid in this type of system can expand, but as it is heated its vapor pressure increases. Water in a pressure cooker responds in a similar manner. The pressure increases as the water is heated and changes to steam (water vapor). Vapor pressure does not increase linearly (unit increase in pressure for each unit of temperature rise). At lower temperature, the increases of vapor pressure for each unit of temperature change is small. The change of vapor pressure at higher temperatures is much greater. The vapor pressure is said to change logarithmically.

thermoplastic elastomers. Thermoplastic elastomers or rubbers combine in a large degree the end-use characteristics of vulcanized rubber with the rapid processing advantages of the thermoplastics. Several types are commercially available, including styrene–butadiene–styrene block copolymers, under the trademark Kraton thermoplastic rubbers; and both block and random copolymers of styrene and butadiene, as well as butadiene homopolymers, under the trademark Solprene elastomers. Thermoplastic elastomers are supplied in pellet or diced form for extrusion, molding, or thermoforming; the Solprene and Kraton products are also available in crumb form for compounding use. While the thermoplastic elastomers are largely processed as such, they can also be blended with other polymers. **(See also rubbers.)**

thermoplastic materials. A thermoplastic material is defined as one that can be repeatedly softened when heated and become firm when cooled. Thermoplastic polymers are comparable with candle wax or cooking grease in their reaction to temperature changes. Unlike wax or grease they do not liquefy to pourable consistency, instead they attain a state of high melt viscosity, requiring suitable pressure to mold or shape them. There are exceptions such as nylons and vinyls; however, in most cases, decomposition occurs before liquefication. Thermosetting resins during processing undergo a chemical reaction that results in a thermally stable, three-dimensional molecular network. Molding compounds and laminating and casting resins are cured by catalysts, often accompanied by heat and pressure. In each instance, conversion to final form involves a chemical and a physical change. Processing involves close control over batch-to-batch uniformity, storage, shelf life, preparation of materials, and reaction time. Thermoplastics are readily softened because they are not polymerized into three-dimensional molecules; instead, they are made up of linear chains that may be parallel, coiled, interconnected slightly, or tangled. All thermoplastics have one feature in common. They contain molecules of extremely great length; these molecules are often referred to as giant or macromolecules. Other thermoplastics are brittle or difficult to melt and process. To overcome these limitations or to provide variations of basic resins, copolymers are made by simultaneously polymerizing two monomers. Resin manufacturers can produce an almost limitless variety of grades of one particular polymer by the method used to effect polymerization and choice of catalysts. **(See also plastics.)**

thermoplastics, low permeability (LPT) (nitrile barrier resins). A group of nitrile polymers that are excellent barriers to the transmission of gases such as carbon dioxide and oxygen. With good mechanical, thermal, optical, and melt-processing properties, the resins find wide application in packaging. Feasible processes include blow molding, film extrusion, laminating extrusion coating, sheet extrusion and calendering, and thermoforming.

tin bronze (see bronzes)

thread milling machines, offset. In these machines both the part and the cutter are rotated. Unlike the eccentric-drum miller, the amount the cutter can be offset is fairly large. These machines handle a wide range of internal and external threads, utilize standard or special, single or multiple-thread mills, and are available with either hand or power collet chucking, air chucking, or with hollow spindles having pot chucks or fixtures. External thread length is only restricted by the particular machine size, a standard 4½-in. (11.4-cm) machine handling diameters up to 4½-in. (11.4-cm) by 36 in. (91.4 cm) in length. A 6-in. (15.2-cm) machine, on the other hand, handles diameters up to 6 in. (15.2 cm) and lengths to 120 in. (305 cm). Thread forms up to 11/32-in. (8.7-mm) deep can be cut on the 4½-in. (11.4-cm) machine while on the 6 in. (15.2 cm), up to 11/16-in. (17.5-cm) depths can be had. Internal threads are more limited, maximum collet capacity on the 6-in. (15.2 cm) machine being 4 in. (10.2 cm), swing over the carriage 9½ in. (24 cm) and over the bed, 16 in. (41 cm). Internal thread depths up to 2¾ in. (7 cm) can be produced with standard equipment on this miller. Other machines, standard and special, allow precision external and internal milling and threading of diameters ranging up to cannon breech-lock threads, and milling of large screws, etc. The primary limitation, however, is that parts must be chucked, colleted, held between centers, or swung in jigs or fixtures for the milling operation.

thread rolling, metal. A cold forming process for producing threads or other helical or annular forms by rolling the impression of hardened steel dies into the surface of a cylindrical or conical blank. In contrast to thread cutting and thread grinding, thread rolling does not remove metal from the work blank. Rather thread rolling dies displace the surface metal of the blank to form the roots and crests of a thread. Dies for thread rolling may be either flat or cylindrical. Flat dies operate by a traversing motion. Methods that use cylindrical dies are classified as radial infeed, tangential feed, through feed, planetary, and internal.

tier. A set of storage locations having a common elevation.

tile. Forms of tile include roofing, wall and flooring tile for building construction, and drain tile for conveying liquids. Roofing tile requires careful clay selections, forming and burning and made in many forms and shapes. Floor tile is made by the "dust pressed" process and wall tile by either this or the "plastic" process. The essential difference in these processes is the moisture content of the materials at the time they are pressed in molds. Glaze may be applied to wall tile after burning; floor tile is left unglazed. Drain tile is usually made from a red-burning clay by the stiff mud process. It is burned in a manner that increases porosity by avoiding glazing and vitrification. Sewer pipe must be water-tight so it is glazed by the addition of salt during the burning process. Sodium vapor reacts with the clay to produce a hard, dense, non-absorbing "salt glaze."

tin. Tin is obtained from the tin oxide cassiterite (SnO_2). Tin oxide is reduced to tin by smelting in a blast furnace with coal and limestone. The tin obtained from the blast furnace is refined by electrolysis or by reheating. The ultimate tensile strength of tin in the annealed condition is about 2200 psi (1.5×10^4 kPa), the yield strength is about

1300 psi (9×10^3 kPa), and the elongation in 2 in. (5.08 cm) is about 45%. Cold working has a temporary effect on tin, since normal temperatures have an annealing effect. The high resistance of tin to corrosion makes it suitable for plating. Tin is also used for making solder. Copper and lead are the most common metals used with tin to produce alloys. The word bronze has been used for a number of alloys, but primarily designates copper–tin alloys. Copper–tin alloys may be strengthened by cold working. Gun metal is one of the best known bronzes and contains about 10% tin. Another bronze called Government bronze contains about 88% copper, 10% tin, and 2% zinc. It is commonly used for valves and gears. Phosphor bronze is bronze with phosphorus added in amounts up to 4%. Phosphorus is added to improve corrosion resistance in applications such as turbine blades and gears. Phosphor bronze, with 20% tin makes a suitable material for bearings subjected to high pressures and low speeds. Tin alloys of antimony and copper are also suitable for bearings and are known as babbitt metal.

tin alloys. One of the most important alloys of tin is in the manufacture of bearings. The bearing alloys are essentially tin, antimony, and copper and are referred to as babbitts, although this term is also applied to tin-base bearing alloys that contain lead. The composition varies, but most of these alloys fall within the range of $83\frac{1}{3}$–91% tin, $4\frac{1}{2}$–$8\frac{1}{3}$% antimony, and $4\frac{1}{2}$–$8\frac{1}{3}$% copper. The bearing functions by supporting the load on hard crystals of copper–antimony and tin–antimony compounds embedded in the relatively softer matrix of copper and antimony dissolved in tin. The babbitt metals are also suited for die casting. The principal applications are in the fields where the bearing qualities will be useful or where greater accuracy in the finished dimension is required without machining, and in certain food-processing industries. The cost of tin-base die castings is much greater than that of zinc-base castings. An alloy of tin called pewter contains about 80% tin, 5–15% antimony, 0–3% copper, and 0–15% lead. This alloy is used to make extruded ornamental molding, and also vases and pitchers. Many colonial houseware articles were made of pewter. Solders are alloys of tin and lead with the addition at times of antimony and cadmium. The composition of solders varies considerably ranging from 5% tin and 95% lead, through 60% tin and 40% lead, to 95% tin, and 5% antimony. Lead and tin are combined with other elements to form low-temperature melting or "fusible" alloys. Two well-known alloys are Wood's metal and Rose's metal. Alloys of lead, tin, and antimony are used as type materials or metals. An alloy of 20–25% tin and 80–75% lead is called "terne" metal and is used in terne plating. This plating is used to protect steel sheets from erosion.

titanium. The first step in the extraction of titanium from its ore is to chlorinate a TiO_2 and carbon mixture to titanium tetrachloride, which is then reacted with magnesium metal in a heat-resistant steel vessel at red heat, under an inert gas blanket to prevent oxidation. The resulting products are spongy titanium, known as titanium sponge, and magnesium chloride, $MgCl_2$, the latter drained out of the reaction chamber as a liquid. Chlorine and magnesium are recovered from the molten magnesium chloride in electrolytic cells. The subsequent melting of the titanium sponge into massive ingots takes place, with the complete exclusion of oxygen and nitrogen under inert argon or vacuum. Titanium is a silver-gray metal, classified as one of the so-called light metals, being 60% heavier than aluminum but only 56% as heavy as alloy steel. Titanium-base alloys are very strong and in addition very ductile, easily drawn or hammered down to thin sections. It has a low specific gravity (4.5), a high corrosion resistance to acids, and a high strength to weight ratio, especially at elevated temperatures. The yield strength in the form of hot-forged bars is 75,000 psi (5.2×10^3 kPa). This strength-for-weight ratio is high but does not exceed that of aluminum at ordinary temperatures. Above 200°F (94°C), the strength of aluminum falls off rapidly, whereas titanium continues to retain a high proportion of its strength, thereby making it more desirable than aluminum. However, for operation at temperatures above 800°F (427°C), steel alloys have higher strength–weight ratios. The corrosion resistance of titanium is good, and sea water has practically no adverse effects. Titanium is being used as a corrosion-resistant material in the petroleum, food-processing, and other chemical industries. The metal can be forged, hot rolled, or cold rolled. It can also be brazed and welded to itself. The greatest use of titanium metal and its alloys has been in airframe and jet engine parts, and for certain military applications.

titanium alloys. The corrosion resistance and mechanical properties of titanium alloys compare favorably with those of austenitic stainless steel. In general, the titanium alloys are used as substitutes for stainless steel, particularly the austenitic grades, in applications where the lower specific gravity of titanium alloys justified the much higher cost. About 90% of titanium alloy production is used in aerospace applications. Titanium alloys occur as three different structures: (1) alpha, which has a hexagonal-close-spaced structure and cannot be hardened by heat treatment; (2) beta, which has a body-centered-cubic structure and can be age hardened; and (3) alpha–beta mixtures, which also can be hardened by heat treatment. Better than 60% of titanium alloy production is in the form of mixed alpha–beta structures. Aluminum is a major alloying addition for titanium alloys. It functions as an alpha stabilizer, dissolves in the alpha, and strengthens alpha while in solution. Fabrication of titanium alloys is much more difficult than for aluminum-base alloys. Machinability is about the same as for stainless steel, which is difficult. Titanium alloys are used for such applications as aircraft gas turbine compressor blades, forged airframe fittings, missile fuel tanks, and structural parts operating for short times up to 1100°F (593°C).

toluene. An aromatic hydrocarbon somewhat similar to benzene but of higher boiling point produced in coking of coal and also by petroleum processing. A solvent secured from coal tar, which like benzene is ring-shaped and found sparingly in crude oil; also known as toluol.

tool center point (TCP). The user of a robot is primarily concerned with the position of a given point on an end effector, such as the center of a pair of gripper jaws or the tip of an arc welding gun. Basically, the robot arm is a means of moving the TCP from one programmed point to another in space. The control philosophy of the robot is built up around TCP concept. The control directs the movement of the TCP in terms of direction speed and acceleration along a defined path between consecutive points. This method of control, termed a controlled-path system, utilizes the mathematical ability of a computer to give the operator coordinated control of the predefined TCP in a familiar coordinate system during the teaching operation. It also controls the TCP in terms of position, velocity, and acceleration between programmed points in the replay or automatic modes of operation. The TCP is predefined by the operator who enters a tool length into the control. This represents the distance between the end of the robot arm and the TCP. In the controlled-path system the coordinates of the TCP are stored as x, y, and z coordinates in space and not as robot axis coordinates. During the automatic mode of operation the position of the TCP and the orientation of the end effector relative to the TCP are known at all times by the control system.

tooling plate. A cast or rolled product of rectangular cross section of thickness 0.250 in. (6.4 mm) or greater, and with edges either as-cast, sheared, or sawed, with internal stress levels controlled to achieve maximum stability for machining purposes in tool and jig applications.

torch cutting (air-carbon-arc). This process is widely used for a variety of metalcutting, beveling, piercing, and metal-removal chores ranging from washing and trimming iron and steel castings and scarfing billets to preparing weld-joints and removing defective welds. Unlike oxyfuel flame cutting the air-carbon-arc process uses an arc generated by a consumable carbon–graphite electrode to melt the base metal. Simultaneously, a flow of compressed air sweeps molten metal away from the operator and the workpiece. Air-carbon-arc equipment consists of four major components: the torch (electrode holder), the carbon–graphite electrode, the power source, and the compressed air source. Selection should match application to obtain maximum performance and economic benefits. Air-carbon-arc electrode holders are made in two basic versions; manual and automatic. A typical manual torch consists of an insulated holder with jet ports to channel compressed air toward the arc for fast metal removal. The air orifices are situated in the clamping device of the holder so that air flow is always aligned with the electrode. Electrodes from 1/8 to 1 in. (0.32 to 2.54 cm) in diameter can be used. Automatic torches come in three varieties: a manually actuated electrode feed; a spring-loaded, mechanical electrode feed; and a voltage-controlled electrode feed. The last of these can hold groove depths to a tolerance of 0.025 in. (0.64 cm). **(See also flame cutting.)**

transducers. A number of input and output devices are utilized in conjunction with the circuits found in the different branches of electronics. Such units usually convert one form of energy to another and therefore are known as transducers. Thus, a microphone is a transducer since it converts sound-wave energy into electric signals. Similarly, a loudspeaker is a transducer because it converts electric signals to sound waves. There are a number of other transducers utilized in automation, industrial control, computer systems, and other allied branches of electronics.

transfer devices. Transfer devices can be grouped into two main categories: bulk transfer devices, which transfer randomly placed parts, and orientation transfer devices, which hold the workpiece in transfer devices. Conveyor belts, sheet-metal chutes or guides, and simple part-slide mechanisms are all orientation transfer devices. The familiar tote bin, often assuming different names, is the most basic form of bulk transfer device. The overhead conveyor, which is a simple type of orientation device, can be configured in one of three ways: indexing from position to position, continuously moving, or powered and free. In a powered and free conveyor system, handling hooks or racks move continuously around a closed loop. When a work station needs a hook at a given point, it can divert the hook from the main conveyor system and rigidly lock it into position. After the hook is loaded with a product, it returns to the main loop, from which it can move to a different work station. Another type of orientation transfer device is the pallet conveyor. This device is simply a normal conveyor equipped with pallets for individual parts. Parts are placed into the fixture of a part pallet and then transferred to another work station, either by continuous motion or by indexing in controlled steps. By using shot pins and other mechanical devices, a robot can locate a palletized part within 0.005 in. (0.127 mm). A lift and carry device, which uses a mechanical linkage system to index parts from one work station to the next, is similar to a pallet conveyor. Other types of orientation transfer devices include the bulk hopper feeder with orienter and iron-man-type devices. The bulk hopper automatically feeds loose, randomly oriented parts from a bowl-like vessel by elevating individual parts to a drop-off position.

transfer mechanism. Any device that transfers objects onto or off of a conveyor line or from one conveyor line to another.

transfer molding (see molding, transfer)

transformer. When two coils are brought into close proximity so that the magnetic lines of force of one coil link with those of the others, the device is known as a transformer. A transformer usually consists of a primary winding to which the ac energy is applied, and a secondary winding. The ac energy can be the 60 Hz, 110 V derived from the power mains, or it can be in the form of audio- or radio-frequency signal energy. With ac applied to the primary, a voltage is induced into the secondary, and a current flow will occur if the secondary is a close circuit. The induced voltage is generated by the magnetic lines of force cutting the coil turns of the secondary winding. Thus, energy is also drawn from the primary winding and this is the energy that is dissipated in wattage across the load resistor. The only other resistance that would consume energy in a circuit of this type is the internal resistance of the coil windings. Coil reaction does not consume power. Transformers are extensively used in all electric and electronic applications. Transformers are employed in instances where it is necessary to have a higher voltage than from a given ac source or in such cases where it is desirable to convert an existing ac voltage to one of lower value. A transformer is also useful to couple the signal energy in one amplifier stage to that of another amplifier stage.

transistors. Transistors are made of materials such as silicon and germanium. Such materials are called semiconductors. They have greater resistance to the flow of current than metallic conductors, but not as great a resistance as insulators. Semiconductor crystals used in the manufacture of transistors can be either *n*-type or *p*-type. The *n*-type is one which, by the addition of a donor impurity, possesses a quantity of free electrons. It is therefore a material which depends upon the flow of free electrons for its conductivity. The *p*-type is one, which by the addition of an acceptor impurity, is left with a quantity of holes. These holes are electrical charge carriers similar to electrons but possessing instead a positive charge. Therefore, the *p*-type is also a material which depends on the flow of positively charged holes for the conductivity. There are five principal classifications of semiconductors: (1) diode, (2) bipolar or junction transistors, (3) silicon-controlled rectifier (SCR), (4) unijunction transistors, (5) unipolar or field effect transistor. When utilized in circuits appropriate to the characteristic encountered, function in a fashion similar to vacuum tubes and gas tubes in their ability to amplify, generate and process electric signals and switch dc and ac as required. Solid-state units can be substituted for tubes in virtually all electronic circuitry. There are many types of transistors available. Some are designed for low-frequency signal service; others are for high frequencies ranging through the microwave regions. Special types are also available for pulse signals, control circuits, and computer logic systems. Transistors can be formed by using *p*- and *n*-region combinations to produce either *pnp* transistors or *npn* types. In either of these, a triode (three element) unit is formed with amplifying and signal-generating characteristics comparable to the vacuum tube.

trap rock. Rocks of greater density and finer grain than granite. They split irregularly and are used as crushed stone for railway ballast or coarse aggregate for concrete. A small amount of trap rock is processed for use as paving block or curbing.

tread plate. Sheet or plate having a raised figured pattern on one surface to provide improved traction.

trepanning (machining). A machining process for producing a circular hole or groove in, or a disk, cylinder or tube from, solid stock by the action of a tool containing one or more cutters (usually single point) revolving around a center. The process is used in at least four distinct types of applications: the production of round disks, large shallow through holes, circular grooves, and deep holes.

tribology. Based on the Greek word *tribus*, meaning sliding, is the name that applies to the subject of wear, friction, and lubrication as wholly interdisciplinary—embracing physics, metallurgy, chemistry, mechanical engineering, mathematics, etc.

tricot. From the French word *tricoter* (to knit). A fabric made on a machine that knits from a warp rather than from cones, usually characterized by a thin texture as the machine handles single or fine yarns. These fabrics are usually porous, will not run (might "ladder" for very short distances), and have varying degrees of stretch, usually greater in the width than the length.

triode tube. Similar to the diode tube except that it has one additional electrode. This is the grid, which is placed between the cathode and the plate. The effect of the grid in a triode is to make it possible for the tube to act as an amplifier; that is, small changes in voltage on the grid will cause very substantial changes in the current flow from the cathode to the plate.

truck, hand. A general group of nonpowered industrial trucks for operation by a pedestrian.

truck, hand/rider. A dual-purpose powered industrial truck that is designed to be controlled by a pedestrian but which may also be controlled by a riding driver.

truck, high lift. A self-loading industrial truck equipped with an elevating mechanism designed to permit tiering. Popular types are high lift fork truck, high lift ram truck, high lift boom truck, high lift clamp truck, and high lift platform truck.

truck, high lift platform. A self-loading industrial truck equipped with a load platform, intended primarily for transporting and tiering loaded skid platforms.

truck, industrial. A wheeled vehicle, primarily intended for the movement of objects or materials and usually associated with manufacturing, processing, or warehousing, but not including vehicles intended primarily for earth-moving or over-the-road hauling. **(See specific types.)**

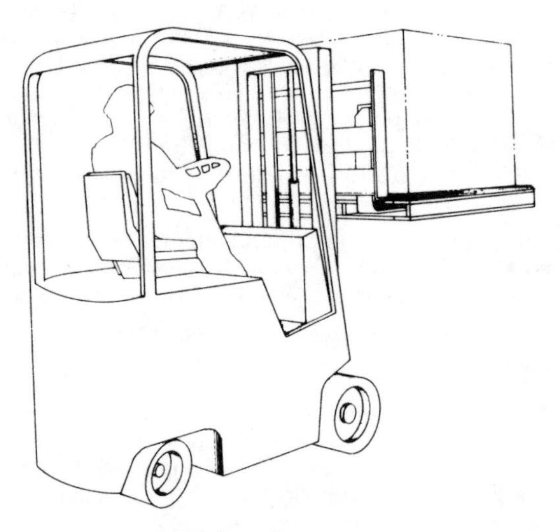

Industrial truck.

truck, industrial lift. Also called a powered industrial truck. A mobile, self-propelled truck used to carry, push, pull, lift, stack, or tier material. **(See also fork lift truck.)**

truck, low lift. A self-loading industrial truck equipped with an elevating mechanism designed to raise the load only sufficiently to permit horizontal movement. Popular types are low lift platform truck and pallet truck.

truck, low lift platform. A self-loading industrial truck equipped with a load platform intended primarily for transporting loaded skid platforms.

truck, order picker high lift. A high lift truck controllable by the operator stationed on a platform movable with the load-engaging means and intended for (manual) stock selection. The truck may be capable of self-loading and/or tiering.

truck, order picker low lift. A low lift truck controllable by an operator when stationed on, or walking adjacent to, the truck, and intended for (manual) stock selection. The truck may be capable of self-loading.

tube. A hollow wrought product that is long in relation to its cross section, which is round, a regular hexagon, a regular octagon, elliptical, or square or rectangular with sharp or rounded corners, and that has uniform wall thickness except as affected by corner radii.

tube, alclad. Composite tube composed of an aluminum alloy core having on either the inside or outside surface a metallurgically bonded aluminum or aluminum alloy coating that is anodic to the core, thus electrolytically protecting the core against corrosion.

tube, brazed. A tube produced by forming and seam-brazing sheet.

tube, butt-seam (see tube, open-seam)

tube, drawn. A tube brought to final dimensions by drawing through a die.

tube, extruded. A tube formed by hot extruding.

tube, lock-seam. A tube produced by forming and mechanically lock-seaming sheet.

tube, open-seam. A shape normally produced from sheet of nominally uniform wall thickness and approximately tubular form but having a longitudinal unjointed seam or gap of width not greater than 25% of the outside diameter or greater over-all dimension.

tube, seamless. Tube which does not contain any line junctures resulting from the method of manufacture.

tube, sized. A tube that, after extrusion, has been cold drawn a slight amount to minimize ovalness.

tube, stepped drawn. A drawn tube whose cross section changes abruptly in area at intervals along its length.

tube, structural. Extruded tube, which may contain an extrusion seam, suitable for applications not involving internal pressure.

tube, welded. A tube produced by forming and seam-welding sheet longitudinally. Butt-welded tube is a welded tube the seam of which is formed by positioning one edge of the sheet against the other for resistance welding. Helical-welded tube is a welded tube produced by winding the sheet to form a closed helix and joining the edges of the seam by welding. Lap-welded tube is a welded tube the seam of which is formed by longitudinally lapping the edges of the sheet for welding.

tumbling. Finishing operation for small articles by which gates, flash, and fins are removed and/or surfaces are polished by rotating them in a barrel together with wooden pegs, sawdust, abrasive pellets and polishing compounds.

tungsten. Tungsten which melts at about 6170°F (3410°C), has a higher melting point than any other metal. Tungsten is also very stiff, having a modulus of elasticity of 50,000,000 psi (3.45×10^8 kPa). In fine wire (18 mils) the tensile strength of tungsten is 264,000 psi (1.8×10^6 kPa). In wire having a diameter of 3.96 mils, its tensile strength is 483,000 psi (3.33×10^6 kPa). Because of its high melting point and high atomic number, tungsten is used as a target material in x-ray tubes. Tungsten is an important alloy addition to steels; its hard persistent carbide is effective in resisting wear and softening at high temperatures. It is also used in the cobalt-base alloys to produce good high-temperature properties. Cemented carbide tools consist of hard carbides of tungsten cemented together by cobalt or nickel. Because of its high melting point, casting and melting are impracticable. Tungsten is mined as an ore containing the minerals wolframite, $FeWO_4$, or scheelite, $CaWO_4$. This is converted to tungstic acid, H_2WO_4, which is reduced to tungsten powder by hot hydrogen. The powder is compressed to form a briquet and sintered in a hydrogen atmosphere at about 2000°F (1094°C) using the technique of powder metallurgy. Following sintering, the dense, solid bar is mechanically worked in order to improve its properties and convert it to wire filaments.

turbidimeter. A turbidimeter continuously monitors a process stream and instantly signals a change to control pollution or product quality.

turret lathe. Usually classified as hand-operated or automatic. In both instances the distance traveled by the cutting tool is controlled by stops or cams. The essential difference is that the hand turret lathe is manipulated by the operator, whereas the controls are preset and operated automatically on the automatic turret lathe. Horizontal turret lathes may be further identified by the construction of the hexagon turret. Those having fixed saddles and movable slides upon which the hexagon turret is mounted are called ram turret lathes. The saddle type has its hexagon turret mounted directly on the saddle, so that the entire saddle moves when actuated by a handwheel. The saddle-type turret lathe is sometimes equipped with a hung saddle that operates off the front way and thus permits greater swing capacity. The ram-type turret requires that the saddle be positioned and clamped to reduce the ram overhang. The saddle-type turret lathe is constructed so that the entire saddle moves. This makes for rigid construction and eliminates the overhanging slide. Many turret lathes are equipped with automatically controlled spindle speeds.

twill weave. Weave whereby the filling yarns are interlaced with the warp yarns in such a way as to form diagonal ridges across the fabric.

U

ultrasonic hardness tests, steel. The "Sonodur" hardness tester operates on a completely different principle, which makes it very useful for certain conditions. A magnetostrictive, diamond-tipped rod, vibrating at ultrasonic frequencies, is brought into contact with the metal at a load of 1½ lb (6.7 N). Tip penetration, which is related to hardness of the material, changes the vibration frequency of the rod. This change in frequency is read on a meter scale in terms of either Rockwell or Vickers numbers. The equipment is portable, and the shallowness of the indentation, less than 0.0005 in. (0.013 mm), makes it particularly suitable for use where a larger indentation would be objectionable. Its use is restricted primarily to steel. **(See also hardness testing.)**

ultrasonic testing. A nondestructive test applied to sound-conductive materials having elastic properties, for the purpose of locating inhomogeneities or structural discontinuities within a material by means of an ultrasonic beam. **(See also contact scanning.)**

ultraviolet curing (see curing, ultraviolet)

undershoot. The degree to which a system response to a step change, in reference to input, falls short of the desired value.

unit load. Any load configuration handled as a single item.

universal impact testing machine, metals. A test machine, which is important to manufacturing processes. A series of tests may be performed with this machine. They are the Charpy, Izod, and tensile impact tests. The test consists of raising a hammer to a fixed height, holding the specimen in a vise, and releasing the hammer that causes an impact. The specimen absorbs some of the energy of the swinging hammer. The loss of energy by the hammer means that it will not return to its original height. The more energy absorbed by the specimen, the less is the return swing of the hammer. This loss of energy by the hammer is recorded by the machine as foot-pounds of energy absorbed. The Charpy test requires that the prepared specimen be held at the ends. It is loaded as a simple beam. The specimen in the Izod test is held at one end. The test is a cantilever type. The tensile impact test uses a specimen similar to the static tensile specimen. The impact loads the specimen in tension.

unscramblers, bar. These machines eliminate the need for manual handling of strapped bundles of steel bars as received from the mill. The bundle is dropped into the machine and the straps are cut. The unscrambler slings raise and lower, automatically, feeding bars in a single layer down an inclined ramp. One or two bars at a time (as required) are released to a horizontal feeder, which loads a machine tool, press, or shear. Flow of the bars is stopped by a hold back arm until additional bars are called for.

up-ender. A device to rotate an object from a position on its side to a position on its end. **(See also down-ender.)**

urea-formaldehyde resins. Urea and formaldehyde react under alkaline conditions to produce either monomethylol or dimethylol urea. Aqueous solutions of urea and formaldehyde with an alkaline catalyst such as calcium, sodium, or ammonium hydroxide are reacted together in nickel-clad, stainless-steel, or glass-lined reactors. The reaction is stopped by the addition of acid; and a filler, either alpha cellulose or wood flour, is added and the water evaporated. When alpha cellulose is used, it is no longer visible in the molded product but is apparent dispersed in a state of molecular divisions, or perhaps combined chemically. Colorants, mold lubricants, and acid catalyzers are added to the molding powder. If acids are added, the final cross-linking proceeds rapidly. Usually latent acid catalysts, which react only at temperatures above ordinary room temperatures, are used. Examples are aniline hydrochloride, dimethyl oxalate, and ammonium chloride. A small amount of urea or thiourea is also added to react with any excess formaldehyde and to form cross-links. The urea resins are readily molded by compression, setting up thermally in the mold. They have good tensile strength and hardness but are inferior to some phenolics in impact strength. They have good resistance to weak alkalies and most organic solvents, but not to acids or strong alkalies. Their high dielectric strength and nontracking property make them suitable for many electrical uses. The urea resins are transparent. The alpha cellulose filled material is translucent and can be colored in a wide variety of pastel colors.

urethane rubbers. The reaction product of certain organic diisocyanates and polyglycols are rubbery products known as polyurethanes. These compounds are speciality rubbers and possess high abrasion resistance and are useful at high temperatures and with high concentration of solvents and oxygen or ozone. A major use of this type of rubber is in the production of flexible foam and elastic fibers, which result when a diisocyanate is mixed with polyester containing both free hydroxyl and carboxyl groups. The reaction is extremely rapid, with the evolution of gas which serves to expand the mass yielding foams either hard or soft, depending on the reactants and the conditions employed.

VA (see value analysis)

vacuum (industrial gases). A vacuum, as used in industry, must be considered as an atmosphere or gas with special inert properties. Industrial vacuum for process uses is almost always what is called "very high vacua," that is, in the range of 10^{-7}–10^{-3} torr (10^{-3}–10^{-7} mm Hg). At least two vacuum pumps in series are required to reach such low-vacuum conditions. Vacua for such purposes as electron beam welding or the vacuum melting of steel are at the high-pressure end of this range, but vacuum tubes, including x-ray tubes and oscilloscope tubes, must maintain a vacuum of at least 10^{-6} torr if they are to function properly. A wide range of metals react rapidly with oxygen at temperatures above 1000°F (538°C) and must be melted under vacuum conditions. Such metals include titanium, beryllium, zirconium, all the refractory metals, and the semiconductors such as silicon and germanium.

vacuum distillation, petroleum. A method of distillation carried on at a low pressure area or partial vacuum in a still, usually associated with steam distillation to reduce to a minimum decomposition, the lubricating oil being distilled. Such a still is called a vacuum still and the residue remaining in the still is frequently referred to as short residium.

vacuum metalizing. Products to be coated are rotated on "trees" inside a boilerlike sealed vacuum chamber. Arc electrodes carrying small tabs of aluminum are turned on. Aluminum volatilizes and deposits on surface. The extremely bright "silver" effect can be modified by a tinting top-coat (organic) to produce a brilliant shiny gold.

vacuum plating. A method of depositing thin metal films by deposition from the vapor phase under high vacuum. Under high vacuum conditions the metal to be deposited has a high vapor pressure and a low boiling point. The vacuum deposition of cadmium avoids the hydrogen embrittlement problems of electroplated cadmium. Vacuum deposition is performed on plastics as well as metals; capacitors are made by the vacuum deposition of zinc or aluminum on paper. This is the best method for the production of reflective surfaces such as are found in sealed-beam headlights.

vacuum-sealed molding. The essential differences between vacuum-sealed molding for precision steel, iron, or aluminum castings and conventional sand molding is (1) the V-process uses a very thin thermoplastic film, which is heated to its softening point and them vacuum-formed over a pattern on a hollow carrier plate, and (2) the dry, free-flowing, unbonded sand used to fill the special flask set over the film-coated pattern. Slight vibration quickly compacts the fine grain sand to peak density. The flask is then covered with a second plastic film. The vacuum is drawn and the sand becomes rigidized. Releasing the vacuum originally applied to the pattern permits easy stripping. The other half of the mold is fashioned in the same manner. The cope and drag are then assembled forming a plastic-lined mold cavity. Sand rigidity is maintained by holding the vacuum within the mold halves at 200–400 mm Hg. As molten metal is poured in the mold, the plastic film melts and diffuses into the sand leaving a hard-glazed surface. After the metal solidifies and cools, the vacuum is released. Gravity pulls the sand away to disclose a casting with a smooth, sand-free surface. Molding sand can be recycled once cooled.

valency compounds, metals. When one metal is very strongly electropositive and the other only weakly electropositive, they may combine to form a compound in which the ordinary rules of chemical valency apply. Thus, the strongly electropositive metals, toward the left-hand side of the periodic table, form compounds such as these with the very weakly electropositive metals adjacent to the nonmetals on the right-hand side of the table, e.g., Mg_3Sn and Mg_2Pb. In these compounds the range of solubility of any excess metal is generally very small and often they have simple lattice structures like ionic crystals. Sometimes such a compound is so stable that it will have a higher melting point than either of its parent metals.

value analysis (VA). Also termed value engineering, value analysis is a functionally oriented scientific method for improving product value by relating the elements of product worth to their corresponding elements of product cost in order to accomplish the required function at least cost in resources. Value analysis is more concerned with the functional than with the structural characteristics of the product. It uses the word function to describe everything that the product does for the customer, including how well it does it, how often, for how long, and under what conditions. All this is encompassed in the qualification functionally oriented.

vanadium steels. Although vanadium is not a major alloy in many of the special alloy steels, the role it plays as a minor alloying element is very important. Vanadium has a strong affinity for carbon and, when added to steel, it is found partly combined with ferrite but principally with the carbide or cementite. The presence of vanadium greatly increases the strength, toughness, and hardness of the ferrite matrix of steels. The carbide or cementite formed with vanadium is likewise greatly strengthened and hardened and shows little tendency to segregate into large masses. The most important effect of vanadium in steels is to produce a very small grain size and induce grain control in the steel. The vanadium steels are inherently fine-grained, and this fine grain size is maintained throughout the usual heat-treating temperature range. Steels containing vanadium are also shallow hardened because they have a faster transformation rate from austenite to pearlite during the cooling cycle from the hardening temperature. Owing to the grain-control characteristics imparted by the addition of vanadium, steels may be heated over a wide range of temperatures during the hardening operation without danger of developing a coarse and weak structure. Vanadium is effective in steels in amounts as low as 0.05% and in many of the common alloy steels the vanadium content runs under 0.30%. Alloy steels designed for use at elevated temperatures, such as hot work steels and high-speed cutting tools, contain up to 4.0% vanadium.

vaporforming. Basically this process is gas plating of nickel accomplished by allowing vapors of nickel carbonyl to contact a heated surface in a closed chamber. As the vapors decompose, a uniform layer of pure nickel is deposited on the heated mandrel gradually forming a shell that accurately reproduces the surface of the mandrel. Used mainly for making molds, porosity and casting-type shrinkage distortion do not occur in vaporforming, leaving an excellent surface finish and absence of porosity. Vaporformed nickel shell molds rarely need parting compounds, which cause loss of detail, and with less sticking, there is less mold abuse. The accurate surface finish, which adds durability to the mold, also contributes to the precision, close tolerance, reproduction of molded shapes and textures. Deposition of the nickel is smooth and consistent and shells can be grown as fast as 0.010 in. (0.254 mm) per hour. The most outstanding capability of nickel gas plating is its ability to plate and fill uniformly internal corners, yet not build up excessively on projections or develop nodular growths. The important difference between vaporforming of nickel and electroplating deposition of nickel is the uniformity of deposit on internal corners. In practice, the electroform will have 20–25% as much metal in an internal corner as on the flat areas; the vaporformed parts will have 80% or more. Sizes can range up to 30 ft (9.1 m).

varnishes. A liquid composition that converts to a transparent or translucent solid film after being applied as a thin layer. There are two general types of varnishes, oil varnishes and spirit varnishes. Neither contains pigments; the color results from the oils or resins that are present. Oil varnishes are made up of an oil, resin, thinner, and drier. The oil, thinner, and drier may be any of those used in making paints. Resin is defined as a solid or semisolid, water-insoluble organic substance with little or no tendency to crystallize. Originally all resins were natural, coming from pine trees or fossil deposits. Many resins today are synthetic. Spirit varnishes are solutions of resin in a volatile solvent such as alcohol or benzol. Oil varnishes are ordinarily made by melting resin, mixing it with hot oil, and continuing to heat the mixture. The resin is either dissolved or suspended in the oil. Heating causes the oil to polymerize and changes its physical characteristics. The film formed by the drying of this oil is more elastic and water resistant than a paint film. When spirit varnishes such as shellac are applied, the solvent evaporates and leaves a film of resin. Spar varnishes are of the highest grade and are used largely for exterior and marine purposes. They are long-oil varnishes, tung oil being usually employed, although with certain resins linseed and soya oils are used. Many types of natural and synthetic resins go into spar varnishes. These varnishes are characterized by their high resistance to damage through exposure. Shellac is a spirit varnish, the resin being the excretion of the small lac insect of India. It is used as a plaster sizing, as an electric insulation coating, in rubber ink, and in phonograph records. Special varnishes are used to coat the inside of food cans, for electrical insulation, and water-proofing, and as bases for enamels. Rubbing varnishes are hard semi-solids. Some varnishes are baked on industrial and domestic appliances.

vehicle, automatic guided (AGV). A vehicle equipped with automatic guidance equipment, either mechanical, electromagnetic, or optical. Such a vehicle is capable of following prescribed guide paths and may be equipped for vehicle programming and stop selection, blocking, and any other special functions required by the system.

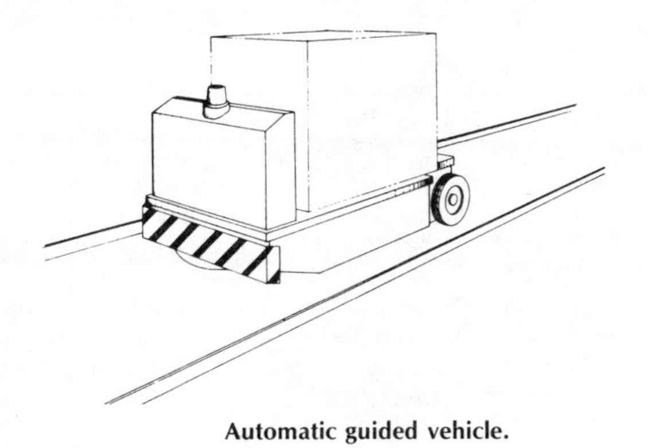

Automatic guided vehicle.

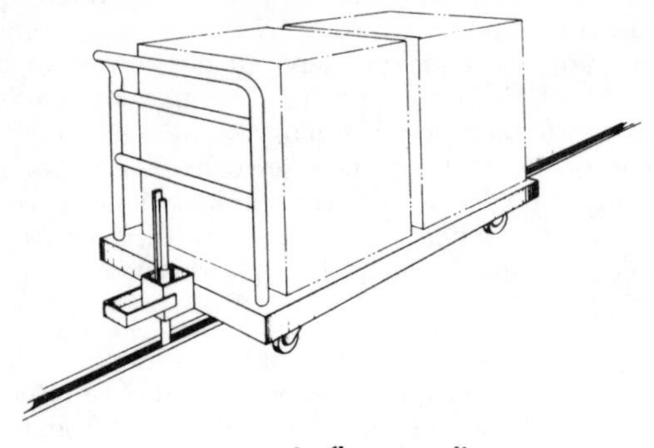

In-floor tow line.

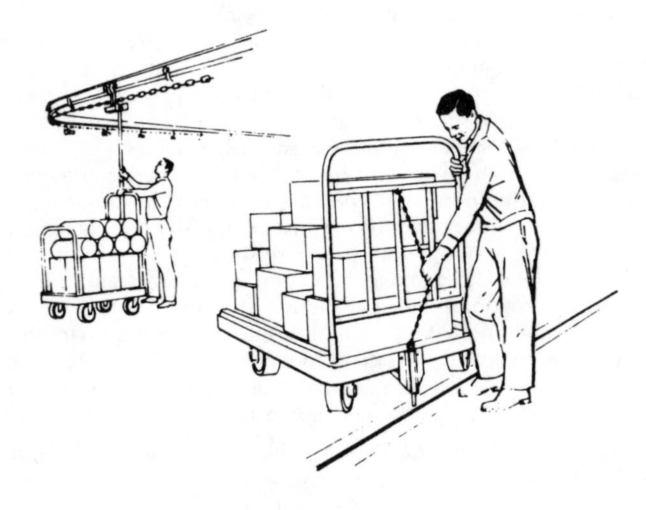

Automatic guided vehicles.

veneer and plywood. A thin piece of wood used primarily to cover another piece of wood or several other veneers. The purpose of a veneer is to give a finer appearance to the surface. Veneers may be cut in a continuous strip by rotating a log in a lathe, or sliced or sawed in strips from a log or other large timber. When three or more layers of veneer are glued, together the product is known as plywood. The grains of alternate layers run at right angles to one another in order to produce properties that are the same in both directions. An odd number of plies is always used to secure "balanced" construction. This reduces the distortion or warp with moisture changes. There is considerable variation as to the number and thickness of the plies and the type of wood. The finished piece may be from $^1/_{16}$ to 3 in. (0.16 to 7.6 cm) thick. Plywood can be made with a high ratio of strength to weight.

vent mark. A small protrusion of a forging or casting resulting from the entrance of metal into die or mold vent holes.

vermiculites (minerals). A group of minerals similar to the micas in properties and appearance. They show the characteristic cleavage, although thin flakes may not be so elastic as those of mica. Hardness is 1.5–3.0. Color ranges from bronze through yellowish brown to black, and the luster from bright to pearly. Nineteen mineral species have been identified in the group, but only two, vermiculite and jefferisite, are common, and vermiculite is generally applied to both. Vermiculite is a hydrous magnesium silicate, with varying amounts of iron and aluminum. The property of vermiculite that makes it of commercial value is its capacity to expand when heated. On exposure for a few seconds to temperature 1472–2012°F (800–1100°C), particles of vermiculite increase by volume by 6–20 times. Expanded vermiculite has a bulk density of only 5–10 lb/ft^3 (80–160 kg/m^3) and a low thermal conductivity. It is chemically inert and relatively refractory. These properties make it especially suitable in many types of thermal and acoustic insulation. It is used as loose-filled insulation in walls, ceilings, and roofs of dwelling and industrial buildings, and also in the insulating spaces of refrigerators, incubators, water heaters, and similar applications. Another major application is in lightweight concrete, monolithic walls, floors and roof decks, and precast structural panels and special shapes. Other applications are in gypsum-vermiculite plaster, insulating cement, and various molded articles.

vertical boring mills (see boring machines)

vertical cold saw (see cut-off machines)

vertical turret lathe (see boring machines)

vibrating screen. A single or multiple deck screen for separating fines from coarse material, usually inclined to the horizontal. The screen decks are vibrated by mechanical, pneumatic, or electrical means.

Vickers hardness test (see hardness test, Vickers)

video image analysis. On-line examination of products, both inside and out, for external surface and geometry according to preprogrammed criteria. The system can be programmed to look for surface imperfections, edge defects, placement of components, precision of manufacture, size, shape, dimensional errors, etc. Visible light or x-ray replaces the human eye in the examination of exterior product surfaces.

vinyl acetate chloride copolymer (plastics). This plastic combines the good qualities of each component polymer, and it is used in the same applications. In the molded form it has remarkable dimensional stability and warping resistance, making it an excellent material for phonograph records. The copolymer plastic is also used as a lining for food containers.

vinylidene chloride. A member of the polyvinyl family, which may exist in either amorphous or crystalline state. In the amorphous form, it is flexible and extrudes easily. Mechanical working crystallizes the material, which then becomes tough and loses its flexibility. The amorphous form is extruded into filaments, which are used as a substitute for rattan fabrics in furnishings, furniture, and seats in public conveyances. Finer filaments are woven into window screening.

vinyl plastics. Ethylene, obtained by thermal cracking of petroleum or from the propane fraction of natural gas, is most widely used because of its importance as a starting chemical for polyethylene, polystyrene, acrylics, and other high-volume plastics. Calendering is a method for producing film and sheet for a variety of end uses such as upholstery materials and waterproof fabrics such as rainwear, shower curtains, and baby pants. Calendered sheets may be transparent, pigmented to any desired shade, embossed, or printed. They are laminated to knitted cotton for upholstery, or applied as facings for nylon or other wide mesh fabrics for tear-resistant tarpaulins. Extruded PVC meets many stringent electrical codes when used as wire insulation. Other uses are extruded medical and chemical tubing, garden hose, refrigerator door gaskets, and vacuum sweeper bumper strips. Injection molded parts are manufactured for use in environments where there are stringent requirements. Depending on formulation, resins may be extremely flexible, semirigid, or rigid. Common examples are bicycle handlebar grips, atomizer-type closures, pipe fittings, and snap-on devices. Blow molding containers are now being used. Transparency, toughness, and crack resistance are properties that enable PVC to compete with other thermoplastics in packaging foods. Fabrics utilizing glass cloth, nylon, cotton, and other webs are coated with PVC on one or both sides for an unlimited number of industrial, commercial and decorative uses.

vinyls, foamed. Produced as either open- or closed-cell structures, depending on the method used. To produce open-cell foam a gas or chemical blowing agent is combined with the plastisol, which is then spread on a conveyor belt or on release paper and carried through a fusing oven. Closed-cell foam is produced by confining the material in molds during heating, which retains the blowing agent in nonconnecting cells. Closed-cell foams are used in flotation devices and shaped cushions. Open-cell foams are used for rug backing, cold-weather apparel, sporting goods, automotive arm rests, visors, and dashboard coverings.

vinyl plastics (organosols). Dissolving a vinyl in a solvent with small amounts of plasticizers results in an organosol. Such solutions are important in textile coating for better penetration into close-weave fabrics, or as top coats for paper and cardboard. Because of lower viscosity, organosols can be extended with fillers such as clays or carbon black to modify properties. Notebook and book covers, lamp shades, and bottle-cap liners are examples of organosol-coated products. **(See also foams.)**

vinyl plastics (plastisols). Polyvinyl chloride can be plasticized with a high proportion of plasticizer into a viscous solution called a plastisol or fusing vinyl. When heated from 330 to 450°F (166 to 232°C), the plasticizer forms a more intimate part of the structure and the liquid becomes firm. Plastisols resemble corn starch in water; unless heated, the starch remains as discrete particles. When heated, the particles expand and coalesce as water penetrates, and a gel is produced. The two components cannot be separated. In the same way, vinyl and plasticizer fuse under heat to produce plastisol. Plastisols are used for dipcoating metal products such as baskets for dishwashers, electroplating racks, tool handles, and chemical ware. They are cast into hollow molds for doll parts, novelties, and a variety of industrial products. Fabric coaters employ plastisols where heavy build-ups, single surface coating, or different formulations of colors are required on opposite sides.

Vinyon (synthetic fiber). A copolymer of vinyl chloride (88%) and vnyl acetate (12%). It is called a copolymer because the two are polymerized together. In manufacturing the copolymerization of vinyl chloride and vinyl acetate is effected by heating, probably in the presence of a catalyst of the aluminum chloride, $A1C1_3$, or boron trifluoride, BF_3, type. The polymer is dissolved in a solvent, acetone, or its homologous methyl ethyl ketone (MEK) and a solution that contains 23% by weight of the copolymer is made. This is filtered and deaerated and then extruded through fine orifices by the pressure of a pump. The "filaments" emerge into a countercurrent of warm air that evaporates the acetone so that the filaments solidify and can be wound. Vinyon yarn has a tenacity of about 3.4 grams per denier and an elongation at break of 18%. It is as strong wet as dry, probably because it is very resistant to the action of water. Because Vinyon is a stretched yarn, the filaments are usually very fine, e.g., a yarn of 40 deniers may have 28 filaments. Vinyon has found very considerable application. Its superlative chemical resistance makes it suitable for use in filter-pad and in protective clothing for chemical workers. Its resistance to water has enabled it to be used for fishing lines and nets. Other uses include felts, sewing-threads and twines, and ladies gloves, which have been made from warp-knitted Vinyon fabric. Its great defect of having such a low-melting point means that it cannot be used for materials and garments that normally are laundered, and because it has very poor absorption of moisture, it is unsuitable for underwear. Most important applications are felts, carpets, and filter fabrics.

viscose rayon. Regenerated cellulose, the cellulose coming from wood, of which it is the major constituent. It is purified, treated with caustic soda, which converts it into alkali cellulose, then treated with carbon disulfide, which converts it into sodium cellulose xanthate, and then dissolved in a dilute solution of caustic soda. This solution is then "ripened," the solution becoming at first less viscous and then increasing nearly to its original viscosity; it is then spun into an acid coagulating bath, which precipitates the cellulose in the form of a viscose filament. The cellulose that constitutes the final filament differs chemically from the original cellulose of the wood in only one respect, that it has suffered some degradation during the manufacturing process; the very long cellulose molecules have been partly hydrolyzed and have been broken down into shorter, although still very long, molecules. When artificial fibers are made by chemical processes from naturally occurring polymers, as in the case of viscose from wood cellulose, or Bember, or cellulose acetate from cotton cellulose, or alginate from seaweed, it is desirable so far as possible to keep the inevitable degradation of the material to a minimum. If the natural long-chain polymers are broken down considerably, there is a significant loss in strength and fibrous properties. In viscose, the cellulose molecules are about one-quarter as long as those in the wood cellulose from which it was made. The uses of viscose rayon are manifold. Practically all the textile field is open to it. Hose, underwear, and dress goods are made from viscose. There is no problem of poor washability. Linings are particularly good in viscose rayon; they are resistant to abrasion, bright, shiny, and slippery; the garment lined with viscose rayon slips readily into position. Rayon staple fiber, made crease resistant, has many uses. It is also used for curtains, chair coverings, transport furnishings, table cloths, cushions, bedspreads, lace, fine fabrics, and underwear.

viscosity, fluid. The viscosity of a fluid expresses its resistance to flow. When viscosity is measured, the nature of the fluid must be known. The viscosity of some liquids varies with the temperature, or changes considerably after the fluid has been shaken or is flowing. Such fluids are generally classified as non-Newtonian. Fluids whose viscosity is constant at any given temperature are classified as Newtonian. Before measurement of fluid viscosity is attempted, the classification of the fluid must be determined. There are several unit systems for expressing viscosity. Absolute viscosity is a measure of the resistance of a fluid to internal deformation. The most common unit for expressing absolute viscosity is the centipoise. Kinematic viscosity is the ratio of the absolute viscosity and the mass density. The common unit of kinematic viscosity is the centistoke.

viscosity index, lubricants. An indication of the rate of change of viscosity with temperature. An oil with a VI of 100 arbitrarily established for an average of paraffin oil, does not change viscosity with temperature as rapidly as an oil with a lower VI. The average VI for a selected naphthene base oil has been set arbitrarily at zero. Oils of low VI thin out more rapidly with temperature increases and thickens with reduced temperatures. A high VI is desirable when temperature changes vary widely and also in internal combustion engines where starting may be at low temperature but steady operation is at high temperature. Additives and special refinement methods may produce indices as high as 150.

viscosity, lubricants. A measure of internal friction of resistance to flow. The Saybolt viscosimeter is most commonly used. In it, the time in seconds required for 60 mm of oil to flow through a standard orifice is reported as Saybolt Universal Seconds or

SUS. The temperature of the oil is closely held at 100–130°F (38–54°C), for lighter oils, and 210°F (99°C) for heavier oils. It is more common now to determine the kinematic viscosity in a Ubbelhode or Ostwald viscosimeter for greater accuracy, and then convert from centistoke to SUS. Viscosity is an indication of the suitability of a lubricant for operation at certain temperature. If the viscosity is not high enough, an oil film cannot be formed to separate the contacting parts; if it is too high, the oil may not flow properly to reach all parts that require lubrication.

vision, machine (see sensors, image processing)

voltmeter. A voltmeter is employed for measuring the voltage drop across a resistor or some other component of a circuit. For this reason the voltmeter is not placed in series with the resistor or other components, but across it. Since the voltmeter is shunted across the place where the voltage is to be read, the meter should have a high internal resistance so that it will not behave as a low-shunt resistor. If the meter resistance is low, some of the current flowing into the resistor will branch into the meter and upset true voltage readings. Voltmeters are constructed by using a basic milliammeter, plus a series resistance, that reduces the current flowing through the meter to a value within its range. The ohmic value of the external resistor will establish the range of the voltmeter.

vulcanization. A term used in the rubber industry to refer to a variety of cross-linking processes used. Natural and synthetic rubbers such as styrene–butadiene (SBR), polybutadiene, acrylonitrile, and others contain a double bond capable of cross-linking. The mechanism of such cross-linking is essentially identical with cross-linking through additional polymerization, such as curing of unsaturated polyesters. However, in a typical rubber, there is only about one to every few hundred chain atoms. Vulcanization can be accomplished by heating raw rubber with some sulfur or some sulfur compounds and accelerating agents. The snarls of the rubber are fastened at certain points by the covalent bonds between sulfur and two carbon atoms associated with the double bond. The more numerous the points of linkage are, the greater restriction of molecular slip exists and the lower the extensibility is until, finally in harder rubber (ebonite) the structure becomes similar to that of a completely cross-linked thermosetting resin such as Bakelite (phenol–formaldehyde resin).

vulcanized fiber. When high alpha cellulose pulp is treated with a 75% zinc chloride solution, the solution penetrates between the cellulose chains, causing swelling and gelatinization. This swollen pulp is commonly pressed into sheets from 0.005 to 2 in. (0.127 mm to 5.08 cm) thick or formed into tubes around mandrels. After thorough washing, it is dried. Vulcanized fiber cannot be molded by the usual methods. It can be machined by such operations as cutting, turning, milling, shaping, drilling, and tapping. The sheets can be formed in punch and hydraulic presses by such operations as bending, drawing, swaging, and cupping. Vulcanized fiber has very good physical and dielectric properties, but very high water absorption. Standard colors are red, gray, and black. It is used in many electrical applications and in silent gears, wastebaskets, textile, bobbin heads, welder's helmets, and gaskets.

wand scanner. A hand-held scanning device used as a contact bar code or optical code reader.

warm heading. A method in which blanks are heated to below the transformation temperature, usually 200 to about 1000°F (93 to 538°C). A valuable method of producing cold-working strength in the hard to deform tougher materials, such as some of the newer high-strength, high-temperature-resistant metals, or obtaining greater deformation with materials that are easy to upset. At elevated temperatures, work hardening is greatly reduced, less pressure is required, die life is increased, and unlike hot heading, the metal's properties are not altered. Methods of rapidly heating the stock or blanks, by means of induction, high-temperature-gas flame, or other means, are necessary for economical production. The correct temperature to be used for warm heading is critical since the strength of some materials is raised with slight increases in temperature.

warp. The yarns which run lengthwise in a woven fabric.

water gas (blue gas). Made in a cyclic process, in which coke is "blown" with air to raise its temperature and then "blasted" with steam. The steam reacts with the hot carbon endothermally as follows: $C + H_2O \rightarrow CO + H_2$. The gas is sometimes called "blue gas" because of the characteristic blue flame with which it burns, the coloring being due to the high percentage of hydrogen and carbon monoxide. A typical analysis of this gas is:

Component	Percent by Volume
CO_2	5.4
CO	37.0
H_2	47.3
CH_4	1.3
N_2	8.3
Higher heating value	28.7 Btu/ft^3 (1010 kJ/m^3

water gas, carburetted. Water gas is often enriched with fuel oil to raise its heating value. The enriched gas is termed carburetted water gas. The process involves the injection of oil into the carburetting chamber during the steam blow; the water gas, passing into the carburetor, picks up the oil vapor, which is then cracked into gases. The composition and heating value of the carburetted water gas varies with the amount of oil used, the latter property varying between 400 and 700 Btu/ft^3 (14,480 and 25,340 kJ/m^3).

water jet cutter. Using a thin, highly focused collimated stream of water pressurized up to 100,000 psi (6.9 × 10^5 kPa) and a computer-controlled pantagraph or a CNC control table, these machines can cut any desired pattern on soft materials such as plastics, wood, leather, paper, corrugated, carpeting, etc. A high-production machine, it can shape, trim, or cut with straight-line speeds from 165 to 2000 fpm (50.3 to 610 m/min). Newsprint can be cut at close to a mile a minute. Because cutting is done with water, there are virtually no airborne particles created. Some systems use long-chain polymer additives to provide jet-stream cohesion.

wattmeter. The unit for measurement of electric power is the watt. One watt is the power expended when a current of 1 A is flowing under the pressure of 1 V. In an electric circuit, the power in watts is equal to the product of the voltage and the amperage. This is true in dc and ac circuits when the voltage and currents are in phase. A wattmeter is contructed in a manner similar to a dynamometer movement, but the circuits for the magnetic field and the moving coil are separate. One of the windings must provide a field proportional to the current in the circuit, and the other must produce a field proportional to the voltage. Since the current windings must be heavy enough to carry the current of the circuit, the current coil is stationary. The moving coil carries the voltage winding because this winding must have a high resistance and is relatively light in weight. The current circuit is connected in series with the load circuit, and the voltage circuit is connected in parallel with the load circuit. A current-limiting resistance is connected in series with the voltage coil. When the circuits are connected in this manner the strength of the staionary field is proportional to the load current, and the strength of the moving coil is proportional to the voltage across the load. The indicating needle moves a distance proportional to the product of the voltage and the amperage; hence the scale may be marked directly in watts.

wavy edge. A rippling departure of an edge from flat.

wax. Esters of long-chain, unbranched fatty acids and long-chain alcohols. Both the acid and the alcohol that combine to form a wax may be 16–30 carbons in length. Both plants and animals produce natural waxes. Waxes usually are mixtures of esters and contain, in addition, small amounts of free acids, alcohols, and even hydrocarbons. The waxes melt over a wide range of temperature 95–212°F (35–100°C).

wax distillate. A neutral oil distillate containing a high percentage of crystallizable paraffin wax obtained in the distillation of paraffin-base or mixed-base crude and on reducing or neutral lubricating stocks. It is a primary base for paraffin wax and neutral lubricating oils.

wax, paraffin. A wax occurring principally in residues of paraffin-base petroleum, although it also occurs in limited quantities in mixed-base crudes and asphalt-base oils. It is removed by pressing, sweating, and filtering from the wax distillate taken from the tar stills. When thoroughly refined, it is a pure white wax having an unctuous surface.

wax press. Mechanical unit consisting of a series of canvas covered plates designed to remove certain forms of paraffin from chilled lubricating oil prediluted with naphtha to prevent the mass from congealing. The paraffin or wax collects between the plates from where it is removed.

wax tailing. In petroleum refining a heavy distillate during the final states of distilling certain mixed base oils down to coke. The product contains anthracene and chrysene produced by cracking. It is used for weatherproofing and waterproofing compounds.

wear. The unintentional removal of solid material from rubbing surfaces. The basic type of wear, which is always present to varying degrees when rubbing action between solid surfaces occurs, is that of adhesive nature. Adhesive wear, referred to as galling, scuffing, or scoring, is characterized by an intensive interaction between two bearing surfaces resulting from a mutual adhesion of the metals at the junction. Such adhesion results in a heavy tearing of the rubbing surfaces, leading ultimately to seizure over a wide area. Galling or scuffing can be defined in terms of the extensive damage done to a surface area, usually penetrating to some depth. Seizing is an intensive galling action, in which the relative motion of the opposing metal parts is stopped because of excessive adhesion. The extent of adhesive wear is controlled by the same factors as those involved in friction between sliding surfaces. Galling and seizing are promoted by an increase in temperature at the rubbing surfaces and also by clean surfaces. Relatively soft metals and solid-solution alloys of homogeneous structure tend to adhere strongly, leading to excessive damage to their surfaces. Dissimilar metals and alloys, particularly those having hard and heterogeneous surfaces, have little tendency to weld, therefore causing only a small amount of adhesive wear. Austenitic and ferritic stainless steel, titanium, platinum, and silver tend to gall severely, whereas cast iron, highly hardened steels, and chromium-plated surfaces show little tendency to gall. The surface finish may also affect galling. The extent of wear depends on the size, shape, and hardness of the particles. Thses make scratches on the wearing surfaces, often irregular size, and cause the removal of surface material, thereby resulting in a gradual wear of that surface.

web. A single thickness of foil as it leaves the rolling mill. Also, a connecting element between ribs, flanges, or bosses on shapes and forgings.

welded tube (see tube, welded)

welding, arc. The source of heat in all arc-welding processes is an electric arc. In some processes the arc occurs between a single electrode and the work. And in others it is formed between two electrodes. The temperatures produced are always well above the melting points of the common engineering metals. The base metal is almost always melted during arc welding. There are wide variations in fluxes and shielding methods for different arc-welding processes. The two principal kinds of arc welding are metal arc-welding and carbon-arc welding.

welding, arc-button or arc-spot. Shielded by gas, metal-arc welding one spot at a time. One form of the operation is programmed to burn through one or more layers, up to 3/8 in. (9.5 cm) thick and then stop, leaving a plug of fused metal through the sheets topped with a button of excess metal on the surface. The main advantage of this method is that it can be done entirely from one side without high pressure. A way to pierce metal is to burn through a piece with a shielded electrode and blow out the molten metal. A process for quantity production utilizes a string of electrodes, along a straight or curved joint. The electrodes are fired in rapid succession, and produce a series of overlapping spot welds.

welding, arc stud welding. Stud welding is an arc-welding process wherein coalescence is produced by establishing an arc between a metal stud and the workpiece until sufficient temperature is produced, and then pressing the stud against the work with enough pressure to complete joining of the stud to the work. The process is usually done without shielding. A special "gun" is used with auxiliary equipment, which controls the establishing of the arc, its duration, and the application of pressure. Direct current is used. The end of the stud is recessed and the recess filled with welding flux. A ceramic ferrule is placed over the end of the stud before it is positioned in the welding gun.

This ferrule is an important part of the process, since it concentrates the arc heat, protects the metal from atmosphere, confines the molten metal to the weld area, and shapes it around the base of the stud. After the weld is completed the ferrule is broken from the stud. Stud welding has eliminated much drilling and tapping of holes in manufacturing.

welding, atomic hydrogen arc. Similar to gas, tungsten arc welding but is done with an arc between two tungsten electrodes in a steam of hydrogen gas. The hydrogen gives protection from the atmosphere. The molecules of hydrogen split into atoms in the heat of the arc, and the gas transfers heat more readily in that state. The atoms recombine away from the arc and liberate heat at the workpiece. Thus, a high temperature is created for the weld. The process is not used often except for deep welds, as in die blocks, and for high-temperature alloys, particularly for surfacing.

welding, braze. In braze welding a filler is melted and deposited in a groove, filet, plug, or slot between two pieces to make a joint. In this case the filler metal is a copper alloy with a melting point below that of the base metal but above 800°F (427°C). The filler metal is puddled into and not distributed in the joint by capillarity as in plain brazing. Metals with high melting points, such as steel, cast iron, copper, brass, and bronze are braze welded. The base metal does not melt, but a bond is formed. Bronze welding may be done by oxyacetylene, metallic-arc, or carbon-arc welding. The filler metal is applied by a rod or electrode together with a suitable flux. The main advantages of braze welding result from the low temperature of the operation. Less heat is needed and a joint can be made faster than by fusion welding. Braze welding joints are not satisfactory for service at over about 500°F (260°C) nor for dynamic loads of 15,000 psi (1.03×10^5 kPa) or more.

welding, carbon arc. In this process a carbon arc acts as the electrode and supplies the dc current. Straight polarity current creates the arc. A filler rod, uncoated, is used. Flux may or may not be used as desired with this rod. The filler rod is consumed and mixes with the base metal. Since only dc is used in this method, splatter may occur because of the magnetic fields created. This is called magnetic arc blow. The arc tends to wander and very careful control must be instituted. Edge welding is almost impossible because of the high concentration of magnetic fields at the edge of the work.

welding, cross wire. A variation of projection welding. Crossed wires touching at a point give the same effect as projection. Large-surface-area electrodes apply pressure to the pattern. As the pressure and current are applied the wires are welded at the point of contact to form the desired mesh. Wire baskets, lampshades, etc., may be joined by this process.

welding, dip brazing. Accomplished by dipping the assembled parts into a molten bath of the filler material. This method is called metal bath brazing. In chemical bath brazing the assembly is dipped into a neutral salt bath with the filler rod in position on the work. The oxidation that takes place is negligible because the parts do not come into contact with the atmosphere. In both processes accurate temperature control is possible. In the chemical bath process no flux is needed.

welding, electron. A fusion welding process in that the mechanism of welding is to join two or more pieces of metal by local melting, which results in coalescence. Filler metal and base metal, or base metal only, may be melted together to produce a joint. The heat required for welding is produced by the bombardment of the work by a dense beam of high velocity electrons. The electron beam is directed upon the area to be fused, and the electrons upon striking the surface of the metal give up their kinetic energy almost completely in the form of heat energy. Since a vacuum in the order of 1×10^{-4} mm Hg is needed to produce a stable electron beam, the welded joints are extremely pure and singularly free from absorbed gases. As a result, electron beam welding is suited for welding the refractory metals (tungsten, molybdenum, columbium, tantalum, etc.), the highly oxidizable metals (titanium, beryllium, zirconium, uranium, etc.), and also the vacuum-melted alloys (M-252, Rene 41, Udimeter 500, hastelloys, etc.). Any weldable metal or alloy, and possibly some ceramics, can be fused by means of the electron beam process.

welding, electro-slag. A form of vertical welding used particularly for joining pieces more than 1½ in. (3.8 cm) thick for turbine shafts, boiler parts, and heavy presses. Welding is initiated on a starting block at the bottom of the vertically positioned joint. Flux poured around the electrode is converted to slag that floats on a layer of molten metal confined in the joint by water-cooled copper shoes that slide on the sides. No arc is visible and a larger part of the heat is said to come from the electrical resistance of the slag. The welding dams and head move upward as weld metal solidifies and new metal is fed in by the wire electrodes. The wire is oscillated across the joint to spread the weld evenly and two or three more wires may be fed at the same time in a wide joint.

welding, flash-butt. Two pieces to be flash welded are clamped with their ends not quite touching. A current with a density of 2000 to 5000 A/in.^{2} is amplified to the pieces and arcs across the joint in a flash that melts the metal at the ends of the piece. A pressure of 5000–20,000 psi (3.45×10^4–1.4×10^5 kPa) is applied suddenly to close the gap.

welding, furnace brazing. Generally uses a controlled atmosphere furnace; an electric furnace and a controlled atmosphere will give the best results. Any of the inert gases may be used to prevent copper braze from oxidizing. If oxide formation is no problem, a controlled atmosphere is not necessary. With this method many parts may be assembled with the brazing material and flux in position. The entire assembly is then placed in the furnace and brazed in one operation. The heating is uniform and relatively free from distortion. Good control of temperature and atmosphere gives good surface conditions.

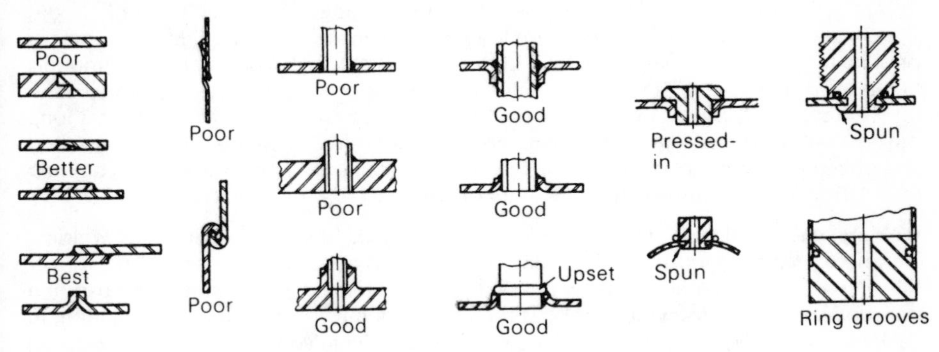

Typical lap joints utilized in designing brazed assemblies.

welding, inert-gas arc. Helium (called heliarc), argon, carbon dioxide are introduced into the chamber that houses either a tungsten or carbon electrode. The gas, which is emitted through an opening, circumscribes the electrode and forms a protective cloud about the arc when aluminum or magnesium are being welded. Both metals oxidize rapidly and need this projection during the welding process. The arcing takes place between the electrode and the work. In the case of mangesium a noncombustible electrode is used in combination with helium and dc current. In the case of aluminum a combustible electrode and argon are used in combination with ac current. Straight polarity dc current may be used to weld steel, stainless steel, copper, or cast iron. Carbon dioxide gas, because of excellent penetration and high speed is at times used to weld carbon steels.

welding, inertia. Inertia welding can be considered a type of machine operation—somewhat comparable to combining a lathe and a heavy press. In operation, workpieces to be joined are clamped in chucks with one piece in a nonrotating holding device and the other in a rotatable, flywheel-equipped spindle. This spindle is accelerated to a predetermined high rpm under no load, power is cut, and the workpieces immediately thrust against one another. Frictional deceleration converts the flywheel's kinetic energy to heat, making the interface plastic but not molten. This action forms a strong, solid-state bond. Rotation ceases when the weld is completed. Metal at the weld interface is hot-worked during the last half-revolution of the workpiece. The process is fast and is considered to be one of the most efficient means of converting applied energy to generate heat for welding. No external heat source is needed, nor is flux, filler material, shielding, or long cooling periods. The parameters of mass, speed, and thrust can be predetermined and set for the specific run of parts. Once set, the operator need only actuate the start button to produce a weld. Inertia welders are available in varying capacities and configurations and can be customized to the degree of automatic operation required for specific applications.

welding, laser. A process which focuses monochromatic light consisting entirely of a pure color into extremely concentrated beams. The heart of laser welding hardware is a high-purity rod, usually a ruby (trivalent chromium in aluminum oxide). The average duration of a laser weld beam is 0.002 s. Because of the short duration, two basic welding methods are used. In one method, the workpiece is rotated or moved fast enough so that the entire joint is welded with a single burst of light. The other method uses many pulses of the laser to cover the joint. With the pulse technique, the weld is comprised of round, solidified puddles each overlapping the previous one by about half a puddle diameter. The pulse method is the one commonly used. Most laser systems pulse an average of 10 times per second so that each point on the weld line melts and solidifies in microseconds. Some substances may be pulsed continuously but at low power, and in practice with a ruby laser 6–10 pulses/min are obtained. The system is only about 1% efficient; most of the loss heats the laser and, even with cooling, limits the rate of pulsing. In spot welding flashes of 1500 J/pulse can be generated, but such large amounts of energy disintegrate and vaporize materials. Quantities of 10 J or less are more useful for welding. Short pulses broadly focused may be applied for shallow welds and long pulses sharply concentrated for deeper and narrow welds. A minute puddle is melted and frozen in a matter of microseconds, and the heat-affected zone is normally confined to 0.010 in. (0.254 mm) or less. **(See also laser.)**

welding, gas metal arc. Called GMA or MIG for metal inert gas, shielded arc welding is done with a wire fed through the welding head to act as the electrode and supply filler metal. Because filler metal is needed in most welding applications, the use of a consumable electrode, which melts at the arc temperature, greatly speeds welding in such cases. Consequently, gas metal-arc welding (gas-shield consumable metal-electrode welding) was developed as an offshoot of gas tungsten-arc welding. With CO_2, helium, or argon as the shielding gas, a special welding gun and feeding mechanism feed the electrode, in the form of coiled wire, as it is consumed. This eliminates the necessity for frequent changes of electrodes, as in the case with stick-type electrodes, and greatly increases the welding speed. There is a further saving of time due to the fact that no slag is formed, which must be removed after the welding. The electrode usually is of the same general composition as the material being welded. With an inert gas, such as helium, or argon, this process can be used for welding almost any material. Because of their cost, when these gases must be used, the process usually is more costly than ordinary shielded electrode arc welding. Therefore, these gases are used primarily when welding aluminum, magnesium, or stainless-steel alloys where a completely inert atmosphere is a requisite, or when excellent appearance is desired, but cheaper carbon dioxide has been popular for a large variety of production operations. The carbon dioxide also dissociates into carbon monoxide and oxygen, and deoxidizers must be added to protect the weld. One form of the process uses a flux-cored wire containing arc stabilizers and deoxidizers. Another variant of the process is called microwire welding and feeds wires as small as 0.030 in. (0.762 mm) diameter at high speeds. With high current densities, up to 150,000 A/in.^{2}, the wire is well preheated and the current is concentrated at the weld.

welding, gas (metallurgy). Heat is normally supplied by an oxygen acetylene flame, with the temperature produced by the flame well above the melting points of any of the common engineering metals. The base metal is invariably melted in the immediate vicinity of the joints. In gas welding of steel, the base metal and weld may be controlled by proper manipulation of the welding torch. The temperatures in the weld and base metal vary between the melting point of the steel and room temperature. Portions of steel heated above the critical range undergo recrystallization and grain growth, and upon rapid cooling, hardening of the base metal may occur. Both coated and bare rods are used to provide filler metal. The coated rods supply a protective liquid slag cover. Fluxes in the form of powders, pastes, or rod coatings are more common for gas welding of nonferrous metals where they afford protection and also flush out impurities. Fairly recently, stabilized methylacetylene propadiene, in conjunction with oxygen, has come into fairly extensive use for gas-flame welding, primarily because it is more convenient to store. However, flame welding is used for only a small percentage of fusion welding, and largely in repair and remote field work because it requires no power source and is both flexible and portable.

welding, metal electrode. Virtually all arc welding is done with metal electrodes. In one type the electrode is consumed and thus supplies needed filler metal to fill in the voids in the joint and to speed the welding process. In this case the electrode has a melting point below the temperature of the arc. Small droplets are melted from the end of the electrode and pass to the parent metal. The size of these droplets varies greatly and the mechanism of the transfer varies with different types of electrodes. As the electrode melts, the arc length changes and the resistance of the arc path varies, due to the change in the arc length, and the presence or absence of metal particles in the path. Arc welding with bare consumable electrodes is difficult because the arc tends to be unstable. In addition, it is necessary to maintain a short arc length to avoid severe oxidation of the hot metal. The instability of the arc makes this difficult. As a result almost no welding is done with bare electrodes, except in the case of stud welding.

welding, plasma-arc. Plasma is high-temperature ionized gas and occurs in any electric arc. If a stream of injected gas and the arc are made to flow through a restriction (a water-cooled copper orifice), the current density of the arc and velocity of the gas are raised. In that way the ionization and temperature of the gas are greatly increased. The excited particles gave up large amounts of energy when they reform into atoms, some at the workpiece surface. In thin materials, the force of the jet opens a hole that is filled with molten metal as the arc is moved along. Penetration is deep and thorough, and much work can be done without filler metal. Any known material can be melted, even vaporized by the plasma-arc process and become subject to welding.

welding, projection. Uses projections to concentrate the current. The localized current and the multiple projections create a multiple spotweld effect. The pieces to be welded are held in position under pressure supplied by the electrodes. The current consummates the weld. The projections at the point of contact should be approximately equal to the thickness of the metal and increase in cross section. The height of the projection should be 10–80% in thickness of the material welded. If pieces of unequal thickness are to be welded, the thicker piece should have the projections. The desired number of spots determine the number of projections pressed into the sheet. The process is fast because a number of spots can be welded in one closure of the press. A metal must have sufficient hot strength to be projection welded successfully. For this reason aluminum is seldom welded by this process and some brasses not at all. Free-cutting steels high in phosphorus and sulfur must not be projection welded because the welds are porous and brittle. Most other steels are readily projection welded.

welding, resistance. The resistance of the workpiece to the flow of current through it is the source of heat in resistance welding. In flash welding, which is a special form of resistance welding, the heat of an electric arc is an additional source of energy. The temperature produced in resistance welding are always above the melting point of the metal being welded. High temperatures are produced over only localized areas, and the amount of metal involved is fairly small. Original properties can be restored, if necessary, by suitable heat treatment. Fluxes and shielding gases are rarely used, nor is filler metal needed for resistance welding. Pressure is required for all resistance welding processes. There are four principal varieties of resistance welding: (1) spot, (2) seam, (3) projection, and (4) flash.

welding, resistance brazing. Requires that an electric current be passed through the parts being brazed. The resistance to the passage of the current generates heat and melts the braze material and the flux. Any of the resistance machines may be used for this type of brazing. Special machines using water-cooled carbon electrodes have been designed for use in resistance brazing. These machines are equipped with accurate time-controlling devices. The size of the work is one of the limitations of this type of machine.

welding rod. A rolled, extruded, or cast round filler metal for use in joining by welding.

welding, seam. A series of spot welds either overlapping or spaced at short intervals. The latter is called a roll spot weld or stitch weld. Seam welding is done by passing the work to conserve the electrodes and cool the work rapidly to speed the operation. Both lap and butt joints may be seam welded. Seam welding is done on metal sheets and plates from 0.0003 to 0.187 in. (0.0076 to 4.75 mm) thick with standard equipment and up to 3/8 in. (9.5 cm) thick with special machines. Seam welding is done mostly on low-carbon alloy and stainless steels, but also on many other metals including aluminum, brass, titanium, and tantalum. Typical seam-welded products are mufflers, barrels, and tanks.

welding, shielded metal arc. Shielded electrodes consist of a metal wire which usually from 1/16 to 3/8 in. (1.6 to 9.5 mm) in diameter, upon which is extruded a coating containing chemical components that provide a number of desirable characteristics to the welding process. These coatings may accomplish all or a number of the following qualities: provide a protective atmosphere; stabilize the arc; act as a flux to remove impurities from the molten metal; provide a protective slag to accumulate impurities, prevent oxidation and slow down cooling of the weld metal; reduce weld metal spatter and increase the efficiency of deposition; add alloying elements; affect arc penetration. Coated electrodes are classified on the basis of tensile strength of the deposited weld metal, welding position in which they may be used, type of current and polarity, and the type of covering. All electrodes are marked with colors in accordance with a standard established by the National Association so they may readily be identified as to type.

welding, spot. The most common form of resistance welding and the simplest. Electrodes with reduced ends are pressed against the work, the current is turned on and off, and the pressure is held or increased to forge the weld while it solidifies. This is commonly repeated in a series of spots along a joint. If a welded spot between two sheets is sectioned, a small mass or nugget of metal is found embedded in and joining the sheets together. Sound welds are obtained consistently by applying the proper current density, timing, and electrode pressure uniformly in each case to clean surfaces.

Metals	Al	St. St.	Br.	Cu	Gal. Fe	St.	Pb.	Mon.	Ni	Nich.	Tin Pl.	Zn	Ph. Br.	Ni Sil	Te.Pl.
Aluminum	B	E	D	E	C	D	E	D	D	D	C	C	C	F	C
Stainless Steel	F	A	E	E	B	A	F	C	C	C	B	F	D	D	B
Brass	D	E	C	D	D	D	F	C	C	C	D	E	C	C	D
Copper	E	E	D	F	E	E	E	D	D	D	E	E	C	C	E
Galvanized Iron	C	B	D	E	B	B	D	C	C	C	B	C	D	E	B
Steel	D	A	D	E	B	A	E	C	C	C	B	F	C	D	A
Lead	E	F	F	E	D	E	C	E	E	E		C	E	E	D
Monel	D	C	C	D	C	C	E	A	B	B	C	F	C	B	C
Nickel	D	C	C	D	C	C	E	B	A	B	C	F	C	B	C
Nichrome	D	C	C	D	C	C	E	B	B	A	C	F	D	B	C
Tin Plate	C	B	D	E	B	B		C	C	C	C	C	D	D	C
Zinc	C	F	E	E	C	F	C	F	F	F	C	C	D	F	C
Phosphor Bronze	C	D	C	C	D	C	E	C	C	D	D	D	B	B	D
Nickel Silver	F	D	C	C	E	D	E	B	B	B	D	F	B	A	D
Terneplate	C	B	D	E	B	A	D	C	C	C	C	C	C	D	B

*A = excellent, B = good, C = fair, D = poor, E = very poor, and F = impractical.

Spot welding metal combinations and weldabilities.

welding, submerged arc. In submerged arc welding an arc is maintained beneath a granular flux. Either alternating or direct current may be used as the power source. The flux is deposited just ahead of the electrode, which is in the form of coiled wire, copper coated to provide good electrical contact. The arc is completely submerged by the flux so that only a few small flames usually are visible. The granular flux provides excellent shielding of the molten metal so that high-quality welds are obtainable. In addition, because the pool of molten metal is relatively large, there is good fluxing action so as to remove impurities. The welding head is moved along the work automatically, or the work moved past the head. A portion of the flux is melted and solidifies into a glasslike covering over the weld. This, with the flux that is not melted, provides a thermal coating that slows down cooling of the weld area and helps to provide a soft, ductile weld.

welding, thermit. In thermit welding heat is provided by a chemical reaction between iron oxide and powdered aluminum. The temperature produced by this reaction is far above the melting point of steel, which is the only metal joined by the process. Iron oxide is reduced to iron by the aluminum to supply filler metal. Sufficient carbon is provided in the powdered mixture so that actually molten steel, rather than iron, is the filler. Within certain limitations it is also possible to add other alloying elements to produce alloy steels of high strength. Thermit welding is particularly suited to welding rather heavy sections such as rails, large crankshafts, heavy steel castings, and similar objects. The parts to be joined are positioned in a fixture that holds them a certain distance apart. The distance separating the parts depends on their size. This space is then filled with wax, which is given the shape of the joint to be made. A form corresponding to a molding flask is placed around the joint, into which the joint is rammed. Heat is then applied to melt out the wax and leave a void of the desired shape. Additional heating serves to dry out the same mold and preheat the pieces being joined, so that better fusion is obtained. In the meantime, the correct amounts of iron oxide and aluminum and the necessary alloying elements are placed in a bottom-pour crucible. The powder is ignited, and after the reaction is completed, the molten metal is poured into the mold.

welding, torch brazing. The most common method used for brazing. A neutral flame is directed at both parts so as to get uniform heating of both pieces before the braze melts. The fuels used may be oxyacetylene, oxyhydrogen, or natural gas. Oxidization and distortion may be factors in this type of operation which must be dealt with if the operator is not skilled.

welding, torch cutting (see flame cutting)

welding, tungsten inert gas (TIG). Developed originally for welding magnesium, this process employs a tungsten electrode in a special holder, through which a gas may be supplied at low pressure so as to provide sufficient flow to form a shield around the arc and the molten pool of metal, thus shielding them from the atmosphere. For most work, the inert gases argon or helium are used, but in welding steel CO_2 may be substituted. Because the tungsten is virtually not consumed at arc temperature in these inert gases, the arc length remains constant so that the arc is stable and easy to maintain. To make it easier to start and maintain the arc, a high-frequency, high-voltage current usually is superimposed on the regular ac or dc welding current. In the nonconsumable-electrode process, any additional filler metal that is needed is supplied by a separate filler rod as is done in gas-flame welding. In one modification of tungsten arc welding, the needed filler metal is supplied automatically in the form of a continuous fine wire, and the wire is heated by the passage of an electric current so that it melts into the weld puddle behind the arc. This hot-wire procedure gives a four- to sixfold increase in the metal deposition rate. It cannot be used with copper or aluminum because of their lower resistivities. Because no flux of any kind is used in tungsten arc welding, the resulting welds are very clean, and no special cleaning or slag removal is necessary.

welding, ultrasonic. A process by which the welder converts electrical energy to high-frequency mechanical vibrations. These vibrations are delivered to the parts being welded in such a way that dynamic stresses are induced in a limited area near the interface between the workpieces. When these dynamic stresses reach sufficient intensity, they cause local plastic deformation of the interface material in such a way that adhered moisture, organic, and oxide films are broken up and dispersed. The irregularities of the original interface material surfaces are also eliminated in order to create an area of intimate nascent metal contact. This results in a true metallurgical bond, formed in the solid state, with no melting of the material being joined.

welding, upset-butt. Consists of pressing two pieces of metal together end to end and passing a current through them. Current densities range from 2000 to 5000 A/in.^{2}. The resistance of the contiguous surfaces under light pressure heats the joint. The pressure is increased. This helps metal from the two parts to coalesce and squeeze some metal out into a flash or upset. The current may be interrupted one or more times for large areas. Final pressures range from 2500 to 8000 psi (1.7×10^4 to 5.5×10^4 kPa) depending on the materials. Most metals can be upset welded. Common applications are the joining of wires and bars end to end, welding of a projection to a piece, and welding together the ends of a loop to make a wheel rim. Pipe may be upset-butt welded.

welding wire. Wire for use as filler metal in joining by welding.

wettability. The degree to which a metal surface may be wet to determine the absence of or the amount of residual rolling or added lubricants or deposits on the surface.

whip marks. Markings on a sheet surface generally running normal to the direction of rolling, resulting from a whipping of the sheet as it enters the rolling mill.

white metals. The white metals are low-melting alloys in which lead, tin, or antimony predominate. Zinc, aluminum, and magnesium base castings are sometimes erroneously called white-metal castings. The three most important classes of white metals are (1) fusible alloys, (2) type metals, and (3) bearing alloys. Of these, the most important from an engineering standpoint are the bearing alloys. These are known as babbits. True babbits are tin-base alloys strengthened by the presence of antimony and copper, and occasionally containing lead. Lower-cost lead-base babbits are also used in place of true babbits. The microstructures of all these alloys are similar. They consist of hard primary cube-shaped particles in a soft matrix. When the primary particles solidify, they tend to separate by gravity segregation, sinking to the bottom of the casting. Segregation is prevented by the use of copper, which solidifies first as an interlocking network of spiny crystals. This produces a structure in which the hard, cube-shaped particles are uniformly distributed in the matrix. The babbit bearing alloys are quite readily cast in the shop.

whiteware. Generic term for ceramic products which are usually white and of fine texture. These are based on selected grades of clay bonded together with varying amounts of fluxes and heated to a moderately high temperature in a kiln 2192–2732°F (1200–1500°C). Because of the different amounts and kinds of fluxes, there is a corresponding variation in the degree of vitrification among whitewares, from earthenware to vitrified china. These may be broadly defined as follows: Earthenware, sometimes called semivitreous dinnerware, is porous and nontranslucent with a soft glaze. Chinaware is a vitrified translucent ware with a medium glaze which resists abrasion to a degree; it is used for nontechnical purposes. Porcelain is a vitrified translucent ware with a hard glaze which resists abrasion to the maximum degree. It includes chemical, insulating and dental porcelain. Stoneware, one of the oldest of ceramic wares, may be regarded as a crude porcelain, the raw material being of a poorer grade and not so carefully fabricated. Sanitary ware, formerly made from clay, now made with a vitreous composition. Prefired and sized vitreous grog is sometimes included with the triaxial composition. Whiteware tiles, are generally classified as floor tiles, resistant to abrasion and impervious to stain penetration, glazed or unglazed, and as wall tiles which also have a hard, permanent surface.

wire. A solid wrought product that is long in relation to its cross section, which is square or rectangular with sharp or rounded corners or edges, or is round, a regular hexagon or a regular octagon, and whose diameter or the greatest perpendicular distance between parallel faces (except for flattened wire) is less than $^3/_8$ in. (9.5 mm). If a wire is covered with insulation, it is properly called an insulated wire; while primarily the term wire refers to the metal, nevertheless, when the context shows that the wire is insulated, the term wire will be understood to include the insulation.

wire, Alclad. Wire having on its surface a metallurgically bonded aluminum or aluminum alloy coating that is anodic to the core to which it is bonded, thus electrolytically protecting the core alloy against corrosion.

wire, cold heading. Wire of quality suitably uniform in structure for use in the manufacture of cold-headed products such as rivets and bolts.

wire drawing. Wire drawing involves pulling the stock through a tungsten carbide die having a tapered bore. Large quantities of rods, tubes, and wires are produced by this technique. Usually the reduction of the rod area is limited to about 30% so that for a fine wire, the process consists of a long series of successive wire drawing operations through subsequent dies of suitable diameters.

wire, drawn. Wire brought to final dimensions by drawing through a die.

wire, flattened. A solid section having two parallel flat surfaces and rounded edges produced by roll-flattening round wire.

wire, flattened and slit. Flattened wire that has been slit to obtain square edges.

wire forming. Known today by the general term "four-slide machines," these machines range from relatively simple single-slide machines to the plain or combination four-slide types useful in producing wire and ribbon parts. Essentially, the basic four-slide machine consists of a stock straightening arrangement, a feed mechanism for either wire or ribbon-metal, a cutoff unit, four horizontal slides, and a vertical forming attachment with a stripping mechanism. Normally the stock is drawn from a horizontally positioned reel through the straightener rolls by the positive feed mechanism and is fed to the desired length as preset. The front tool, commonly referred to as the "binder," then advances to hold the part against the tool or stationary forming post on the head. The cutoff slide then functions, the horizontal slides following in the preset forming sequence, and lastly, where necessary, the vertical forming attachment and/or the stripper operate. All the slides employed are usually cam actuated while the feed mechanism is usually crank or toggle actuated in conjunction with cam-operated feed and holding grip clamps.

wire insulation, rotating machinery. The insulation that is applied to a wire before it is made into a coil or inserted in a machine.

wire, rivet (see wire, cold-heading)

wood. The structure of wood is cellular. A piece of wood basically is a bundle of thin-walled tubes. This structure explains the anisotropy in the properties of wood. The chemistry of the walls of these wood cells is of great industrial importance especially for the pulp and paper industry. From 55 to 70% of the cell wall is cellulose, a long-chain fibrous polymer resistant to acids. It is estimated that the cellulose polymer is composed of 5000–10,000 monomer units. From 15 to 35% of the cell wall is lignin. This is the chemical that produces the brown color of corrugated cardboard cartons, and causes the yellowing of newsprint with age. Secondary constituents of wood include volatile (distillable) materials, gums, and ash. The property of anisotropy is a design consideration in the use of wood. The tensile strength of wood on a plane prependicular to the grain is about 10 times better than the strength parallel to the grain. Shear strength parallel to the grain is poor, while shear strength perpendicular to the grain is high. The dimensional changes in wood due to variations of moisture content are likewise influenced by the cellular or tabular structure of wood. Changes in length parallel to the grain as a result of moisture variation are negligible. There are numerous kinds of commercial woods. They may be classified in two groups: hardwoods and softwoods. Softwood usually comes from trees with needlelike leaves (mostly cone bearing), while hardwoods come from trees with broad leaves. The main softwoods are pine, cedar, fir, spruce, hemlock, and cypress. The principal hardwoods are oak, maple, walnut, hickory, ash, and poplar. Trees may also be classified according to the manner of growth as exogenous or endogenous. The exogenous tree's growth occurs by the formation of rings between the old wood and the bark, while in endogenous trees the new fillers grow intermingled with the old. Palm, bamboo, and yucca trees are examples of endogenous trees.

workability. The relative ease with which various alloys may be formed by rolling, extruding, forging, etc.

work hardening, metals. The ability of metallic materials to be hardened and strengthened by mechanical deformation is one of the most important metallic characteristics. The application of a stress to a metallic material will tend to deform it. The resulting strain can be either elastic, in which case it is very nearly proportional to the stress and will disappear completely in time when the stress is removed, or plastic, producing thereby a permanent deformation. Within certain temperature ranges, the metal will be hardened and strengthened by any permanent deformation. This effect is known as work or strain hardening and since it is related uniquely to the metallic type of atomic structure, it probably will never be found or produced in any other type of material. At normal atmospheric temperatures, most metals and nearly all commercial alloys are susceptible to strain hardening.

work-in-process (WIP). Product in various stages of completion throughout the plant, including raw material that has been released for initial processing and completely processed material awaiting final inspection and acceptance as finished product or shipment to a customer. Many accounting systems also include semifinished stock and components in this category.

workstation, graphics. This word is used to describe diverse graphics displays, terminals, subsystems, and turnkey systems. A graphics workstation is a hardware–software combination that gives a user local graphics input, graphics display, and graphics processing. **(See also CADD.)**

wrought iron. A material in which fibrous particles of siliceous slag are distributed through very pure iron. The result is a ductile iron that has certain properties, particularly corrosion resistance, which makes it of importance as an engineering material. Wrought iron formerly was made by the puddling process. The Aston process has virtually replaced the puddling method. In the first step of this process, pig iron is tapped into ladles where it is desulfurized. It then is purified to a high state in a Bessemer converter. Simultaneously, the second step preparation of the slag is carried out in an open-hearth furnace, by melting together iron oxide and siliceous materials. In the third step the refined iron is poured into a ladle containing the molten slag. The iron solidifies rapidly with the liberation of dissolved gases. The gas liberation is sufficient to shatter the solidified metal into small fragments. These fragments cohere, as they settle to the bottom of the ladle, and are impregnated with liquid slag. The surplus slag is poured off and the metal mass is squeezed in the same manner as the puddler's balls were handled. Wrought iron obtained by either the puddling or Aston process is about 98% pure iron, with the carbon content being about 0.02%. The remainder is slag.

wrought product. A product that has been subjected to mechanical working by such processes as rolling, extruding, forging, etc.

Y

yarn. A generic term for continuous strands of textile fibers or filaments in a form suitable for knitting, weaving, or otherwise intertwining to form a textile product. It may be composed of a number of fibers twisted together; a number of filaments laid together without twist; a number of filaments laid together with some twist; a single filament, either with or without twist (a monofilament).

yarn, continuous filament. Yarn made from filament having an extreme length not readily measured. **(See also filament.)**

yarn, filling. The yarns running back and forth in a woven fabric at right angles to the warp yarn.

yarn, monofilament. A single filament of sufficient size to function as a yarn in normal textile operations.

yarn, multi-filament. A yarn made from several filaments plied together.

yarn, spun. Yarn made by twisting together thousands of small fibers 1¼ in. (3.2 cm) to about 4 in. (10.2 cm) long into a yarn suitable for knitting and weaving.

yellow brass (see brasses)

yield point. The stress at which a marked increase in strain occurs without an increase in stress. Structural steel exhibits this property to a marked extent, but the number of materials that show a true yield point is limited. Brittle metals such as cast iron, have stress–strain diagrams that rise continuously until rupture occurs. Such materials are not considered to have a yield point.

zero defects program. An industrial motivational program aimed at the development within each employee of a personal enthusiasm for doing his or her job without error. A zero defects program must be custom fitted to the plant and environment in which the program is to work. The program is designed to motivate people toward the pursuit of excellence and pride in craftsmanship. It is a positive program which rewards people for good performance by enlisting the aid of all employees in eliminating causes for error. It is a program which achieves higher and higher levels of quality through constantly improving levels of human performance.

zinc. Zinc is used for many important alloys such as the various bronzes and brasses. The zinc sphalerite (ZnS) is the most important zinc ore. Zinc may also occur in the form of carbonate or oxide. The method of manufacturing zinc is similar to the procedure used for other metals and consists of roasting, reducing, and refining. After grinding and heating the ore in a furnace, zinc sulfide is reduced to zinc oxide. Oxygen is removed from the zinc oxide by heating with carbon, usually in the form of coal. In this reducing process the carbon combines with oxygen and the oxygen is removed in the form of gas, while the oxide is reduced to zinc. With the temperatures used, the zinc is in the gaseous state and is collected and condensed in cooling chambers. There are a number of copper–zinc alloys, known as brass, which vary in the relative amounts of copper and zinc. Manganese bronze is an alloy of copper and zinc with small amounts of tin, iron, and manganese. Naval brass is a zinc alloy similar to manganese bronze but it does not contain iron. For use in die casting a series of zinc alloys known as Zamak have been developed. In addition to zinc, these alloys contain aluminum, magnesium, iron, and copper. Zinc is used also for galvanizing, in batteries and in the manufacture of paint.

zinc alloys. The most important alloy additions for zinc are magnesium and aluminum, which act to improve corrosion resistance and strength in zinc-base die-casting alloys. Copper is sometimes also added to these alloys to stabilize their dimensions. Zinc-base die castings are compared with those of other metals. The alloys of zinc used in die casting contain approximately 4% aluminum, 0.1% magnesium, and as much as 2.5% copper. The zinc must be of high purity and relatively free of lead, antimony, and tin. Such alloys have the trade name "Zamak". Die-cast articles have numerous uses, particularly in automobiles. Such parts as radiator grills, hardware, and carburetors are zinc die castings. An alloy with much the same composition as the die-casting alloys is used to make stamping dies. This alloy is also called ''Zamak'' or ''kirksite''. These dies are used for forming aluminum or magnesium parts when the number required is not large enough to justify the cost of steel or cast iron dies, but is still too large to be economically hand formed. The zinc alloy has a low melting temperature and is easily formed into the desired shape. When wear becomes excessive or production is completed, the die can be melted and the material rescued. The punches used are most often of rubber or lead but kirksite may also be used.

zinc graphite permanent mold casting. The Zn-12 alloy is used nearly exclusively in the zinc graphite permanent mold (Zn-GPM) process. Zn-12 was the first Zn-Al gravity casting alloy developed recently. Zn-8 and Zn-27 were introduced within the last few years and have also been successfully cast in graphite dies. The zinc/graphite process incorporates more refined methods than conventional permanent mold casting with ferrous dies. Iron dies are often mounted on home-built, hand-operated mechanisms individually fabricated for each tool. The shops that have adopted the zinc/graphite process, on the other hand, incorporate standardized production equipment which use semiautomatic casting machine principles. These principles include mounting dies (usually vertically) on electropneumatically controlled platens with built-in ejector systems. The machines operate much like a simplified die-casting machine. The operator starts the casting cycle by pushing the start button on his or her control panel after molten metal has been hand ladled into the die. A thermocouple in the mold provides a visual temperature display for the operator to check die performance. Ejector pins push the castings off the die during the open cycle, and the operator removes the parts by hand. Tongs are not required. After part removal the die automatically closes and is ready for the next cycle. Careful control of the casting parameters sometimes allows for immediate trimming by hand while gates are still soft. In most cases, however, gates are sawed off and ingate areas are ground to blend smoothly with the casting profile. **(See also permanent mold casting.)**

zirconium. A reactive metal, with extraction by converting the ore to zirconium chloride and then reducing the chloride with magnesium. Like titanium, molten zirconium reacts with crucible materials, reducing their oxides to metal. Zirconium melts at 3366°F (1852°C). It is not characterized by good hot strength and oxidizes readily above 1000°F (538°C). Fine zirçonium powders or matching chips may even ignite spontaneously, though the metal is not hazardous in massive form. Crystal structure of zirconium is hcp at room temperature, with a bcc phase above 1590°F (866°C). In sheet form, zirconium somewhat resembles stainless steel, though it has somewhat lower specific weight. It has a low thermal conductivity, like stainless steel, but a much lower coefficient of expansion, 0.000004 in./in. Like stainless steel its corrosion resistance is excellent, though not as good at high temperatures. Small amounts of carbon, oxygen, nitrogen, and hydrogen greatly reduce the corrosion resistance and strengthen and embrittle the metal. Most zirconium is consumed in the manufacture of fuel rods for nuclear reactors. The only alloys of zirconium are zirconium–2.5% niobium (i.e., columbium) and the zircaloys. Zircaloy-2 contains 1.2–1.7% tin with small amounts of iron and nickel. Both tin and niobium have low neutron absorption and offset the detrimental influence of impurities such as nitrogen. Zircaloy-2 has a tensile strength of 66,000 psi (4.6×10^5 kPa) and elongation of 20% in 2 in. (5.08 cm).

Select Bibliography

Altan, Taylan, Oh, Soo-Ik, Gegel, Harold L., *Metal Forming: Fundamentals and Applications.* Metals Park, OH: American Society for Metals, 1983.

Alting, Leo, *Manufacturing Engineering Processes,* New York, NY: Marcel Dekker, Inc., 1982.

American Welding Society, *Welding Handbook.* New York, NY. Current edition.

Anderson, Robert Clark, *Inspection of Metals Volume 1: Visual Examination.* Metals Park, OH: American Society for Metals, 1983.

ASM, *Metals Handbook, Ninth Ed.,* Vol. 1. Properties and Selection: Irons and Steels; Vol. 2. Properties and Selection: Nonferrous Alloys and Pure Metals: Vol. 3. Properties and Selection: Stainless Steels, Tool Materials and Special-Purpose Metals; Vol. 4. Heat Treating; Vol. 5. Surface Cleaning, Finishing, and Coating; Vol. 6. Welding, Brazing, and Soldering. Metals Park, OH: American Society for Metals, 1983.

ASM, *ASM Metals Reference Book, 2nd Ed.* Metals Park, OH: American Society for Metals, 1983.

ASM, *1983 SAE Handbook, Vol. 1: Materials.* Metals Park, OH: American Society for Metals, 1983.

Battista, O. A., *Synthetic Fibers in Papermaking.* Melbourne, FL: Krieger Publishing Co., Inc., 1964.

Baumeister, Theodore, ed., *Marks' Standard Handbook for Mechanical Engineers, 8th Ed.* New York, NY: McGraw-Hill.

Beatty, John C., Booth, Kellogg S., *Computer Graphics, 2nd Ed.* Silver Spring, MD: IEEE Computer Society, 1982.

Becher, Paul, ed., *Encyclopedia of Emulsion Technology. Basic Theory.* New York, NY: Marcel Dekker Inc., 1983.

Bell, David A., *Electrical Instrumentation and Measurements.* Reston, VA: Reston Publishing Co., 1983.

Berry, Robert W., Hall, Peter M., Harris, Murray T., *Thin Film Technology.* Melbourne, FL: Krieger Publishing Co., Inc., 1979.

Bikales, Norbert M., ed., *Molding of Plastics.* Melbourne, FL: Krieger Publishing Co., Inc., 1971.

Bolz, Roger W., *Production Processes—The Productivity Handbook. 5th ed.,* New York, NY: Industrial Press, 1977.

Brady, George S., Clauser, Henry R., *Materials Handbook, 11th Ed.,* New York, NY: McGraw-Hill, 1977.

Brooks, Charlie R., *Heat Treatment, Structure and Properties of Nonferrous Alloys.* Metals Park, OH: American Society for Metals, 1982.

Brown, Robert Goodell, *Materials Management Systems.* New York, NY: John Wiley & Sons, 1977.

Budinski, Kenneth, *Engineering Materials.* Reston, VA: Reston Publishing Co., 1983.

Campbell, J. E., Underwood, J. H., Gerberich, W. W., *Application of Fracture Mechanics for Selection of Metallic Structural Materials.* Metals Park, OH: American Society for Metals, 1982.

Carey, Howard B., *Modern Welding Technology.* Englewood Cliffs, NJ: Prentice-Hall, 1979.

Carlson, Harold, *Spring Designer's Handbook.* New York, NY: Marcel Dekker, Inc., 1978.

Carlson, Harold, *Spring Manufacturing Handbook.* New York, NY: Marcel Dekker, Inc., 1982.

Childs, James J., *Principles of Numerical Control.* New York, NY: Industrial Press Inc., 1982.

Considine, Douglas M., *Encyclopedia of Instrumentation and Control.* Melbourne, F: Krieger Publishing Co., Inc., 1981.

Considine, Douglas M., *Encyclopedia of Instrumentation and Control.* Melbourne, FL: Krieger Publishing Co., Inc., 1982.

Corey, E. Raymond, *Procurement Management: Strategy, Organization, and Decision-making.* Boston, MA: CBI Publishing Company, Inc., 1978.

Cutcho, Marcia, ed., *Adhesives Technology. Developments Since 1979.* Park Ridge, NJ: Noyes Data Corp., 1983.

DeLollis, Nicholas J., *Adhesives, Adherends, Adhesion.* Melbourne, FL: Krieger Publishing Company, Inc., 1980.

der Schadensfalle, Das Buch, trans. by Naumann, Friedrich Karl, *Failure Analysis: Case Histories and Methodology.* Metals Park, OH: Dr. Riederer-Verlag and American Society for Metals, 1980.

Drozda, T. J., Wick, C., eds., *Tool and Manufacturing Engineers Handbook, Fourth Ed., Vol. 1: Machining.* Metals Park, OH: American Society for Metals, 1983.

Duley, W. W., *Laser Processing and Analysis of Materials.* New York, NY: Plenum Publishing Corp., 1983.

Einspruch, Norman G., Treatise Ed., VLSI Electronics—Microstructure Science, Vol. 6: Einspruch, Norman G., Larrabee, Graydon B., eds., *Materials and Process Characterization.* New York NY: Academic Press, Inc., 1983.

Engleberger, Joseph F. *Robotics in Practice.* New York, NY: Amacom Divisions, American Management Associations.

Enrick, Norbert L., *Market and Sales Forecasting, 2nd rev. ed.* Melbourne, FL: Robert E. Krieger Publ. Co., 1980.

Evans, Ulick R., *An Introduction to Metallic Corrosion.* Metals Park, OH: Edward Arnold Ltd. and American Society for Metals, 1981.

Farago, Francis T., *Handbook of Dimensional Measurement, 2nd ed.* New York, NY: Industrial Press, 1982.

Feigenbaum, Armand V., *Total Quality Control.* New York, NY: McGraw-Hill, 1983.

Freeman, Herbert, *Interactive Computer Graphics.* Silver Spring, MD: IEEE Computer Society, 1980.

Gibbons, Robert C., ed., *Woldman's Engineering Alloys, 6th Ed.* Metals Park, OH: American Society for Metals, 1979.

Grady, George S., Clauser, Henry R., *Materials Handbook, 11th Ed.* New York, NY: McGraw-Hill, 1977.

Graham, A. Kenneth, ed., *Electroplating Engineering Handbook, 3rd Ed.* New York, NY: Van Nostrand Reinhold, 1971.

Griff, Allan L., *Plastics Extrusion Technology, 2nd Ed.* Melbourne, FL: Krieger Publishing Co., Inc., 1976.

Gunn, Thomas G., *Computer Applications in Manufacturing.* New York, NY: Industrial Press Inc., 1981.

Hamer, D. W., Biggers, J. V., *Thick Film Hybrid Microcircuit Technology.* Melbourne, FL: Krieger Publishing Co., Inc., 1983.

Harrington, Joseph Jr., *Computer Integrated Manufacturing.* Huntington, NY: Robert E. Krieger Publishing Company, 1979.

Harvey, Philip, ed., *Engineering Properties of Steel.* Metals Park, OH: American Society for Metals, 1982.

Hawley, George F., *Automating the Manufacturing Process.* New York, NY: Reinhold Publishing Corp., 1959.

Higgins, Raymond A., *The Properties of Engineering Materials.* Melbourne, FL: Krieger Publishing Co., Inc., 1979.

Hine, Charles R., *Machine Tools and Processes for Engineers.* Melbourne, FL: Krieger Publishing Co., Inc., 1982.
Huebner's Machine Tool Specs, 2nd Ed., 8-Vol. set, Solon, OH: Huebner Publications, Inc., 1982.
Jensen, John E., *Forging Industry Handbook.* Cleveland, OH: Forging Industry Association.
Juran, J., ed., *Quality Control Handbook.* New York, NY: McGraw-Hill, 1974.
Kazmerski, Lawrence, L., ed., *Polycrystalline and Amorphous Thin Films and Devices,* New York, NY: Academic Press, Inc., 1980.
Kennedy, Clifford W., Rev. & ed. Andrews, Donald E., *Inspection and Gaging.* New York, NY: Industrial Press, 1977.
Kinney, G. F., *Engineering Properties and Applications of Plastics.* Melbourne, FL: Krieger Publishing Co., Inc., 1957.
Klar, Erhard, ed., *Powder Metallurgy: Applications, Advantages, and Limitations.* Metals Park, OH: American Society for Metals, 1983.
Koistinen, Donald P., Wang, Neng-Ming, *Mechanics of Sheet Metal Forming.* New York, NY: Plenum Publishing, 1978.
Krauss, George, *Principles of Heat Treatment of Steel.* Metals Park, OH: American Society for Metals, 1980.
Lambert, Douglas M., *The Development of an Inventory Costing Methodology: A Study of the Costs Associated with Holding Inventory.* Chicago: National Council of Physical Distribution Management, 1975.
Laue, Kurt, Stenger, Helmut, trans. from German, *Extrusion,* Metals Park, OH: American Society for Metals, 1981.
Lee, C. S. G., Gonzalez, R. C., Fu, D. S., *Robotics.* Silver Spring, MD: IEEE Computer Society, 1983.
Lester, Ronald H., Enrick, Norbert L., and Mottley Jr., Harry E., *Quality Control for Profit.* New York, NY: Industrial Press Inc., 1977.
Levitt, Albert P., *Whisker Technology.* Melbourne, FL: Krieger Publishing Co., Inc. 1970.
Licari, James J., *Plastic Coatings for Electronics.* Melbourne, FL: Krieger Publishing Co., Inc., 1981.
Machining Data Handbook, 2 Vols., 3rd Ed. Cincinnati, OH: Machinability Data Center, 1981.
Makridakis, Spyros, and Wheelright, Steven C., *Forecasting—Methods and Applications.* New York, NY: John Wiley & Sons, 1978.
Manko, Howard H., *Solders and Soldering, 2nd Ed.* New York, NY: McGraw-Hill, 1979.
Martens, Charles R., *Emulsion and Water-Soluble Paints and Coatings.* Melbourne, FL: Krieger Publishing Co., Inc., 1964.
Martens, Charles R., ed., *Technology of Paints, Varnishes, and Lacquers.* Melbourne, FL: Krieger Publishing Co., Inc., 1974.
Mascia, L., *Thermoplastics: Materials Engineering.* Englewood, NJ: Applied Science Publishing Ltd., 1982.
Mather, Hal, and Plossl, George W., *The Master Production Schedule—Management's Handle of the Business.* 2nd ed. Altanta, GA: Mather & Plossl, Inc., 1977.
Mayofis, T. M., *Plastics Insulating Materials: Chemistry, Properties and Applications.* Melbourne, FL: Krieger Publishing Co., Inc., 1967.
McGonnagle, Warren J., ed., *International Advances in Nondestructive Testing.* New York, NY: Gordon and Breach Science Publ. Inc., 1983.
McGreivy, Denis J., Pickar, Kenneth A., *VLSI Technologies—Through the 80s and Beyond.* Silver Spring, MD: IEEE Computer Society, 1982.
Metaxes, A. C., Meredith, J. R., *Industrial Microwave Heating.* Piscataway, NJ: Inspec. Dept., IEEE, 1983.
Mohler, Rudy, *Practical Welding Technology.* New York, NY: Industrial Press Inc., 1983.
Mohr, Gilbert J., Oleesky, Samuel S., Meyer, Leonard S., Shook, Gerald D., *SPI Handbook of Technology and Engineering of Reinforced Plastics/Composites, 2nd ed.* Melbourne, FL: Krieger Publishing Co., Inc., 1981.
Morton, Maurice, *Rubber Technology, 2nd ed.* Melbourne, FL: Krieger Publishing Co., Inc., 1981.
Murarka, S. P., *Silicides for VLSI Applications.* New York, NY: Academic Press Inc., 1983.
Norton, F. H., *Fine Ceramics: Technology and Applications.* Melbourne, FL: Krieger Publishing Co., Inc., 1978.
Oberg, Erik, Jones, Franklin D., Horton, Holbrook L., *Machinery's Handbook, 21st ed.* New York, NY: Industrial Press Inc., 1979.
Open Die Forging Manual, 3rd Ed. Cleveland, OH: Forging Industry Association, 1982.
Orlicky, Joseph, *Material Requirements Planning.* New York, NY: McGraw-Hill Book Company, 1975.

Parmley, Robert O., *Standard Handbook of Fastening and Joining.* New York, NY: McGraw-Hill, 1977.

Peckner, Donald, Bernstein, I.M., eds., *Handbook of Stainless Steels.* New York, NY: McGraw-Hill, 1977.

Plossl, George W., *Manufacturing Control—The Last Frontier for Profits.* Reston, VA: Reston Publishing Company, Inc., 1973.

Plossl, George W., and Welch, W. Evert, *The Role of Top Management in the Control of Inventory.* Reston, VA: Reston Publishing Company, 1979.

Potts, Daniel L., Arcuri, James, eds., *International Metallic Materials Cross Reference, 2nd Ed.* General Electric Company, 1983.

Pusztai, Joseph, Sava, Michael, *Computer Numerical Control.* Reston, VA: Reston Publishing Co., 1983.

Reed, Richard P., Clark, Alan F., eds., *Materials at Low Temperatures.* Metals Park, OH: American Society for Metals, 1983.

Rice, Rex, *VLSI Support Technologies (Computer-Aided Design, Testing, and Packaging).* Silver Spring, MD: IEEE Computer Society, 1982.

Rigney, David A., Glaeser, W. A., eds., *Source Book on Wear Control Technology,* Metals Park, OH: American Society for Metals, 1978.

Riley, Frank J., *Assembly Automation.* New York, NY: Industrial Press Inc., 1983.

Roberts, George A., Cary, Robert A., *Tool Steels, 4th Ed.* Metals Park, OH: American Society for Metals, 1980.

Sachs, G., Voegeli, H. E., *Principles and Methods of Sheet Metal Fabricating—Second Ed.* Melbourne, FL: Krieger Publishing Co., Inc., 1966.

SAE/ASTM, *Unified Numbering System for Metals and Alloys, 2nd Ed.* Warrendale, PA: Society of Automotive Engineers, 1977.

Scheflan, Leopold, Jacobs, Morris B., *The Handbook of Solvents.* Melbourne, FL: Krieger Publishing Co., Inc., 1973.

Schey, John A., *Tribology in Metalworking: Friction, Lubrication and Wear.* Metals Park, OH: American Society for Metals, 1983.

Schweitzer, Philip A., ed., *Corrosion and Corrosion Protection Handbook.* New York, NY: Marcel Dekker, Inc., 1982.

Skiest, Irving, ed., *Handbook of Adhesives, 2nd Ed.* New York, NY: Van Nostrand Reinhold, 1977.

Skinner, Wickham, *Manufacturing in the Corporate Strategy.* New York, NY: John Wiley & Sons, 1978.

Smith, Bernard T., *Focus Forecasting—Computer Techniques for Inventory Control.* Boston, MA: CBI Publishing Company, 1978.

Sorensen, O. Toft, ed., *Nonstoichiometric Oxides.* New York, NY: Academic Press, Inc., 1981.

Stuart, R. V., *Vacuum Technology, Thin Films, and Sputtering—An Introduction.* New York, NY: Academic Press, Inc., 1983.

The Institute for Interconnecting and Packaging Electronic Circuits, *General Requirements for Electronic Interconnections.* Evansville, IL.

Tompkins, James A., Smith, Jerry D., *Automated Material Handling and Storage.* Pennsauken, NJ: Auerbach Publishers, 1983.

Tu, K. N., Rosenburg, R., eds., *Treatise on Materials Science and Technology (Preparation and Properties of Thin Films).* New York, NY: Academic Press, 1982.

Turner, William C., *Thermal Insulation Handbook, 2nd Ed.* Melbourne, FL: Krieger Publishing Co., Inc., 1981.

Tver, David F., Bolz, Roger W., *Robotics Sourcebook and Dictionary.* New York, NY: Industrial Press Inc., 1983.

Tver, David F., *Dictionary of Dangerous Pollutants, Ecology and Environment.* New York, NY: Industrial Press Inc., 1981.

Uhlmann, D. R., Kreidl, N. J., eds., *Glass—Science and Technology, Vol. 1: Glass-Forming Systems.* New York, NY: Academic Press Inc., 1983.

Uhlmann, D. R., Kreidl, N. J., eds., *Glass—Science and Technology, Vol. 5: Elasticity and Strength in Glasses.* New York, NY: Academic Press Inc., 1980.

Walton, Charles F., Opar, Timothy J., eds., *Iron Castings Handbook.* Rocky River, OH: Iron Castings Society, Inc., 1981.

Wieser, Peter F., ed., *Steel Castings Handbook, 5th Ed.* Steel Founders' Society of America, 1980.

Wight, Oliver, *Production and Inventory Management in the Computer Age.* Boston, MA: Cahners Books International, Inc., 1974.

Willardson, R. K., Beer, Albert C., Semiconductors and Semimetals, *Vol. 19: Deep Levels, BaAs, Alloys, Photochemistry.* New York, NY: Academic Press, Inc., 1983.

Wilson, Robert, *Metallurgy and Heat Treatment of Tool Steels.* New York, NY: McGraw-Hill, 1975.